Air & Water:

An Introduction to the Atmosphere and the Hydrosphere

Walter Martin
Associate Professor
Department of Geography and Earth Sciences
University of North Carolina at Charlotte

Nelson Nunnally
Professor Emeritus
Department of Geography and Earth Sciences
University of North Carolina at Charlotte

KENDALL/HUNT PUBLISHING COMPANY
4050 Westmark Drive Dubuque, Iowa 52002

Front cover photo: Banff National Park, Alberta, Canada (1984) by Walter Martin

Back cover photo: The Atchafalaya, Louisiana (1982) by Walter Martin

ISBN 0-7872-7743-6

Printed in the United States of America
10 9 8 7 6 5 4 3 2 1

To the memory of Jim Clay and Don Steila who inspired and mentored many students of the natural environment. We remember fondly their leadership and counsel.

Table of Contents

Acknowledgements

Several individuals are owed our personal gratitude for their contributions. Jeff Simpson, Michael Wilcox, and Eric Swinson provided indispensable assistance with several maps and graphic figures. Others, including David Gay, Ken Burrows, Jamie Strickland, and Matt Hanson, have made helpful suggestions to improve form or content. To these people and many others, we express our gratitude. Any errors or omissions in graphics, text, or content, however, are our own.

Charlotte, North Carolina

October 11, 2000

Walter Martin

Nelson Nunnally

Preface

This book is intended to support an introductory course in earth surface systems. Our goal is to aid the reader in understanding the fundamentals of atmospheric and hydrologic systems and processes. Many aspects of resource use and global change involve patterns of matter and energy exchange between these systems. The implications for environmental quality and wise resource use are vast. We believe that it is timely and essential to understand these linkages and the implications they pose for the future. The purpose of this book is not merely to help students understand principles of the fluid earth, but also to involve them in the excitement of current discoveries in physical geography, climatology, meteorology, hydrology, and global change studies.

Chapter **1**

Perspective

It is easy for the members of post industrial societies, snug in their thermostatically controlled worlds, to ignore the symphony of interacting processes taking place all around them in the natural environment. It may be easy, but it is certainly not very smart. Basic education for most of our ancestors meant acquiring enough folk wisdom about the local environment to follow accepted hunting, fishing, or farming practices. The survival of each generation depended upon the accurate transmission of environmental knowledge from the preceding generation. The penalty for ignorance or defiance of those practices was severe: starvation or death by thirst, disease, or esposure. In each of the great agricultural civilizations of the past, such events as droughts or floods were not only life threatening events in their own right, but they frequently led to local crop failures, spoilage losses, food shortages, and famine. Although today, we are somewhat insulated from the exigencies of local environmental events by international supply and distribution systems, we cannot pretend that natural systems are unimportant.

There is still a penalty for environmental ignorance in the post–industrial world. Though subtle and pervasive, the penalties are just as costly as those our ancestors faced. In many instances our smug ambivalence toward natural environmental systems has profound implications for the quality and safety of our lives and the lives of our children. The leading free market democracies of Western Europe, North America, Japan, and Australia have been more successful in generating a high–quality lifestyle for their citizens than any other economic-political system the world has yet seen. Yet, despite considerable success, democracies like the United States will increasingly rely on a scientifically enlightened electorate as we face the technical challenges of the twenty-first century. As we confront difficult environmental questions such as ozone pollution, global climate change, or reductions of stratospheric ozone, it is essential that personal and political choices are made with insight and awareness of natural systems. Ignorance about our environment will not absolve us from the responsibility to make environmental decisions, but it will contribute to sluggish and ineffectual environmental management. However, the perspective presented here is neither ecologic nor economic. The purpose of this book is to provide students with a basis for understanding the role of atmospheric and hydrologic systems within the functional unity of natural systems.

There are five major global patterns: climatic, physiographic, biotic, edaphic, and demographic. The climatic deals with patterns of heat and cold, of moisture and aridity. The major physiographic features such as rugged

Figure 1.1 A Paradigm of Geography

Defining Geography

Greek Origin: GEO = "earth" + GRAPHIA = "to write about, to describe"

Brief Definition: "The science of the earth's surface" (life layer, human habitat)

Conceptual Definition "The study that seeks to provide scientific description of the earth as the home of humankind through the accurate, orderly, and rational description, analysis, and interpretation of physical and cultural gradients."

Physical Environment
(Natural Elements)

Meteorology
Climatology
Geomorphology
Biogeography
Hydrology
Pedology & Edaphology
Economic Geology

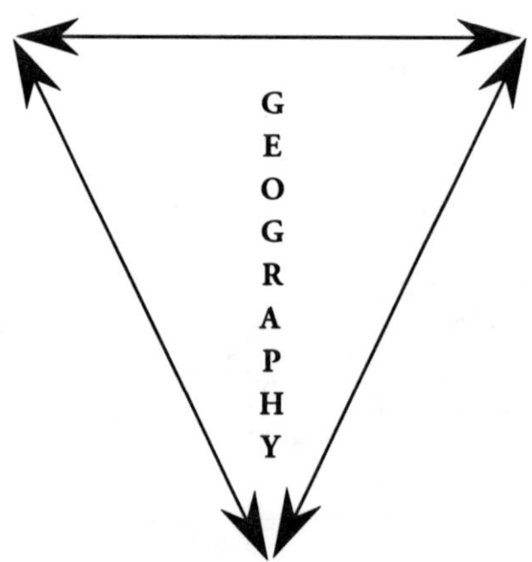

Cultural Environment
(Human Elements)

Demography
Settlement/Land Use
Economic Geography
Medical Geography
Political Geography
Transportation and Communication

Environmental Science

Air & Water Quality
Natural Resource Management
Environmental Disturbances

Geographic Analysis

1. Where is it? What is the distribution?
2. Why is it there? What are the processes?
3. So What? How are other phenomena interrelated?

Tools and Techniques: Cartography, Image Analysis, History, Statistics, GIS

Employment: Environmental, Urban, Land Use, and Transportation Planning; Retail Location; Cartography; Environmental Science; and Geographic Information Systems mountains, bald prairies, and high rocky plateaus result principally from tectonic forces relocating the earth's crustal plates. The erosional and depositional influences on minor land surface forms, however, are dramatically climate driven. Prior to the eighteenth century, naturalists believed that valleys controlled the form and location of streams and rivers. Nothing could be further from the truth. It is the erosive power of moving water, the stream itself, operating across vast stretches of time that has carved the length and breadth of common valleys. The third pattern, the biotic, refers to the range of natural vegetation (flora) and natural animal life (fauna). Natural vegetation largely reflects differences in a response to available light, heat, and moisture. It provides the habitat that supports animal populations. Because most animals are herbivorous, they are generally confined to their specific food producing habitat. Even carnivorous animals are similarly limited to the range of their prey. Together, geology, climate, land surface form, flora, and fauna have been principal forces in creating types and patterns of soils. Most soils evolve from youthful to mature in response to the availability of moisture and heat. Their fertility and productivity are mainly a product of geochemical, climatic, and biologic processes operating over long stretches of time. Each of these global patterns and the matter and energy systems that support them are linked and interrelated to climate. As an over-arching theme, atmospheric processes are useful in understanding the interdependent structure of environmental systems and their regional character. Lastly, the demographic pattern reflects the saga of human migration from our origins in East Africa to all the presently inhabited reaches of the earth. The growth, form, and even the prosperity of those settlements are intertwined with the way in which climatic, physiographic, edaphic, and biological resources are perceived and utilized.

Since the time of the ancient Greeks, naturalists, physical geographers, earth scientists, and climatologists have sought principles that unify the complex matrix of earth surface processes. As specialization has fragmented scientific inquiry during the last two millennia, the need for synthesis and integration has increased. The scientific niche, or functional role, that this text addresses is to integrate, unify, and associate specialized knowledge generated by field researchers in meteorology, climatology, earth science, and physical geography. Meteorologists are atmospheric scientists who use the laws of physics and chemistry to forecast short-term changes in the weather. Climatologists may study similar phenomena but differ from meteorologists in being more interested in the long-term atmospheric characteristics. Many earth scientists have explored physical changes in the earth's atmosphere to better understand processes which form and sculpt the earth's surface. Physical geographers also seek answers to questions about processes operating in the natural environment, but they are particularly interested in the spatial distribution of those phenomena. It is the spatial perspective that identifies physical geography and links it with other physical sciences. We will adapt from geography a useful paradigm that identifies overlapping physical, cultural, and environmental elements (Figure 1.1).

A comprehensive view of our environment involves all of these interacting elements and systems. We are utilizing here an introduction to atmospheric science as the principal component in an integrated structure to survey earth systems. As a home for human beings, the earth surface environment involves each of these disciplines and the interrelated systems and perspectives they offer. For example, hurricane forecasts projecting the time and location of landfall trigger a sequence of responses by public safety officials and the general public. The efficacy of the these forecasts ultimately depends on the land surface form of the coastal region, antecedent soil moisture, how densely populated the region is, how effectively transportation systems can evacuate the resident and tourist populations, how vulnerable communication systems are to severe storms, and even the degree to which municipal water and sewer services will be disrupted in the aftermath. Atmospheric processes play a central role in earth systems and therefore serve as an appropriate introduction to earth surface systems. Elements of the physical environment are listed on the left side of Figure 1.1, elements of the cultural environment are listed to the right, and environmental elements are listed beneath.

Geographic inquiry cannot be defined on the basis of subject matter alone. It assumes the philosophical premise that rational factors explain spatial distributions and gradients. Geography can be thought of as a study of gradients. Gradients are defined as a change in the magnitude of some property across distance (normally horizontal distance) or time. That property might be elevation, atmospheric pressure, annual precipitation, groundwater pollution, temperature, wind speed, or soil moisture. The first question a physical geographer seeks to answer is: What is the spatial distribution of the property? (Where is it?) Usually, the answer to this question is a map or a series of maps with the property carefully plotted. Examination of the spatial distribution frequently leads to a second set of questions: Why is this property located "there" and not "here"? What makes "there" different from "here?" or What process(es) can explain this distribution? Process is change through time. Hypotheses are formulated and tested against observed data. Processes are identified. The final and most important set of questions asks: So what? Who cares? Why is the distribution of this property (and the processes associated with it) important? How is this phenomenon related to other aspects of human activity?

Much of current geographic research and analysis is used to plan for regional and metropolitan growth, to minimize the friction (cost) of distance on human activity, and to attenuate environmental pollution and problems of resource management. However, many people enjoy geography because it helps them better appreciate the nature of each place within a larger milieu. Because no other element of the natural environment interacts more pervasively with other earth surface systems than climate, the atmospheric and hydrologic components of physical geography provide a meaningful starting point for exploring the natural environment. The destination is an awareness of environment which may remind us of T. S. Eliot's lines in "Little Gidding:"

> "We shall not cease from exploration
> And the end of all our exploring
> Will be to arrive where we started
> And to know the place for the first time.
> Through the unknown, remembered gate
> When the last of earth left to discover
> Is that which was the beginning..."

Chapter **2**

Places, Spaces, and Gradients: Maps

Planet Earth

Compared with the incomprehensible size of the universe, with its profusion of galaxies and the countless stars with their planetary systems, earth is an infinitely minute entity. From a human perspective, however, earth is a massive celestial body with resources and environmental conditions propitious to sustaining life.

Earth possesses three distinct attributes that render it suitable for human habitation. These are not known to exist on any other planet in our solar system. They are: (1) an abundance of oxygen in the lower atmosphere, (2) a moderate temperature range suitable for the survival of earth's organic life forms, and (3) large standing bodies of water.

Figure 2.1 The Earth is an Oblate Spheroid

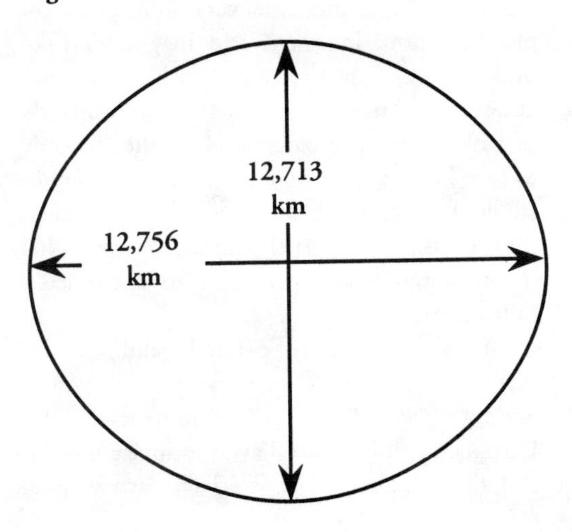

Figure 2.2 The Landmasses of Africa, Asia, and Europe

The diameter of the earth at the equator is 12,756 km (7,926 mi) and its maximum circumference is 40,075 km, (24,900 mi). Note in Figure 2.1 that the equatorial diameter is slightly greater than the polar diameter. This variation from a perfect sphere is less than one percent. It is a result of deformation in response to the centrifugal force exerted upon the earth, particularly along the equator where the rotational speed at the earth's surface exceeds 1600 km/hr (1000 mi/hr). This difference in equatorial and polar diameters means the earth is not a perfect sphere, but rather an **oblate spheroid**. Nonetheless, its deviation from a perfect sphere is very minor, amounting to only about 0.3 percent of the earth's diameter.

Other earth surface variations include landmasses with topographic irregularities that are interspersed with oceans. The large landmasses amount to approximately 29 percent of the earth's surface area and are referred to as **continents**. The smaller bodies are called islands. The usage of both terms is dictated by custom rather than logic. Thus, although Africa, Asia, and Europe constitute one continuous landmass, we arbitrarily refer to the three as separate entities (Figure 2.2). Less spectacular from a planetary viewpoint than the massive land and water bodies, but equally impressive to earth's inhabitants, are the earth's regional topographic irregularities that locally impart deviations from its spheroidal shape. The planet's highest elevation is atop Mt. Everest at 8,850 m (29,000 ft.) above sea level. The Marianas Trench of the Pacific Ocean is recorded to have a bottom at 10,915 m (10.9 km or 6.8 miles) below sea level, amounting to a total difference in elevation between the highest and lowest points of 19,800 m (19.8 km or 12.3 miles). Although these surface irregularities might appear substantial by human standards, they are relatively insignificant in relation to global dimensions. Portrayed on a scale model of the earth that is 30 cm (12 in) in diameter, neither the height of Mt. Everest nor the depth of the Marianas Trench would amount to the thickness of the paper upon which these words are printed. For general purposes, the earth may be considered a perfect sphere. **Geodesy** (science of measurement of the earth), nonetheless, requires all minor deviations from spheroidal (no matter how small) to be considered.

Figure 2.3 A Grid System. This simple grid uses alpha-numeric codes to identify a location at the point of two intersecting lines. Point "X", for example, can be located by identifying the intersection of vertical line number 5 and the horizontal line E. This form of grid is often used on maps displaying city streets.

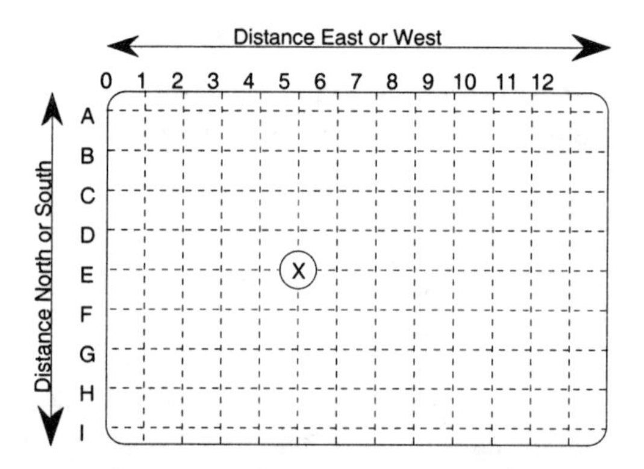

Earth Location

Atmospheric processes vary from place to place in form, intensity, and magnitude. To understand "where" and "why" certain atmospheric interactions occur, it is necessary to have a reference base to locate the "where" factor. Thus, before we enter into an in–depth study of the atmosphere, we will briefly discuss the concept of earth location and its representation upon two-dimensional surfaces, called maps.

If the earth were perfectly uniform in exterior appearance, identifying a location upon it would be a monumental task. Fortunately, it does not have the appearance of a billiard ball and does have astronomic

Figure 2.4 A-B. The Latitudinal Component of the Geographic Grid.

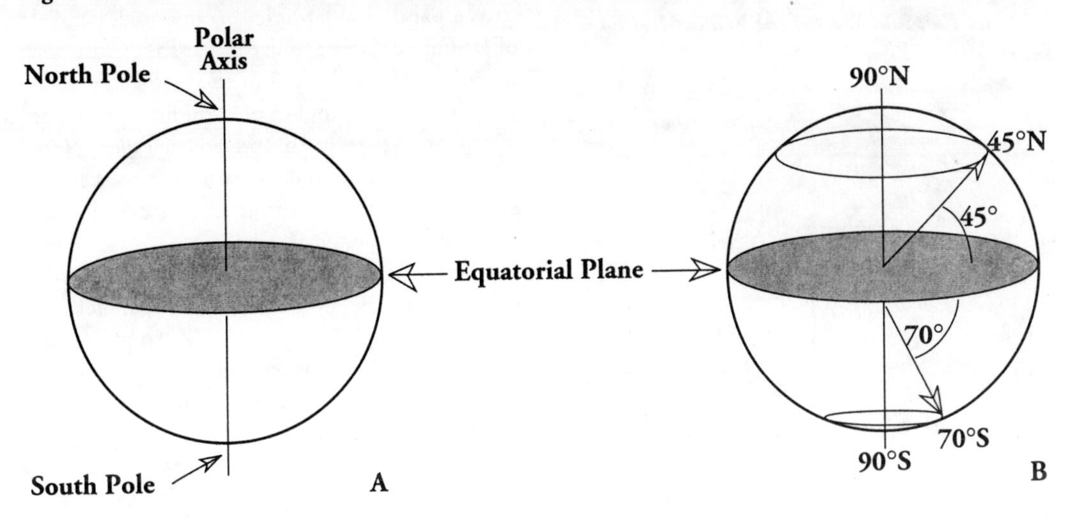

reference points. A given location on earth may be determined by both relative and absolute techniques. **Relative location** refers to the identification of a place upon the earth's surface in relation to a known location or locations. Consider the following example. It is Spring Break at Boston College and a freshman questions a senior class student about directions to reach Daytona Beach, Florida. The senior responds by saying, "Haul out to Interstate 95, head south, and keep going until you see the exit sign: Daytona Beach." The senior provided relative location information which would be useless if the freshman were located virtually anywhere other than Boston. If the student were in St. Louis, Missouri, for example, traveling to Interstate 95 would involve the loss of a couple of days during the break period. The concept and use of relative location is beneficial to each of us in the conduct of daily affairs, but does not allow for the detail required to define legal property boundaries or to navigate air and water craft.

To resolve problems that need definition of **absolute location**, that is identifying with mathematical precision a particular spot on the earth's surface, a network of intersecting lines referred to as a **grid system** is employed. A multitude of such systems have been devised and a simple one of east-west and north-south intersecting lines is illustrated in Figure 2.3. Note in this example that while each point is precisely identified by the intersection of two lines, the individual lines of the grid system do not all have mathematical values and the system lacks reference to known locations on the earth. Thus, a more sophisticated system is needed. That system is the **geographic grid** which is based upon reference points where the earth's imaginary axis of rotation intersects the surface at the North and South poles (Figure 2.4A). Through mathematical division of the polar

Figure 2.5 Meridians of Longitude Measure Distance East and West of the Prime Meridian (dark vertical line)

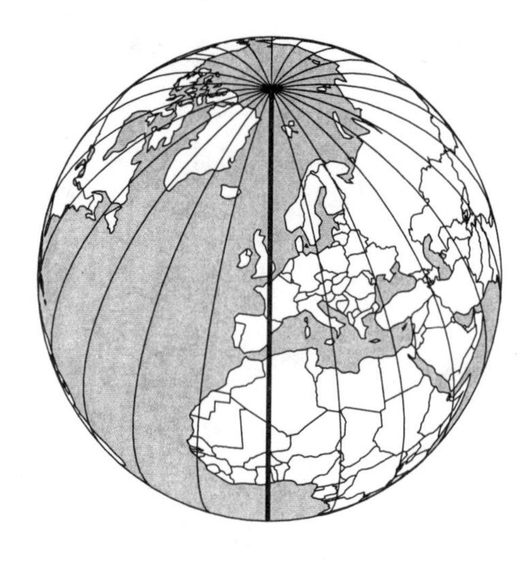

axis' length, a midpoint can be established to create a theoretical plane that separates the Northern Hemisphere from the Southern . The boundary of this plane at the earth's surface is referred to as the **equator**.

Distance north or south can be specified with respect to angular distance from the equatorial plane (Figure 2.4B). This locational trait is referred to as **latitude** and is designated with the capital letter N for Northern Hemisphere points (for example 45°N) and the capital letter S for Southern Hemisphere points (such as 70°S). Because the earth is spherical and a given latitudinal angle can be rotated about its apex at the center of the earth, a series of points can be generated for any given latitude. When the points are connected by a line, they define yet another circular plane that is parallel to the equatorial plane. The outer boundary of this plane defines all points at a given angular distance from the equator, is parallel to the equatorial plane, and is referred to as a **parallel** of latitude. Because each parallel of latitude defines angular north-south distance for locations on all sides of the earth, it is clear that the greatest number of latitudinal degrees is 90°, that is the North and South Poles. Note that parallels of latitude are true east-west lines, even though they measure distance north or south of the equator.

True north-south imaginary lines that converge at the poles and cross parallels of latitude at right angles are referred to as meridians, and each is equivalent to one-half of a great circle. One special meridian, called the **prime**, or **Greenwich**, meridian, serves as reference from which distance east and west may be measured (Figure 2.5). In 1884, the International Meridional Conference convened in Washington, D.C. and reached an arbitrary agreement that the prime meridian (0° longitude) would pass through the Royal Observatory at Greenwich (England) located a short distance southeast of central London. Prior to that time it was common for nations to construct maps with the prime meridian through their capitals. To this day, locations on the earth's surface are identified not only by angular distance north or south of the equator, but also by angular distance east or west from the Greenwich meridian. Distance east or west from the prime meridian is known as longitude. Longitude is measured east and west from the prime meridian (0°) to a maximum of 180°.

When a location's latitude and longitude are combined, for example 35°N and 80°W, any earth location can be precisely defined in mathematical terms based on the angular distance from a standard referent such as the prime meridian (0 longitude) and the equator (0 latitude) (Figure 2.6). As a final note, it is important to remember that the earth is

Figure 2.6 Parallels and Meridians form a Global Geographic Grid.

spherical and in cross-section exhibits the shape of a circle. Hence, because a circle contains 360 degrees, the meridian of greatest longitude within a system that measures both east and west from the prime meridian (0 longitude) is 180 degrees.

The equator is a special boundary that is referred to as a Great Circle. A **Great Circle** is the line of intersection between the surface of a sphere and a plane that passes through the center of the sphere. The shortest distance between any two points on the surface of a sphere is always a segment of a great circle. To find the shortest distance between any two points on the globe, all that is required is a section of twine of sufficient length. When the twine is drawn taunt between two points, its path will be a segment of a great circle and the shortest distance between the two points.

Earth Representation

Planet Earth is spherical and can be represented truly only in its three-dimensional form. Just as the rind of an orange cannot be peeled and laid flat without tearing, neither can any large portion of the earth's surface be represented without distortion on a two-dimensional plane known as a map. Nonetheless, maps are vital tools, not only to geographers who study the relationships among phenomena that are distributed upon, beneath, and above the earth's surface, but also to individuals who depend upon them for myriad purposes. How often have you used a road map or viewed a weather map on television? From agricultural development to urban planning or for identifying strategic sites for bombing raids during wartime, accurate maps are critical to each task's success. For our purposes, we are interested in maps because they can display atmospheric and terrestrial data in a readily comprehensible manner. **Maps** may be defined as selective and systematic representations of all or part of a three-dimensional surface (such as the earth) on a two-dimensional plane. They are used normally to depict various features, where they are found, their distance from some other point, and sometimes, their elevation above or below sea level or their quantity or value. As an analytical tool a map is valuable for determining how something exists and why it is there, and for questioning the significance of its location. So something is located there, so what?!

Many contemporary issues are now effectively investigated through spatial or cartographic analysis. Recent advances in computer technology have brought digital cartography within the budgets of many government agencies, businesses, utilities, and universities. Computer software specifically designed to interrelate large informational databases with large spatial databases and photographic imagery has emerged to provide cartographers, geographers, and other spatial analysts with a powerful new tool. This marriage of maps and databases has rapidly gained widespread acceptance where policy makers, planners, and other spatial decision makers need to understand interactions between multiple geographic patterns. The use of computer cartography to link points, lines, and polygons with numerical attributes is called a **geographic information system** (GIS). By providing a view of several superimposed patterns, GISs can answer many questions about spatial conflicts or complementarity. Many urban planning and engineering decisions are amenable to GIS analysis.

All maps including GIS maps should contain the following elements: **title**, **date**, **legend**, **location reference**, **direction reference**, **scale**, and **source**. The **title** of a map is like the title of a book. It should permit the reader to know about the subject matter that is to be addressed. You would not go to the library to check out a book on World War II history to learn of agricultural practices in Germany. Similarly, maps are devised for special purposes and require appropriate titles: Average Monthly Temperature of the United States in January, Number of Frost-free days in Canada, Global Temperature Patterns, etc. Maps are **dated** to provide the reader with information regarding when the data for its compilation were collected and analyzed. A map entitled Distribution of Tornadoes in the United States would have little value if the reader could not determine whether its date of production was 1890 or 1990. **Legend** refers to an interpretation key which is normally found within a "boxed-in" area in some corner of the map. It helps explain any symbols the map may contain, such as interstate highways, secondary roads, urbanized areas, and state capitals. **Location reference** provides information to identify where upon the earth the mapped data are applicable. This reference is most frequently addressed either by superimposing upon the map appropriate parallels of latitude and meridians of longitude, or by an inset map showing the surrounding region. **Direction reference** is indicated either by a North arrow or by specific parallels and meridians. These provide a frame of reference for the map's compass direction. Note that maps frequently combine both location and direction reference.

Scale provides information on the ratio between map distance and earth surface distance. Because maps are normally smaller than the earth surface they represent, it is important to know the relationship between the two. Consider the following: a freshman at North Carolina State University says to his buddy, "Let's drive down to Florida; it looks real close on the map. In fact it's only a couple of inches." In this case the freshman doesn't know if that distance will involve driving one mile, one hundred miles, or one thousand miles. Looking at the scale resolves the problem. Scale can be **verbal**: "One centimeter represents one hundred kilometers," or "One inch represents one hundred miles." (Note: The foregoing are not equivalent.) It can be expressed as a **representative fraction**: for example, 1:100 (read as "1 to 100"), wherein the numerator (usually 1) is separated from the denominator by a colon, and any system of measure may be used. One centimeter of map distance represents 100 centimeters of earth distance, or one inch of map distance represents 100 inches of earth distance. The idea is that one map unit of distance represents one hundred comparable units of earth distance.

As illustrated in Figure 2.7, map scale also can be presented in graphic form. This is called a linear scale and consists of a graduated line or bar drawn on the map. The line is marked off with specific map distances that are proportionate to distances on the earth. To use such a scale, you first measure distance between two points on the map (such as the distance between two cities), or you can lay the edge of a straight piece of paper on the two points of interest and mark their locations on the paper. You then place the ruler with the measured distance or the paper with the marked points along the linear scale to determine the distance between the two points on the earth's surface. The linear scale is especially useful on maps that are to be photographically enlarged or reduced, as the map scale will change proportionally to the change in map size. Maps are constructed at a scale that allows the desired level of detail to be portrayed. **Large scale** maps have a representative fraction greater than 1:1,000,000.

Figure 2.7 Map Scale: The Ratio between Map Distance and Earth Surface Distance

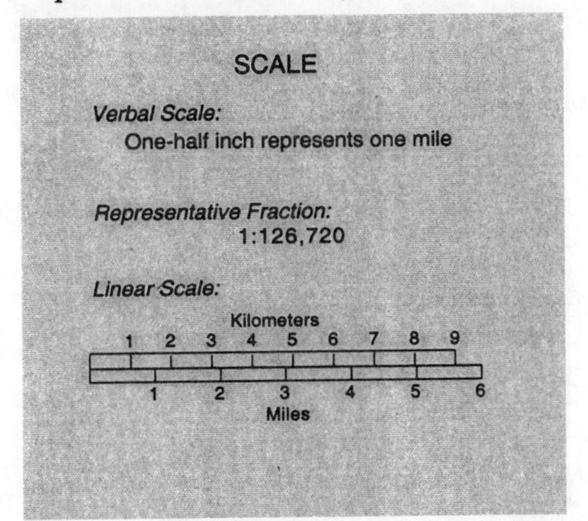

These maps portray a small portion of the earth, but can provide detailed information. **Small scale** maps, for example with a representative fraction of 1:1,000,000 or smaller, display a larger area, but consequently, cannot be as precise and detailed. Suppose your home is to be located on a map with a ratio of 1:10,000. Assume the house is 30 meters (approximately 100 feet) in length. If the house location is identified upon a map scale where one centimeter represents 10,000 centimeters earth distance, the house would occupy map space of less than one-third a centimeter (about 0.30 centimeters or roughly 0.12 inches). At a scale of 1:1,000,000 the house length on a map would be 0.003 cm (0.001 in.) and quite impossible to display in a fashion that would be intelligible.

Source establishes a measure of map reliability. It identifies who collected the data from which the maps were compiled and the date that the data represent. In some cases it may be useful to know the age and source of data in order to have confidence in the map.

Now that the essential components that should accompany each map have been addressed, let's look at the map itself. How can the image of a three-dimensional earth be transferred to a two-dimensional plane (map) with minimum distortion. The technique used to accomplish this task involves the concept of projecting the spherical surface of the geographic grid and associated earth features onto another geometric surface that can be displayed upon a flat plane. The only geometric surfaces that allow this are cylinders, cones, and planes. As illustrated in Figures 2.8 – 2.10, each projection appears different. Each is most realistic and representative of the earth shape at

Figure 2.8 The Mercator Projection

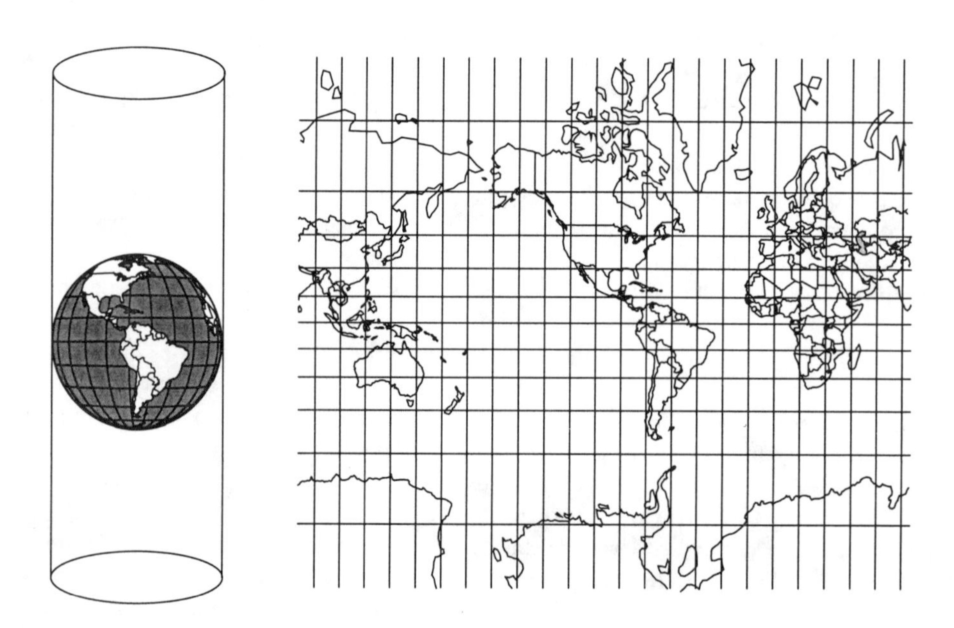

The Mercator is a cylindrical projection characterized by increasing areal distortion toward the poles. Parallels and meridians are straight lines that meet at right angles. Meridians are equally spaced. Parallels are proportionally farther apart as latitude increases. Places poleward of 82° latitude are rarely shown. Because course directions (Rhumb lines) appear as straight lines, Mercator maps are frequently used as base maps for navigation charts. Mercator maps are conformal.

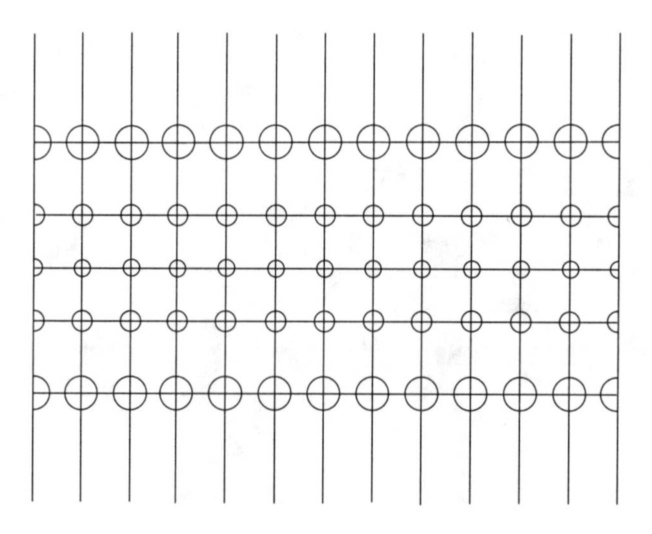

Figure 2.9 The Albers Projection

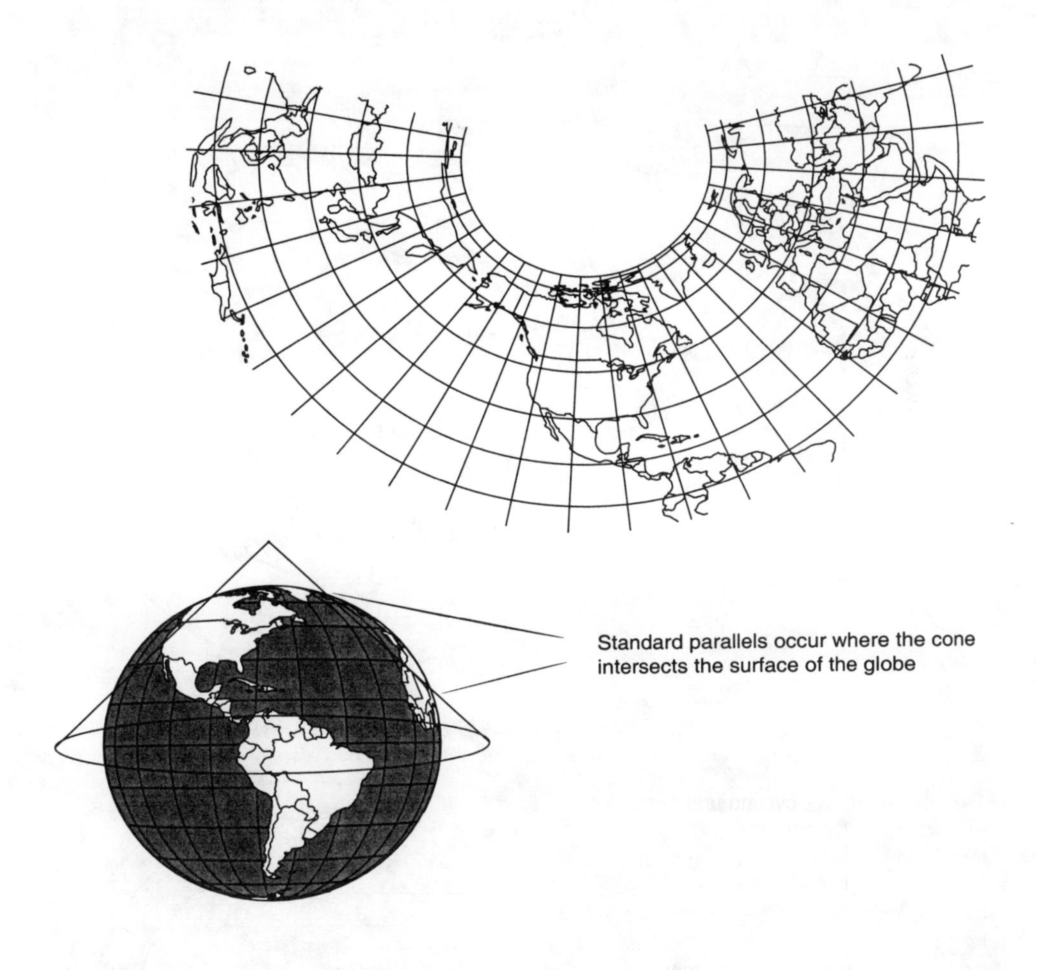
Standard parallels occur where the cone intersects the surface of the globe

The Albers is a conic projection characterized by equivalence. Parallels are concentric arcs and meridians diverge from the poles as straight lines. Accuracy is greatest along the each of the two standard (secant) parallels. These maps are used to show political and statistical data within the middle latitude areas. They are frequently used to depict the United States, central Europe, Russia, and the northern hemisphere.

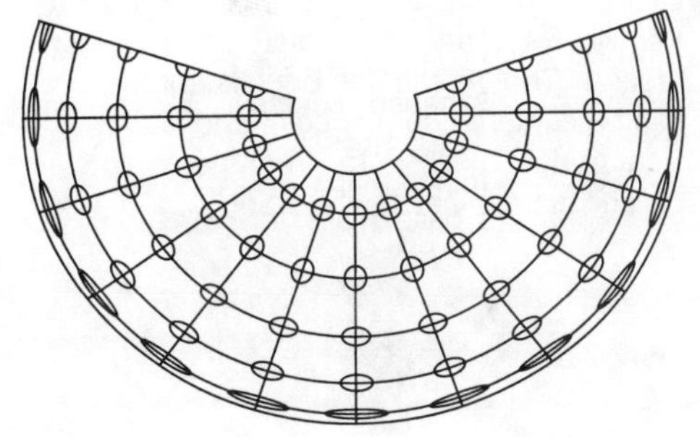

Figure 2.10 The Azimuthal Projection

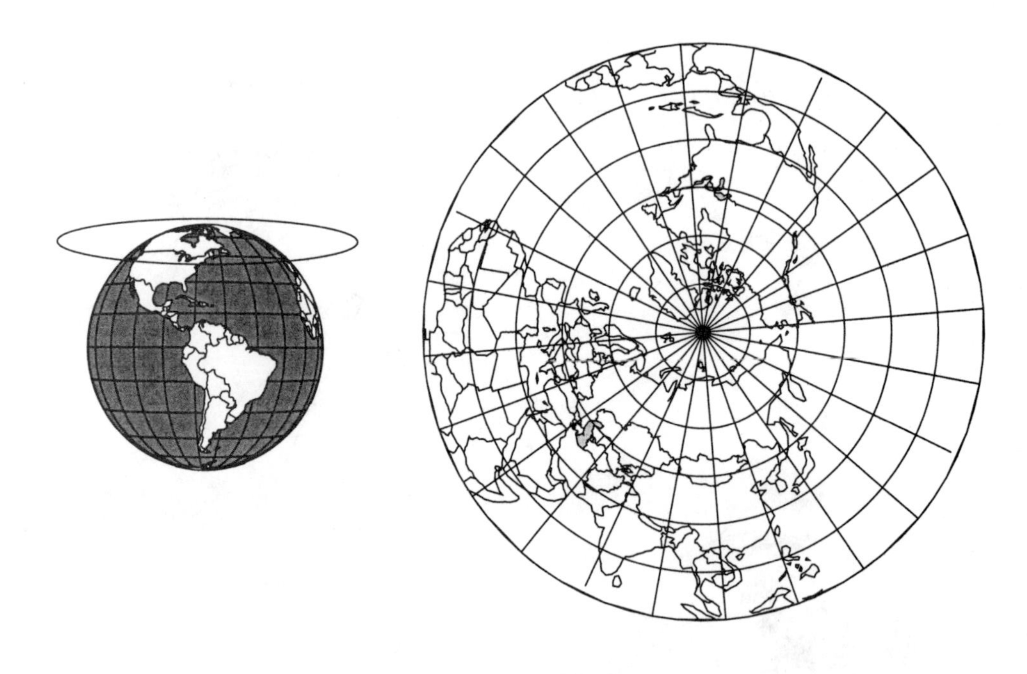

The Azimuthal Equidistant is a plane projection characterized by increasing distortion toward the periphery. Meridians are shown as radii from the central point of tangency. Parallels are concentric circles. Azimuthal projections are frequently used to depict polar areas. The Azimuthal Equidistant can be centered on any point and is frequently used to compare distances from that central point.

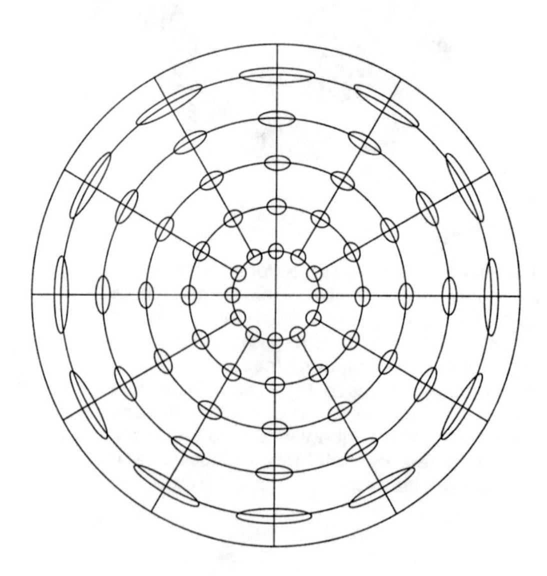

points where the projected surface is tangential to earth, but loses realism as distance from the tangential points increases.

In theory, these maps are produced by projecting the earth's image onto the forms as illustrated. Imagine a pinpoint of light radiating outward from the center of a translucent globe and projecting the image to each different geometric form. In practice, such imagery is mathematically derived. Many mathematical map projections can be devised to serve special purposes. The use of all maps requires an appreciation of the rationale for which they were constructed.

Some maps can be designed to specifically represent the shape of landmasses, but represent area inconsistently across the breadth and width of the map. These maps are called **conformal** and are used to identify locations such as cities or to determine compass directions between points. Several conformal maps are used as navigation charts for pilots and seamen (Figure 2.8). The Mercator projection produces a conformal map and it has been widely used for navigation. One attribute of this projection is that a line of constant compass direction may be drawn with a straight edge. Lines of constant heading are called **Rhumb lines**. This characteristic made plotting and navigating the oceans more straightforward than with other projections. The Mercator projection exhibits massive size distortion in the high latitudes and for that reason should not be used to display comparitive areal data. They are useful and popular when displaying the standard time zones.

Other maps are developed to display earth area in equal proportions and have a property called **equivalence**; that is, each unit area of the earth surface is equally represented (Figure 2.9). These maps are used for an unlimited number of comparative purposes. They may be used to assess, for example, the relative extent of areas having suitable temperature and precipitation for producing a specific crop (e.g. wheat) in various parts of the world (e.g. North America versus Eurasia), to compare cropland in the United States with that in Russia, or to demonstrate global impacts of air pollution.

Another desirable property of maps is **true distance**. Maps designed to represent true distance are called **equidistant**, an example of which is the Azimuthal Equidistant (Figure 2.10). Distances measured radially from the central point of tangency to other locations are accurate and comparable. Azimuthal projections are created by projecting earth surface features onto a plane.

Only the globe can accurately display all four desirable map properties: equivalence, conformality, true distance, and direction. Properties of both equivalence and conformity cannot be uniformly presented on the same flat map. Thus, the use of a conformal map to illustrate the comparable sizes of land masses or of an equivalence map to identify missile targets not only would be misleading, but could result in disastrous consequences. The wise map-users are those who understand the characteristics of the maps with which they work.

The Isoline Map

We need to devote brief attention to a special map that is used constantly in the study of the atmosphere and earth – the **isoline map**. The isoline map provides an orderly presentation of numeric data in graphic form. On these maps, cumbrous volumes of recorded values are converted to comprehensible lines of *equal value* (iso means equal). The best way to understand isoline maps is to provide an example. In this case, our choice will be a

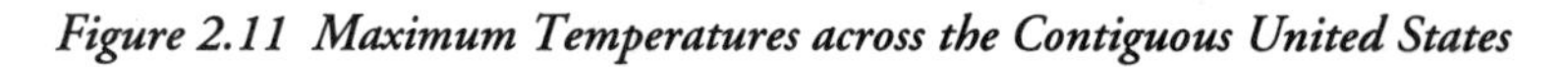

Figure 2.11 Maximum Temperatures across the Contiguous United States

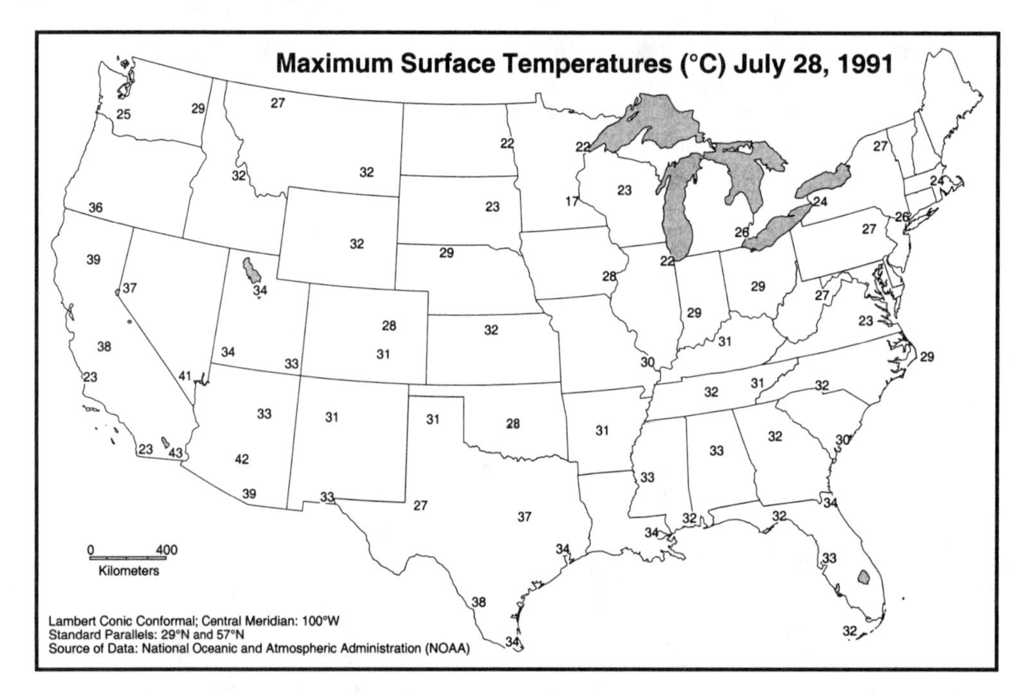

special isoline, the **isotherm**, (therm refers to temperature). An **isothermal map** displays lines of equal temperature. Figure 2.11 is a map of the contiguous United States on which temperature data are plotted at the location where they are recorded. The array of data is not quickly interpreted, unless you find a specific location and read its actual temperature. How does this relate to all the other temperature values on the map?

Construction of isolines helps to make sense out of the array of information. The task is easy. You simply reenact your childhood activity of drawing a picture by connecting dots number-by-number. There are only three minor differences between that activity and isoline construction: (1) rather than connecting points of ascending order, e.g. 1, 2, 3, to 50, you only connect points of equal value, such as all points having a temperature of 30°C (86°F); (2) isolines may never cross one another; and (3) you frequently have to make a decision where a given value exists between two-points of known-value, neither of which matches the value for which you are searching. Item number three is called **interpolation,** which is easy to understand with some examples.

Suppose you have two data points on a map that are spaced ten centimeters apart. The first value is 5°C and the other is 15°C. Where would a value of 10°C be found? The interpolation procedure is based upon the premise that values change in a constant linear fashion. Temperature should, then, increase 1°C for each centimeter of distance between the 5°C data point and the 15°C data point. Thus, the 10°C isotherm value will lie halfway between the data recording points. Figure 2.12 provides an illustration of how the interpolation procedure is operationalized. Now, let us return to the data set provided in Figure 2.11. When we interpolate our temperature values at 5°C isotherm intervals, the formerly unorganized data now reveal distinct patterns as shown in Figure 2.13. Because **cartographers** (professional mapmakers) remove the data points from the final map to make it more readable, isoline maps can illustrate complex data in a comprehensible form. There are many forms of isoline maps that show atmospheric data. In addition to isothermal maps, there are **isobaric maps** (depicting lines of equal atmospheric pressure), **isohyetal maps** (containing lines of equal precipitation amounts), and many more.

Figure 2.12 The Interpolation Procedure for Deriving the 10° C (50° F) Isotherm

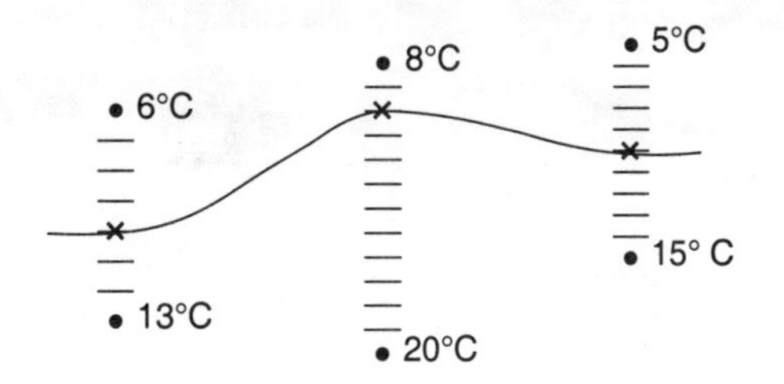

Figure 2.13 Isothermal Patterns of Maximum Temperature across the Contiguous United States

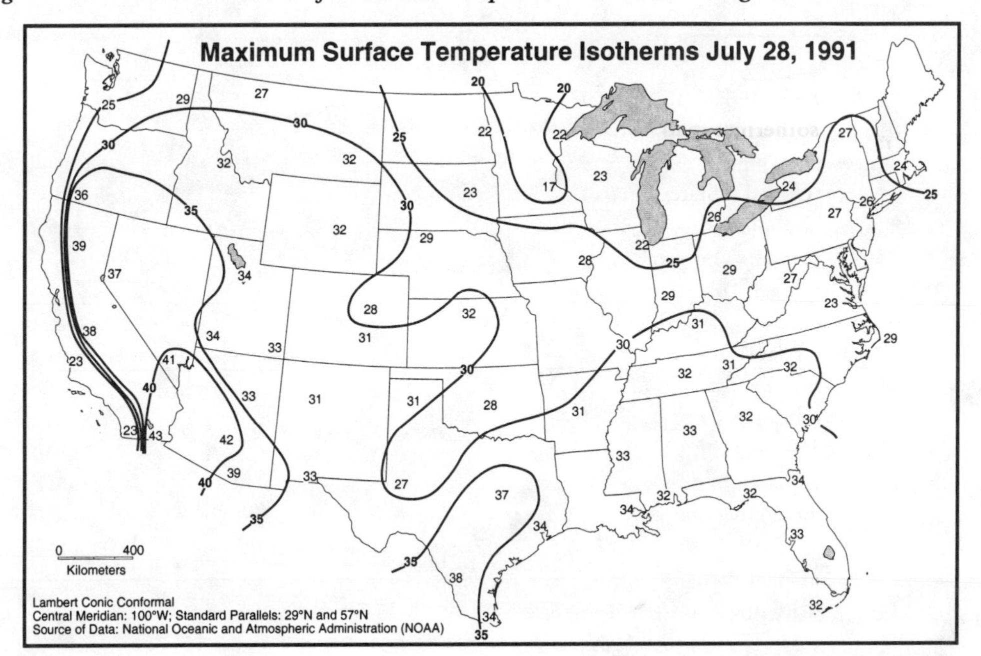

Activity: Location of Earth Features

OBJECTIVES

Knowledge of place locations is essential in understanding the frequent references made in physical geography to real world phenomena. In this activity you will review the location of major earth features. Examples of the types of features that you will locate include: oceans and seas, continents, mountain ranges, rivers and lakes, deserts, U.S. states, Canadian provinces, countries and regions. When you have finished you should be able to locate these features on an outline map. You should also know what types of information are contained in an atlas and how to use atlases and maps to find information.

MATERIALS

You will need this textbook, *Air & Water;* world and regional outline maps (provided in the Appendix B: Outline Maps); a comprehensive atlas such as the National Geographic World Atlas and/or the National Geographic North America Atlas. There are many atlases in any library devoted to many different themes or areas of the world, and any one of several others might also be useful for this review. If one is in use, find another and work with it for a while.

INSTRUCTIONS

Locate as many of the places listed in this activity as you can on the outline maps in Appendix B. If you are uncertain of a location, consult the Atlas of Placenames and Regions in Appendix A. Use the alphabetical list of placenames to find the appropriate map. Check your answers. Also you may use either a Goode's World Atlas, a National Geographic World, or National Geographic North America atlas as a supplement in locating each of the remaining features listed in this activity. Good atlases have a sizeable alphabetized list in the back. Use this index to find the page number and location of each feature. Plot and label each feature on the maps provided. Your maps need not be perfectly neat. It is your knowledge of the locations that is important.

Feature	***Appropriate Map***
Continents	
Africa	(World Map)
Antarctica	(World Map)
Australia	(World Map)
Eurasia	
Europe	(World Map)
Asia	(World Map)
North America	(World Map)
South America	(World Map)
Oceans and Seas	
Arabian Sea	(Middle East)
Arctic Ocean	(World Map)
Atlantic Ocean	(World Map)
Baltic Sea	(Europe)
Bay of Bengal	(India)
Black Sea	(Europe)
Caribbean Sea	(Caribbean)
Caspian Sea	(Russia)
Gulf of Mexico	(N. America)
Hudson Bay	(N. America)
Indian Ocean	(World Map)
Mediterranean	(Europe)
North Sea	(Europe)
Pacific Ocean	(World Map)
Persian Gulf	(Middle East)
Red Sea	(Africa)
Sea of Japan	(China)
South China Sea	(SE Asia)
North Atlantic Drift	(World Map)
Gulf Stream	(World Map)
California Current	(World Map)
Canary Current	(World Map)
Kuroshio Current	(World Map)
Labrador Current	(World Map)
Peru Current	(World Map)
Mountain Ranges	
Alps	(Europe)

Feature	***Appropriate Map***
Andes	(S. America)
Appalachians	(N. America)
Appenines	(Europe)
Atlas	(Africa)
Brazilian Highlands	(S. America)
Brook's Range	(N. America)
Carpathians	(Europe)
Cascade Range	(N. America)
Caucasus	(Russia)
East African Highlands	(Africa)
Great Rift Valley	(Africa)
Guiana Highlands	(S. America)
Himalayas	(India)
Pyrenees	(Europe)
Rocky Mountains	(N. America)
Sierra Madre Occidental	(Mexico)
Sierra Madre Oriental	(Mexico)
Transylvanian Alps	(Europe)
Tibetan Plateau	(China)
Tien Shan	(China)
Urals	(Russia)
Deserts	
Atacama	(S. America)
Gobi	(China)
Great Australian	
Great Sandy	(Australia)
Great Victoria	(Australia)
Gibson	(Australia)
Kalahari	(Africa)
Namib	(Africa)
Patagonia	(S. America)
Sahara	(Africa)
Sonoran	(N. America)
Thar	(India)
Rivers	
Amazon River	(S. America)
Angara River	(Russia)
Colorado River	(N. America)

Feature	***Appropriate Map***
Columbia River	(N. America)
Congo River	(Africa)
Danube River	(Europe)
Euphrates River	(Middle East)
Ganges River	(India)
Hudson River	(N. America)
Hwang-ho (Yellow)	(China)
Indus River	(India)
Lena River	(Russia)
Mackenzie River	(N. America)
Mekong River	(SE Asia)
Mississippi River	(N. America)
Missouri River	(N. America)
Murray River	(Australia)
Niger River	(Africa)
Nile River	(Africa)
Ob (Obi) River	(Russia)
Ohio River	(N. America)
Orinoco River	(S. America)
Po River	(Europe)
Rhine River	(Europe)
Rhone River	(Europe)
Rio Grande	(N. America)
Rio Paraguay	(S. America)
Rio Parana	(S. America)
St. Lawrence River	(N. America)
Tennessee River	(N. America)
Tigris River	(Middle East)
Volga River	(Russia)
Yangtse-Kiang River	(China)
Yukon River	(N. America)
Zambezi River	(Africa)
Lakes	
Great Lakes	(N. America)
Lake Erie	(US)
Lake Huron	(US)
Lake Michigan	(US)
Lake Ontario	(US)

Feature	***Appropriate Map***
Lake Superior	(US)
Great Salt Lake	(N. America)
Lake Baikal (Baykal)	(Russia)
Lake Malawi	(Africa)
Lake Maracaibo	(S. America)
Lake Tanganyika	(Africa)
Lake Titicaca	(S. America)
Lake Victoria	(Africa)
Countries & Regions	
Africa	
Algeria	(Africa)
Egypt	(Africa)
Ethiopia	(Africa)
Ghana	(Africa)
Kenya	(Africa)
Libya	(Africa)
Morocco	(Africa)
Nigeria	(Africa)
Somalia	(Africa)
South Africa	(Africa)
Tanzania	(Africa)
Zaire	(Africa)
Asia	
East Asia	
China	(China)
Japan	(China)
Korea	(China)
Manchuria	(China)
Mongolia	(China)
Taiwan	(China)
Southeast Asia	
Indonesia	(SE Asia)
East Timor	(SE Asia)
Kampuchea	(SE Asia)
Malaysia	(SE Asia)
Myanmar	(SE Asia)

Feature	Appropriate Map
Phillipines	(SE Asia)
Viet Nam	(SE Asia)
Southern Asia	
Bangladesh	(India)
India	(India)
Nepal	(India)
Pakistan	(India)
Sri Lanka	(India)
Tibet	(India & China)
Southwestern Asia	
Turkmenistan	(Russia)
Uzbekistan	(Russia)
Tajikistan	(Russia)
Kirghizia	(Russia)
Europe	
Armenia	(Russia)
Austria	(Europe)
Azerbaijan	(Russia)
Bosnia	(Europe)
Byelorussia	(Europe & Russia)
Croatia	(Europe)
Czech Republic	(Europe)
Denmark	(Europe)
Estonia	(Europe & Russia)
Finland	(Europe)
France	(Europe)
Georgia	(Russia)
Germany	(Europe)
Greece	(Europe)
Hungary	(Europe)
Ireland	(Europe)
Italy	(Europe)
Kosovo	(Europe)
Latvia	(Europe & Russia)
Lithuania	(Europe & Russia)
Moldavia	(Europe & Russia)
Netherlands	(Europe)
Norway	(Europe)
Poland	(Europe)

Feature	Appropriate Map
Portugal	(Europe)
Romania	(Europe)
Russia	(Europe & Russia)
Serbia	(Europe)
Slovakia	(Europe)
Spain	(Europe)
Sweden	(Europe)
Switzerland	(Europe)
Ukraine	(Europe & Russia)
United Kingdom	
England	(Europe)
Northern Ireland	(Europe)
Scotland	(Europe)
Wales	(Europe)
Yugoslavia	(Europe)
Greenland	(N. America)
Iceland	(Europe)
North America	
United States	(N. America)
all 50 states	(US)
Aleutian Islands	(N. America)
Canada	
Canadian Provinces:	
Alberta	(N. America)
British Columbia	(N. America)
Labrador	(N. America)
Manitoba	(N. America)
New Brunswick	(N. America)
Newfoundland	(N. America)
Northwest Territories	(N. America)
Nova Scotia	(N. America)
Ontario	(N. America)
Prince Edward Island	(N. America)
Quebec	(N. America)
Saskatchewan	(N. America)
Yukon	(N. America)
Baffin Island	(N. America)
Mexico	(Mexico)
El Salvador	(Mexico)

Feature	***Appropriate Map***
Central America	
Honduras	(Mexico)
Guatemala	(Mexico)
Belize	(Mexico)
Costa Rica	(Mexico)
Panama	(Mexico)
Caribbean Sea	
Bahama Islands	(Caribbean)
Haiti	(Caribbean)
Dominican Republic	(Caribbean)
Puerto Rico	(Caribbean)
Grenada	(Caribbean)
Barbados	(Caribbean)
Cuba	(Caribbean)
South America	
Argentina	(S. America)
Bolivia	(S. America)
Brazil	(S. America)
Chile	(S. America)
Colombia	(S. America)
Ecuador	(S. America)
Peru	(S. America)
Venezuela	(S. America)
Middle East	
Bahrain	(Middle East)
Iran	(Middle East)
Iraq	(Middle East)
Israel	(Middle East)
Jordan	(Middle East)
Kuwait	(Middle East)
Lebanon	(Middle East)
Saudi Arabia	(Middle East)
Syria	(Middle East)
Turkey	(Middle East)
Oceania	
Australia	(Australia)
New Guinea	(Australia)
New Zealand	(World Map)

Feature	***Appropriate Map***
Cities	
Akron, OH	(US Metro)
Albuquerque, NM	(US Metro)
Anchorage, AK	(US)
Asheville, NC	(US Metro)
Atlanta, GA	(US Metro)
Austin, TX	(US Metro)
Barrow, AK	(US)
Baton Rouge, LA	(US Metro)
Beijing, China	(China)
Berlin, Germany	(Europe)
Bombay, India	(India)
Bonn, Germany	(Europe)
Boston, MA	(US Metro)
Buffalo, NY	(US Metro)
Cairo, Egypt	(Africa)
Charleston, SC	(US Metro)
Charlotte, NC	(US Metro)
Chicago, IL	(US Metro)
Cincinnati, OH	(US Metro)
Cleveland, OH	(US Metro)
Cochin, India	(India)
Columbia, SC	(US Metro)
Columbus, OH	(US Metro)
Dallas, TX	(US Metro)
Dayton, OH	(US Metro)
Denver, CO	(US Metro)
Detroit, MI	(US Metro)
Durham, NC	(US Metro)
Fort Worth, TX	(US Metro)
Fresno, CA	(US Metro)
Grand Rapids, MI	(US Metro)
Greensboro, NC	(US Metro)
Halifax, Nova Scotia	(N. America)
Hartford, CT	(US Metro)
Hong Kong	(China)
Houston, TX	(US Metro)
Indianapolis, IN	(US Metro)
Iquitos, Peru	(S. America)

Feature	***Appropriate Map***
Jacksonville, FL	(US Metro)
Kansas City, MO	(US Metro)
Kisangani, Zaire	(Africa)
Las Vegas, NV	(US Metro)
London, England	(Europe)
Los Angeles, CA	(US Metro)
Louisville, KY	(US Metro)
Memphis, TN	(US Metro)
Miami, FL	(US Metro)
Milwaukee, WI	(US Metro)
Minneapolis, MN	(US Metro)
Mobile, AL	(US Metro)
Montreal, Quebec	(N. America)
Moosonee, Ontario	(N. America)
Moscow, Russia	(Russia)
Murmansk, Russia	(Russia)
Nashville, TN	(US Metro)
New Orleans, LA	(US Metro)
New York, NY	(US Metro)
Norfolk, VA	(US Metro)
Oklahoma City	(US Metro)
Orlando, FL	(US Metro)
Ottawa, Ontario	(N. America)
Paris, France	(Europe)
Philadelphia, PA	(US Metro)
Phoenix, AZ	(US Metro)
Pittsburgh, PA	(US Metro)
Portland, OR	(US Metro)
Providence, RI	(US Metro)
Quebec, Quebec	(N. America)
Raleigh, NC	(US Metro)
Reykjavik, Iceland	(Europe)
Richmond, VA	(US Metro)
Rochester, NY	(US Metro)
Sacramento, CA	(US Metro)
Salt Lake City, UT	(US Metro)
San Antonio, TX	(US Metro)
San Diego, CA	(US Metro)
San Francisco, CA	(US Metro)

Feature	***Appropriate Map***
San Jose, CA	(US Metro)
Seattle, WA	(US Metro)
Shreveport, LA	(US Metro)
Singapore	(SE Asia)
St. Louis, MO	(US Metro)
St. Paul, MN	(US Metro)
St. Petersburg, FL	(US Metro)
St. Petersburg, Russia	(Russia)
Tampa, FL	(US Metro)
Timbuktu, Mali	(Africa)
Toronto, Ontario	(N. America)
Tokyo, Japan	(China)
Vancouver, British Columbia	(N. America)
Verkhoyansk, Russia	(Russia)
Vladivostok, Russia	(Russia)
West Palm Beach, FL	(US Metro)
Wilmington, NC	(US Metro)
Winnipeg, Manitoba	(N. America)
Winston-Salem, NC	(US Metro)

Activity: The Grid System and Time

OBJECTIVES

Given a globe or flat map of the world, you will be able to: (1) correctly determine longitude and latitude of any point on the earth's surface; (2) calculate distance between any two points on the earth lying due east or west from each other at any latitude; (3) calculate distance in miles between any two points on the earth lying due north or south from each other; (4) calculate differences in time and date between any two point locations on the earth; and (5) determine time differences between major places within the United States.

MATERIALS

Time chart and globe

GRID SYSTEM

The Earth's geographic North and South poles (defined as the points of intersection of the earth's rotational axis with the earth's surface) have been designated as reference points upon which the most widely used grid system is based. The equator is a line drawn midway between the North and South polar axis. Examine a globe to familiarize yourself with these features.

Latitude is the measurement of distance north or south from the equator. Lines of latitude run east and west around the earth and are called parallels because they are parallel to the equator and to each other. If viewed from the poles, they appear as a series of concentric circles. The line of latitude of largest diameter is the equator. This is an important reference line on the globe for it divides the earth into two halves, the northern and southern hemispheres. The equator is assigned a value of 0° latitude. Between the equator and the poles there are 90°. This allows one to find one's position between either pole and the equator as somewhere between 0° and 90° north or south of the equator. A latitude of 45° North would be half way between the pole and the equator in the northern hemisphere. Use your atlas or globe to find the approximate latitude, in degrees north or south, for the locations listed in question #1 below.

Longitude is the measurement of distance east or west. Lines of longitude also are called meridians. Meridians begin and end at the poles. The Prime Meridian (which serves as the starting or 0° meridian) passes through Greenwich, England (within greater London) and connects the North and South Poles. Opposite the Prime Meridian is the 180° meridian. If you put two opposing meridians together, a great circle will be formed. A great circle divides the earth into hemispheres. It is the largest circle that can be drawn on the earth's surface, and it is important because the shortest distance between any two places follows the arc or path of a great circle. An infinite number of great circles can be drawn on the globe. Just as the equator divides the earth into a northern and a southern hemisphere, there is a great circle that divides the earth into an eastern and a western hemisphere. This great circle is formed by the Prime Meridian and the 180° meridian together. Locate these two meridians on a globe. The Prime Meridian passes through Greenwich, England by international agreement.

There are 360° of longitude on the globe: 180° are east of the Prime Meridian, and 180° are west of it. When establishing the location of a place (its coordinates), it is important to know whether it is east or west of the Prime Meridian. Use a globe to note that Raleigh, North Carolina is about 35° north latitude and 78° west longitude. This means that Raleigh is 35° north of the equator and 78° west of the Greenwich 0° line. The correct way to write these coordinates is: 35°N, 78°W. Complete question #1 by finding the latitudinal and longitudinal position of each of the places listed.

1. Map Coordinates

   a. Raleigh, North Carolina

   b. Dublin, Ireland

   c. Manila, Philippines

   d. Sydney, Australia

   e. Pt. Barrow, Alaska

Now let's see what distance on the earth's surface is represented by 1° of latitude and 1° of longitude. The calculation of actual ground distance from degree measurement is not always straightforward. First examine Figure 2.1.

2. Do the distances represented by travel lines #1 and #2 appear to be the same? __________

Figure 2.1

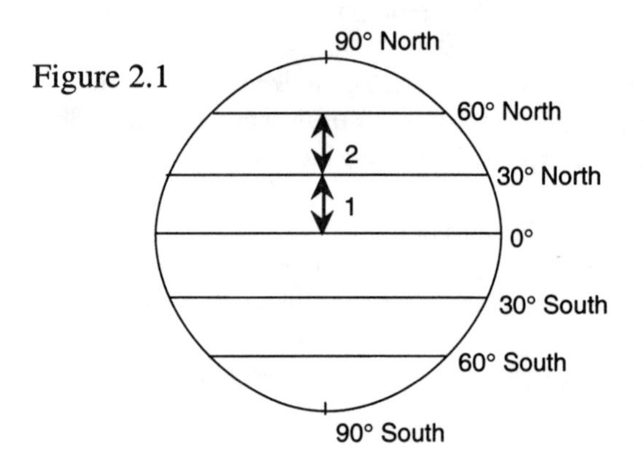

Figure 2.2

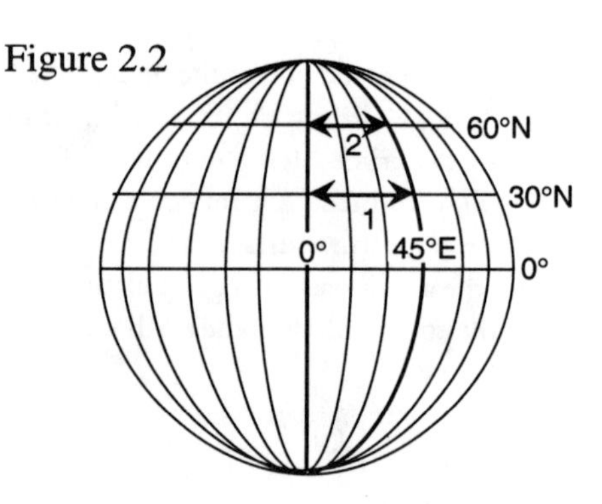

For distance #1, travel in a north-south direction between 0° and 30° North, calculate the number of degrees and the number of miles travelled. Record your calculations in the space provided below. Then, calculate the distance along travel route #2, between 30° and 60° degrees latitude.

3. Calculate the distance in miles between the equator (0°) and a point 30 degrees north latitude.

4. Calculate the number of miles between 30 degrees north latitude and 60 degrees north latitude.

5. Do your calculations agree with your earlier observations? Are the distances the same?______

From this, we may conclude that the number of miles in any degree of latitude (travelling in a North-South direction) is constant. This distance is equal to 69 miles because parallels are always equidistant and the shortest distance between them is along a great circle.

Now let's consider a similar problem, but this time travelling 45° of longitude in a due easterly or westerly direction. Figure 2.2 shows two travel routes. Travel route #1 is along the 30° parallel or line of latitude. Travel route #2 is at 60° North latitude. Do the actual travel distances appear to be the same? The answer is no because meridians converge at the poles. Consequently, although we may be travelling through 45° of longitude at each latitude, we are actually travelling a shorter distance at 60° North. The closer we get to the poles, the shorter distance per degree. Compare travel routes #1 and #2 for a confirmation. At the poles where lines of longitude converge, neither east nor west travel is possible.

The degree to which the lines of longitude converge as we proceed poleward from the equator is described by a trigonometric function, the cosine. The cosine value at the equator is 1. The cosine value at the poles is 0. At latitudes between the equator and the poles, the cosine value lies somewhere between 1 and 0. Note also that the cosine becomes progressively smaller from the equator poleward. This cosine function identifies the portion of 69 miles that is equal to the width of a degree of longitude at various latitudes.

Examine travel route #1 in Figure 2.2 at 30° north between the two lines of longitude that are 45° apart. First decide how many miles are in a degree at 30° North. This is done by multiplying 69 miles (the number of miles in a degree at the equator) by the cosine of the latitude (30°). The cosine of 30° is 0.866.

$$69 \text{ miles/degree} * \cos 30^\circ = 69 \text{ miles/degree} * 0.866 = 59.8 \text{ miles/degree}$$

This means that at 30° North (or South), there are 59.8 miles in each degree of longitude. You would like to know how many miles there are in 45° of east-west travel at that latitude. So now all you need to do is multiply:

$$59.8 \text{ miles/degree} * 45^\circ = 2{,}689 \text{ miles}$$

6. a. Calculate the number of miles between the prime meridian (zero degrees) and a point 45 degrees east along the 30 degree parallel (cos 30° = 0.866).

   b. Calculate the number of miles between the same number of degrees of longitude (45°) along the 60 degree parallel (cos 60° = 0.5).

TIME ZONES

You are already aware that at different longitudinal positions on the earth at a given instant, the clock times and even the calendar dates may differ. You know that on the California coast it is earlier in the day than it is on the east coast of the United States. When you think for a moment of the earth's rotation on its axis, the reason is clear. If you could look down upon the earth at the North pole, you would see that the earth is rotating in a counter clockwise direction.

Assume that you are standing at point "X" on the earth. The sun is directly overhead so it is 12:00 noon. If suddenly you could transport yourself in an easterly direction to a point "A" east of point "X" would the time there be earlier or later? It would have to be later because the sun would appear to be lower in the western sky. Conversely, if you could suddenly transport yourself to a point "B" to the west, it would be earlier in the day because the sun would appear to be lower in the eastern sky than when you were at point "X".

A chronometer is a very accurate clock set to Greenwich Mean Time (GMT) which is used to determine longitude in degrees east or west of the Prime Meridian (zero line of longitude running through Greenwich, England). For every 15° longitude away from the Prime Meridian, there will be a one hour difference in time. As you travel east or west away from the Prime Meridian, you allow the chronometer to remain set for the correct time along the Prime Meridian. To find your longitude, all that is necessary is to determine noon for wherever you might be. You do this by setting your own watch at 12:00 Noon whenever the sun reaches its highest point in the sky. Let's assume that you observe noon at your longitude, at the same moment it is 3:00 p.m. in Greenwich. This gives a discrepancy of 3 hours between the two time pieces. This means that you are 3 times 15° longitude away from Greenwich, or at a 45° line of longitude (either 45°E or 45°W). You would be 45° west of Greenwich because it is three hours earlier where you are than it is in Greenwich.

7. Briefly describe the relationship between the rotation of the earth and the passage of time.

8. Because there are 24 hours in a day and 360° in a circle, how many degrees of longitude pass beneath the sun in one hour?

STANDARD TIME ZONES

For each 15° of longitude, there is a 1 hour difference in the position of the sun (differences in the apparent time of day). If you are going east, the time advances one hour or if you are going west, the time will be one hour earlier for each 15° of travel. This allows us to divide the earth's surface into 24 time zones that correspond, generally, to lines of longitude that are 15° apart. These special lines of longitude are called standard meridians. Each of these standard meridians functions as the "central" meridian for a time zone. Central meridians occur every 15° of longitude starting from the Prime Meridian and extending east and west. There are 24 of these central or standard meridians. Standard time zones extend approximately 7.5° on either side of each standard meridian. Examine a time and date map and record the values for the standard meridians for the time zones listed below. Note that the standard meridian is at the approximate center of each time zone and that the time zone boundaries are not always straight. Convenience and political reasons account for the irregular paths. Study the time zone map of the United States (below) and be able to determine differences between places such as states or major cities.

9. Identify the standard meridian for each related time zone:

   a. Eastern Time Zone (Example) 75°W

   b. Central Time Zone ______________________________

   c. Mountain Time Zone ______________________________

   d. Pacific Time Zone ______________________________

   e. Alaska Time Zone ______________________________

   f. Hawaii-Aleutian Time Zone______________________________

Use the globe, an atlas, and/or the global time zone map in this volume to complete the time problems in question #10.

10. Time Problems. If it is 2:00 p.m in Charlotte, North Carolina, what time is it in:

    a. Denver, Colorado (39°N 105°W) ______________________________

    b. Anchorage, Alaska (61°N 150°W) ______________________________

    c. Dublin, Ireland (53°N 6°W) ______________________________

    d. Cairo, Egypt (30°N 31°E) ______________________________

e. Rio de Janeiro, Brazil (23°S 43°W) ______________________________

f. San Jose, California (Pacific Time) ______________________________

THE INTERNATIONAL DATE LINE AND THE MIDNIGHT MERIDIAN

The International Date Line (IDL) was selected as the boundary for separating dates. The date line was chosen because its location (approximately that of the 180° meridian) is half way around the earth from the 0° meridian. It should be remembered that the IDL is the standard meridian for a time zone as well as being a change of date line.

Assume that you are crossing the date line while traveling in a westward direction. Find the dateline on a globe or map and see if you can determine the change in date. Will it become a day later on the calendar or a day earlier?

When you cross the IDL traveling west, the calendar will be adjusted forward a day. That is, it will be a day later. For example, if it is Monday before you cross the IDL, it will be Tuesday after you cross the IDL. The time (hour) will be the same. When you cross the IDL traveling east, the calendar will be adjusted back one day. That is, it will be a day earlier. For example, if it is Monday before you cross the

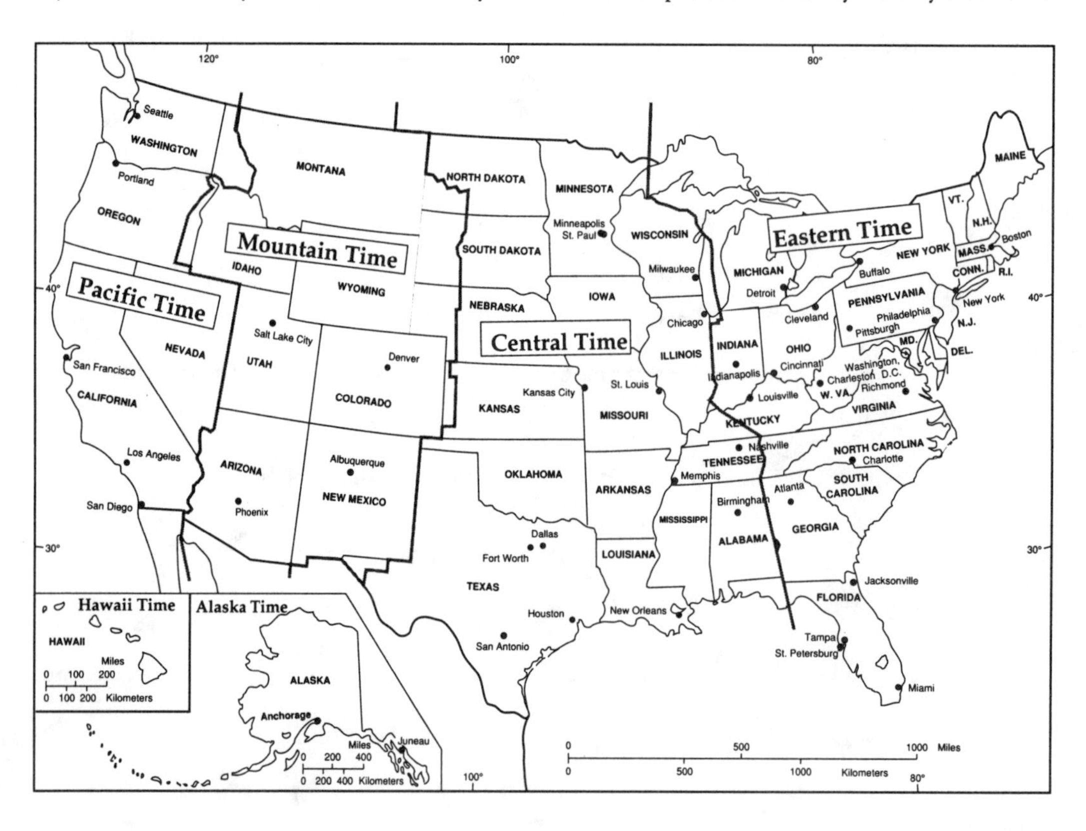

IDL, it will be Sunday after you cross the IDL. The hour will remain the same.

The other meridian that separates calendar dates is the midnight meridian (Figure 2.3). The midnight meridian is a line that connects all the places that are experiencing midnight. Crossing the midnight meridian traveling east coincides with changing the calendar one day forward. This should come as no surprise; it happens every night. Crossing the midnight meridian traveling west changes the date to the previous day. Crossing the IDL and the midnight meridian in the course of a single trip cancels the effect of each and results in no change of date. In calculating date differences between places be sure to consider whether the IDL, the midnight meridian, or both have been crossed.

11. If it is 9:00 a.m. on Friday, December 16 in Los Angeles, California, what time is it in Tokyo, Japan? Is it the same date?

12. If it is 2:00 p.m. on June 2 in Sydney, Australia, what is the time and date in Fairbanks, Alaska?

Figure 2.3

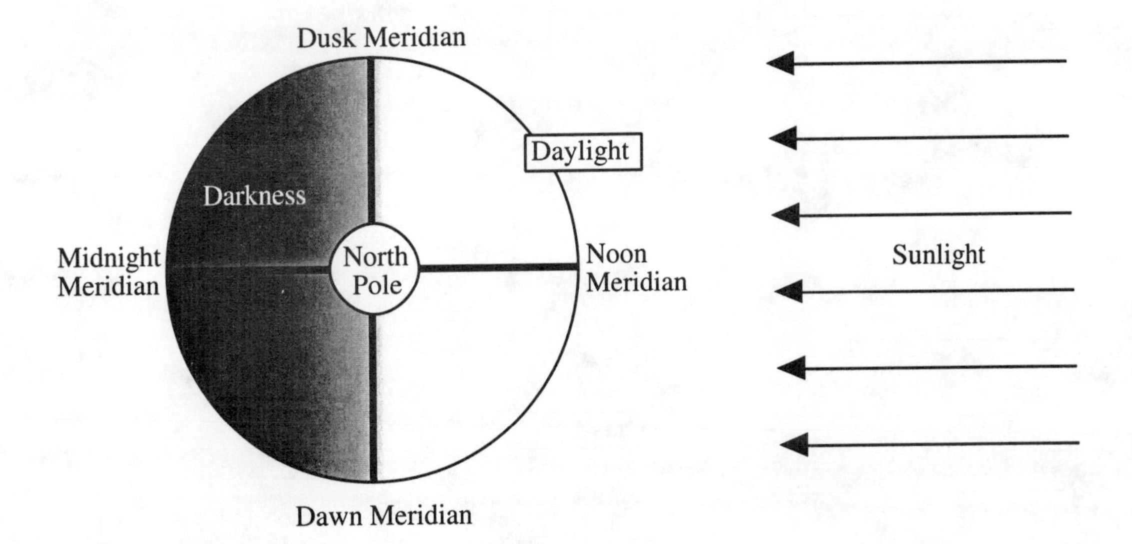

Chapter 3

The Composition and Structure of the Atmosphere

The atmosphere is the envelope of gases that surrounds the solid earth. It plays a key role in many earth processes. It is closely linked with the oceans by the cycling of water vapor and energy and by winds which provide the driving force for much of the ocean's circulation. It is linked to the solid earth by weathering and erosion processes that depend on the atmosphere for the water and heat that provide chemical reactions and by interchanges of numerous substances such as carbon, nitrogen, sulfur and dust. Organisms of the biosphere are dependent on the atmosphere for carbon dioxide and oxygen that they need and for a favorable growth environment. It is for these reasons that we begin with a discussion of the composition and structure of the atmosphere.

The importance of weather to human activities has made the study of the atmosphere a major science, and though much remains to be learned about it, we have learned a great deal in the last half century. In recent years scientists have developed sophisticated techniques for exploring the atmosphere. Radars are used to probe for information about precipitation and tornadoes; balloons and rockets send back temperature, wind, and pressure data; and satellites photograph storms and measure infrared radiation. Meteorologists compile thousands of bits of data from these and other sources and analyze the data with complicated computer models to predict weather events and atmospheric behavior.

Composition

Earth is not unique among planets because it has an atmosphere. Its envelope of gases, however, is different from that of other planets in our solar system. Some differences among planetary atmospheres, such as air density and seasonal temperature contrasts, are due to: (1) variation in sizes of the planets and the gravitational force they impart upon their atmospheres and (2) distance and angular relationships that each planet has with the Sun. These concepts will be reserved for discussion later in the text. Our present concern is with the chemical composition of earth's gaseous envelope which is radically different, particularly in its oxygen content, from other planets of the solar system.

The present atmosphere had its origins in the primordial earth that was much hotter than the one we know today. Gases were released from superheated rock material in the same fashion as today's active volcanoes emit water vapor, carbon dioxide, nitrogen, and small amounts of sulfur compounds. During the early accretionary stage of the earth's formation, the primordial atmosphere is thought to have been 60 – 70% water vapor, 10 – 15% carbon dioxide, and 8 – 10% nitrogen although many uncertainties remain. Such a steam atmosphere would be transient and cool quickly as the accretionary stage ended. The primitive atmosphere could have contained 10 bars of carbon dioxide and carbon monoxide plus 1 bar of nitrogen during the first several million years of earth history.[1] Surface temperatures beneath such an atmosphere would have been near 85°C. Most of the water vapor in the original atmosphere would have condensed into huge water molecules as the earth and atmosphere cooled. These subsequently formed raindrops that fell to earth under the force of gravity to accumulate as the world ocean. In the process much of the original atmospheric carbon dioxide was absorbed by the falling rain to subsequently combine with soluble earth minerals and form vast deposits of rock, such as calcium and magnesium carbonates. Nitrogen, which is relatively inert, probably accumulated gradually. Evidence from microfossils and stromatolites in ancient sediments suggests that the earth supported life as early as 3.5 billion years ago. Major increases in the level of free oxygen began between 1.9 and 2.2 billion years ago and resulted from photosynthesis. As oxygen was gradually released from plants, atmospheric levels remained at around 3% of present levels as oxygen was being stored first in the surface layer and subsequently in the deep ocean. As these oxygen sinks filled, oxygen concentrations in the atmosphere gradually rose to those observed in the present atmosphere.

The current atmospheric composition is vastly different from the primordial atmosphere, but some of the formative processes remain unchanged (Table 3.1). Oxygen, upon which human beings are totally dependent for survival, originates from plants as a by-product of photosynthesis. In this process plants dissociate water molecules consisting of two hydrogen and one oxygen atoms (H_2O) and respire oxygen to the atmosphere. At the same time, they assimilate carbon dioxide, water, and solar energy to produce carbohydrates. Thus, as plant life evolved and became established over large expanses of the land surfaces, atmospheric oxygen content gradually increased to the levels we currently monitor. Despite the importance of oxygen in life processes, it is not the most abundant substance in the gases surrounding the earth. There is nearly four times as much nitrogen as oxygen.

There are three basic states of substances: gases of various types, water in liquid and solid forms, and mineral and organic particles. The latter are referred to as aerosols (particles small enough to be suspended in the gaseous matrix due to their buoyancy). Each of these three substances influences atmospheric processes by its different effects on, and responses to, energy inputs. In this section the major constituents of air are described. Though some of the more important effects that certain substances have on atmospheric processes are mentioned in this section, the processes themselves are not treated in detail at this time.

Gases

Nitrogen, oxygen, and argon are present in relatively constant quantities, and they constitute over 99.99% of the volume of air within the lower 16 kilometers of the atmosphere. Water vapor may vary from nearly 0% over cold deserts to as much as 3% in the humid tropics. Carbon dioxide is present everywhere, but the concentration varies from place to place and from season to season. Amounts of some gases, such as surface ozone, carbon monoxide, or sulfur dioxide, are largely the product of human activities and are present in significant quantities only in urbanized and highly industrialized areas.

[1]*See James F. Kasting, "Earth's Early Atmosphere," Science 259 (February 12, 1993): 920-926.*

Table 3.1 Composition of the Earth's Atmosphere

Gases That Occur in Relatively Constant Amounts		Volume (%)	Concentration (ppm)
Nitrogen	(N_2)	78.08	780,840.00
Oxygen	(O_2)	20.95	209,460.00
Argon	(Ar)	0.93	9,340.00
Neon	(Ne)	0.0018	18.18
Helium	(He)	0.0005	5.24
Krypton	(Kr)	0.0001	1.14
Hydrogen	(H_2)	trace	0.50
Xenon	(Xe)	trace	0.09
Gases That Occur in Variable Amounts			
Water vapor	(H_2O)	0.01 - 3.0	0.1 - 30,000.00
Carbon dioxide	(CO_2)	0.035	350.00
Methane	(CH_4)	0.0002	1.67
Nitrous oxide	(N_2O)	trace	0.30
Carbon monoxide	(CO)	trace	0.19
Ozone	(O_3)	trace	0.04
Ammonia	(NH_3)	trace	0.004
Nitrogen dioxide	(NO_2)	trace	0.001
Sulphur dioxide	(SO_2)	trace	0.001
Nitric oxide	(NO)	trace	0.0005
Hydrogen sulfide	(H_2S)	trace	0.00005

Particulates (Aerosols)

- Dust
- Smoke
- Salt nuclei
- Water droplets
- Ice crystals

Sources: Godish, Thad, 1991: Air Quality. Lewis Publishers, Inc.: Chelsea, Michigan; NOAA, 1976: U.S. Standard Atmosphere.

Even gases that are present in small quantities may be significant in their impact. Ozone accounts for only a tiny fraction of atmospheric mass. In fact, if all atmospheric ozone were concentrated at sea level, it would constitute a layer of air only a few millimeters thick. Yet, an ozone rich layer occurs between 15 and 50 kilometers and shields biologic life from harmful ultraviolet solar radiation. On the other hand, sulfur dioxide, which is produced by burning fossil fuels containing sulfur, may be sufficiently concentrated in some industrial areas to corrode building materials and statuary over a period of time.

Aerosols

Colloidal–sized particles, cloud droplets, or ice crystals suspended within the air are called aerosols. Aerosols serve several functions. Some, like salt, are important as nuclei upon which cloud droplets may form. Without sufficient numbers of them, little precipitation could occur regardless of the amount of water vapor present. In addition, solid particles absorb and scatter light, thereby reducing the amount of sunlight that reaches the earth's surface.

Solid particles such as dust and smoke originate near the earth's surface and are confined largely to the lower atmosphere. These particles are heavier than air and, despite the lifting effects of turbulence, eventually settle out. The settling velocity depends on particle shape, size, and density. Most aerosols spend a relatively short time (a few weeks or less) in the atmosphere.

Water vapor, carbon dioxide, and dust are good absorbers of the radiation emitted by the earth (called long-wave radiation). The concentration of particulates near the surface and the greater density of air molecules in the lower atmosphere give the lower atmosphere a greater heat capacity and higher temperatures.

Structure of the Atmosphere

Air Pressure

The gaseous matter of our atmosphere has mass (matter), just like any other substance in our solar system. As we walk about each day, we rarely think about the weight exerted upon our bodies by the air surrounding us. Yet this force amounts to about 1.03 kg/cm^2, or 15 $lbs./in^2$. If this seems insignificant, try to quickly estimate the atmospheric pressure that is now being exerted upon your hand. From the wrist to fingertips your authors' hands are about 21 cm (8 inches) long and 11.5 cm (4.5 inches) wide, a surface area of 242 cm^2 (36 in^2). Average pressure (or weight) on this one hand is, therefore 242 cm^2 x 1.03 kg/cm^2 (36 in^2 x 15 lbs/in^2) or equal to 249 kg (540 lbs.). Bear in mind, your hand represents a very small portion of your total body area. If you were to extend this mental exercise one step further and take into account your entire body's surface area, the results would indicate that atmospheric pressure (weight) upon your body is equivalent to many tons. Fortunately the internal structure and outward pressure exerted by your body are normally balanced with the external atmospheric pressure. If this were not the case we would be crushed by atmospheric pressure or hemorrhage (if the external pressure suddenly became nonexistent).

Beside having weight, the gases of the atmosphere (like all gases) are compressible (capable of being concentrated in a smaller space). This means that, in some circumstances, more gas molecules may exist in a given volume of space than in other conditions. Current technology provides numerous examples of utilizing compressed gas. These include compressed propane in tanks for home heating and barbecue grills, and compressed oxygen both for hospital use and scuba–diving gear, and certainly compressed air to power pneumatic tools.

Mass and compressibility of gases have important implications for understanding the nature and behavior of earth's atmosphere. Because atmospheric gases have mass they are subject to gravitational attraction (the force of attraction between the earth and all matter surrounding it). Gravitational attraction, or gravitational force, is inversely proportional to the square of distance between two bodies having mass. It is mathematically expressed as:

$$F = G\frac{m_1 m_2}{r^2}$$

where m_1 and m_2 are the masses of the two attractive bodies, G is a universal constant equal to 3.3^{-11} when F and m are in pounds and feet respectively, and r is the distance between the two masses. At the earth's surface, the mass of the other objects can be ignored and the equation becomes:

$$F = G\frac{m_1}{r_1}$$

Gravity at the earth's surface represents an acceleration of 32.15 feet per second per second or 9.8 meters per second per second. This means that within a vacuum where the frictional resistance of air does not exist, any falling object regardless of its size or weight will descend 9.8 meters during the first second, 19.6 meters during the next second, 29.4 meters during the third second, and so on.

Mass, compressibility of gas, and gravitational force all are responsible for the altitudinal variation in atmospheric pressure. Clearly, gases near the earth's surface will experience greater gravitational attraction from the earth than will gas molecules at the outer fringe of the atmosphere. Also, because all air molecules above those at the surface are being pulled toward the earth's surface by gravity, there is a cumulative downward force that causes near-surface air molecules to be more compressed and dense (more molecules per unit volume) than is characteristic of air distant from the surface.

The balance between air pressure and gravity is called hydrostatic equilibrium. The result is high atmospheric pressure near the ground with pressure decreasing at an increasing rate vertically. At sea level average or standard pressure (force) of the atmosphere is 1013.2 millibars.[2] At 5.6 km (18,500 feet or 3.5 miles), the approximate elevation of the highest human settlement in the Andes, air pressure is approximately 500 millibars. This means that,

Figure 3.1 Atmospheric Pressure and Altitude. Half of the Atmosphere Lies Beneath 6 Kilometers

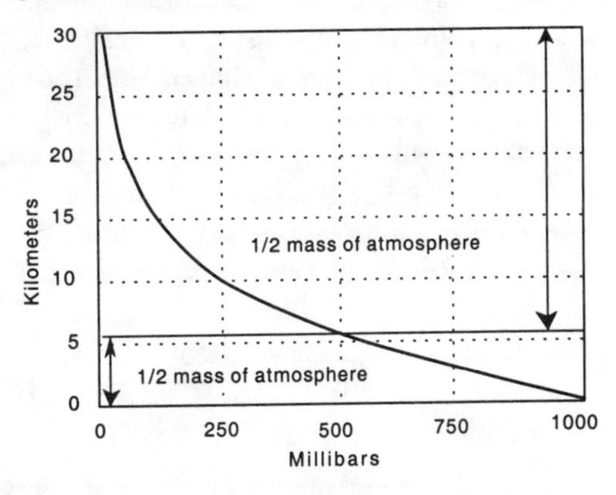

[2] A millibar is a unit of force equal to 1,000 dynes/cm². A dyne is the force necessary to accelerate one gram (0.035 oz.) one centimeter (0.39 in.) per second

for practical purposes, half the atmospheric mass is below that altitude. Figure 3.1 illustrates the vertical change of atmospheric pressure in millibars.

Air Temperature

In addition to significant decrease in atmospheric pressure relative to distance from the earth's surface, our atmosphere also has a distinct, although more complicated, change in temperature. Prior to addressing the nature of this change, however, it is worthwhile for us to address the various ways temperature values are recorded. Temperature is a measure of the mean kinetic energy of the air; that is, an indicator of the velocity at which air molecules travel and bombard one another. It is common in the United States to use the Fahrenheit Scale, whereon 32°F represents the freezing point and 212°F identifies the boiling point of distilled water. Most countries in the world utilize the metric temperature scale, and it is likely the United States will adopt this scale in the future. The metric scale, also called Celsius, has a freezing temperature of 0°C (equal to 32°F) and a boiling point of 100°C (equal to 212°F) for distilled water. Another temperature scale, widely used in science, is absolute, or Kelvin, temperature. This temperature scale has a lower value of 0K (-273°C), at which matter is cool enough to stop internal molecular motion. Freezing and boiling temperatures of distilled water on the Kelvin scale, therefore, are 273K (equal to 0°C or 32°F) and 373K (equal to 100°C or 212°F). In practice, it is simple to determine Kelvin degrees by adding 273 to the Celsius temperature. Figure 3.2 illustrates equivalence between all three temperature scales in the range of common atmospheric temperatures.

Vertical Temperature Structure

Because temperature changes with altitude, it is possible to identify distinct layers, or strata, of air within the atmosphere (Figure 3.3). The lowest layer of the atmosphere is called the troposphere. Elliptical in shape, the troposphere is deepest at the equator (approximately 20 km or 12 mi) and shallowest at the poles (about 10 km or 6 mi). It contains practically all atmospheric water vapor and particulate matter, is the region in which our weather is generated, and is the most dynamic portion of the atmosphere.

Throughout the troposphere, average temperatures are warmest near the earth's surface and normally decrease as altitude increases, the result of the atmosphere being heated primarily

Figure 3.2 The Three Temperature Scales: Celsius, Fahrenheit, and Kelvin

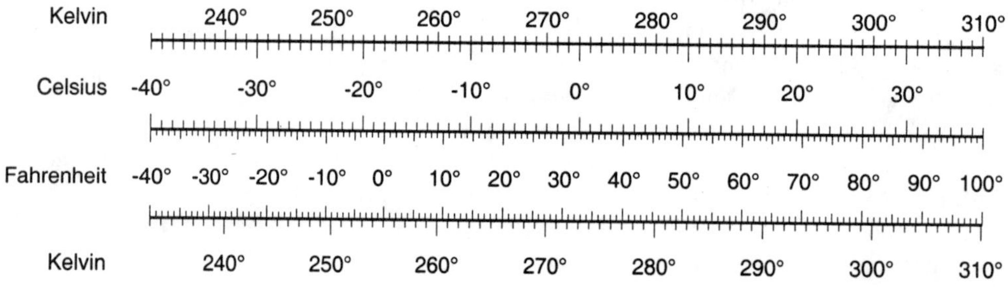

Kelvin 240° 250° 260° 270° 280° 290° 300° 310°
Celsius -40° -30° -20° -10° 0° 10° 20° 30°
Fahrenheit -40° -30° -20° -10° 0° 10° 20° 30° 40° 50° 60° 70° 80° 90° 100°
Kelvin 240° 250° 260° 270° 280° 290° 300° 310°

Figure 3.3 Structure of the Atmosphere Based on Temperature Strata

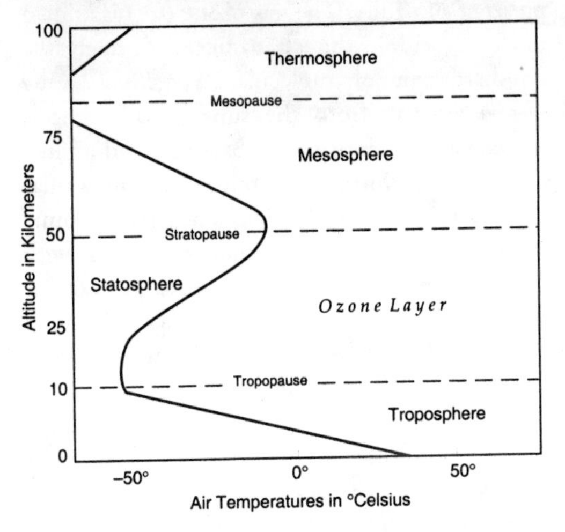

from the earth's surface. The last statement may seem to be a contradiction because it is the sun's energy that heats and animates the atmosphere albeit indirectly. Actually however, both of these statements are true. The sun is the dominant energy supplier. However, solar energy that is emitted at a very high temperature arrives at the atmosphere as shortwave radiation[3] — a form of energy that earth's envelope of gases has little capacity to absorb. This energy travels through the atmosphere as though the air were essentially transparent. The surface of the earth, being opaque, absorbs most of the sun energy reaching it. Thus the surface experiences a temperature increase and, in turn, is capable of radiating energy. Being at a much lower temperature than the sun's surface (average earth temperature is 12°C (54°F) while the sun's is 6,000°C (10,000°F), the earth emits long-wave radiation — a form of energy the atmosphere, particularly its carbon dioxide and water vapor components, can absorb. Hence, the atmosphere is heated primarily by the earth's radiation, warming the air from the surface upward.

The sequence of energy absorption by our atmosphere has often been likened to the processes that heat greenhouses, even though the analogy is not perfect. Shortwave radiation from the sun may pass through the glass of a greenhouse as if it were transparent. It then heats interior surfaces (soil, boxes, wooden shelves, etc.) which in turn become radiators of long-wave energy. It is not easy for this long-wave radiation to escape through the structure, except through conduction via the glass panes. Thus, energy is absorbed by air within the greenhouse and temperature increases. A parallel process is in operation when one opens a closed car on a sunny day. The interior of the car is found to be considerably warmer than ambient conditions outside.

Just as more heat energy is felt when one stands directly in front of a fireplace, so the lower atmosphere is warmest nearest the radiating surface of the earth. Although the rate of temperature change with altitude varies spatially and temporally, the long-term global average amounts to about 6.5°C/1000 m (3.5°F/1000 ft.). This average rate is commonly referred to as the normal temperature lapse rate or the **normal environmental lapse rate** (Figure 3.4). Clear skies and bright sunlight during the day can increase the lapse rate as a result of strong surface heating. Conversely, clear skies at night can permit sufficient surface cooling to actually reverse the lapse rate and create a surface layer that is cooler than air immediately above. This atypical and temporary reversal, is called a

[3]Short-wave radiation is referred to herein as the sun's radiation, which includes energy wave lengths primarily in the range from 0.2 to 3.0 micron. Long-wave radiation emitted by the earth is largely in the range from 3.0 to 50 microns.

Figure 3.4 Examples of Typical Temperature Lapse Rates

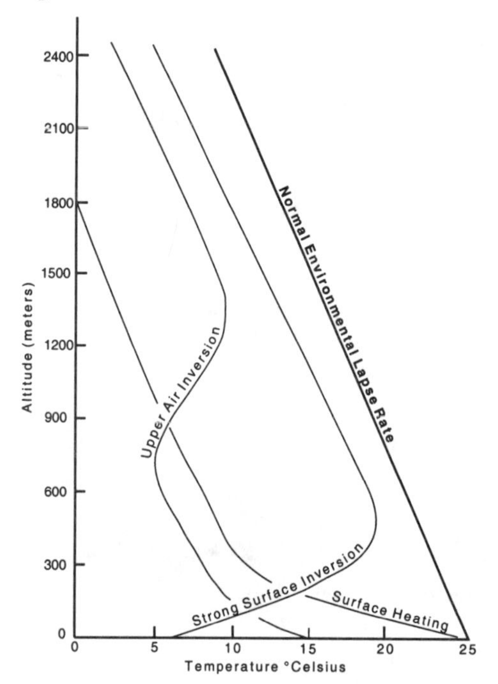

temperature inversion. The normal temperature lapse rate, characteristic of the troposphere, does not continue throughout the entire atmosphere, but is interrupted by a layer in which warming occurs. The level in the upper atmosphere at which temperatures first cease dropping is called the tropopause and defines the upper limit of the troposphere. At this level average temperatures are about -60°C (-70°F).

Beyond the troposphere is the stratosphere, a region of air featuring a low-lying isothermal layer (where temperature remains constant with increasing height). Beyond the isothermal layer, to a distance of approximately 50 km (30 mi) from the surface, temperatures gradually rise to about 0°C (32°F) at the stratospheric boundary (the stratopause). Reversal of the thermal pattern from that of the troposphere is explained by the presence of ozone. The gas ozone (O_3) is known to form within the stratosphere through the bombardment of molecular oxygen (O_2) by ultraviolet rays from the sun. In this process ultraviolet rays are largely absorbed, and atomic oxygen (O) is formed by splitting the molecular oxygen ($O_2 \longrightarrow O + O$). The freed atomic oxygen (O) is unstable and combines with remaining molecular oxygen (O_2) to form ozone (O_3), that is, $O_2 + O \longrightarrow O_3$. Formation and decay of ozone are involved in this process. While this action is taking place, the absorption of ultraviolet radiation adds heat to stratospheric gases which, in turn, results in warmer temperatures than are found in the upper reaches of the troposphere. Ozone gas is small in amount but extremely important. If all atmospheric ozone were located at sea-level, it would form a layer only about 2.5 mm (0.1 in) thick. Its ability to absorb harmful ultraviolet radiation and prevent its reaching the earth's surface, however, is critical to the health and survival of organisms, including human beings. The greatest concentration of ozone occurs near an altitude of 25 km (15 mi), a chemical stratum known as the ozone layer. A discussion of this layer's critical significance and the possible impacts of alterating of its characteristics is provided in the following section: The Post–Industrial Atmosphere.

The mesosphere is the next highest atmospheric stratum. In many respects it parallels the troposphere in temperature characteristics. Extending outward to approximately 80 km (48 mi), temperature in the mesosphere decreases to its minimum, approximately -83°C (-120°F), at the mesopause. Beyond the mesosphere lies a realm of extremely rarefied air, the thermosphere.

The Post–Industrial Atmosphere

Since the industrial revolution, a net increase of several preexisting trace gases plus a few new compounds has been observed in the global atmosphere. Although they account for a minuscule fraction of the total atmosphere by volume, several of these substances have the potential to perturb matter and energy transfers within the atmosphere and other interacting environments. The following table lists several of these gases, their sources, and potential or suspected effects.

Health risks to local populations posed by some of these effluents are dwarfed by the threat of global climate change and/or depletion of the ozone shield. In recent years the suggestion that human activities could actually trigger any significant change in the global atmosphere has emerged from speculation to unverified theory to the reality of stratospheric ozone depletion over the South Pole during the austral spring. Although many meteorologists and climatologists agree that human activities are adding greenhouse gases to the atmosphere, and that those anthropogenic (generated by human activity) gases may contribute to global warming, persuasive empirical evidence that global warming has occurred remains elusive (Figure 3.5). Evidence of reduced concentrations of ozone over the Antarctic, however, are widely accepted, and many (but not all) atmospheric scientists conclude that several anthropogenic gases are being transported into the stratosphere where ozone destroying chlorine (or bromine) is released. Measured as a vertical column in the vicinity of the South Pole in 1987, total ozone levels were found to be almost 50 percent of their 1979 levels (Figure 3.6). These reduced concentrations were associated with chlorine and chlorine monoxide, thereby providing strong evidence that ozone depletion was caused by a combination of chlorine chemistry and the unique polar meteorology. But the story really began years earlier.

During the 1970's practically every household contained a variety of spray cans, ranging from mosquito repellents to deodorants.

Figure 3.5 Anthropogenic Emissions to the Atmosphere

Pollutant	***Major Sources***	***Potential Effect***
Carbon Dioxide	Burning coal, oil, gas, biomass	Global warming, faster plant growth
Carbon monoxide	Burning coal, oil, gas,biomass	Global warming, health risk
CFCs (Chlorine)	Refrigerants, polystyrene, solvents	Loss of stratospheric ozone, global warming
Carbon Tetrachloride	Solvents	Loss of stratospheric ozone, health risk
Halons (Bromine)	Fire Extinguishers	Loss of stratospheric ozone
HCFCs (Chlorine)	Refrigerant	Loss of stratospheric ozone, global warming
Methane	Agriculture, anaerobic bacteria	Loss of stratospheric ozone, global warming
Methyl chloroform	Solvents	Loss of stratospheric ozone
Nitric oxide	Burning coal, oil, gas	Loss of stratospheric ozone, global warming
Nitrous oxide	Agriculture	Loss of stratospheric ozone, global warming
Nitrogen dioxide	Burning coal, oil, gas, biomass	Acidity, global warming
Sulphur dioxide	Burning coal, oil	Acidity, global cooling
Ozone (tropospheric)	Burning coal, oil, gas, biomass	Global warming, health risk

Figure 3.6
Trend of Antarctic Ozone 1979-1994

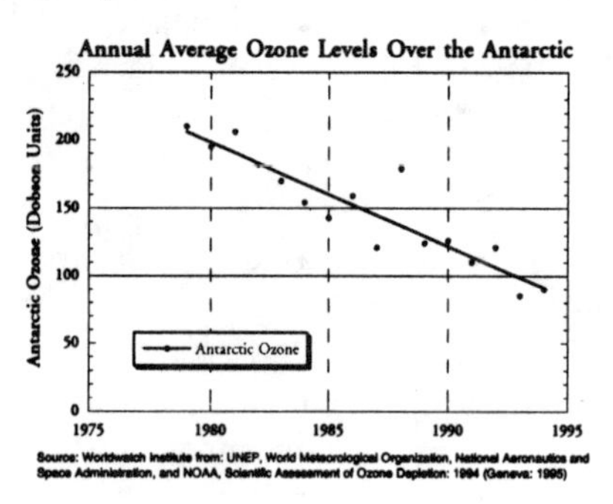

Source: Worldwatch Institute from: UNEP, World Meteorological Organization, National Aeronautics and Space Administration, and NOAA, Scientific Assessment of Ozone Depletion: 1994 (Geneva: 1995)

These aerosol cans were traditionally pressurized with chlorofluorocarbons (CFCs). CFCs are a combination of chlorine (Cl_2), fluorine (F_2), and carbon (C), that is essentially inert. Inert substances will not readily react with other chemicals at surface temperatures and pressures. This property was once considered to be an advantage. Because these chemicals do not react with gases in the atmosphere or with chemicals in surface waters, they appeared to be harmless and simply accumulated in the environment. In 1974, however, researchers at the University of California proved that under ultraviolet radiation, chlorofluorocarbons dissociated (split up into atoms). They felt that this discovery could have serious implications for the survival of the ozone layer.

If CFCs are released to the lower atmosphere in abundance, some scientists postulated that they will also be carried to higher levels, including the ozone layer, by air circulation systems. At the ozone level, where ultraviolet radiation is intense, the freed CFC atoms, especially the chlorine atom (Cl) are available for a series of chemical reactions with ozone (O_3) to produce molecular oxygen. The reaction sequence theoretically decreases the amount of ozone present and also the number of free oxygen atoms readily available to form more ozone. Relating the incidence of skin cancer to ultraviolet radiation exposure, it is estimated that each one percent decrease in stratospheric ozone is accompanied by a two percent increase in ultraviolet radiation reaching the earth's surface, a rise in the incidence of basal cell skin cancer by two to five percent and of squamous cell skin cancer by four to ten percent. In the United States, approximately 500,000 persons contract these diseases each year. Decreases in ozone could cause organic mutations and threaten the existence of all unprotected organisms. Consequently CFCs were banned from aerosol sprays in the United States during late 1978.

Industrial scientists have argued that tests performed with CFCs at surface temperatures are unrealistic when applied to the much colder upper atmosphere, and that the dangers perceived in the University of California studies are nonexistent. Other arguments have pointed out that emissions from supersonic aircraft and refrigerators pose a more serious threat to the ozone layer than the CFCs released by aerosol cans. Yet a more recent NASA study suggests that methane (CH_4) may be more important to ozone destruction than CFCs. Atmospheric concentrations of methane are increasing by one percent/year and are believed to have doubled in total quantity during the past century. Sources of methane production are difficult to control and are numerous, including decaying wood, decomposing rice, and the digestive processes of animals ranging from termites to cattle. The controversy over the threat to the ozone layer has sharpened public awareness of ozone's importance to surface life and has resulted in governmental programs aimed at further defining risks to the earth's vital ultraviolet radiation screen.

A ray of hope has emerged from the ozone controversy. International recognition of the problem has resulted in several agreements to control the impact of future CFC releases. In September 1987, 47 nations pledged to reduce CFC production by 50 percent within the following ten-year period. This was followed in June 1990, by over 70 countries agreeing to totally end CFC use by the year 2000.

The amount of ozone in the atmosphere is monitored by several ground- and satellite-based sensors. Each instrument estimates the total amount of atmospheric ozone in a vertical column by precisely comparing the intensities of two wavelengths in the solar spectrum in the region of partial ozone absorption (0.30 to 0.33 mm).[4] These instruments are called Dobson spectrometers and measure total column ozone in Dobson units (DU). A Dobson unit is a hundredth of a millimeter and refers to the thickness of ozone that would result if all column ozone were measured at standard atmospheric temperature and pressure.[5] Two of the best sources of time series data on levels of total ozone are from the Total Ozone Mapping Spectrometer (TOMS) on board NASA's Nimbus-7 satellite and from the network of 22 ground based Dobson instruments. Data are available from both sources since 1979. Additional data from two other satellite instruments the NOAA–11 SBUV/2, operational since 1989, and another TOMS instrument aboard the Russian Meteor–3 spacecraft, launched in 1991, help to verify accuracy of the older instruments and lend credibility to the time series. Although total column ozone varies by season, by periodic biennial shifts in upper air wind patterns, and by the current phase of the solar cycle, the first 14 years of this series suggest a decline in global average total column ozone.[6]

Suppressed levels of ozone also were recorded across the mid latitudes of the northern hemisphere during the spring of 1993. Reductions were moderate across most of the United States and Europe with somewhat greater losses over the western United States, and the greatest losses across Alaska, northern Canada, Greenland, Scandinavia, and Siberia. Although the complete mechanism responsible for low ozone values is unknown, anthropogenic emissions such as chlorine from CFCs and other industrial emissions are suspected as partial causes, along with natural aerosols from the 1991 eruption of Mount Pinatubo interacting with industrial emissions.

[4]*Refer to Chapter 4: Solar Radiation for more information about wavelength and ultraviolet radiation.*

[5]*The International Standard Atmosphere is defined as a mean sea level temperature of 15°C at 1013.25 millibars with a lapse rate of 6.5°C per kilometer up to 11 kilometers above which the temperature is assumed constant at –56.5°C.*

[6]*J.F. Fleason, P.K. Bhartia, J.R. Herman, R. McPeters, P. Newman, R.S. Stolarski, L. Flynn, G. Labow, D. Larko, C. Seftor, C. Wellemeyer, W.D. Komhyr, A.J. Miller, and W. Planet. Record Low Global Ozone in 1992. Science 260, (April 23, 1993): 523-526.*

Chapter **4**

Earth – Sun Geometry

The maximum solar radiation intensity available at any part of the earth's surface is determined by geometric relationships between the earth and sun. These relationships contribute directly to the diurnal and seasonal patterns of heating and cooling and indirectly to patterns of precipitation. They are fundamental in our understanding of heat and moisture.

Rotation

Viewed from above the north pole, the earth rotates counterclockwise on a polar axis. Rotation has several important effects including the diurnal march of daylight and night. The sun illuminates one-half of the earth at all times and the line separating the daylight hemisphere from the night hemisphere is called the circle of illumination (Figure 5.1). From our earthbound perspective, the earth's west to east rotation causes the perceived east to west motion of the sun, moon, and stars across the sky. This apparent motion of celestial bodies has provided a frame of reference upon which our ancestors devised a means to measure time. Marking the passage of time according to the local position of the sun is called local time and was widely practiced prior to the late 19th century. Prior to the development of accurate timekeeping apparatus, settlements were often widely separated and it became traditional for each location to establish their own local time. Typically, local time was based upon the occurrence of the sun's highest elevation in the sky. This time of day is referred to as local noon. Division of a complete day into 24 units called hours is an ancient, albeit arbitrary, choice.

Figure 4.1 The Circle of Illumination on an Equinox Date

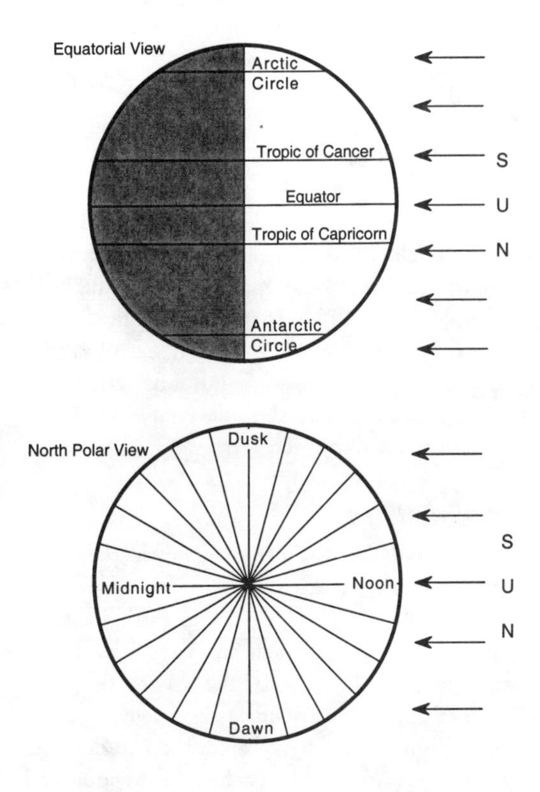

Development of railroads and the demand for train schedules forced adoption of standard time within meridional zones. Each standard time zone centers on a standard meridian and spans an average width of 15° longitude. Time zones for the United States and the world are shown in Figures 5.2 and 5.3. Note from the foregoing diagrams that time zones rarely align perfectly with meridians of longitude. The irregular shape of time zone boundaries results from conventions to minimize the disruptive effects of the time line. Time zone boundaries frequently follow political boundaries such as state borders and avoid political entities such as cities.

Because the axis of rotation is inclined 23.45° from a perpendicular to the plane of the earth's elliptical orbit, all parallels are not illuminated equally. If the axis were not tilted, the earth would be illuminated at all times from pole to pole, and as the earth rotated, each place on the surface would move through the dark hemisphere for 12 hours and through the lighted half for the remaining 12 hours of each 24-hour rotation. Under those conditions all places would have 12 hours of daylight and 12 of darkness each day of the year.

Revolution

Viewed from above the north pole, the earth's counterclockwise movement around the sun is called revolution. As the earth journeys around the sun once each 365 days 5 hours, 48 minutes, and 46 seconds, the axis of the earth maintains a constant alignment, called parallelism (Figure 5.4). In addition to defining a year, revolution also defines a hypothetical plane in space known as the plane of the ecliptic. During the course of tens of thousands of years, the axis exhibits a slight wobble, however, it is not large enough to cause an annual effect. The condition described above, that of the earth being lighted from pole to pole, does occur twice each year, in the spring and in the fall. The dates, March 21 or 22 and September 21 or 22, are called the spring or vernal equinox and the fall or autumnal equinox (equal nights) respectively. At other times of the year, the North Polar region is within the illuminated portion and the South Polar region is in darkness, or vice versa. The combined effect of the axial tilt, parallelism, and revolution produces seasonal temperature change.

The greatest potential for receipt of radiant energy occurs when a receiving surface is perpendicular to the path of the incoming radiation. As parallel rays of radiant energy from the sun strike the sphere of the earth, the incident angle is perpendicular at only one point, known as the subsolar point. The subsolar point and areas nearby hold the greatest potential for solar heating of any place on earth. At the moment of the spring equinox, the subsolar point is at the equator, but, as the earth continues its path around the sun, the subsolar point migrates into the northern hemisphere (Figure 5.4). The subsolar point reaches its most northern latitude, the Tropic of Cancer (23° 27'N), around June 21, the summer solstice. Increases in the receipt of solar energy per unit of surface area generate warmer temperatures, and summer arrives in the northern hemisphere. After the summer solstice, the earth continues its revolution. During the next six months, the subsolar point migrates south to its most southern latitude, the Tropic of Capricorn (23°27'S) about December 22, the winter solstice. Now the northern hemisphere, with lower sun angles experiences winter. Seasonal changes in the southern hemisphere result from the same earth-sun relationship, but six months out of phase from the northern hemisphere.

During the northern hemisphere summer,

Figure 4.2 Standard Time Zones within the United States 1992

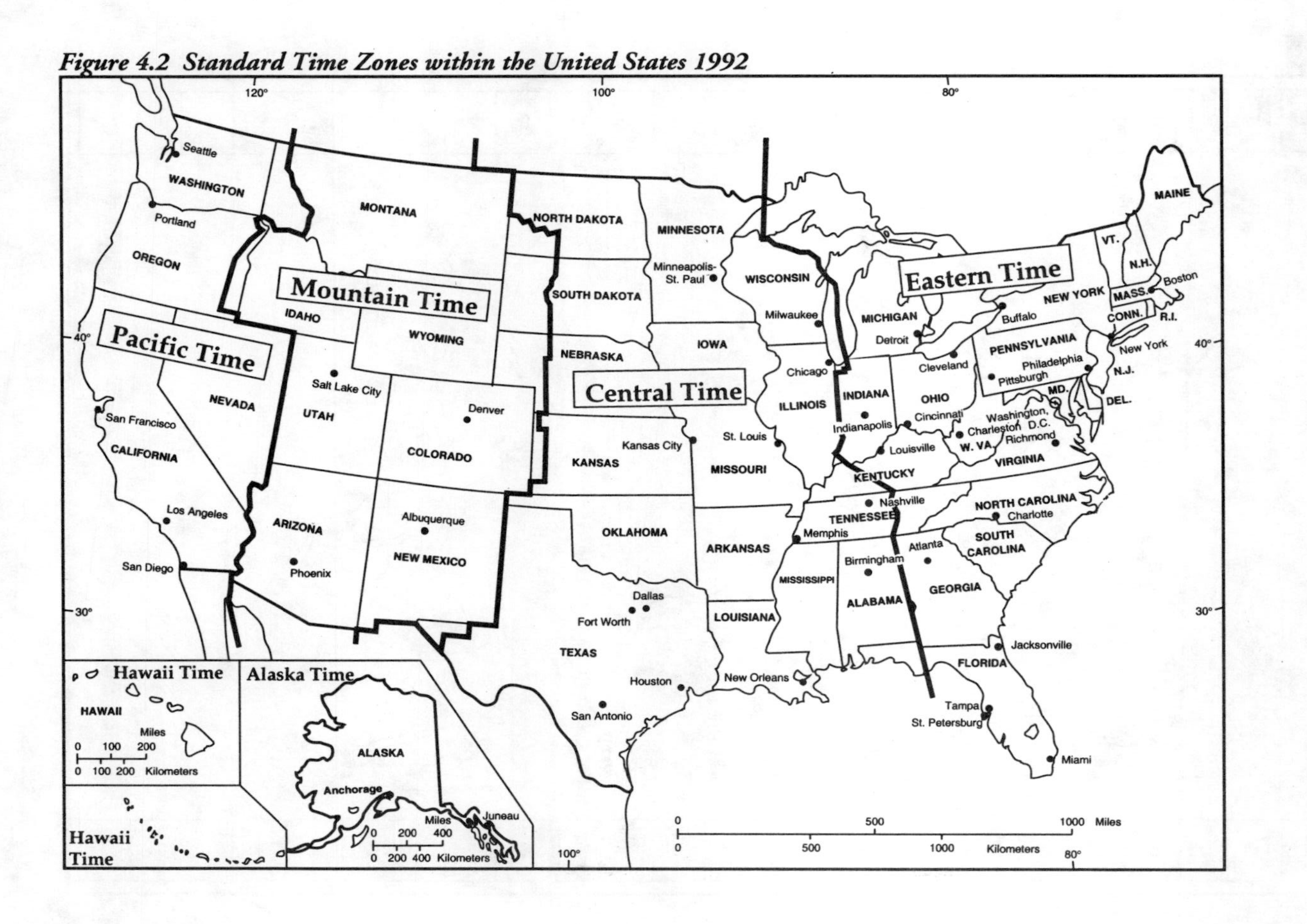

Figure 4.3 Standard World Time Zones in Hours Ahead (+) or Behind (-) Greenwich Mean Time 1992

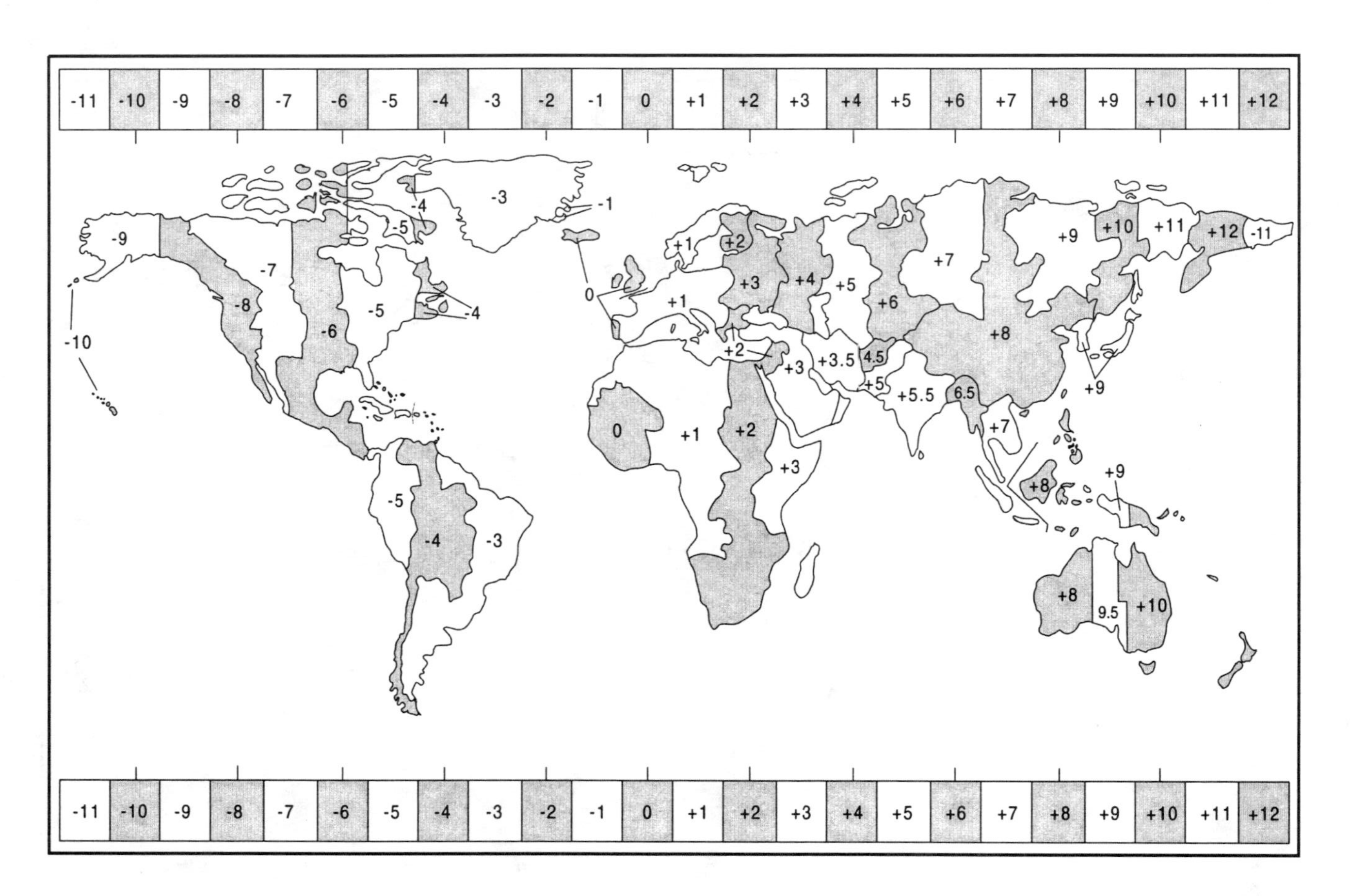

Figure 4.4 The Causes for Seasonal Change: Migration of the Subsolar Point, Parallelism, and Revolution

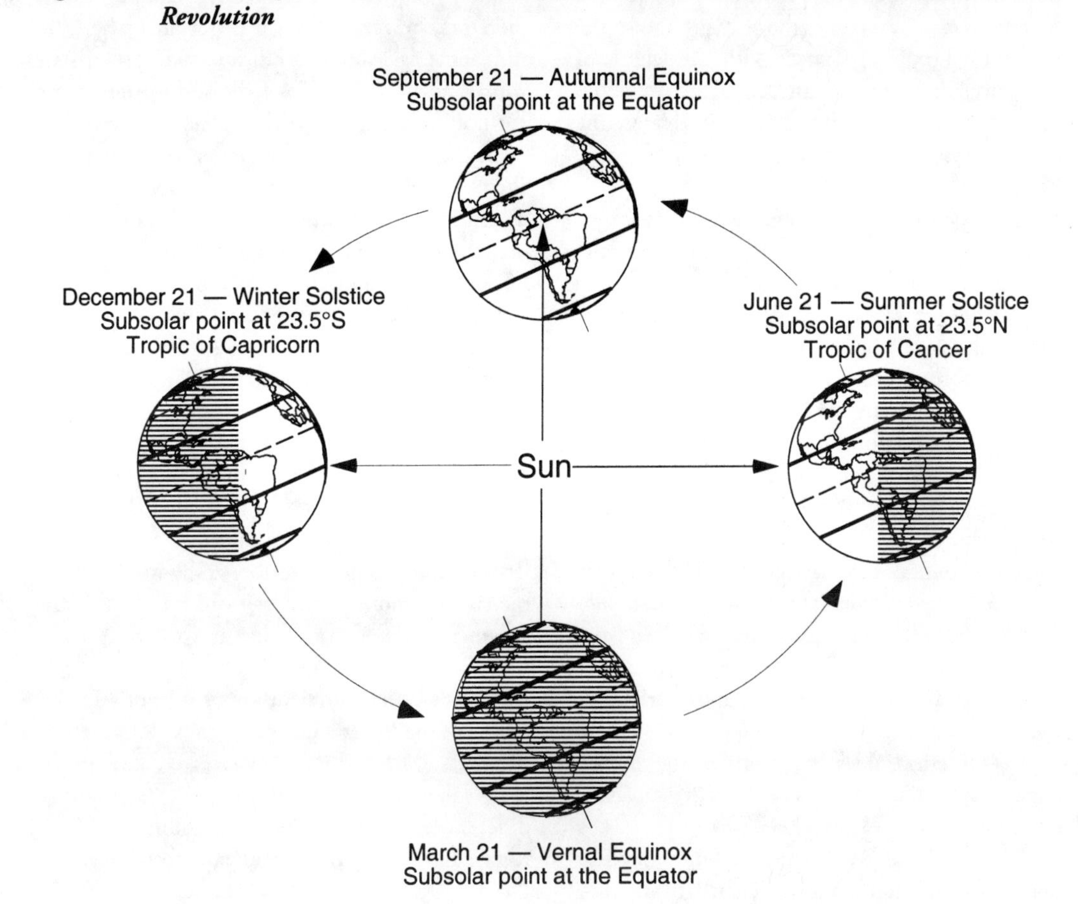

when the North Pole is illuminated continuously, more than half of each parallel north of the equator lies within the lighted hemisphere. Compare the length of the unshaded Tropic of Cancer with the length of the unshaded Tropic of Capricorn at the summer solstice (Figure 5.4). Clearly, as the earth rotates, these parallels are in the lighted portion more than 12 hours, causing the daylight period to be longer than the night period. The higher the latitude, the more of the parallel is within the lighted hemisphere and the longer the day. In fact, inside the Arctic Circle, the parallels are completely lighted on June 21, indicating continuous daylight on the summer solstice. The Arctic and Antarctic Circles at 66.5° North and South mark the latitudes where the sun's rays are tangent during the summer and winter solstices. The Tropic of Cancer and the Tropic of Capricorn mark the limits of the solar declination (the latitude of the subsolar point). These latitudes are often considered to be the approximate boundary between the tropics and the subtropics, and the sun's rays are never vertical poleward of the band marked by these parallels.

What effect does variation of the earth's angular orientation with respect to the sun have on heating the earth? The first effect is a function of the angle at which the noonday sun makes with the earth at a particular latitude. For example, at the latitude of Charlotte, NC (35°N), if you were to record the solar angle at noon each day of the year and plot the points, it would look like the bell shaped curve in Figure 5.5. At 35° North, the noon sun appears 78.5° above the southern horizon at the summer solstice, but climbs only 31.5° above the horizon at the winter solstice.

When the sun's rays strike vertically, a smaller area is illuminated than when the rays strike obliquely. The same energy must, therefore, be spread over a larger area when the sun angle is low, a condition that prevails in winter. Hence, the rays of the sun in the winter are not as warming as in summer, a fact easily confirmed by personal experience. Figure 5.6 shows why the sun angle is such a strong determinant of surface heating. It is not the fact that the lower sun angle causes more light to be reflected, although this may be a contributing factor over water, snow, and ice surfaces. Rather, the same amount of energy is spread over a larger area resulting in less energy input per unit area.

The second effect is that days and nights are of unequal length in winter and summer everywhere except at the equator. Days are increasingly longer in summer in higher latitudes, and nights are longer in winter. This means that more sun energy can be absorbed in summer and more can be lost in winter.

The increasing length of daylight in high latitudes is not sufficient to offset entirely the effects of the lower angle of illumination, even in summer. In the winter they receive considerably less solar energy, resulting in substantial differences in the relative amounts of heat received between polar and equatorial regions. Despite this, the relative differences in temperature between the equator and the polar regions remain more or less constant. By meridional heat transport, the atmospheric circulation system prevents the polar regions from growing colder and the equatorial areas from getting hotter.

All of us who live outside the tropics annually experience the effects of seasonal changes in rates of energy gain and loss. The

Figure 4.5 Angle of the Noon Sun at 35° North Latitude

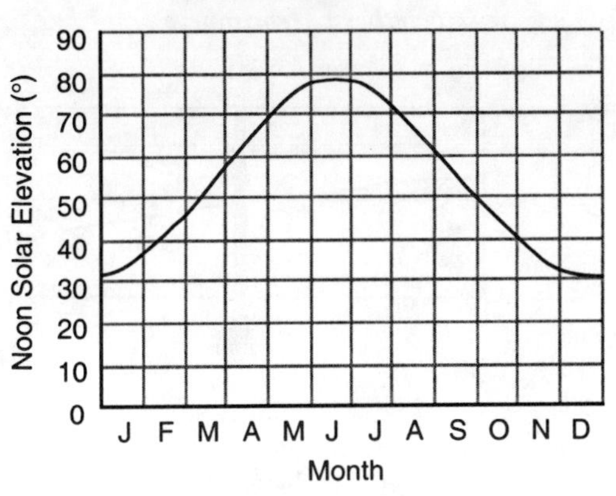

higher sun angle and the longer days of summer provide considerably more input of solar energy and result in high summer temperatures. The short nights of summer do not permit the earth system to lose as much heat as it receives during the day so that there is a gradual accumulation of heat. There is a lag between the absorption of energy by the earth and its heating of the atmosphere, caused by the time required for the earth to store energy, become hotter, and emit an amount of heat equal to what it receives from the sun. It is for this reason that we experience our hottest weather in the northern hemisphere during the latter part of July and in August – despite the fact that we receive the most energy on June 21 or 22.

During the long nights of winter, more heat escapes than can be replaced during the short days. The same lag effect observed in summer occurs in winter with the coldest temperatures normally occurring in late January and February, although the longest nights occur in December.

Figure 4.6 Energy Distribution in Relation to Latitude. Note that the amount of energy in cyclinder A equals the amount of energy in cyclinder B. The energy in cyclinder B is concentrated in heating the part of the earth's surface shown as area C. The energy in cyclinder A, although equivalent to that in cyclinder B, is distributed over area D, a larger than area C. When the sun's rays are vertical (or nearly vertical), energy is concentrated in a small area and heating potential is great. When the zenith angle is great (the sun appears low in the sky), the energy is expended over a large area, and heating intensity (energy per unit area) is low.

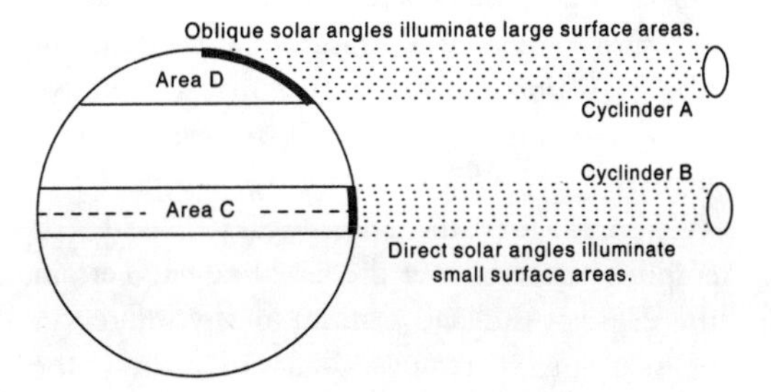

In summary, on the vernal and autumnal equinoxes the sun is vertical at the equator, days and nights are 12 hours long at all latitudes, and the northern and southern hemispheres receive equal amounts of heat. At all other times, the hemispheres are heated unequally, and the summer hemisphere not only has a higher sun angle, but longer days as well.

On June 2l the North Pole is tilted toward the sun and the sun's rays are tangent at 66.5° North and South latitude. Everything north of the Arctic Circle (66.5°N) receives continuous sunlight on June 21, while places south of the

Antarctic Circle (66.5° S) are in continuous darkness on that day. The situation is reversed during the northern hemisphere winter six months later.

Net radiation is incoming radiation minus outgoing radiation. A latitudinal profile of net radiation for the Northern Hemisphere is illustrated in Figure 5.7. Note that a balance occurs at approximately 37°N latitude. Lower latitudes experience surplus radiation, whereas higher latitudes have a deficiency. This should mean that latitudes poleward of 37°N ought to be getting colder while those equatorward become warmer. There is no evidence that this is occurring. How then is this latitudinal energy imbalance accommodated? The simple answer is that surplus energy is transferred to deficient areas by winds and ocean currents, a topic we will discuss in greater detail in forthcoming chapters.

Several other factors exert local or temporary control over the heat budget. Terrain (the shape of the land surface) locally influences the solar incident angle. Vineyards along the Rhein River in Germany are located on the south facing slopes to improve the incident angle for better light and heat. Ski resorts, located in regions where climatic conditions are marginal, build runs and trails on the north facing slopes to retard melting and extend the season. Atmospheric conditions which block solar radiation, such as cloudiness or pollution, can affect local heat budgets. In the aftermath of the 1991 Gulf War, several groups of atmospheric scientists gathered to collect data about the effects of massive oil fire plumes on the regional heat budget. Beneath the thick canopy of smoke from fires that burned 4.6 million barrels of oil per day, temperatures were much cooler. Surface

Figure 4.7 Latitudinal Distribution of Mean Annual Net Radiation in the Northern Hemisphere

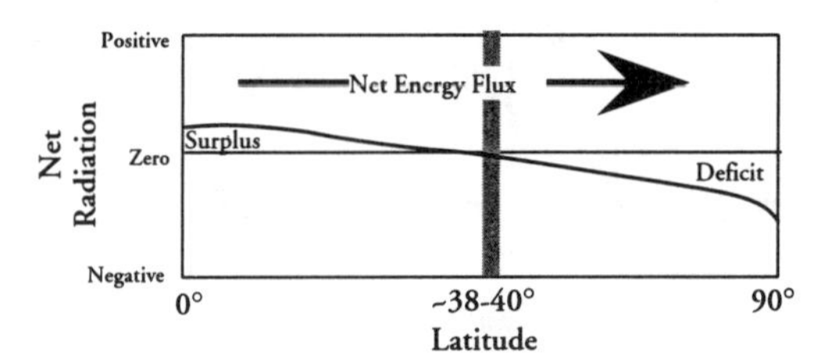

albedos which determine the proportion of radiation reflected and absorbed also account for much local variability in heat budgets. The Kuwaiti oil fires left parts of the desert coated with a black surface film responsible for record high local temperatures. Finally, outgoing radiation is strongly affected by variations in surface composition. The presence or absence of water, with its high specific heat and its tendency to evaporate, and of vegetation are the most pervasive surface factors influencing changes in energy output and storage. When water and vegetation are abundant, a large portion of the heat striking a surface is transferred to the atmosphere by evaporation or transpiration, resulting in lower air temperatures over that surface. Water also stores more energy, thereby reducing the amount available for transfer to the atmosphere. On the other hand, the presence of atmospheric moisture increases the interception of terrestrial radiation and returns more counterradiation to the surface below.

Energy absorbed by surface materials is temporarily stored, then emitted as longwave radiation. If input remains constant for a long period of time then the system attains thermal equilibrium and a constant temperature exists. If, however, energy inputs change, temperature must increase or decrease.

The changing angle of sun during the day causes input to increase until noon and decrease thereafter. There is a short lag, however, between maximum input and maximum output so that maximum surface temperatures are not achieved until some time after noon. As a result, maximum air temperatures do not occur until early to mid afternoon in most locations. This occurs when outgoing radiation exactly equals incoming radiation. Students frequently have difficulty in understanding why maximum temperatures do not coincide with the time of maximum radiation intensity. Think back for a moment to our reference in an earlier chapter to the hot oven. No matter how hot the heating element is in the oven, it takes time to absorb heat before it becomes a radiator of energy. Likewise, when you turn off the heating element, the oven does not immediately return to room temperature. Rather, it gradually radiates its stored energy. The earth behaves the same way. During daylight hours, surplus energy is stored in the earth materials, radiated to the atmosphere, and finally raises air temperatures. Note in Figure 5.8 that once the noontime peak insolation is past, there is still surplus energy added to the system until mid-afternoon. Thus, to experience the hottest daytime temperature during the summer you would go outdoors about 3:00 to 4:00 P.M. If on the other hand, you desire either to get a sun tan or avoid getting a sunburn, you would want to know when local noon occurs at your location, for it is then that the sun's rays are most intense.

At night, input is zero, but heat loss continues until sunrise, and minimum temperatures usually occur about this time. Exceptions occur when cold air masses move into an area so that horizontal heat transfer is negative.

Surface albedos influence the relative amounts of solar radiation that are reflected and absorbed. Surfaces which are good emitters are good absorbers, and this means that rough, dark surfaces absorb more solar radiation than light colored, smooth surfaces, all other things being equal. Anyone who has ever stepped, barefoot, onto an asphalt or tar surface on a bright midsummer day can confirm this principle. Air temperatures above such surfaces are usually several degrees higher than those over lighter colored surfaces.

The pattern of heating and cooling we've examined for an idealized day is also applicable to seasonal temperature changes. In the Northern Hemisphere our maximum radiation intensity occurs on the date of the summer solstice (June 21), but the highest monthly

Figure 4.8 Diurnal Profile of Incoming and Outgoing Radiation on an Equinox Day

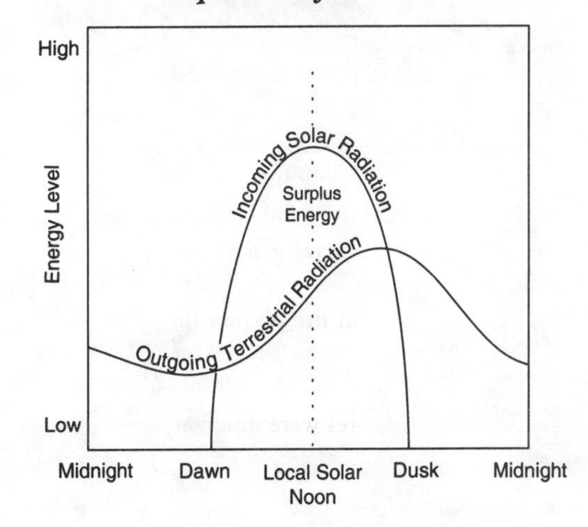

temperature normally is associated with July. Why? The temperature lags behind the seasonal insolation maximum until outgoing exceeds incoming radiation. The date of lowest intensity of radiation is December 22 (Winter Solstice), yet our coldest month is typically January. Again, the lag is caused by the dynamics of the radiation balance.

Activity: The Solar Angle of Incidence

OBJECTIVES

Upon completion of this lab you will be able to: (1) understand the variability of insolation and the causes for seasonal change; (2) explain the relationship between sun angle and solar heating potential; (3) calculate the zenith angle, the angle of the sun, and determine solar declination; (4) list the dates for the beginning of each season; (5) define solar irradiance, direct and diffuse solar radiation, solar constant, azimuth, solar altitude, perihelion, aphelion, circle of illumination, parallelism, solstice, equinox, zenith angle, angle of the sun, and analemma; and (6) calculate the maximum potential surface insolation densities by applying the sine law.

INTRODUCTION

Seasonal, diurnal, and spatial differences in surface temperatures result largely from variations in receipt of radiant energy from the sun. The density or strength of incoming solar radiation, insolation, varies depending on four important factors. In order of decreasing importance they are: 1) angle of incidence; 2) length of daylight; 3) transparency of the atmosphere; and 4) variation in solar irradiance. Angle of incidence and length of daylight vary as a function of the earths rotation on its axis and revolution around the sun. The earth-sun relationship which controls these two factors is the principal theme of this exercise. Atmospheric transparency, mostly clouds and pollution, shadow the earths surface and reduce the insolation received at the surface. Solar irradiance is the intensity of insolation as measured at the outer edge of our atmosphere. It varies ± 3.3% mainly because of changes in the earth-sun distance. The average annual solar irradiance is known as the solar constant, although that term is being used less frequently because solar irradiance is not constant. Solar irradiance varies from 1300 to 1400 watts/m3. While transparency and solar irradiance are each important, the key to understanding variations in earth surface temperatures is understanding earth-sun geometry.

Two concepts are particularly important in understanding insolation. First, the earth-sun relationship determines the solar angle. Second, solar angles determine the potential for surface heating the length of night and day, latitudinal temperature gradients, and seasonal change.

EARTH-SUN RELATIONSHIPS

Four concepts are particularly important to our understanding earth-sun relationships. First, the earth makes one complete rotation on its axis every 24 hours, giving us day and night. Second, the earth revolves around the sun. One complete revolution takes 365.242 days. Third, the rotational axis of the earth is inclined 23.5 degrees from a perpendicular to the plane of the orbit. Fourth, the alignment of the earth's axis does not change during the year. The earth-sun relationship controls the length of night and day, latitudinal temperature gradients, and seasonal change.

During the earth's revolution around the sun we are closest to the sun, perihelion, (147 million kilometers) during the first week of January and farthest from the sun, aphelion, (152 million kilometers) during the first week of July. Consequently direct sunlight is about 6.5% more intense on January 1 than it is on July 1. Yet, in the northern hemisphere we have winter when the earth is closest to the sun and summer when the earth is farthest from the sun. Change in the earth-sun distance does not cause the seasons. Record, in the spaces below, the earth to sun distances at the perihelion and the aphelion.

Perihelion: _______________ kilometers Aphelion: _______________ kilometers

The earth's rotational axis is inclined 23.5° from a perpendicular to the plane of the earth's orbit around the sun. Regardless of its position in solar orbit, the earth's axial angle remains constant. In addition, the position of the earth's axis at any given time is always parallel to its position at any other time during the year. This effect is called parallelism. Inclination and parallelism of the earth's axis combined with the earth's revolution around the sun are responsible for the changing seasons.

As the earth revolves around the sun, changes in the angle of the suns rays with the earths surface drive seasonal change. One of the most important measurements borrowed from the study of optics is the angle of incidence. When applied to sun angles, it is the angle between the sun's rays and a line perpendicular to a receiving surface, in this case the earth's surface. That point in the sky directly perpendicular to the earth's surface is frequently called zenith, and the angle of incidence is also referred to as the zenith angle (Figure 1). In the workbook label the angle of incidence in Figure 1. The angle of the sun is an angle complementary to the zenith angle. At sunrise the angle of the sun would be 0° and the zenith angle would be 90°. Because the earth rotates on its axis, the position of the sun (as well as the moon and stars) appears to migrate across the sky from east to west. Through the hours of the day and the seasons of the year, the angle of the sun's rays progressively changes to produce cycles of diurnal and seasonal temperatures.

Figure 1 Sun Angle Definitions

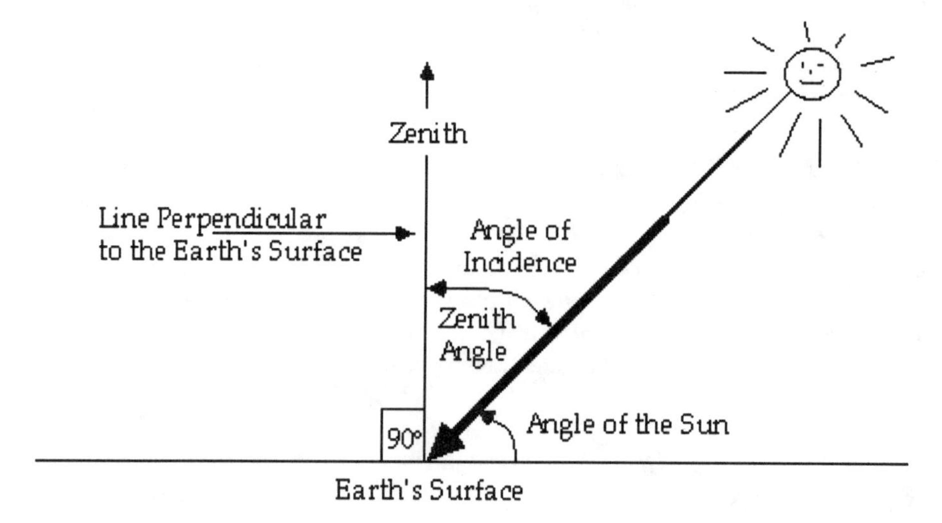

THE SUBSOLAR POINT

Because the earth is nearly spherical, parallel rays of incoming solar radiation can strike the earth precisely at a right angle at one and only one place. That place is called the subsolar point. The zenith angle is useful in determining distance north or south from the latitude of the subsolar point. Consider the geometric relationship between the earth and incoming rays of sunlight at noon. In Figure 2, angle ZXY is the zenith angle at point X. Angle ZAB is equal to the latitudinal distance between point X and the subsolar point. Angle ZXY and angle ZAB are corresponding (and therefore equal) angles formed by a diagonal line intersecting two parallel lines. Therefore, the noon zenith angle is always equal to the latitudinal distance between the observation and the latitude of the subsolar point. Since ancient Greece, this relationship has provided travellers a way of calculating distance. An optical instrument called a sextant was used for centuries by navigators to measure solar angles and distance. Today, navigation within the grid system of the earth has been made simple and very accurate by the use of radio beacons, radar, and satellite communication, but finding a specific position on the grid system may still be done accurately with older and simpler equipment.

To determine distance from the subsolar point:

(1) Determine the noon sun angle.
(2) Subtract this from 90° to get the zenith angle.
(3) The zenith angle is equal to the latitudinal distance between the observer and the subsolar point.

Figure 2 The Subsolar Point

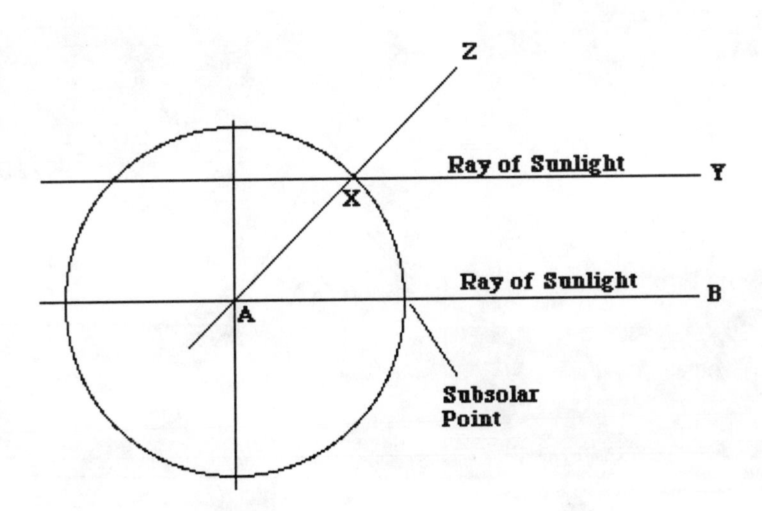

THE SINE LAW

The subsolar point is important because the heating effect of the sun is greatest when the receiving surface is oriented perpendicular to the sun's rays. When the sun's rays are intercepted by a surface tilted at a large incident angle, the energy from the sun is distributed over a much larger area. For example, if you were to shine a flashlight directly (perpendicular) at the desktop, the illuminated area would be a small circle of bright light. If the flashlight were then aimed obliquely, you would see the illuminated area become larger and dimmer. In each case, the same total amount of light strikes the desktop, but the light is less concentrated when it is spread over a larger area.

This same relationship can be applied to solar radiation as it strikes the earth's surface. In Figure 3 direct incoming solar radiation (sunlight and other forms of radiant energy) may be concentrated on a perpendicular surface, such as surface A, or it might strike a surface that is not perpendicular and be spread over a larger surface, such as surface B.

The conservation of energy requires that the illumination throughout area B (see Figure 6) is equal to the illumination of radiation over area A times the quotient of area A divided by area B:

$$I_B = (I_A)\left(\frac{A}{B}\right)$$

This calculation can be simplified, to produce the sine law:

$$I_B = (I_A)(\sin acb)$$

Figure 3 The Sine Law

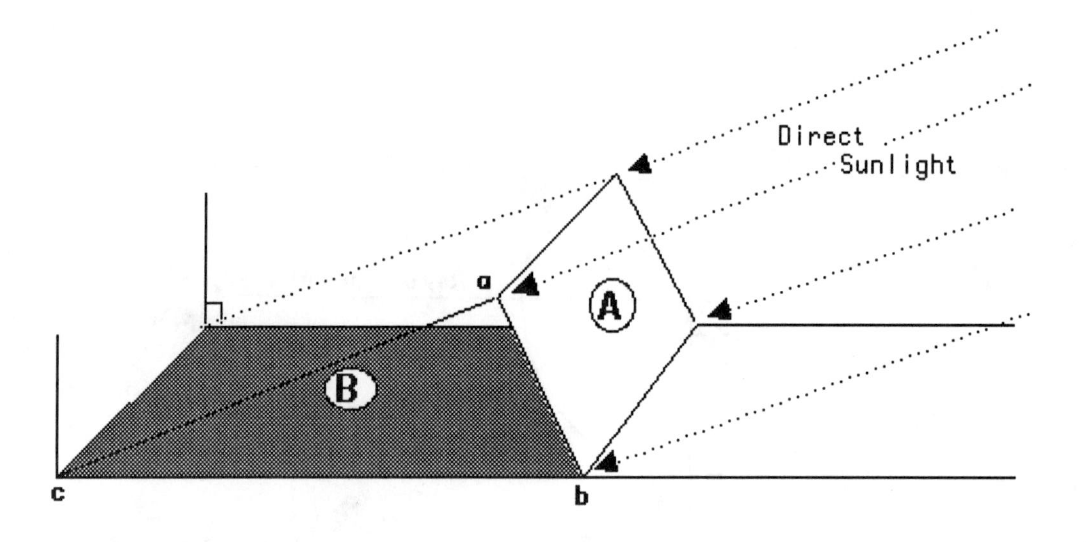

The relationship between the sun angle and radiation intensity can be summarized by the sine law. Where IB is the intensity of radiation across each unit area of B; (IA) is the intensity of radiation across each unit area of A; and sin acb is the sine of angle acb.

Generally, IA is treated as a constant value (the solar constant) for the earth as a whole. The solar constant is the maximum potential intensity of radiation available to the earth from the sun at the outer limits of the atmosphere. The average global value of the solar constant is 1.94 calories per square centimeter per minute. In radiation balance climatology one calorie per square centimeter per minute is called a langley, so the solar constant in langleys is 1.94. The solar constant can also be expressed as 1,370 watts per square meter or as 428 BTUs/hr/ft^2. The maximum potential brightness of one square meter of sunlight is as bright as thirteen 100 watt light bulbs. The solar constant is an average between slightly higher insolation values near perihelion and slightly lower values near aphelion. The actual radiation reaching the earth's surface cannot exceed the solar constant unless the earth were to move closer to the sun or the sun were to begin radiating more energy. Thankfully neither of these conditions is very likely. The actual radiation reaching the earth's surface is reduced by conditions which 1) reduce the transparency of the atmosphere (such as clouds or dust) and/or 2) reduce the angle of the sun.

On approximately June 21 (the summer solstice), the angle of the sun at noon here in Charlotte is 78.50°. Using the sine law calculate the maximum potential (clear day) intensity of radiation in langleys at noon on that day.

__________________Langleys

On approximately December 21 (the winter solstice), the angle of the sun at noon here in Charlotte is 31.5°. What is the maximum potential intensity in langleys at noon on that day?

__________________Langleys

Compare the difference between the intensity of radiation at noon on December 21 with that on June 21 here in Charlotte.

____

____

Just as Charlotte experiences variation of sun angles throughout each day and season, each geographic location experiences a range of different sun angles. Because the earth is nearly spherical, insolation always strikes the high latitude places obliquely. The zenith angle at noon in the tropical areas is relatively small but quite large in the high latitude locations. The maximum potential intensity of radiation striking the earth's surface at noon at any latitude may be easily estimated by applying the sine law.

Calculate the maximum potential radiation density in Langleys for Bergen, Norway (latitude 60° North) on December 21 (angle of the noon sun = 6.5°) and on June 21 (angle of the noon sun = 53.5°).

December 21__________ Langleys June 21__________Langleys

What is the difference between June and December ? ________Langleys _______% difference

What is the difference in maximum potential radiation intensity between Bergen and Charlotte on December 21?

Charlotte: ________Langleys Bergen: ________Langleys Difference:_______Langleys

LOCAL SOLAR ANGLES

The sun path diagram describes the sun's position in the sky for a specific latitude, for example 35° N in the case of Charlotte. The path of the sun across the sky is shown by a series of lines which start at the eastern edge of the circle (sunrise) and finish on the western edge (sunset). Take a moment to locate the North, East, South, and West sides of the chart. The northernmost of these lines represents the sun's path on June 21 and the southernmost line represents it on December 21. A label of 23.5° S for the southernmost sun-path line refers to the declination of the sun on that date. Selected declinations and their corresponding dates are listed at the bottom of the page to aid in identifying each of the sun-path lines. The sunpath line marked 0° for example, describes the track of the sun across the sky at 35° North latitude when the solar declination is zero (on the equator). The shorter more vertical lines that cross the sun's path represent the hours of the day. The sun-path diagram describes the angle of the sun (also called the solar altitude) and compass direction (also called azimuth) as they appear to an observer at 35° North latitude. Imagine yourself facing north. Turn to your right and raise your arm until you are pointing directly at the sun. Azimuth is the rotation of your body in angular degrees and solar altitude is how far you raised your arm in degrees. The angle of the sun is measured from the perimeter to the center of the graph, from 0° to 90°. The azimuth is measured by the radii, 90 = East, 180 = South, 270 = West, and North = 0.

For example, to find the sun's path in Charlotte (35° N) on March 21:

A. Find the subsolar point on March 21. The table at the bottom of the graph provides a few selected declinations. On March 21 the SSP is on the equator.

B. Find the sun path line that corresponds to 0° declination.

Notice the seasonal and diurnal variations in solar angles.

Using the sunsight determine the angle of the sun here at UNCC. Take the sunsight and a piece of stiff white paper outside to an area where you can observe the sun unobscured by trees or buildings. If you can observe the sun through a window, you can also try this part of the experiment from inside. Orient the small cylinder of the sunsight precisely toward the sun and read the angle of the sun beneath the plumb. Do not look directly through the cylinder at the sun, but rather find the proper orientation by holding a page of white paper beneath the sunsight.

Date of Observation_________ Time of Day_________ Zenith Angle __________

Angle of the Sun __________Angle of Incidence ___________

Sunlight entering the earth's atmosphere is scattered in a diffuse pattern by water vapor, dust, pollution, even the oxygen and nitrogen molecules of the atmosphere itself (Figure 4).

Figure 4 Direct and Diffuse Sunlight

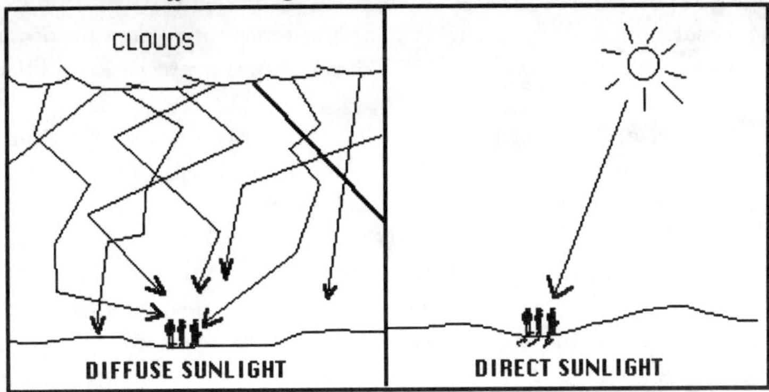

Because the atmosphere scatters sunlight, direct insolation at the earth's surface is always less than the solar constant. On a clear day, 90% of the insolation is direct and 10% is diffuse radiation. On days cloudy enough to obscure the sun, 100% of the insolation at the earth's surface is diffuse. Direct insolation casts shadows but diffuse radiation does not. Take the solar radiation meter and the collimator tube outdoors to an open area unobscured by buildings or trees. Aim the solar cell on top of the meter toward the sun. Keep the meter's photocell perpendicular to the sun's rays. Orient the sensor until you get the highest reading on the meter. Record and label the intensity of solar radiation in BTUs/hr./sq. ft.

Intensity of all solar radiation ______________________BTUs

Now hold the collimator tube over the solar cell, keep the meter's cell perpendicular to the sun's rays and again record the intensity of the sun's rays. How much difference between the two reading could be detected?

Intensity of direct solar radiation with collimator tube? __________BTUs

Difference between total and direct radiation___________BTUs

According to your observations what percent of the insolation is diffuse? (To find the percentage, subtract the direct insolation value from the total insolation value and divide by the total insolation value.) ________%

HEATING POTENTIAL

The greatest density of radiation received at the surface occurs when the receiving surface is perpendicular to the incoming radiation. The greatest potential for incoming solar radiation exists near the subsolar point. The subsolar point always occurs at noon sun-time, but the latitude of the SSP varies throughout the year. Notice in Figure 5 that the SSP (indicated by an arrow) strikes a different latitude during each season.

Migration of the SSP is caused by several factors: inclination of the axis, revolution, and parallelism. The northern hemisphere axis always points to Polaris, the North Star, regardless of the earths location in its path around the sun. The combined effect of these three factors causes an annual migration of the subsolar point between 23.5° N and 23.5° S latitude. Northern hemisphere summers occur because the subsolar point migrates into the northern hemisphere, winters when the subsolar point is in the southern hemisphere, while fall and spring are intermediate seasons when the subsolar point is at the equator. In Figure 5, label the beginning dates of the seasons for the northern hemisphere as Vernal Equinox (March 21-23); Autumnal Equinox (September 21-23); Summer Solstice (June 21-23); or Winter Solstice (December 21-23). The precise latitude of the SSP is called the declination of the sun. The solar declination corresponding to each day of the year is plotted on a special type of graph called an analemma. Refer to the large globe in the lab. Analemma is the term for the figure-eight patterns you will see plotted across the equator. Find the latitudinal position of today's date on the analemma. That is the latitude of the perpendicular sun today. In the spaces provided below, record the latitudes of the sun's perpendicular rays for each of the following dates:

	Vernal Equinox	Summer Solstice	Autumnal Equinox	Winter Solstice
Solar Declination	___________	___________	___________	___________

In studying the relationships between the earth and sun on a seasonal basis, it is helpful to know the locations of important parallels of the earth's grid system. Use one of the globes in the lab to locate each of the following special parallels. In the spaces below label each with the correct latitude before continuing with this lab.

Location of Important Parallels:

Arctic Circle: _________ Equator: _________ Antarctic Circle: _________

Tropic of Cancer: ___________ Tropic of Capricorn: ___________

SUNPATH CHART
SOLAR ALTITUDE AND AZIMUTH 35° N

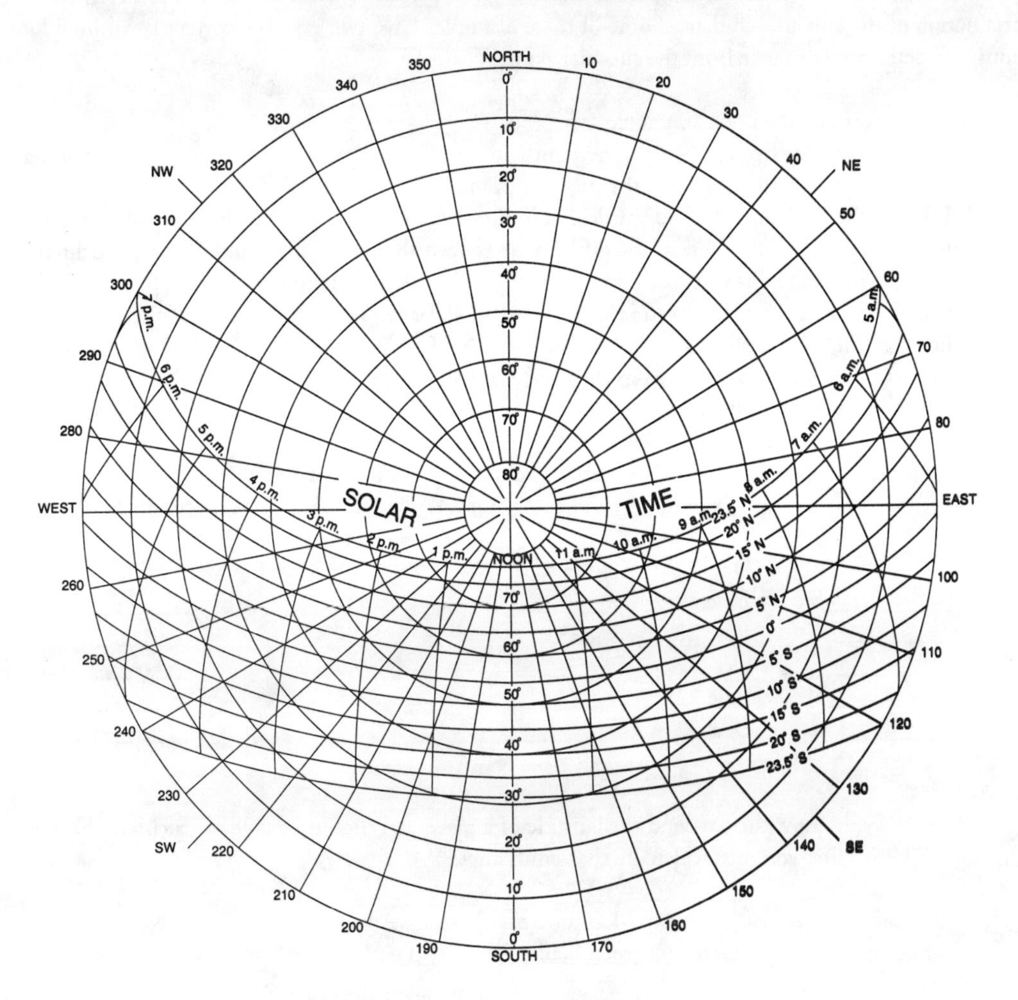

Declination of the Sun	*Dates*
23.5°N	*June 22*
20°N	*May 21, July 24*
15°N	*May 1, Aug. 12*
10°N	*Apr. 16, Aug. 26*
5°N	*Apr. 3, Sept. 10*
0°	*Mar. 21, Sept. 23*
5°S	*Mar. 8, Oct. 6*
10°S	*Feb. 23, Oct 20*
15°S	*Feb. 9, Nov. 3*
20°S	*Jan. 21, Nov. 22*
23.5°S	*Dec. 22*

Calculate the latitudinal location from the data given below. Use the analemma to determine the declination of the sun at each date. Some of these examples have two possible correct latitudinal locations. To determine distance from the subsolar point:

(1) Determine the noon sun angle.
(2) Subtract this from 90° to get the zenith angle. The zenith angle is equal to the latitudinal distance between the place and the subsolar point.
(3) Use the analemma to find the declination of the sun.
(4) Find the absolute value of the difference between the zenith angle and the solar declination (Δ L). This is the distance from the subsolar point in degrees latitude.
(5) If the noon sun is observed in the southern half of the sky, locate the parallel that is Δ L north of the solar declination.
(6) If the noon sun is observed in the northern half of the sky, locate the parallel that is Δ L south of the solar declination.

	Latitude of Place		Date	Latitude of Perpendicular Sun	Angle of the NoonSun	Noon Zenith Angle
a.	40	N	Dec. 22	23.5 S	26.5°	63.5°
b.	________	S	June 22	______	0.0°	______
c.	________	N	June 22	______	47.0°	______
d.	________	S	Dec. 22	______	79.5°	______
e.	________	N	June 22	______	78.5°	______
f.	________	N	Dec. 22	______	17.5°	______

Likewise if you know the date and the latitude of a place you should be able to calculate the noon sun angle. (Hint: first you must calculate the zenith angle.)

Latitude of Place	Date	Latitude of Perpendicular Sun	Angle of the NoonSun	Noon Zenith Angle
40° N	Mar. 4	______	______	______
25° S	Aug. 10	______	______	______
25° N	Aug. 10	______	______	______
35° N	Feb. 8	______	______	______

Figure 5 Migration of the Subsolar Point

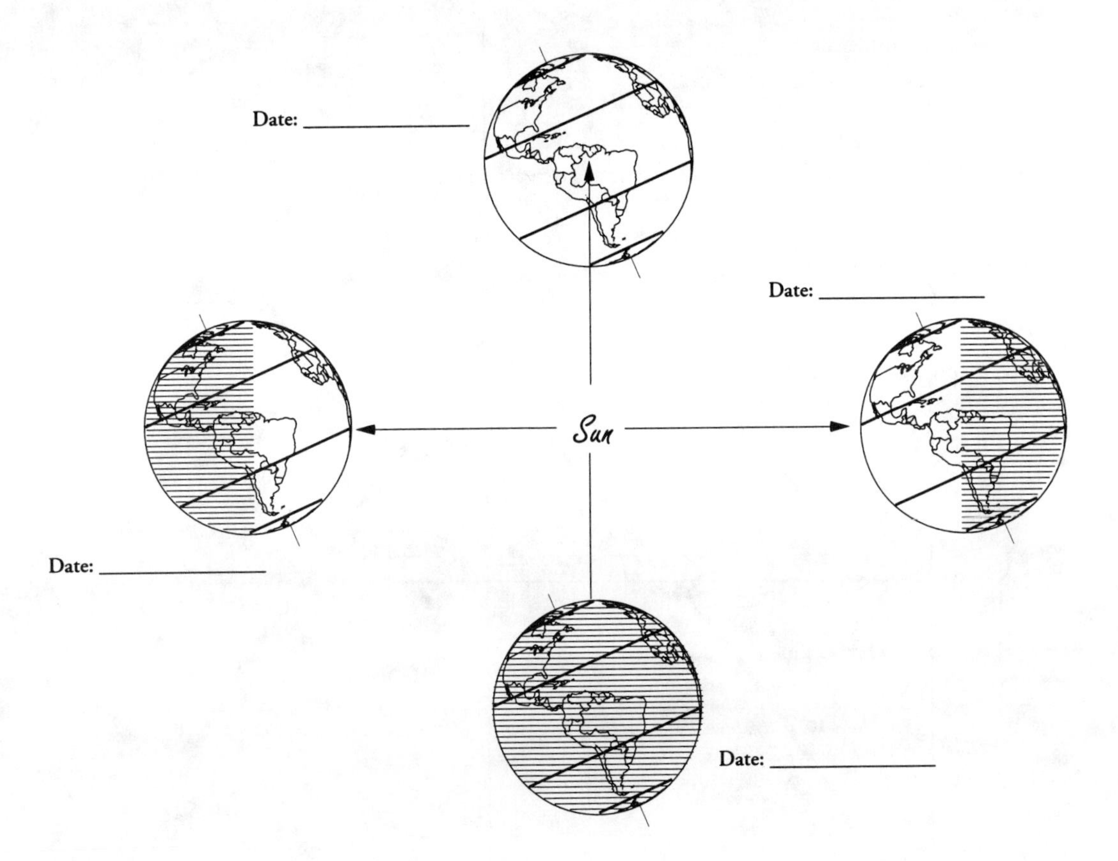

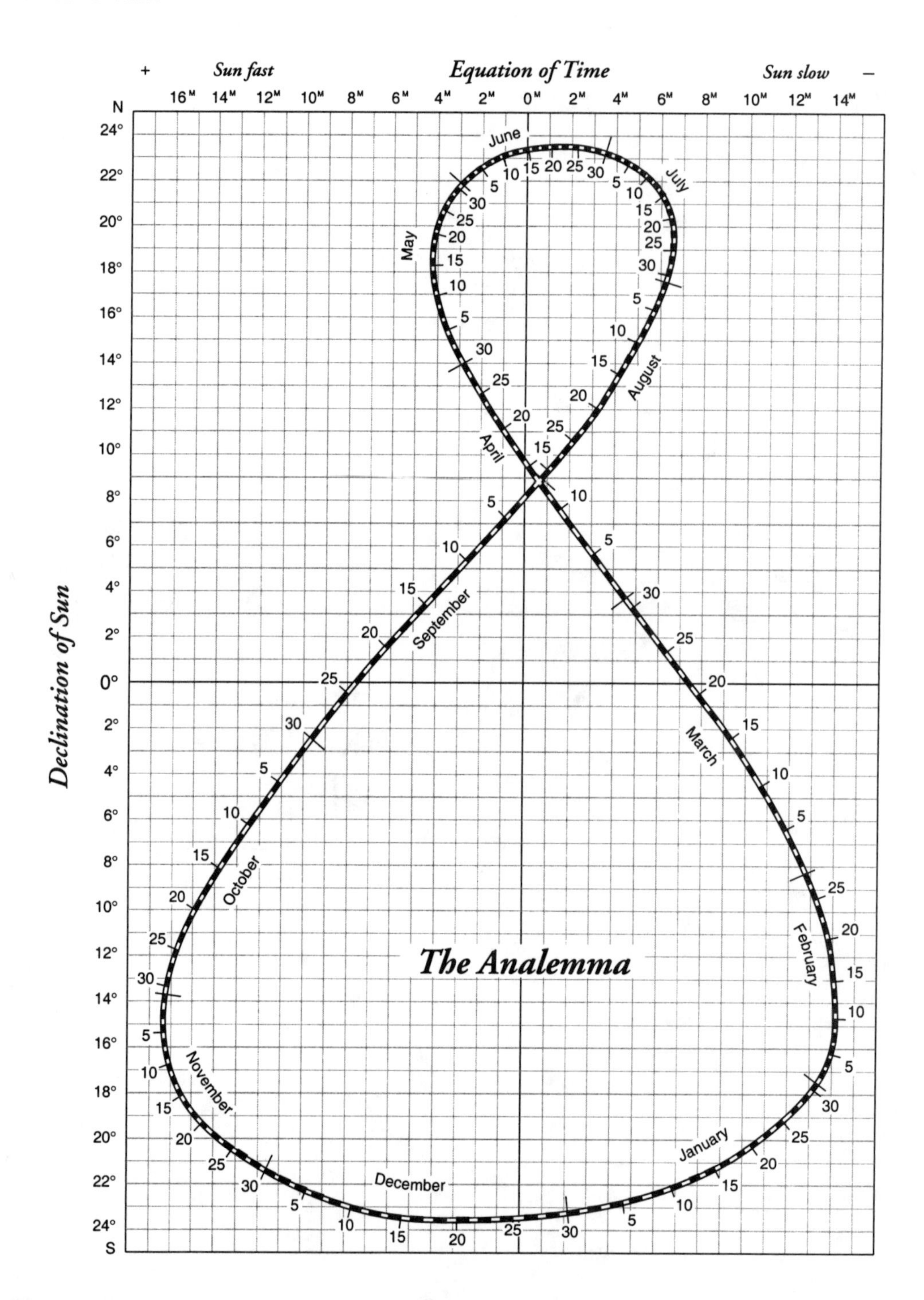
+ Sun fast
Equation of Time
Sun slow –
16M 14M 12M 10M 8M 6M 4M 2M 0M 2M 4M 6M 8M 10M 12M 14M
N
24° 22° 20° 18° 16° 14° 12° 10° 8° 6° 4° 2° 0° 2° 4° 6° 8° 10° 12° 14° 16° 18° 20° 22° 24°
S
Declination of Sun
June
July
May
August
April
September
March
October
February
November
January
December
The Analemma

Chapter 5

Solar Radiation

Solar radiation is the principal unifying component of earth surface systems, for it is the flow of energy and associated matter that ties the atmospheric processes together with vegetation, soils, and land surface form. Therefore, even though our major objective is to see how the spatial and seasonal distribution of radiation drives the atmospheric circulation system and influences our weather and climate, the same energy distribution has an impact on related systems. Energy to run the atmosphere originates from the sun and is transferred to earth by radiation. Radiation is not only an energy transfer process, but it is a basic form of energy that needs to be well understood because of its central role in physical processes.

Electromagnetic Radiation

During the early 19th century, physicists working with light explained its behavior by postulating that it consisted of particles transferred through a substance they called ether. As they learned more about light, this explanation did not adequately explain its behavior under certain circumstances, so that a wave model was suggested for light. In the wave model light consists of two perpendicular vectors, one electrical and one magnetic, that oscillate together. The oscillations are similar to waves propagated in water. The distance from crest to crest is **wavelength** and the number of waves passing a particular spot over a given time period is **frequency** (Figure 4.1). Because all light travels at a speed of 300,000 km (186,000 mi) per second, more short waves must pass a stationary point in a given time than longer waves. Thus, short waves are associated with high frequencies and long waves with low frequencies. Short wavelengths possess greater energy than do long wavelengths. As wavelength becomes exceedingly short, radiation behaves more like a stream of particles and less like a wave. These discrete packets of energy are referred to as **photons**. Photons of ultraviolet and x–ray wavelengths are energetic enough to penetrate human tissue, and both are disruptive enough to damage DNA and introduce errors in the genetic code of cells.

Figure 4.1 The Distance from Crest to Crest Is Wavelength.

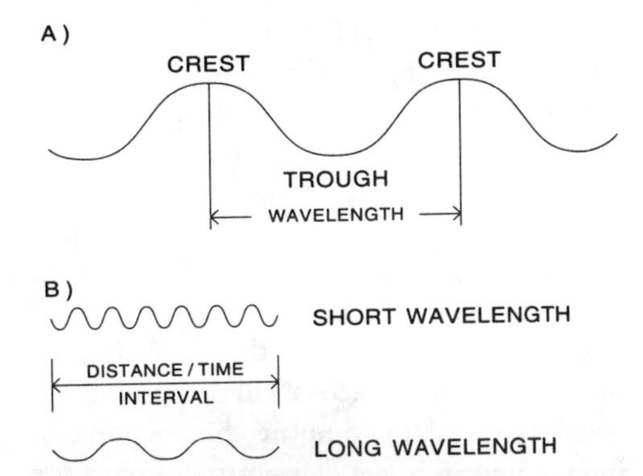

The number of waves passing a stationary point over a given time period is frequency.

The Electromagnetic Spectrum

There is a continuum of wavelengths of electromagnetic radiation ranging from low energy, long wavelength radiation, to very short wavelength, high energy x–rays and gamma or cosmic radiation. Figure 4.2 shows the spectrum with the names for the various wavelengths and frequencies of radiation. Radio waves are longest followed by microwaves, infrared, visible light, ultraviolet, x–rays, and gamma rays.

Many effects of radiation are familiar to us. X–rays are used by dentists and physicians to penetrate organic matter and to reveal the health status of internal parts of the body; ultraviolet (UV-A) radiation is responsible for the summer tan that most college students aspire to achieve; however, UV-B is also responsible for sunburns and basal and squamous cell skin cancer. Electromagnetic radiation in the range of about 0.4 to 0.75 micrometers[1] in wavelength is selectively suited to stimulate the sensation of vision in human beings. The warmth from a fire in the hearth or a friendly body resting close to you is carried by infrared radiation. These very long waves (>1 meter) pass through the atmosphere relatively unhindered and have been coaxed by electronic engineers to provide us with the enjoyment of radio. Broadcast television, weather radar that provides for improved weather forecasting, air traffic control, and highway patrol radars all employ microwaves. Microwaves can also be used to heat any substance containing water. In this instance, the energy of microwaves is absorbed

[1]A micrometer (μm) is sometimes referred to as a micron. It is equal to 0.001 millimeters or 1,000 nanometers.

Figure 4.2 Electromagnetic Radiation

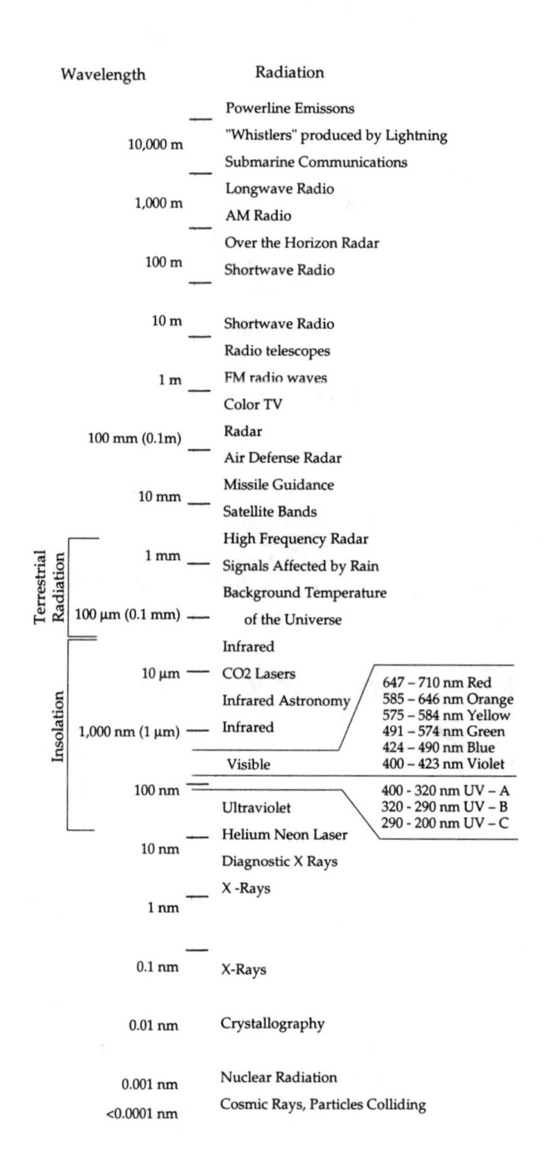

by water molecules in organic substances where rapid heating takes place, temperature increases, and the cooking time of food is reduced radically.

The Stefan-Boltzmann Analysis

As we have seen, electromagnetic radiation serves a number of useful purposes, as well as being capable of harming earth's life forms. In addition to the phenomena discussed, electromagnetic radiation from the sun provides planet earth with energy that imparts motion to the atmosphere. When energy from the sun is absorbed by the earth and its atmosphere, the heat increases the internal molecular motion (also known as **kinetic energy**) of substances, and their temperatures increase. We should note here that molecules, atoms and ions of all substances with temperatures above **absolute zero** (0 Kelvin) possess kinetic energy. Armed with this information, we can now address the laws that explain the properties of radiation and the results of radiation on atmospheric processes.

A basic radiation law is: any body capable of absorbing radiation must also radiate energy. Let us begin with a simple example. Suppose you are planning to broil chicken in an electric oven. You set the temperature, turn on the heating element, and place the chicken inside. As the electric current raises the temperature of the heating element, more energy is radiated to the chicken and the contents inside the oven.[2] As they absorb energy, their temperatures increase and even the walls of the oven become hot and also radiate heat that contributes to cooking the chicken. The increased kinetic energy of the hot surfaces causes an increase in the radiant energy. Upon finishing your cooking chores and turning off the heating element you find the oven still radiates energy, that is, it is still hot. It will continue to radiate energy until its kinetic energy establishes **equilibrium** (balance) with its surrounding environment. Thermal equilibrium will be achieved when the oven has lost a sufficient amount of energy to return it to room temperature.

The amount of radiation emitted from a substance is related to its temperature, because temperature is an indirect measurement of the kinetic energy a substance possesses. This relationship between the temperature of a substance and the electromagnetic radiation it emits is described mathematically as the Stefan-Boltzmann Law, named after its two discoverers, Joseph Stefan (1835-1893) and Ludwig Boltzmann (1844-1906). The Stefan-Boltzmann Law states that the rate at which an object (e.g. the earth) radiates heat is proportional to the fourth power of its absolute temperature. In equation form:

$$E = \sigma T^4 \varepsilon$$

where E is the intensity of radiation in units of Watts per square meter, σ is the Stefan-Boltzmann constant of $5.67 \bullet 10^{-8}$ Watts per square meter, T is the temperature in Kelvins, and ε is the emissivity ratio of the object. Note: an object that absorbs (and radiates) all radiation that strikes it is called a blackbody or perfect radiator and has an ε of 1.0. The surface of the earth approximates the definition of a black

[2]*Direct molecular heat transfers (conduction) and transfers through mixing (convection) greatly reduce cooking time in a real oven.*

body with emissivity rates between 0.84 and 0.99; therefore, the ε factor is sometimes ignored in general radiation temperature studies and the Stefan-Boltzmann equation becomes:

$$E = \sigma T^4$$

At this point you might ask — "so what?" or "who cares?" Basically, this is important in meteorology and climatology because locations on the earth may temporarily experience temperature disequilibrium because of daytime heating and nighttime cooling, warm and cold seasons, or between tropical and polar regions, but long-range temperature equilibrium must exist for planet earth. Otherwise, the global temperature would continuously increase or decrease.

The Stefan-Boltzmann equation provides a measure of the long-range balance of energy absorbed and radiated from the earth-atmosphere system[3]. Given that the average earth temperature as seen from space is -18°C (255K), then:

$$E = \sigma T^4$$
$$E = (5.67 \cdot 10^{-8} \text{ Watts/m}^2)(255\text{K})^4$$
$$E = (5.67 \cdot 10^{-8} \text{ Watts/m}^2)(4.23 \cdot 10^9\text{K})$$
$$E = 239 \text{ Watts/m}^2$$

[3]Although the Stefan-Boltzmann equation may be used with various units of measurement, in this calculation we use Watts per square meter. A Watt is a unit of power (a rate of energy flow or flux). It is 1 joule per second. A joule is the energy needed to move 1 newton 1 meter. A newton is the force needed to accelerate 1 kg 1 meter/sec. A newton = 0.2248 pounds force.

We may also use the equation to calculate energy emitted from the sun by inferring its temperature, which has experimentally been determined to be 5800K in the following fashion:

$$E = \sigma T^4$$
$$E = (5.67 \cdot 10^{-8} \text{ Watts/m}^2)(5800\text{K})^4$$
$$E = (5.67 \cdot 10^{-8})(5.8 \cdot 10^3\text{K})^4$$
$$E = (5.67 \cdot 10^{-8})(1{,}131.6 \cdot 10^{12})$$
$$E = 64{,}161{,}720 \text{ Watts/m}^2$$

You have identified a monumental difference between energy emitted from the sun versus that emitted by earth. Yet, we are taught that solar radiation supplies energy and motion to earth's atmosphere and that solar energy is not diminished by travelling through the vacuum of space. Therefore, the earth should receive an average of 64,161,720 Watts/m^2 of solar energy and not 239 Watts/m^2. This apparent discrepancy is explained by two facts: (1) the distance of the earth from the sun and (2) the spheroidal curvature of the earth. Solar radiation is increasingly dispersed over a larger area as distance from the sun increases. At a distance of $149.5 \cdot 10^6$ km (~ 93 million miles), the intensity of solar radiation striking a perpendicular surface averages 1,368 Watts/m^2 suggesting the area over which the radiation is distributed has expanded tremendously.

This average value of radiation intensity at the earth's distance from the sun ($149.5 \cdot 10^6$ km) is called the **solar constant**. It is now estimated to be 1,368 Watts/m^2. This still does not equal the average radiation emission of the earth (239 Watts/m^2) that we calculated from the Stefan-Boltzmann equation. We can explain this difference in two ways. Consider how earth's energy receipts are distributed over our planet (Figure 4.3). The energy intercepted by the earth is proportional to the planet's cross-sectional area (a circle), mathematically expressed as $\pi r^2 \cdot 1{,}368$ Watts/m^2. This energy is distributed, however, over a rotating sphere with area dimensions equal to $4\pi r^2$, thus reducing the per square meter receipt of 1,368 Watts to 1/4 of its original intensity, or equivalent to 342 Watts/m^2.

Our first estimate of the earth's radiation balance is 239 Watts/m^2 or only 70% of this value. Why? Nearly 30% of the insolation that strikes the earth-atmosphere system is not absorbed and does not affect total radiative emissions to space. Therefore. only 70% of the 342 Watts/m^2 drive emissions from the earth, and that is 239 Watts/m^2.

The earth emits radiation just as the Stefan-Boltzman equation predicts and most natural surfaces have emissivities close to unity, but certain gases within the atmosphere block the loss to space of some terrestrial radiation and force the average surface temperature above the value it would have without the atmosphere. This **greenhouse effect** results in higher temperatures at the surface of the earth. The average surface temperature is 15°C (288K) or 33°C warmer than would be possible without this effect. Exchanges of radiant energy between the surface, the atmosphere, and space that support the greenhouse and other effects are useful components in atmospheric models that help us understand changes in weather and climate. But that is another story best left to the chapter on solar heating.

Figure 4.3 The Area of a Sphere Is 4 Times Greater Than That of a Circle

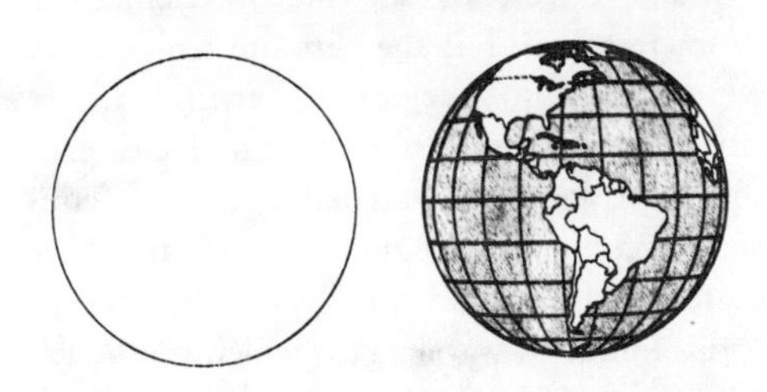

Wien's Radiation Balance Law and Atmospheric Heating

Emissivity is defined as the ratio of energy emitted by a real object to the energy emitted by an idealized perfect radiator of the same temperature. Radiated energy is transmitted directly through empty space. When the radiation comes into contact with matter, one of several kinds of interaction may occur. Some of the energy may be reflected from the surface of large particles or scattered by smaller ones. **Reflected** radiation has been diverted from its original path of motion, but is otherwise unchanged. That which is not reflected or **scattered**, passes into the medium and is subsequently absorbed by the material or **transmitted** through it. Absorption refers to energy that has been transferred to a receiving surface. How much is reflected depends on a number of factors: the angle at which radiation strikes the surface (angle of incidence), the smoothness of the surface, the color of the surface, and even the wavelength of the energy.

Figure 4.4 demonstrates light striking a water surface. The angle of incidence and the angle of reflection are always equal. With water surfaces, the nearer the incidence angle is to 0, i.e. vertical, the less light is reflected and the more is absorbed.

The ratio of light transmitted to light absorbed depends on the transparency of the medium, its thickness, and the wavelength of light. In the case of water for example, light penetrates clear water much further than it does turbid water (water with heavy concentrations of suspended sediment). There is no light in water bodies below a depth of several hundred meters. Furthermore, all wavelengths of light are not transmitted equally. Longer wavelengths are absorbed much more readily by water than short wavelengths. In part, this is what gives water its blue-green color. It also explains why long wavelength infrared radiation is so readily absorbed by water vapor in the atmosphere. Different substances absorb different wavelengths of light because of variations in their molecular structure. Figure 4.5 shows absorption of solar radiation within the atmosphere including those for ultraviolet, visible, and infrared radiation. The selective absorption of certain wavelengths over others can be attributed to the presence of specific molecules in the atmosphere which are stimulated by particular wavelengths (Figure 4.6).

Figure 4.4 The Angles of Incidence, Reflection, and Refraction

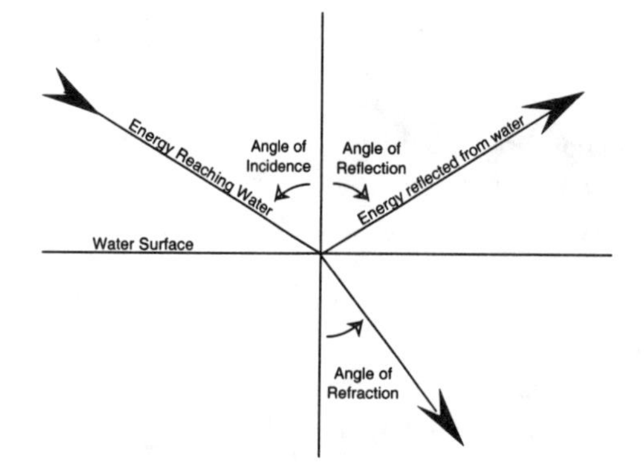

Absorption of radiation by a solid substance varies with the material composition, color, and surface roughness and with the wavelength of the radiation reaching it. As objects absorb radiation they get hot and begin to emit (radiate) energy themselves. The hotter they get the more they radiate, in accordance with Stefan's law. Warm objects not only radiate more of each wavelength, but they shift to higher energy or shorter wavelengths to help relieve the heat load. This wavelength relationship is referred to as Wien's Law and is mathematically expressed as:

$$L_m = \frac{2,898}{T}$$

where L_m stands for the wavelength in micrometers, and T is the absolute temperature (temperature in degrees Kelvin). Since temperature is the denominator, Lm decreases as temperature increases. A good example of this process occurs when the unit of an electric stove is heated. At first, as the unit gets warm it begins to radiate infrared wavelengths which can be felt but not seen. As it gets hotter it glows dull red. If more energy were passed through the material it would turn yellow, then blue. Finally, if it got hot enough it would give off large quantities of

Figure 4.5 Absorption of Solar Radiation by Wavelength

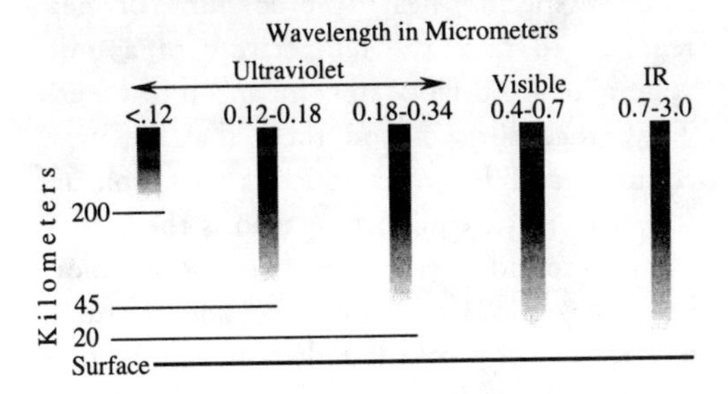

all wavelengths and appear white. Of course, the latter does not occur with the energy supplied to the electric stoves found in homes.

If we apply Wien's Law to determine the optimal wavelength emissions of the sun and the earth, we find the following:

$$L_m(sun) = \frac{2,898}{5,800K} = 0.50\mu m$$

$$L_m(earth) = \frac{2,898}{288K} = 10.1\mu m$$

As you will note in Figure 4.7 these values realistically depict the efficiency of wavelength radiation at specific temperatures. Also notice that 10,000 nanometers[4] corresponds to infrared radiation wavelengths. If our eyes were sensitive to these wavelengths, we would be able to see the earth "glowing at night." Technological advances have allowed scientists to develop film and night scopes, however, that are sensitive to infrared radiation. One special application of this technology is the ability to photograph atmospheric weather systems in the

[4]A nanometer is a unit of distance equal to one billionth of a meter.

Figure 4.6 Absorption of Incoming Solar Radiation by the Atmosphere

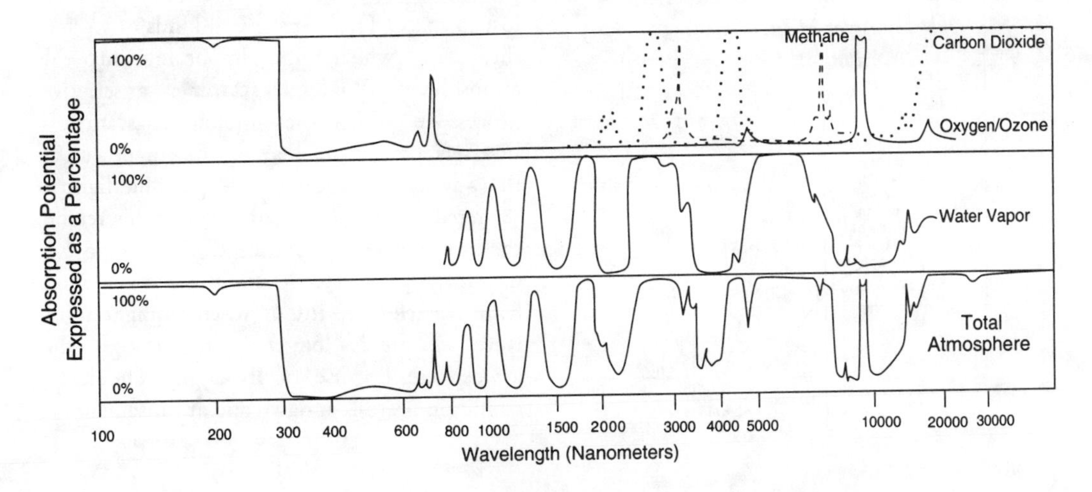

dark of night. This has led to improvements in short-term weather forecasting.

Emission is a process that prevents objects from getting hotter and hotter and which allows them to develop a thermal equilibrium with their environment. The **Second Law of Thermodynamics** dictates that all energy is converted to heat ultimately. **Entropy**, the degradation of energy into heat, is maximized by emission. Factors that affect the absorption and storage of heat in real objects include: 1) angle of incident radiation; 2) length of exposure to incoming energy; 3) color; 4) surface roughness, and transmissivity; 5) surface to volume ratio; 6) heat conductivity of the absorbing material; and 7) specific heat (the amount of heat required to raise the temperature of a unit-volume of a substance in comparison to water). Thus, road surfaces and rocks that are more compact and better conductors than soil are slower to reach equilibrium than is the surface of the ground. Because they can store more heat, they stay warm longer at night. This is one reason why cold blooded animals such as snakes and lizards are found lying on the road and among rocks at night in the early spring and fall.

Figure 4.7 Wavelengths of Solar and Terrestrial Radiation

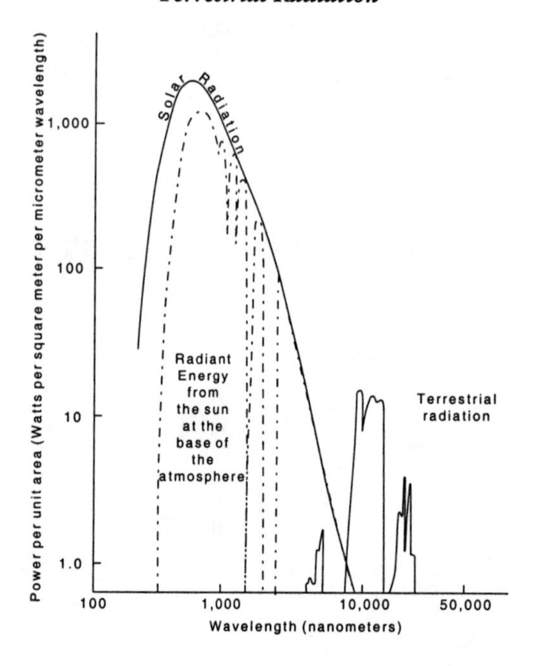

Solar Radiation Scatter

Scattering of solar radiation occurs when gas molecules and atmospheric aerosols deflect sun energy from a direct path toward the earth. This results in photons being redirected from their intended movement in a variety of directions that may result in their finally reaching the earth's surface, being absorbed by gas molecules while being redirected, or even lost to space. There are several kinds of scatter, three of which are important in the atmosphere. **Rayleigh scatter** is a selective process in which the amount of scatter is inversely proportional to the fourth power of the wavelength. Short wavelength blue light is scattered much more than red. For this reason the sky appears blue except when there are unusually large concentrations of dust or other large particles present or when sunlight must travel a much longer path through the atmosphere because of its angle. The latter condition prevails at dawn and at sunset, and, if there is a lot of dust in the air, produces brilliant orange and red sunsets.

Mie scatter results when incoming radiation of a specific wavelength strikes a molecule with a diameter approximately the same as the wavelength of light. The effect of Mie scatter is more uniform than Rayleigh scatter, affecting all wavelengths of light. When abundant water vapor is present, however, its effect on longer wavelengths is increased.

Nonselective scatter occurs when large spherical particles reflect light from their curved surfaces causing it to go in all directions. Since all wavelengths are susceptible to the process, the scattered light appears white. Clouds and fog are good examples.

Energy from the sun is affected by all of the foregoing processes. Radiation emitted by the sun travels approximately 150,000,000 kilometers through space before striking the earth. The small fraction of that energy intercepted by the atmosphere is subjected to reflection, scatter, and differential absorption. The power of this energy as measured at the base of the earth's atmosphere is estimated to be 1,370 watts on each square meter measured at right angles to the solar beam. Radiation absorbed at the earth's surface is transformed to longwave heat radiation and emitted upward. Solar and terrestrial radiation are absorbed, reflected, and scattered as they pass through the atmosphere. Some of the longwave radiation is absorbed by the atmosphere, but large amounts are lost to space.

A Final Note

Several electromagnetic radiation concepts have been introduced in this chapter. You may at this point be wondering what they have to do with the subject. Rest assured, we have not discussed these topics idly. As we will see in forthcoming chapters, these principles relate to the so called "Greenhouse Effect" and to assumptions about the balance of global heating. Do we, or do we not, have an atmospheric energy balance? A negative answer should result in a shift in Wien's predictive wavelength emissions and Stefan-Boltzmann's energy (heating) potential. And what about scattering? If pollutants either from industrial processes or from volcanic activity increase scattering, the earth's surface could "theoretically" receive less energy and, thus, become subject to a cooling trend. If, however, emissions of greenhouse gases increase, earth heating can be induced. Another possibility is that increased scattering and increased greenhouse gases can counterbalance each another and maintain present global climate patterns. Each of these alternatives implies a very different future for the world's ecosystems. These topics will be discussed in the next chapter.

Activity: The Greenhouse Effect

OBJECTIVES

Upon completion of this activity you will be able to: (1) describe the difference between the wavelength of radiation received by the earth and the wavelength of radiation emitted from the earth; (2) relate these differences in radiation wavelengths to the role of the earth's atmosphere; (3) use Wien's Law to determine the wavelength of the most intense radiation that a body emits, given the temperature of the body; (4) use the Stefan-Boltzmann Equation to determine the predicted temperature of a surface, given the universal constant for that body and the intensity of radiation coming in to that surface; (5) calculate the average balance surface temperature of any planet by applying the Stefan-Boltzmann Equation; (6) describe the greenhouse effect in terms of: (a) insolation; (b) temperature; (c) wavelength of radiation received; (d) wavelength of radiation emitted; (e) effects of the atmosphere; and (f) effects of greenhouse gases; (7) list three important greenhouse gases; (8) describe and give examples of both long- and short-wave radiation (especially, ultraviolet, infrared, and visible wave length radiation).

MATERIALS

Solar test box, 1 black absorber sheet, 2 glass sheets, solar radiation meter, flood lamp or direct sunlight, thermistor, 2 temperature probes, and perhaps a calculator.

INTRODUCTION

This activity may be performed either outside under a bright sunny sky or in the lab beneath an artificial light source. It is more realistic to do this outdoors but you must be very careful transporting the glass plates. They may be carried securely inside the solar box.

The total intensity for solar radiation is spread over a range of comparatively short wavelengths. Most of these pass easily through the earth's atmosphere and are absorbed at the earth's surface. As the earth's surface warms up, it in turn reradiates the absorbed energy back toward space.

The earth radiates heat energy at much longer wavelengths, and this radiation is effectively blocked by the earth's atmosphere. Water vapor, ozone and carbon dioxide are the major components of the atmosphere which permit passage of shorter waves of solar radiation, but interfere with the passage of longer wavelengths (heat) radiating from the earth's surface.

The ability of the earth's atmosphere to admit most of the incoming solar radiation and retard the re-radiation from the earth is called the greenhouse effect. As we burn more and more fossil fuels, abnormally large amounts of carbon dioxide enter the earth's atmosphere, potentially causing a change in the global heat balance. Some scientists are concerned that this will eventually lead to regional shifts in global climatic patterns, including shifts in agricultural productivity, a partial melting of the polar icecaps, and an increased rise in sea level.

In 1900 a physicist named Max Planck discovered what is called the black body radiation law. In trying to explain mathematically how fast an object could radiate heat, he formulated the relationship between wavelength, temperature, and maximum rates of radiation. Part of Planck's law gives the maximum amount of radiation that an object can give off as a function of its temperature. An object that radiates heat at this maximum rate is called a black body, because such objects also absorb all the light (or heat) that shines on them. Planck's law is actually a combination of two older laws. Both of these laws are important because they help us understand energy flows from the sun to the earth and consequent interactions between the atmosphere and other earth surface systems. Each law tells us about the flow of energy relative to temperature, an easily measured atmospheric parameter.

The Stefan-Boltzmann Law:

One of the most important concepts in meteorology is the **Stefan-Boltzmann law.** The Stefan-Boltzmann law states that the rate at which an object radiates heat is proportional to the fourth power of the absolute temperature. The principal of the Stefan-Boltzmann Law can be applied to the earth by the equation:

$$T^4 = \frac{E}{\sigma}$$

Where T is the absolute temperature (Kelvin), E is the intensity of incoming radiation in units of calories per square centimeter per minute (Langleys), and **σ** is a universal constant equal to 8.17×10^{11} derived for the earth from the Stefan-Boltzmann constant and the surface area of the earth in square meters. Versions of this law can be used to determine the balance temperature of an object so long as the intensity of incoming solar radiation is known. The Stefan-Boltzmann law actually tells how much heat a body radiates away and not how much it receives. Where atmospheric temperatures are neither rising nor falling, incoming and outgoing radiation are in balance. While earth surface temperatures may change daily or seasonally, the average temperature of the earth has been virtually stable over the last several decades. Any time that the Stefan-Boltzmann law is used to calculate the temperature of a planet, what is really being computed is an average balance temperature.

Wien's Law:

The second of these laws is **Wien's law** (after Wilhelm Wien, 1864-1928) It states that, as the temperature of an object increases, the wavelength of the most intense radiation that it emits decreases.

$$\lambda_m = \frac{a}{T}$$

Where λ_m stands for the most intense wavelength in micrometers (microns or μm), T is the absolute temperature (temperature in degrees Kelvin), and a is a constant that has a value of 2898 if λ_m is expressed in micrometers. Since temperature is in the denominator of the equation for Wien's law, as the temperature increases the fraction decreases. This means that the wavelength decreases. Red is the color with the longest wavelength and red is produced by cooler temperatures than blue which has a much shorter wavelength. This is true of all objects. As they become hotter the wavelength of the radiation they emit becomes shorter and shorter. As an electric stove-top heating element is switched on and begins to heat, it radiates relatively longwave infrared radiation which is too longwave for the human eye to see. As the temperature increases the wavelength decreases and it begins to glow dull red barely visible to the eye. As the temperature increases it glows orange. If it could be heated further the wavelength would change to yellow, and even blue.

Initially as radiant energy enters the earth's atmosphere, it is either reflected, transmitted, or absorbed. **Reflection** is the process of sending back a portion of the incident radiation without altering either the reflecting surface or the radiation. The earth does not absorb all the incoming radiation because some of this energy is reflected back to space. The percentage of incoming radiation that is reflected back from the tops of clouds, snow, ice and other surfaces is called **albedo**. Albedo has not been included in the version of the Stefan-Boltzmann law above. **Absorption** is the ability of an object to assimilate energy from incident radiation. The atmosphere reflects some of the insolation back to space, but transmits most to the earth's surface. As the surface absorbs insolation the atmosphere is heated generally from the ground up. In this series of exercises you will observe these processes by measuring incoming radiation and temperature changes.

Orient the test box and the floodlamp (or the sun) so that the angle of incidence is 0°. Insert the black metal plate on the lowermost slot and place the solar radiation meter sensor on top of it being sure to locate the photovoltaic cell directly beneath the light source.

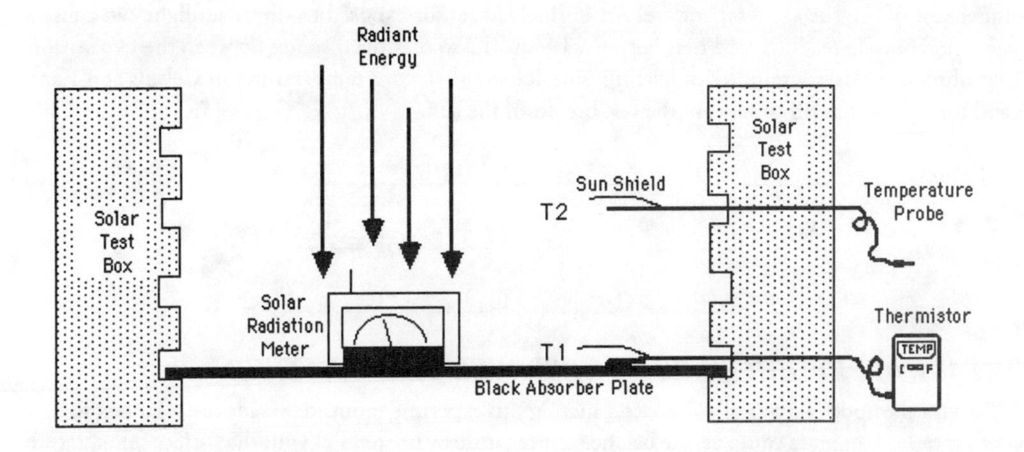

MEASURING INSOLATION

With the solar radiation meter, measure the direct incoming radiation. Be sure to keep the solar radiation meter sensor at 0° angle of incidence. Record the data and **REMOVE** the radiation meter. Convert to langleys using either of the suitable conversion factors: 1 Langley = 1 calorie/square cm/ minute = 220.618 Btu/hour/square foot = 697.422 Watts/square meter.

Direct incoming radiation___________Btus/hour/square foot

Direct incoming radiation___________Watts/square meter

Direct incoming radiation __________ Langleys per minute

PREDICTING TEMPERATURE

Use this observed rate of incoming radiant energy and the Stefan-Boltzmann Equation above to predict the average balance temperature of a perfect radiating body receiving this amount of radiation. Record your calculations here:

Predicted temperature _______ ° Kelvin _______ ° Celsius _______ ° Fahrenheit

$F = 1.8\,C + 32°$; $C = .55\,(F - 32°)$; and $K = C + 273°$

MEASURING ACTUAL TEMPERATURE

Now place a surface thermistor probe on the black metal plate. Be sure it is flush on surface to insure an accurate reading. Shield the temperature sensors with a slender strip of white tape to get an accurate measurement of the surface temperature. An unshielded sensor exposed to direct sunlight can cause a higher than accurate reading. Add tiers between T1 and T2 so that the distance between the two sensors is 10 centimeters. After 5 minutes of heating, simultaneously record temperatures in Celsius at T1 and T2 and turn the lamps off or remove the test box from the sun.

Minutes	1	2	3	4	5
T1	______ ° C	______ ° C	______ ° C	______ ° C	______ ° C
T2	______ ° C	______ ° C	______ ° C	______ ° C	______ ° C

COMPARING PREDICTED WITH OBSERVED TEMPERATURE

Turn off the flood lamp or, if you are conducting this experiment outside, shade the solar box so that heat can escape. Compare your average balance temperature estimate (T2) with the surface temperature (T1) you just observed at the end of the test. Is the observed temperature of the black metal plate (T1) higher or lower than the temperature predicted by the Stefan-Boltzmann law? ______________

____

Why are they different ? ____

____

The Greenhouse Effect

Ozone, carbon dioxide, and water vapor in the earth's atmosphere transmit incoming solar radiation (insolation) but reflect and absorb outgoing radiation from the earth. These so-called greenhouse gases are frequently compared with glass which is also transparent to insolation (e.g. visible light) and translucent to radiant energy from the earth's surface (e.g. heat). To simulate this aspect of the earth's atmosphere place a glass plate in the uppermost slot of the solar test box as shown below.

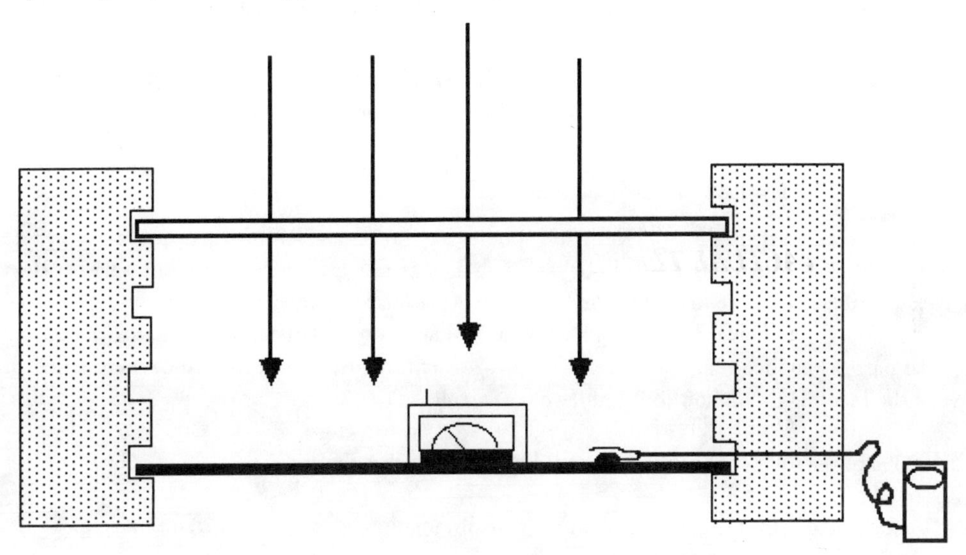

Switch on the flood lamp, or place the solar test box in direct sunlight and record the incoming radiation level beneath the glass. **REMOVE THE METER** and replace the glass. Are the levels similar to those without the glass?________

A. Incoming radiation without the glass cover_______________ ___BTUs/hour/square foot

B. Incoming radiation beneath the glass cover __________________BTUs/hour/square foot

What percentage of the radiation is reflected by the glass? (1 - B/A) _____ reflected

Which is equivalent to how many BTUs? (A-B) __________BTUs

What percentage of the radiation reaches the black plate through the glass? _______%
What percentage of the radiation would penetrate to the surface if three layers of glass were used?

________%

Although the principle is correct, the term "greenhouse effect" is actually a misnomer. The main reason that greenhouses on earth are so hot is that the glass prevents the heated air from rising. Greenhouses made of rock salt (which allows infrared radiation to escape) are almost as hot as greenhouses made of glass.

THE AVERAGE BALANCE TEMPERATURE OF THE EARTH

In this section you will find the average balance temperature of the earth. However, in order to find the balance temperature of a planet we must find the average intensity of radiation and not the intensity of direct radiation. Direct sunlight falls on only one point of a planet at a time. Half of the planet receives no light at all (night) at any given instant, while the rest of the planet receives sunlight obliquely. Although it may seem like a very difficult task to find the average intensity of sunlight it is actually quite easy.

All planets are essentially spherical, and spheres have an area given by the formula, $A=4\pi r^2$. This is 4 times the area of a circle that would be receiving direct sunlight. Average sunlight on any planet has an average intensity of 0.25 to that of direct sunlight (the solar constant).

One fourth of the earth's solar constant (1.94 Langleys) is 0.485. Also, just as the glass plate reflected part of the incoming radiation in the experiment, the earth and the atmosphere reflect about 33% of the incoming radiation (33% of 0.485 = 0.160).

If 33% of the solar constant is reflected, how much strikes the surface? _______%

What is the maximum average potential value of radiation at the earth's surface?

_________Langleys.

Using the Stefan-Boltzmann equation, what then would you expect the average temperature of the earth to be? Solve for T and show your work.

____

____

Average temperature of the earth: _____° Celsius ______° Fahrenheit ______ Kelvin

Does this seem reasonable?________
The actual average temperature of the earth is 15° C or 288 K.

Suggest reasons why your answer differs from the true average temperature of the earth.

____

____

____

In the earth's atmosphere the carbon dioxide (CO_2), water vapor (H_2O), ozone (O_3), methane (CH_4), and cloud droplets heat up when they absorb longwave radiation from the ground. These heated molecules then collide with the other air molecules, thus heating the rest of the atmosphere as well. Once these absorbing gases have heated up, they too radiate heat. Some of this radiation escapes to space, but some heads back toward the ground. This extra radiation (in addition to the solar radiation) received at the earth's surface produces the greenhouse effect. The greenhouse effect on earth produces a net warming of about 37° C. Add 37° C to your Stefan-Boltzmann estimate to account for the greenhouse effect.

Now what do your calculations suggest is the average temperature of the earth?

_______° Celsius _______° Fahrenheit ________ Kelvin

Since the earth radiates almost like a black body all the time, why doesn't it glow at night? Use the average temperature of the earth and Wien's law in your answer.

____

____

____

____

____

Chapter 6

Solar Heating

Introduction

The previous chapter focused upon the nature of electromagnetic radiation and the characteristics of several forms of radiation. Now we have the opportunity to examine the manner by which energy is distributed in the earth-atmosphere system. We've learned that when a substance absorbs more energy than it radiates, its temperature will rise. Conversely, if it radiates more energy than it receives, temperature will decrease. We can liken this relationship to a household budget where income has to be balanced with expenditures. When income exceeds expenditures, surplus money is available to be stored in a bank; similarly, when absorbed radiation exceeds emitted energy, surplus energy is stored in the earth-atmospheric system and temperatures rise. If your monthly income suddenly drops below your expenditures, however, your savings will soon disappear and indebtedness will ensue, a situation most college students can readily relate to. The earth-atmosphere system also experiences times when energy receipts are low and, as the system releases its stored energy, temperatures decline. We become aware of these energy imbalances, for example, in the difference between day and nighttime temperatures and in the changing of seasons. Each of these imbalances varies in intensity and magnitude. Yet, when the positive and negative imbalances are averaged for the entire globe for a long length of time, scientists believe they mathematically cancel each other: i.e. energy absorbed – energy re-emitted = 0. Otherwise, major global climate change would have to occur.

To define the balance of earth-atmospheric heating, it is possible to evaluate the separate energy components, as illustrated in Figure 6.1. For the sake of simplicity, we will assume that solar radiation at the top of the troposphere is 100 units or 100 percent. (Note: If we wish to establish the exact amount in either calories/cm^2 or watts/m^2 of energy received, it is a simple matter to multiply the average radiation intensity times a specific period of time, e.g. one year.) Remember, according to Wien's Law, radiation emitted by the sun reaches planet earth as **shortwave radiation.**[1] The distribution of this energy is illustrated in Figure 6.1A. For every 100 units entering the atmosphere, 6 units are scattered and deflected back to space. This is caused by air molecules and aerosols altering the energy's direct path toward the earth. Another 20 units are reflected back to space from the tops of clouds, and 4 units reach the earth's surface, only to be lost to space. Thus, a total of 30 input units of energy are lost to space as shortwave radiation and are not effective in

[1]Shortwave radiation is used as a general term in radiation balance climatology to refer to radiation with a wavelength less than 2,000 nanometers.

Figure 6.1 Generalized Diagram of the Global Heat Budget

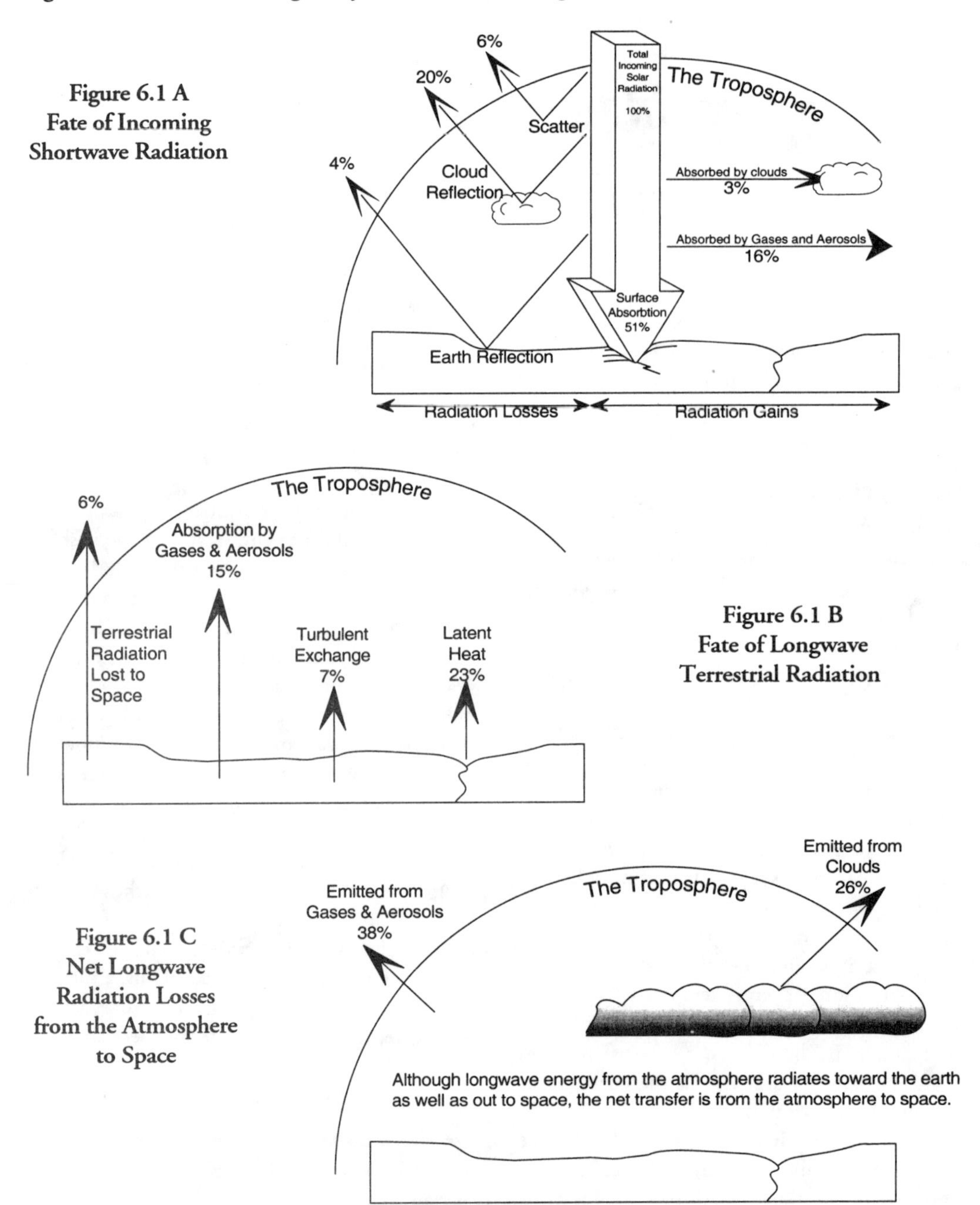

Figure 6.1 A
Fate of Incoming Shortwave Radiation

Figure 6.1 B
Fate of Longwave Terrestrial Radiation

Figure 6.1 C
Net Longwave Radiation Losses from the Atmosphere to Space

heating either the earth or atmosphere. These 30 units, constitute the earth–atmosphere system **albedo** (reflectivity of its surfaces).

It is worthwhile to note that all surfaces have some degree of reflectivity. Have you ever taken an airplane trip on a day when it was raining and skies overcast? If so, you may recall that when the plane rose above the clouds the sunshine was brilliant and if you looked downward the glare from reflected sunlight off the cloud tops was blinding. Reflectivity levels of 75 – 95 percent have been measured above dense clouds. A well mowed lawn averages about 20 percent reflectivity and fresh snow approximately 80 percent. Even you reflect solar energy as you walk about outdoors.

Of the remaining 70 units of incoming radiation, 51 pass directly through the atmosphere without being absorbed, as if the atmosphere didn't exist. When this energy reaches the earth's surface, it is absorbed and used to heat the earth itself or to transform water at the earth's surface into a gaseous form that is diffused to the atmosphere. We have learned that our atmosphere has a low capacity to absorb shortwave radiation; thus, the atmosphere is essentially transparent to much of the energy traveling through it. This is not true for earth bound surfaces which are opaque. Solar radiation cannot pass through the earth, so it must be absorbed.

You may be wondering by this time whether there is any direct sun energy that actually heats our atmosphere. The answer is yes, but only the remaining 19 units that have not yet been accounted for. Of these 19 units, approximately 3 units are absorbed by clouds. The last 16 units are absorbed by aerosols (dust sized organic and mineral particles) and certain gases, among which the most important are: water vapor and carbon dioxide. Bear in mind that this latter absorption is effective primarily throughout the depth of the troposphere. This means that its contribution to the air temperature you experience at the earth's surface on a hot, sunny day is indeed minor.

Our discourse on the energy budget began with the statement that energy receipts and losses must balance. Thus far, we've seen that of 100 original units of energy, 30 units were lost for heating purposes through scattering or reflection, 51 units heated the earth and 19 units were absorbed by the atmosphere. It is obvious that something must happen to the 70 absorbed units. Recall Wien's Law stating: any substance that has a temperature greater than absolute zero (0°K) will also radiate energy. Because the atmosphere and earth surface absorb radiation and have a temperature much greater than absolute zero, they then must emit radiation. Referring again to Wien's Law, wavelengths emitted by the earth and its atmosphere must be relatively longwave radiation[2] because they originate from surfaces much cooler than that of the sun.

Figure 6.1B illustrates how longwave radiation is distributed in the earth-atmosphere system. Since the earth's surface absorbed 51 units of energy it must likewise radiate 51 units. Earth radiation emits 21 units. Of these, 6 units escape directly to space because they are emitted at wavelengths where atmospheric absorption potential is slight. The other 15 units are directly absorbed by air in contact with planetary surfaces.

A second form of longwave energy exchange is the evaporation-condensation process. As solar energy flows toward earth, we must bear in

[2]*Longwave radiation is used as a general term in radiation balance climatology to denote radiation with a wavelength greater than 2,000 nanometers.*

mind that three-fourths of it is directed to water surfaces. This, due to the fact that approximately 75 percent of the planet's surface is covered by water bodies. A significant portion of this energy is used to evaporate water, meaning to change its state from a liquid to a gas. It is important to understand that the energy used in evaporating water fulfills the purpose of altering its molecular structure from water as a liquid to water as a gas or vapor. The foregoing statement has been emphasized to stress that energy involved in the evaporation process does not heat the water; rather, it converts water to, and maintains it in, the gaseous state. For this reason, energy tied up in the evaporation process is referred to as latent heat. The word latent suggests something has potential, but does not yet exist. In the case of latent heat, the energy exists, but is not heating anything as it presently exists. Yet, it has the potential to be released eventually and to heat the atmosphere. This concept may seem strange at this point, but hopefully will soon become obvious. Before going further, however, we should point out that the energy required to evaporate water is relatively substantial, amounting to about 580 calories per gram, a value generally accepted as the latent heat constant.

Once water vapor is mixed with other atmospheric gases, it is subject to vertical and horizontal motions. Specific combinations of these motions can cause the air to cool and lose the ability to hold the water vapor it contains. This results in condensation, the change in state of water from a gas to a liquid. You commonly witness this phenomenon when you see clouds in the sky or drive through fog. What is amazing about this transformation is that the latent heat which kept the water in gaseous form is now released to the atmosphere and available for heating it. It is difficult to comprehend the enormous amount of energy that warms the atmosphere from the countless clouds that form throughout the atmosphere each day. An examination of Figure 6.1 indicates that latent heat (through the evaporation-condensation process) is foremost in atmospheric heating, providing 23 units and surpassing earth radiation (15 units) and shortwave absorption (19 units).

A third form of longwave radiation interaction is called turbulent exchange (7 units). This is a simple process. When air moves over a warm earth surface, heat is transferred from the surface to the air. The intensity of this radiation is well demonstrated, in a qualitative way, by the heat you feel emanating from an asphalt parking lot during the summer. At the contact between the air and the heated surface, heat is transferred by conduction (a transfer of heat between adjacent molecules). This is a reversible process so that as warm air moves over a cold surface, the air loses heat to the surface. This particular form of heat exchange between the atmosphere and the earth's surface would not be so important without turbulent (meaning restless or agitated) airflow. Turbulent airflow helps to exchange air in the contact layer and facilitates the transfer of heat. Although this energy exchange takes place in both directions, the net result is a positive increase in energy to the atmosphere so long as the surface temperature is greater than the air temperature.

This lengthy discourse on the global energy budget has demonstrated the various means that the earth-atmosphere system gains and loses energy. When we started this discussion, it was pointed out that these relationships should balance. The units of gains and losses related to Figure 6.1 are provided in Table 6.1. Compare the two and establish whether a balance actually exists.

The budget just presented utilized net radiation flows. It does not take into account counterbalancing exchanges which occur between the earth and the atmosphere. Much more energy is actually radiated from the earth to the atmosphere only to be absorbed and radiated back toward earth again. Energy radiated back to earth is called **counterradiation**. Budgets which include counterradiation values are quite different from the one presented here because counterradiation must be added to both input and output sides of the budget equation. Minor discrepancies may be found between the values used in various energy budget calculations because of differences in estimated albedos employed by different investigators.

Greenhouse Effect

A careful analysis of the energy budget components provided in Table 6.1 will reveal that about 70 percent of atmospheric heat absorption takes place from longwave radiation exchanges that take place at the earth's surface. This phenomenon has often been referred to, although imperfectly, as the greenhouse effect. To explain this analogy, let us begin with the principles that make a greenhouse work. Shortwave radiation passes through the clear glass, as if it were transparent, and penetrates to interior surfaces of soil, shelves, and miscellaneous objects. These substances, after being heated, radiate longwave energy that does not easily penetrate the glass, but remains trapped inside the greenhouse to be absorbed by the air and raise its temperature. If this concept seems remote to your experience, here's another approach. It's a sunny afternoon, you open your car door and feel a blast of hot air. Why? shortwave radiation has entered through the car windows, as if they didn't exist. Inside, it heats the dashboard, upholstery, steering wheel, etc., until they emit longwave radiation, producing heat (which increases air temperature) that is trapped inside the car. Both of these examples

Table 6.1 The Global Heat Budget. Values are percentage units of short and longwave radiation

	Shortwave	+	Longwave	=	Total
Earth Surface					
Gains	51	+	0	=	51
Losses	0	+	-51	=	-51
					0
Atmosphere					
Gains	19	+	45	=	64
Losses	0	+	-64	=	-64
					0
Total Earth-Atmosphere System					
Gains	100	+	0	=	100
Losses	-30	+	-70	=	-100
					0

emphasize the role of longwave radiation in heating our atmosphere. So what? Although these processes have been going on for millennia, it was not until the twentieth century that we have come to understand the global heat budget and its implications for human activities.

Let us offer two hypothetical conditions related to human impact on atmospheric resources. First, suppose the world's population were to increase the percentage of atmospheric gases that are efficient energy absorbers (called greenhouse gases). If this were to happen, global temperature should rise and climatic changes should occur. Some scientists believe this process is already underway and they have largely blamed carbon dioxide (CO_2) generated from the burning of fossil fuels for causing this effect. In addition to using fossil fuels, however, the clearing and burning of tropical forests has recently accelerated CO_2 emissions. We also bear the major responsibility for producing other greenhouse gases, such as methane, chlorofluorocarbons (CFCs), soot, ozone, and halons, that take up long-term residence in the atmosphere. Halons and methane have been used for a variety of purposes for several decades in the wealthy industrial countries and, despite efforts to find price competitive substitutes, their use is still increasing worldwide.

Up until 1950, data collected by scientists for the previous three-quarters of a century showed an increase in carbon dioxide concentrations from about 285 to 315 ppm (Figure 6.2). During the same period, temperature records for different locations showed average increases of 0.5°C generally, with the highest increases being associated with midlatitude locations. Many scientists interpreted these data to mean that increases in average temperatures were caused by increased concentrations of CO_2 and projected dire consequences if trends in fossil fuel consumption continued.

Between 1950 and 1970 average annual temperatures declined appreciably. Some scientists attribute the decline in temperatures to increased concentrations of dust in the atmosphere and suggested that, if those trends continued, we would be on the verge of a new ice age. Volcanic eruptions (as great as that of Mt. Pinatubo) or dust and smoke produced by industrial and land modification activities increase the percentage of solar rays that is subject to scattering and lost from the atmosphere for heating purposes.

Since 1970, average annual temperature has risen sharply and now rivals or exceeds the warm years of 40 years ago. Because levels of greenhouse gases have been continually increasing since the beginning of the industrial revolution, some atmospheric scientists expect global temperatures to continue to increase. Evidence of global warming and related changes in the observed climate record are currently being carefully scrutinized. The decade of the

Figure 6.2 Global Atmospheric Carbon Dioxide Levels

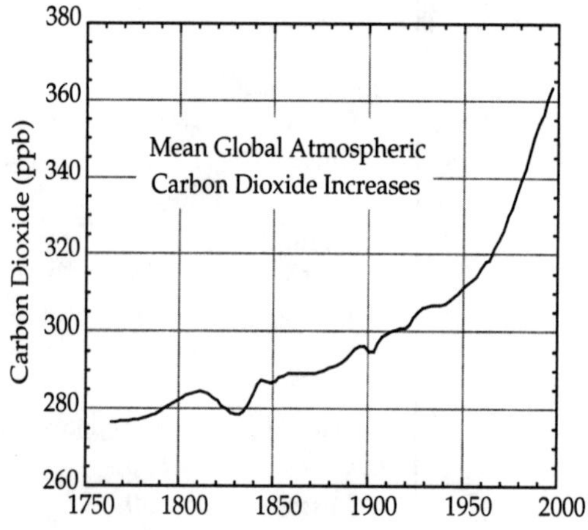

1980s witnessed several record breaking warm years followed by even warmer years in the 1990s. Land based instrumental records identify the warmest globally averaged 10 years of the past century as 1998 followed in descending order by 1997, 1995, 1990, 1991, 1988, 1987, 1983, 1981, and 1980. Many meteorologists and climatologists are cautious, however, and resist assertions that global warming has actually begun. Many climatic parameters exhibit stable oscillations and the current period of warm temperatures may not yet be clearly different from previous warm periods. Of those who are persuaded that warming has begun, many are skeptical that human activities are responsible for the change. Finally, some atmospheric scientists concede that global warming is occurring and that human pollution is to blame.

Divergent perspectives are natural considering the extremely complex nature of the atmosphere. For example, consider the role of natural and anthropogenic tropospheric sulfate aerosols in reflecting solar radiation. Sulfate aerosols are added to the atmosphere as coal and other fuels containing sulfur are burned. Therefore, industrialization contributes to increases in the volume of atmospheric sulfate. These aerosols are important climatically because they reflect solar radiation back to space directly and because they increase cloud albedo which results in an additional loss of incoming solar radiation. The combined cooling effect is thought to nearly counterbalance the warming effect of additional greenhouse gases.[3]

There is at least one other explanation for the changes in the earth's mean temperature, and that is the possibility of fluctuations in solar radiation. We know that variations do occur. During periods of increased sunspot activity, such as those during 1989–1990 and during 2000 for example, there are increases in the output of specific wavelengths of light. Sunspot related increases are only 0.1% above typical output and they recur generally every 11 years with the greatest peaks every 22 years. The debate is likely to continue for some time, because it is not known whether variations in solar radiation are of significant magnitude to affect the earth's temperature.

Energy - Temperature Patterns

We have repeatedly mentioned that the flow of energy reaching earth from its source (the Sun) is steady and also that variation in solar radiation intensity, at a given location, is strongly influenced by latitude. From this general statement, it is possible to assume that any point along a given parallel of latitude has the potential to receive exactly the same amount of energy as every other point. This means that if the earth's surface were a homogeneous featureless plain, its temperature patterns would be comparable in an east-west direction and generally decrease from the tropical regions to the poles. Unfortunately, such simplifications rarely occur in nature. The major difference in solar radiation absorption and temperature response on a real, as opposed to a hypothetical, earth is that surface materials are not uniform. The range of surface properties varies widely. Exposed minerals range in composition and in color from black to white, vegetation may, or may not, be present and water clarity can vary from crystal clear to extremely turbid. Thus, we might expect a myriad of different heating potentials along a given parallel of latitude. To simplify our approach to understanding earth-atmosphere heating, however, we will limit our

[3]J. T. Kiehl and B. P. Briegleb. The Relative Roles of Sulfate Aerosols and Greenhouse Gases in Climate Forcing. Science 260 (April 16, 1993): 311-314.

present discussion to the differences associated with the general heating characteristics of the mineral substances of the landmasses and those related to the earth's oceanic bodies.

A summary of the dominant factors that cause differential heating of land and water surfaces is presented in Figure 6.3. There are four:

1) Evaporation Differences. Recall from our discussion of the heat budget that the evaporation process ties up energy as latent heat, i.e., it is unavailable to heat the substance. Since evaporation potential is greater over water bodies than it is over land surfaces, we can expect that less energy is available for heating water bodies than is available for heating land areas.

2) Opaque versus Transparent Substances. Relative to land surfaces, water bodies are transparent. Thus, we can see the bottom of a swimming pool because sunlight penetrates the water. Consequently, this sun energy is distributed throughout a large volume. The same amount of energy reaching soil surfaces is concentrated in a very shallow layer at the atmosphere-earth contact. Consequently, per unit volume, more energy is available to heat the land surface relative to the water. Perhaps you have observed nature's response to these subtle differences. During the summertime when temperatures soar, dogs will frequently try to find a cool spot in which to relax. Their response is to find a shady spot, dig away the surface soil layer, and rest in the cooler subsoil.

3) Relative Motion of Land and Water. Land is static. It doesn't move on a daily basis. Water is turbulent. Ocean currents transport water equatorward from cold, high latitude places and poleward from warm, tropical locations. Vertical currents that upwell (raise) water from the sea floor or density currents common where sea surface temperatures are less than 4°C help to disperse heat and matter from

Figure 6.3 Factors Determining Land and Water Heating Differences

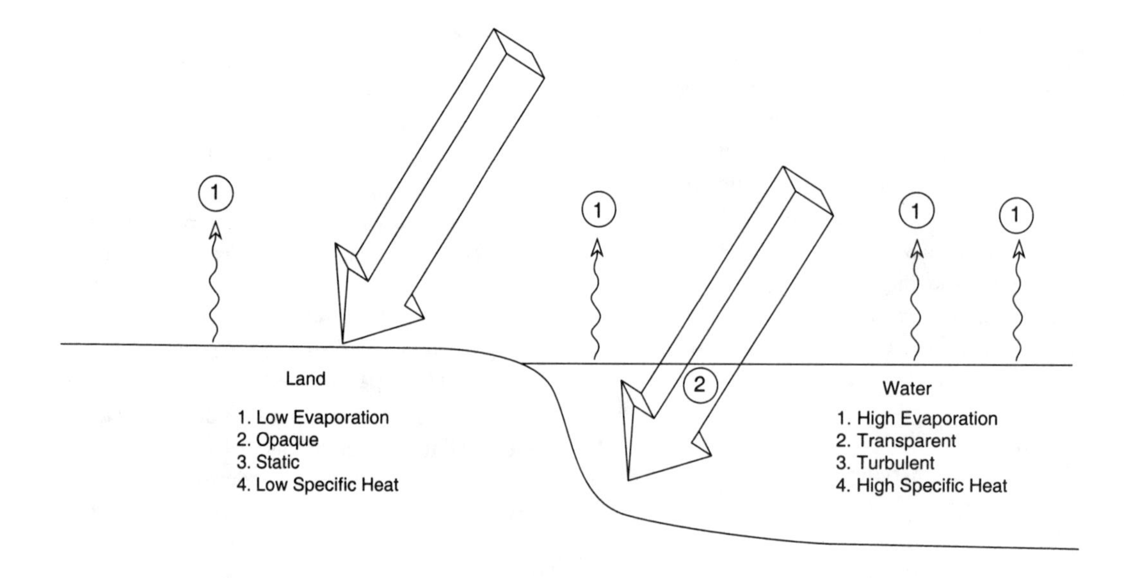

the ocean surface. Energy intercepted by a land location is concentrated in-place, whereas that absorbed by water is subject to vertical and horizontal mixing.

4) **Specific Heat** Relationships. The specific heat of a substance is the amount of energy required to raise its temperature 1° per unit mass. This is true for any system of units, but in Système International (SI) units the specific heat of water in 4.186 Joules/gram. Earlier systems put the specific heat of water at 1.00. For example, it takes 1 calorie[4] to increase the temperature of one gram of water from 14.5°C to 15.5°C. This is approximately 5 times the energy required to increase the temperature of one gram of mineral matter by 1°C. Thus, it takes more energy per unit mass to get a temperature increase in water than an equivalent temperature increase in dry land substances.

Let us now relate the significance of the foregoing to patterns of global heating. Figures 6.4 and 6.5 illustrate the pattern of mean (arithmetic average) sea level temperatures in January and July as **isotherms** (lines connecting points of equal temperature). These two months are usually chosen to illustrate temperature patterns, since for most world locations they represent the coldest and warmest months. Remember that seasons are reversed in the Southern Hemisphere. Several important features can be recognized on these maps:

1) Observe the isotherms in the middle-latitudes of the Southern Hemisphere. Over water bodies, they almost approximate east-west lines. This strongly suggests uniformity of heating along a parallel of latitude. Variation from this simple pattern occurs when the water surface is interrupted by landmasses.

2) There are two massive regions of below freezing temperatures in the Northern Hemisphere during January. These are centered over Canada and Siberia. As will be discussed in following chapters, they serve as source regions for very cold, dense air that frequently moves equatorward. The result is winter storms and sometimes bitterly cold weather that cause crop damage as far south as Florida.

3) Notice the effect of the Gulf Stream on temperatures of the North Atlantic. Temperatures above freezing are found even poleward of the Arctic Circle in January. This clearly illustrates the role of ocean currents in distributing heat from regions having a surplus to those having a deficit.

4) During January when the Northern Hemisphere is experiencing winter, Southwest Africa and Australia have regions where the average monthly temperatures soar above 32°C (90°F).

5) The January and July maps show little variation in temperature conditions along the equator.

6) In July, the Northern Hemisphere cold temperature centers disappear and high temperature regions appear in the southwestern United States and northwestern Mexico, North Africa, and from Saudi Arabia through the Ganges Plain of India. We want to remember these "hot spots," particularly the last, as it plays a dominant role in one of the largest and most energetic wind systems: the monsoon.

Landmasses tend to heat rapidly and intensely with the approach of summer. They likewise quickly lose their meager supply of stored energy as winter nears and intensity of solar radiation decreases. This means the annual range (difference) between summertime and

[4]*A calorie is defined as the heat required to raise the temperature of 1 gram of water from 14.5°C to 15.5°C.*

Figure 6.4 Mean Global Temperatures in January (°C)

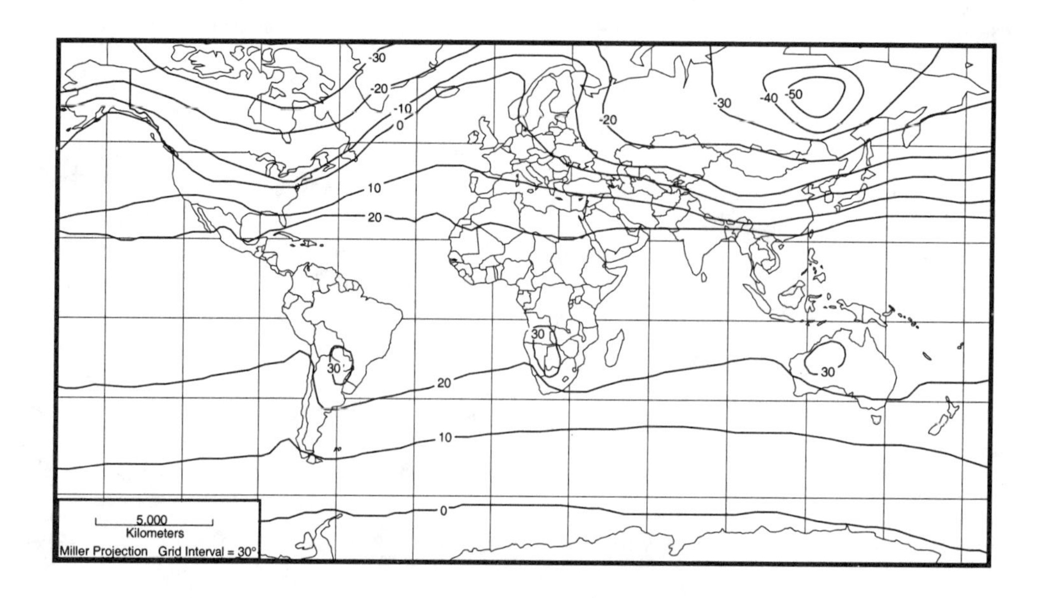

wintertime temperatures is relatively large. We call this feature of landmass heating and cooling **continentality**. As a general rule, the larger the landmass and higher the latitude, the greater will be the effects of continentality. Notice the tremendous continentality of Asia. Verkhoyansk in northeastern Siberia for example, has a range in temperature of about 63°C (115°F), varying from a brisk –49°C (–57°F) in January to a toasty 14°C (+58°F) in July. Look again at Figures 6.4 and 6.5 and examine the temperature range in the vicinity of the equator. The range here is relatively small because of less variation in the altitude of the sun and consequently, radiation intensity. Singapore, which has a tropical location and is minimally affected by the Asian landmass to its north has a mean monthly temperature range that rarely varies by more than 1 to 1.5 degrees. Its average is about 24°C (80°F).

Whereas large landmasses normally experience large temperature extremes, water bodies have a moderating influence on temperature. They heat up more slowly and never get extremely hot. In addition, their stored solar energy is greater than that of landmasses. Thus, as winter nears and landmasses cool quickly, water bodies still have warmth. Therefore, large water bodies have a moderating influence on the temperature range. Small islands and coastal areas where the wind generally blows onshore receive the benefits of this moderating influence, a phenomenon we call the **marine effect**. For example, Rejkavek, Iceland, has a mild temperature range of less than 10°C. Also, the west coasts of North America and Europe have moderate

Figure 6.5 Mean Global Temperatures in July (°C)

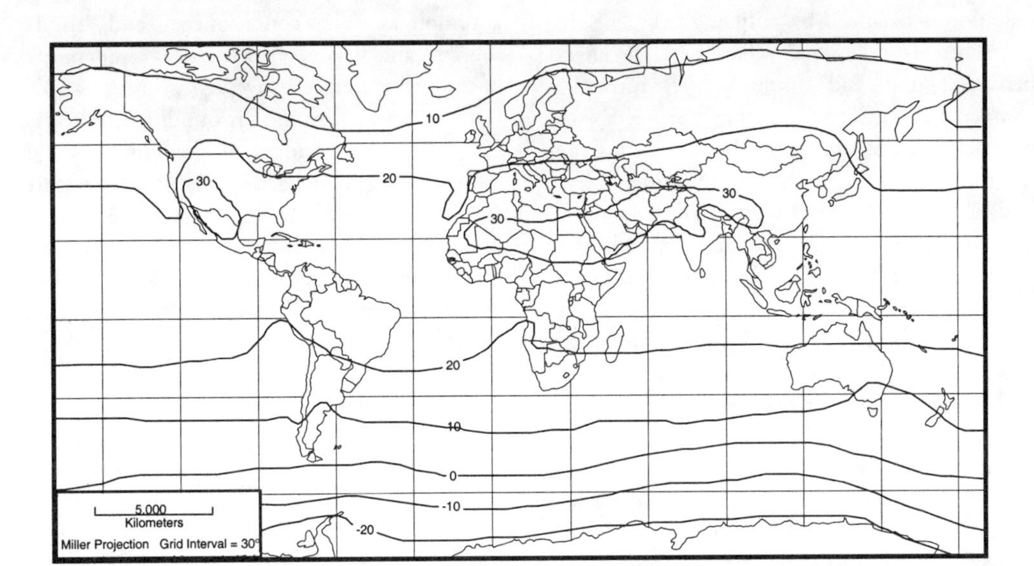

temperature ranges, compared to the interior of landmasses because the dominant wind systems affecting them originate to their west, over adjacent bodies of water. The eastern coasts at the same latitudes are not nearly as mild. Although they are also near a water body, the winds that dominate their climate are again from the west, but in this case from the continental interior.

Summary

The atmosphere receives its energy from the sun — some of it directly, but the bulk of it indirectly from earth. Thirty percent of the incoming radiation is reflected and scattered back to space without having any effect on heating the earth and atmosphere. The water vapor, carbon dioxide, ozone, and dust in the air directly absorb about 19 percent of the incoming radiation and the remaining 51 percent is absorbed by the earth. That 51 percent is lost through direct radiation to space (6 percent), evaporation (23 percent), radiation to the atmosphere (15 percent) and convection and turbulence (7 percent). The atmosphere loses the 64 percent it has absorbed by radiating it to space.

Energy budgets vary according to latitude, season, time of day, and type of surface. In general, higher latitudes receive less energy, and more of it is concentrated in summer when longer days and higher sun angles produce maximum absorption. Land surfaces absorb more than water surfaces and dark surfaces more than light surfaces. Regardless of the exact amount of energy involved or its distribution through time, in the long run the earth-atmosphere system may be thought of as a steady state system in which incoming solar radiation is balanced by outgoing heat loss. In the short term, however, there is a lag effect between input and output which must be compensated for by adjustments in storage, with heat being gained during the day and during the summer and lost at night and in winter. Heat budgets are made up of several components: a general effect which is a function of latitude and which, in turn, influences both the sun angle and the length of day; constant local effects which are determined by such things as exposure and surface albedo; and more random variable effects caused by such things as weather conditions, soil moisture availability, and atmospheric pollution.

The arguments over the nature and causes of climatic variation, whether global or local, center about the causes for variation in annual heat budgets. Variations in solar output, changes in net radiation due to increased albedos, or other causes theoretically can induce climatic changes. The role that human activities play in climatic variations is a much debated topic.

Some of the **anthropogenic** (of human origin) effects on heat budgets and climate at the local level are well illustrated in urban areas. Urbanization changes landscapes by removing much of the vegetation and by replacing naturally permeable soils with impermeable concrete, asphalt, and roof materials. In addition, tall buildings effectively increase total surface area while changing the exposure angle. Early morning and late afternoon sunlight shines more directly on walls than on the ground. With little vegetation and little moisture in the soil, a greater proportion of the more direct solar radiation is transferred as sensible heat. Add to this the energy lost to the environment by factories, buildings, automobiles, air conditioners, and similar sources, and it is easy to see why temperatures in inner city areas may average as much as 10°C (18°F) warmer than the surrounding rural areas.

The energy budget of the atmosphere is most obvious to us as patterns of temperature. We have seen that global temperature patterns vary widely over landmasses and much more conservatively over water bodies. These differences can be effective in determining the movement of air, a topic reserved for the next chapter.

Activity: Temperature

OBJECTIVES

Upon completion of this lab you will be able to: (1) convert from the Fahrenheit scale of temperature measurement to the Celsius scale and to the Kelvin scale measurement and vice versa given the appropriate conversion factors; (2) discuss the different rates of reflectivity (of solar radiation) and absorption of different materials; (3) calculate temperature variations with altitude given the Normal Environmental Lapse Rate (ELR) in degrees Celsius and in degrees Fahrenheit; (4) locate temperature inversion layers on a graph of temperature versus altitude and discuss how and why they occur; (5) discuss the threats to society that temperature inversions pose. When you have *completed* this exercise, you should have a clear understanding of three major factors which influence the temperature of the earth. These are:

1. Albedo - the percentage of solar radiation reflected, or turned back without energy loss, by a surface. How does albedo affect heating and cooling rates of different materials? How do the heating and cooling of these materials affect air temperatures.

2. Vertical Temperature Distribution (Normal Environmental Lapse Rate) - the *amount* or rate of temperature decrease upward through the troposphere. Why are mountain tops generally cooler than adjacent lowlands?

3. Temperature Inversions - reversal of the normal environmental lapse rate. Why do temperature inversions occur? How are temperature inversions related to smog?

TEMPERATURE

Temperature is commonly measured by one of two scales, the English, or Fahrenheit Scale or the Metric, or Celsius (C) scale. Americans are more familiar with the Fahrenheit scale, devised in 1714. By this scale, the temperature at which water boils at sea level is 212°, and the temperature at which water freezes is 32°. The Fahrenheit scale uses the freezing point of a mixture of salt and water as the baseline (0°) temperature. The Celsius scale was devised in 1742 by a Swedish astronomer, Anders Celsius. The temperature at which water freezes at sea level on this scale was arbitrarily set at 0°. The temperature at which water boils is 100°C. The Celsius scale is used nearly everywhere *except* the United States. Even in the U.S., the Celsius scale is used by the majority in the scientific community. The U.S. is moving toward the acceptance of the scale in an attempt to increase uniformity of statistical information throughout the world. Both scales are used throughout these exercises. If you want to convert one measure to another, the necessary formulas are provided. It is better, however, to think what a certain temperature means. Compare the selected equivalent temperatures on the thermometers pictured below.

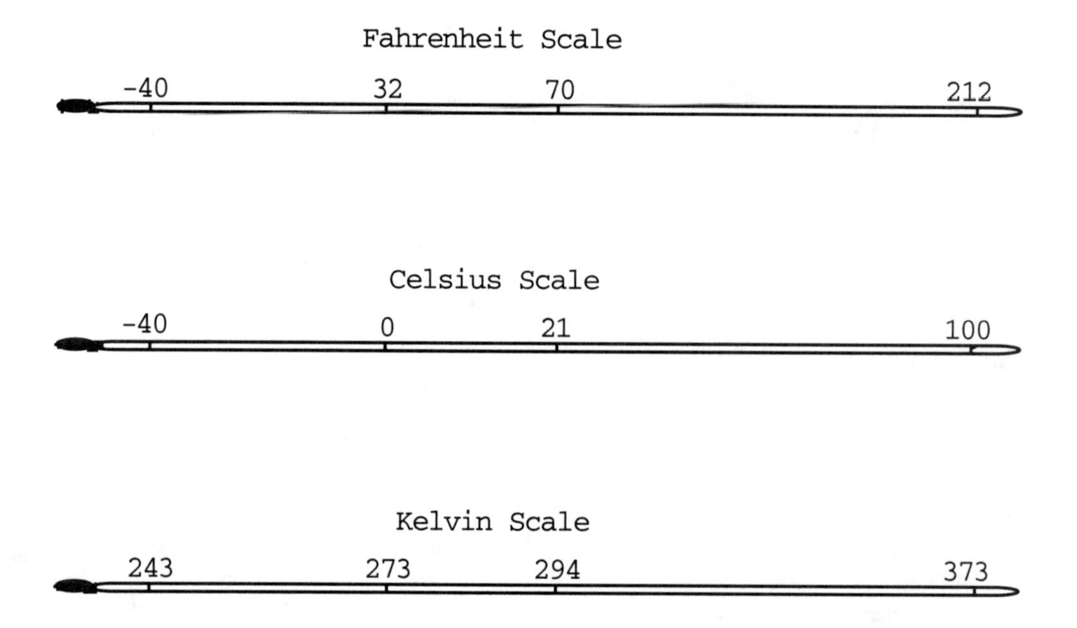

Celsius to Fahrenheit:	$F = 1.8\,C + 32°$
Fahrenheit to Celsius :	$C = .55\,(F - 32°)$
Celsius to Kelvin:	$K = C + 273°$

ALBEDO

Human activities modify our environment and thus produce variations in albedo from place to place. The list below gives average figures for reflectivity of several different surfaces. Following the example given for asphalt, plot the reflectivity/absorption ratio of each material listed below on the graph provided.

Material	Reflectivity	Material	Reflectivity
asphalt	10%	dark soil	15%
snow (fresh)	75%	dense forest	5%
*water	8%	grassy field	10%-20%
dry sand	35%	concrete	25%

*Albedo ranges from 2% to more than 60% for water, depending upon the sun's zenith angle. When the angle of incidence is less than 60°, the albedo for water surfaces is less than 5% but, with an angle of incidence greater than 87°, reflectivity is greater than 60%.

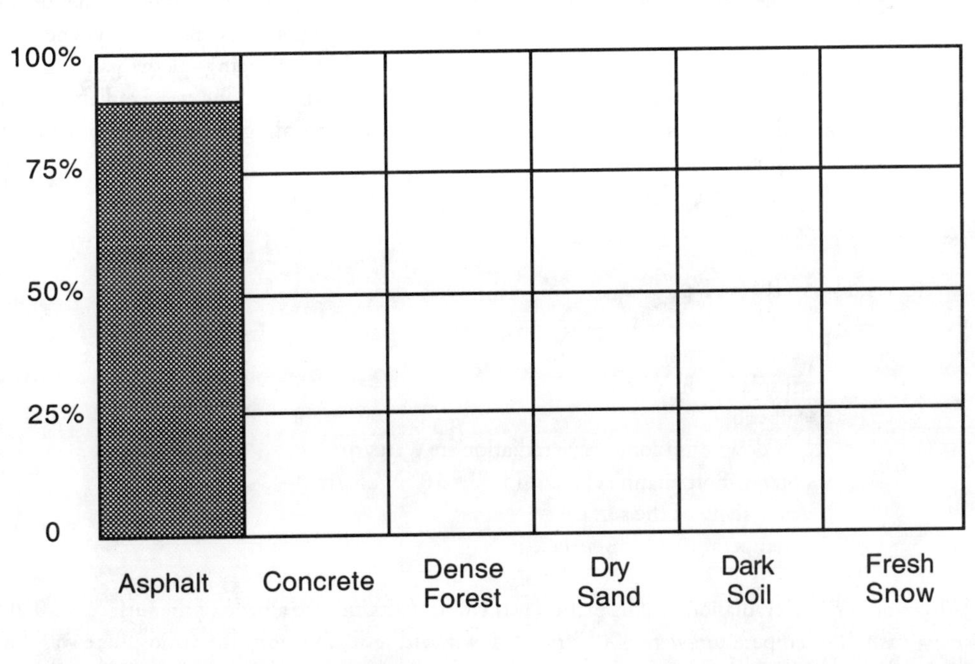

A percentage of the energy that has been absorbed by each of these materials is at some later time radiated into the atmosphere, thus heating the air above. This terrestrial radiation is readily absorbed by the atmosphere and is more important for heating the lower atmosphere than direct insolation. Remember this as you complete the section of this lab dealing with temperature inversion. You might recall that the term insolation is derived from INcoming SOLar radiATION.

NET RADIATION

During the day insolation heats the surface which acts as a heat source to warm the atmosphere. At night, the earth surface radiates this longwave (heat) energy through the atmosphere to space, which results in lower temperatures near the ground. The rate of heat loss at the surface and in the contact layer often exceeds the rate of heat loss from the air aloft. The rate of radiative heating or cooling at the surface oscillates from daytime heating through nighttime cooling each day. Temperature changes at the surface and in the contact layer are a function of **net radiation**, all incoming energy less all outgoing energy.

The rate of net radiative heating or cooling can be calculated by balancing energy gains against losses. Energy is gained at the surface by solar irradiance (I) and by downward long-wave radiation from the atmosphere ($R\downarrow$). Energy is lost by long-wave emitted radiation according to the Stefan-Boltzmann law ($\sigma T^4 \varepsilon$). Net radiation can be calculated as:

$$R_n = I(1-a) + R\downarrow - sT^4\varepsilon$$

where: R_n = net radiation in Watts/m²
I = insolation in Watts/m²
a = albedo
$R\downarrow$ = downward long-wave radiation in Watts/m²
s = Stefan-Boltzmann constant ($5.67 \bullet 10^{-8}$ Watts/m²)
ε = emissivity of the surface
T = temperature of the surface in K

What would the net radiation be if insolation was 1,000 Watts/m²; the albedo of the surface was0.20; emissivity was 0.95; temperature were 300° K; and downward radiation from the atmosphere was 250 Watts/m²?

$$R_n = 1{,}000\,(1-0.2) + 250 - (5.67\bullet 10^{-8}) \bullet 300^4 \bullet 0.95$$
$$R_n = 800 + 250 - 440$$
$$R_n = 610 \text{ Watts/m}^2$$

Now calculate net radiation above a snow covered surface assuming insolation is 1,000 Watts/m^2; the albedo of the surface is 0.95; emissivity is 0.99; temperature is 300° K; and downward radiation from the atmosphere is 250 Watts/m^2. What effect does snow cover have on net radiation?

Calculate net radiation assuming insolation is 0 Watts/m^2 (night); the albedo of the surface is 0.20; emissivity is 0.95; temperature is 300° K; and downward radiation from the atmosphere is 250 Watts/m^2. What effect does night have on net radiation?

ENVIRONMENTAL LAPSE RATES

There are four major zones or spherical layers above the earth's surface that are defined on the basis of temperature change. Although your studies for this lab will concentrate on the lowest sphere, the troposphere, all are important and affect air temperature. The four spheres are, in ascending order: troposphere, stratosphere, mesosphere, and thermosphere. The troposphere extends from the surface to about 18 km (11 miles) at the equator, but only to about 6.5 km (4 miles) over the poles. Temperature generally decreases at a uniform rate with increased altitude within the troposphere, although extreme cold or hot surface temperatures may temporarily distort the rate of temperature change in the lower boundary layer. The troposphere contains about 75% of the total mass of the atmosphere, and almost all the water vapor and clouds.

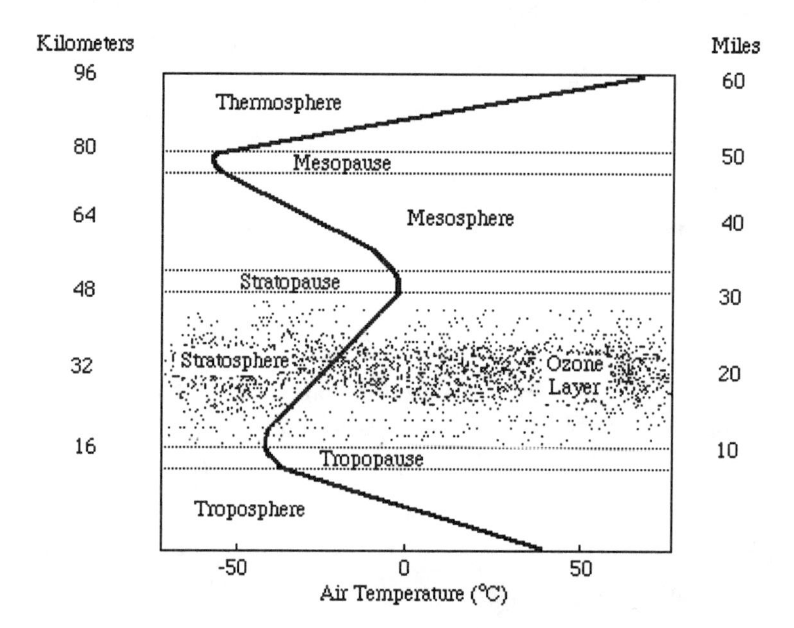

Average temperature decrease is 6.5° C for each 1,000 meters (3.5° F/1,000 ft) increase in altitude. This is called the **Normal Environmental Lapse Rate** or **ELR**. For this exercise use a sea level temperature of 20° C (65° F). Assume average weather conditions at a mid-latitude location, such as Charlotte.

1. On the following graph, calculate and plot the Normal ELR for each altitude listed on the graph, then connect the points. The lapse rate you plot is the result of perfect, or ideal, conditions.

2. To see a more realistic ELR, plot the points that are given, and connect them with a smooth line. Do you see how the ELRs differ?

Temperature (C°)	18°	4.5°	- 7.0°	-18°	-30°	-40°	-50°
Altitude (meters)	sea level	800	1600	3000	4500	6800	10500

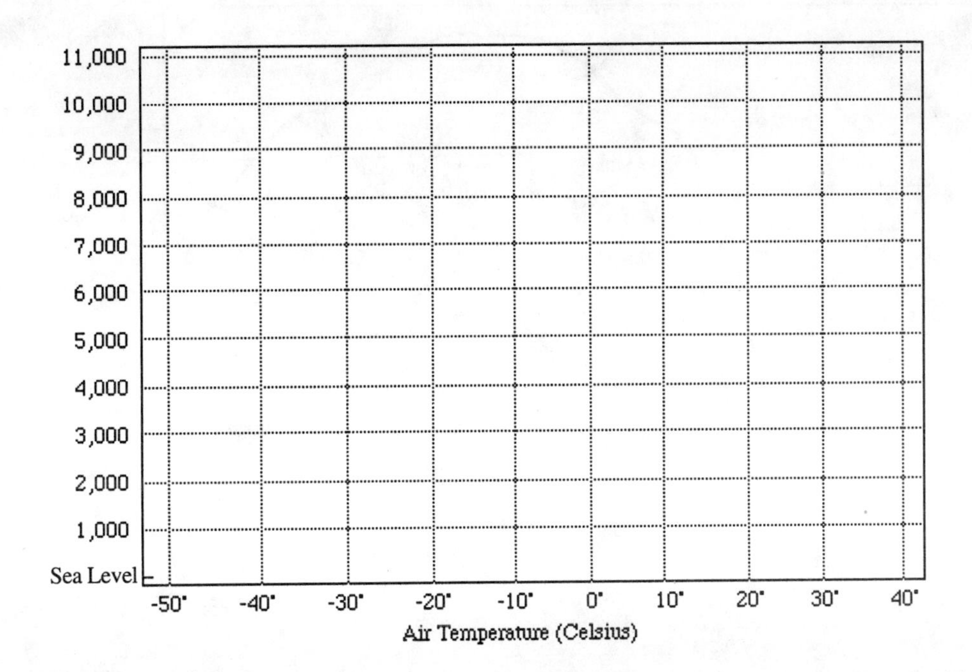

From what you have learned about ELR, compute the temperatures for each of the altitudes on the mountain outline below. Assume a sea level temperature of 20° C. NOTE: this example applies only to temperature decrease with increasing elevation through a *relatively calm* layer of air.

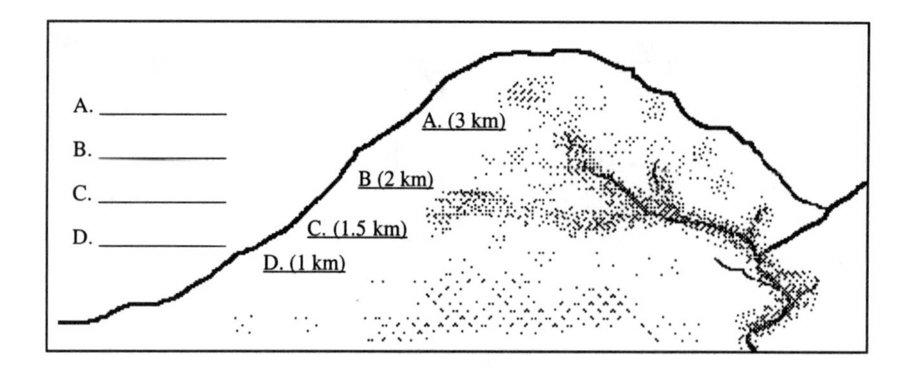

Explain *why temperatures* are higher at the base than at the top of the mountain.

On the graph below, plot environmental lapse rate for three different conditions. First, a normal environmental lapse rate, such as you have already plotted. Second, a low-level inversion. Third, an upper-level inversion. Use a different color for each line, and label each as ELR (normal environmental lapse rate), LLI (low-level inversion), or ULI (upper-level inversion). When this is complete, use a regular pencil to draw a horizontal dashed line marking the level below which temperatures increase. This line identifies the potential "smog zone" of a low-level inversion. NOTE: Plot each *set* of points once only, as you did previously.

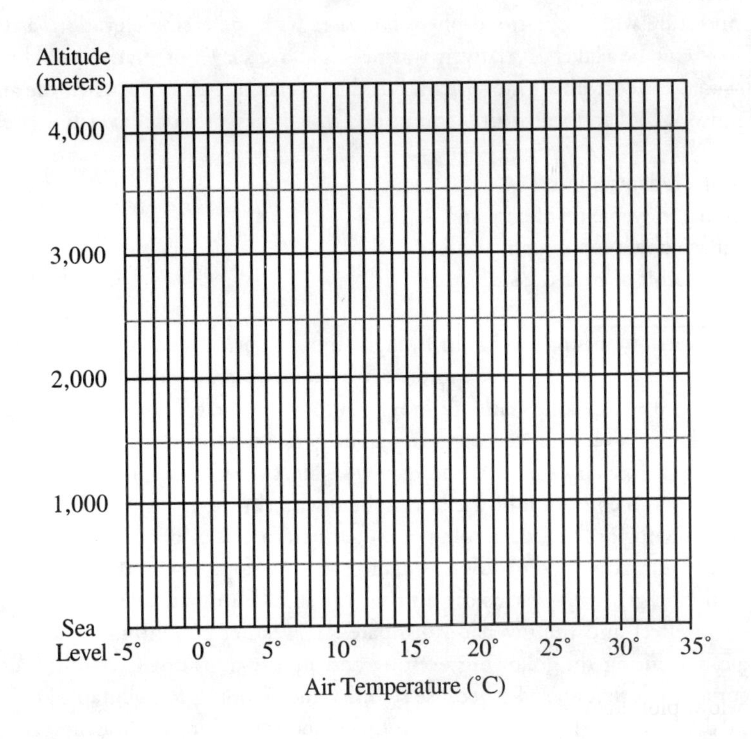

Altitude (meters)	Low Level Inversion 7:00 a.m.	Normal ELR 3:00 p.m.	Upper Level Inversion no given time
Sea Level	10.0° C	30.0° C	31.0° C
500	16.0° C	26.0° C	28.0° C
1000	14.0° C	24.0° C	25.0° C
1500	11.0° C	19.0° C	22.0° C
2000	9.0° C	17.0° C	18.0° C
2500	6.0° C	14.0° C	15.0° C
3000	4.0° C	11.0° C	13.0° C
3500	1.0° C	7.0° C	15.0° C
4000	-1.0° C	4.0° C	12.0° C

TEMPERATURE INVERSIONS

When the temperature within the troposphere increases with increasing elevation, a temperature inversion exists. There *must* be a layer of relatively warmer air above a layer of relatively colder air for this to occur. Inversions may result from air movement, but the inversion itself exists when the air layers are still. There is little mix of cool and warm air. Conditions which may produce inversions are:

1. Night-time radiation losses
2. Advectional movements of cool and warm air
3. Air drainage down topographic valleys
4. Subsidence aloft or near the surface

Among the four inversion causing mechanisms, night-time radiation losses operate most frequently over continental land masses and produce the most short-lived inversion periods. As insolation receipts decline after sunset, surface energy losses chill air near the ground while energy is transferred, warming the layer of the atmosphere above. Thus temperatures actually rise slightly at some distance in altitude. Because the less buoyant surface air cannot rise on its own against the air aloft, the lower atmosphere is said to be stable. Stability is a condition of the atmosphere wherein the lower layers do not mix with layers aloft. Early morning fog or frost, depending upon temperature, is evidence of this stability. Long winter nights and clear, calm air would obviously enhance the development of these inversions. After sunrise, incoming radiation gradually warms the air near the surface so that, by mid-morning, normal lapse rates are again in effect and the inversion dissipates. If weather conditions remain similar, the process may be repeated during the following evening and morning. This mechanism alone is not responsible for major air pollution episodes because stability and trapping of pollutants are interrupted by air movement and dispersion as the inversion breaks up. Good thing, too, because these inversions are quite common.

Advectional movements of large and small air masses can also create conditions that cause less buoyant cooler air to be overlain by more buoyant warmer air. Cold fronts form the leading edge of cooler air masses that vertically displace warm air masses that rise over cooler, more dense air. Departures from normal lapse rates occur, allowing the potential for inversions with frontal passages. Again, foggy conditions are indicators of stability which may punctuate the periods of instability associated with precipitation events. Large-scale events such as these can affect relatively extensive areas for brief periods.

On a somewhat smaller geographic scale, cool maritime air may flow into coastal areas with similar results. Stable conditions associated with inversions occur frequently, making coastlines and their ports some of the foggiest areas of the world. Coastal winds change frequently so these inversions are typically short-term events and thus not associated with serious pollution episodes.

Cool dense air sinking into valleys is also conducive to the development of inversions. The frequent fogs encountered in mountainous areas are products of air drainage. A vivid example of this process occurred at Boise, Idaho, in December, 1985. Ground level (at Boise, 831 meters) temperatures of -18° C were recorded while upslope at 1970 meters, readings of 1° C were observed. Boise's weather forecast included a pollution alert; stagnant, stable air trapped atmospheric pollutants near the surface. In other places, persistent inversions in heavily industrialized valleys with their concentrations of harmful emissions can result in serious pollution episodes. Between December 1 - 5, 1930, the Meuse Valley of Belgium suffered as pollution from coke ovens, steel mills, glass factories, zinc smelters, and sulfuric acid plants was trapped by an inversion. Sixty people died and many respiratory illnesses were blamed on poor air quality. A similar episode occurred in the valley around Donora, Pennsylvania between October 26 and 31, 1948. Approximately 6,000 out of a population of 14,000 suffered severe irritations of the throat and lungs, and 20 people died. Pollution in London has been infamous since King Edward I banned the burning of "sea coale" to reduce air pollution in 1273. In *Hard Times,* Charles Dickens described urban air quality in industrial England when he wrote "Coketown lay shrouded in a haze of its own, which appeared impervious to the sun's rays. You only knew the town was there because there could be no such sulky blotch upon the prospect without a town." In 1911 more than 1,100 Londoners died from the effects of coal smoke, and the author of the report coined the word "smog" to describe a combination of smoke and fog. In addition to fog, which can be heavily laden with sulfuric oxides, another sign of these inversions is the appearance of smokestack plumes which do not rise, but spread horizontally into the air around them. The smoke's failure to rise is, like fog, an indicator of stability.

Atmospheric subsidence associated with high pressure systems is a final inversion-producing mechanism. Air is warmed adiabatically by compression aloft as it sinks through the atmosphere, thus creating a warmer layer above cooler surface air. Because these systems move rather slowly, subsidence can be responsible for the most lengthy inversion episodes. The marked stability of these conditions is associated with aridity in areas where these systems prevail. The deserts of the southwestern United States, Northern Africa, southwest Asia, and Central Australia are notable examples of dry climates associated with subsidence and persistent inversions.

Elsewhere, subsidence may operate seasonally. In the southeastern United States, summertime Three H (Hot, Humid and Hazy) forecasts are products of dominating high pressure, associated with the Azores-Bermuda Subtropical High. Wintertime inversions, on the other hand, frequently occur with the movement of cold, dry, stable polar continental and arctic air masses. Its easy to see that such systems greatly influence not only weather patterns, but regional climates as well.

Severe pollution episodes result when the above mechanisms operate in combination and when sources of atmospheric pollution such as industrial activities, thermal-electric power plants, and automobiles are geographically concentrated. The well-known smogs of Los Angeles result from the unfortunate coincidence of industry, auto exhausts, topography, advection of maritime air, and subsidence from the Pacific Subtropical High Pressure cell. Inversions alone are not the cause of unhealthful air but do result in the stability, which traps the smoke, soot, and chemicals of human activity in the lower troposphere. In short, the inversion layer forms an atmospheric lid, closing stagnant air beneath it. Until this stagnation is broken by instability, air quality is poor.

EXPERIMENTAL PROCEDURE

1. Observe and record temperatures from the upper and lower thermometers in the chamber. Given these temperatures, is the air within the chamber stable or unstable?

2. Place ice cubes or ice water in the chamber tray and observe the temperature decrease in the lower level of the chamber.

3. Postulate two theories that might explain how such a temperature decrease could occur under actual environmental conditions.

4. Identify two geographic settings or areas where the theories you listed above might operate.

5. Recheck thermometers until a difference of 4.5° C is noted. Do you think that air within the chamber would be classified as stable or unstable? Do you think that a column of smoke would rise out of, or stagnate at some level within the chamber under these conditions?

6. Introduce smoke into the chamber and observe its behavior. What happens? What does this simulate?

7. Remove the ice from the chamber tray and heat the lower chamber. Observe what is taking place as temperature changes. Is the air in the chamber remaining stable or becoming unstable? What does this simulate?

8. Refer to your answers to step 3. Explain the events which took place in step 7 as they would relate to the theories you set forth in step 3. What would actually take place?

Sources used in the preparation of this exercise include: Gedzelman, S.D., *The Science and Wonders of the Atmosphere*, New York: John Wiley and Sons, 1980. Lutgens, F.K. and Tarbuck, E.J. *The Atmosphere,* Englewood Cliffs: Prentice-Hall, Inc. 1979. Turk, J. and Turk, A. *Environmental Science* (3rd Ed.) Philadelphia: Saunders College Publishing, 1984.

Chapter 7

The Nature and Behavior of Atmospheric Gases

Introduction

The foregoing chapters on energy and temperature relationships have set the stage for us to embark upon a discussion of atmospheric pressure. This subject of study is the most intriguing, at least from the authors' perspective, of all topics dealing with weather phenomena. In addition to being interesting, it is also one of the most important factors that determine atmospheric behavior. Yet, for most people, it is the least understood of all the weather components we experience on a day-to-day basis. Our bodies can sense most weather components. We can see rain, snow, or clouds; we can smell and sometimes see the results of air pollution; we experience changes in degree of body comfort when temperature or humidity vary over wide ranges; and we can feel variation in wind speed and see its strength demonstrated in the swaying of tree branches. Only a few of us, particularly the aged and/or arthritic, can sense variations in atmospheric pressure. Perhaps you have heard a grandparent saying, "I know it's going to rain, I feel it in my joints." What they are experiencing is the change in air pressure that accompanies the arrival of rain. Most young people do not know this discomfort. Why? The answer is relatively simple, but requires an understanding of how pressure affects the human body. At this point we will avoid any detailed discussion of this issue and just state that the age/pressure sensitivity factor is related to health status. Internal body pressures of the young and healthy adjust to changes in external pressure forces (acting upon the body) much more readily than in older or health-impaired bodies. Hopefully, the information provided in the next two chapters will help unravel some of the mysteries related to atmospheric pressure.

The air surrounding us is a mixture of billions of microscopic gas molecules that are travelling at high speeds and constantly colliding with one another. The distance they travel is incredibly small (about 1/10,000 cm), yet tremendously long when considered relative to the diameter of the gas molecule. For example, near sea level they travel on paths that average one thousand times their diameter. These minute substances also have weight. Thus, they exist in the environment surrounding our solid earth because of gravitational attraction. If air were weightless, its gases would escape the earth's gravitational field and be lost to outer-space.

Air pressure is the result of the impact that gas molecules exert upon an object, including you, and depends upon:

1. The velocity of air molecules.
2. The mass of air molecules.
3. The frequency with which air molecules impact one another.

These concepts are discussed in the following sections of this chapter.

Kinetic Theory of Gases

A gas molecule is at rest only when its temperature is absolute zero (0K, -273°C or -459°F). As temperatures increase above absolute zero, gas molecules travel at increasingly high speeds and in random directions. The average speed of these motions is proportional to the temperature of the gas and represents its kinetic energy (energy of motion). These relationships can be expressed neatly in mathematical terms as demonstrated below:

$T = (0.00004)\ Mv^2$
where M equals molecular weight,
T is absolute temperature (K),
and v represents average velocity in meters per second.

Suppose the air temperature is a mild 10°C (50°F). How fast are the air molecules moving, if the molecular weight of air is 29?[1]

$T = (0.00004)\ Mv^2$
$283 = (0.00004)\ 29v^2$
$v^2 = 243{,}965$
$v = 493.9$ m/sec = 1,778 km/hr = 1,106.4 mi/hr

Compare the foregoing results with the speed of sound (336 m/sec) at the same temperature. As you can see, the average velocity of gas molecules is substantial, even at moderate temperatures, and must be associated with a significant amount of energy. Just how much energy of motion (kinetic energy) is contained within a small amount of gas at sea-level pressure? This may be expressed as:

$KE = 1/2\ m\ v^2$
where KE is kinetic energy in joules
m = mass in kilograms
v = velocity.

Given our former example of air temperature at 10°C (50°F) and velocity 493.9 m/sec, the kinetic energy of 1 kilogram of gas would be:

$KE = 1/2\ m\ v^2$
$KE = 1/2\ (1)\ (493.9)^2$
KE = 121,968 joules = 29,150 calories
(note: 1 joule = 0.239 calories)

Understanding that temperature represents the average molecular speed, it is possible to explain with the first of the above equations why earth's atmosphere has little helium or hydrogen. These elements are common throughout most of the universe. If we keep temperature constant, for example 283K, and substitute into the formula: $T = (0.00004)Mv^2$, the molecular weight of helium (4) or hydrogen (2), it is obvious that their average velocities must increase substantially compared to those of the heavier oxygen (32) and nitrogen (28) molecules. Because these velocities exceed the escape velocity of the earth, these elements migrate into space.

The foregoing illustrations have dealt with two components of atmospheric pressure, the

[1] *The value of 29 is frequently used as an average molecular weight for atmospheric gases.*

speed and mass of gas molecules. Note that as speed and/or mass increase, pressure will increase. Also, if the speed or concentration of molecules increase, the frequency of molecules impacting one another will necessarily increase, therefore increasing air pressure. To express the latter, i.e. concentration of molecules, the density term (ρ) which equals mass (m) divided by volume (v) is employed:

$$\rho = \frac{m}{v}$$

As stated at the beginning of this section, all of the factors discussed above contribute to the magnitude of atmospheric pressure. Now that we know how each component is derived, we can present a combined, yet simplified, equation to define the force we call atmospheric pressure, as:

$$\mathrm{P} = R\rho T \quad or \quad R\frac{m}{v}T$$

where P is atmospheric pressure expressed in millibars,[2] ρ is density in kilograms per cubic meter, R is a constant with a value of 2.87, m is mass, v equals volume and T is absolute temperature in degrees Kelvin.

This equation is known as the **ideal gas law**. What importance is it to us? The answer is that it explains relationships between pressure, density, and temperature of gases. We will see later that these relationships affect air movement. For the present, let's examine Figure 7.1. In this illustration, part A represents temperature and density variation when air pressure is held constant. Note that low temperatures are associated with high densities; whereas, high temperatures have low air densities. When we speak of air density, we are referring to how many molecules of gas (identified as mass) are found in a unit volume of air. Thus, in a relative sense we can think of high densities as representing heavy air and low densities as light, or buoyant, air. Diagram 7.1B compares pressure and density changes when temperature is held constant. Notice that there is a direct relationship: as density increases, so does pressure. Diagram 7.1C presents the temperature and pressure relationships when density is held constant. Notice this is a direct association: as one increases the other does as well.

Figure 7.1 Relationship between Pressure, Density and Temperature

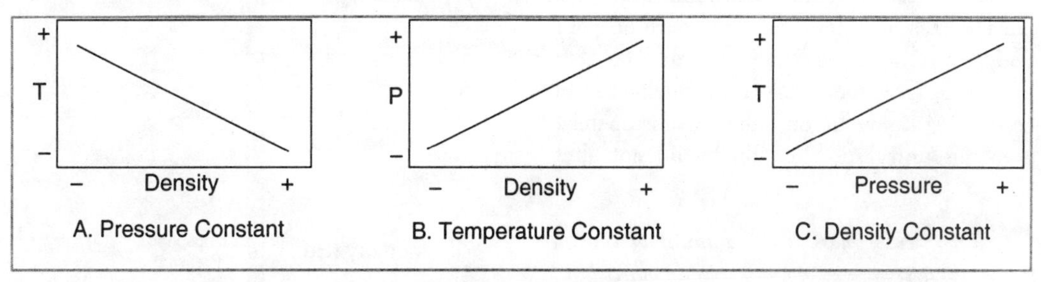

[2]A millibar is defined as a unit of force equal to one-thousand dynes/cm². A dyne is a force under whose influence a body with a mass of one gram (0.035 oz.) would experience an acceleration of one centimeter (0.39 inches) per second.

Pressure versus Force

Pressure

In the foregoing section we discussed that gas molecules have mass (or substance) and that mass is a characteristic of gases that ultimately determine the magnitude of atmospheric pressure. Many students confuse the concept of mass with that of weight. Rather than continue possible confusion, let's pause to explain the differences between the two terms.

Mass is the amount of matter contained within an object.[3] Your body, for example, has the same mass whether you are standing on the surface of the earth or on the surface of the moon. Yet, your weight in these two locations is radically different. Weight is a product of the mass contained within a body times the gravitational attraction (acceleration of gravity) exerted upon it, that is:

Weight (w) = mass (m) • acceleration of gravity (g),
or w = mg; where w is expressed as Newtons,[4]
m is defined in kilograms
and g (for the earth system) is 9.8 m/sec^2.

This means that if the earth did not have a gravitational attraction for the mass of your body, your weight at sea-level would be zero kilograms (zero pounds). This might sound comforting if you are on a diet. Unfortunately, even this unrealistic situation would not alter your mass. The following example may help you understand this difference. Suppose you have a mass of 50 kg (110 lbs.) What is your weight at sea-level? Using the above equation:

$$w = mg$$
$$w = (50)\ (9.8) = 490 \text{ Newtons}$$

Since 1 Newton equals 0.10 kg (0.225 lbs.), your weight is 49 kg or 110.3 lbs. On the moon your mass remains the same, but the acceleration of moon gravity is one-sixth of that on earth; therefore, your weight would be a mere 8.2 kg (18.4 lbs). In conclusion, we can see that weight is a force.

The gases of our atmosphere have mass that is affected by the acceleration of gravity; hence, air must have weight. And, it does! The average weight of the atmosphere at a standardized sea-level altitude is 1.03 kg/cm^2 (14.7 lbs./in^2). This value results from the total mass of gas molecules extending from the earth's surface to

[3]Technically mass is the amount of inertia (resistance to change in motion) that an object has.

[4]A Newton represents the force required to move an object with a mass of 1 kg at a speed of 1 meter per second (m/sec) each second.

Figure 7.2 Diagram of a Mercurial Barometer

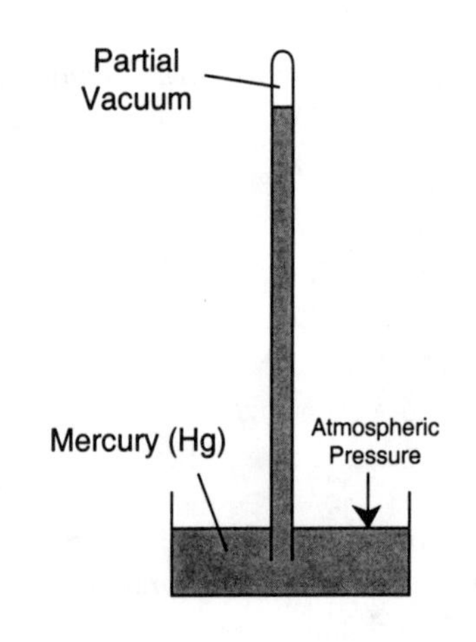

the top of the atmosphere. Its variation through time is measured with an instrument called the barometer.

Barometers

A **barometer** is an instrument used to measure air pressure. There are two primary types: the mercurial barometer and the aneroid barometer. The mercurial barometer is based upon the concept that if the air has weight, then a column of liquid (mercury is commonly used because it has high density) can be balanced against air weight. This barometer consists of a container in which a vacuum or mass of air can be isolated from the atmosphere (Figure 7.2). If it were not isolated, the air inside the container would exchange molecules with the outside air as external pressure changes and remain in equilibrium. By isolating the system the difference between internal pressure and external pressure can be determined by measuring the displacement of a volume of mercury.

The **aneroid barometer** consists of a hermetically sealed and collapsible canister from which air has been evacuated. Springs that keep the canister partially stretched out prevent total collapse. When air pressure increases the canister is compressed. When pressure decreases the springs stretch the canister. This is the type of barometer frequently found in homes. The word aneroid means "containing no liquid."

Figure 7.3 Pressure-Altitude Relationships. Note that pressure and density decrease rapidly in the vertical. This rapid change is considered an inverse-logarithmic relationship.

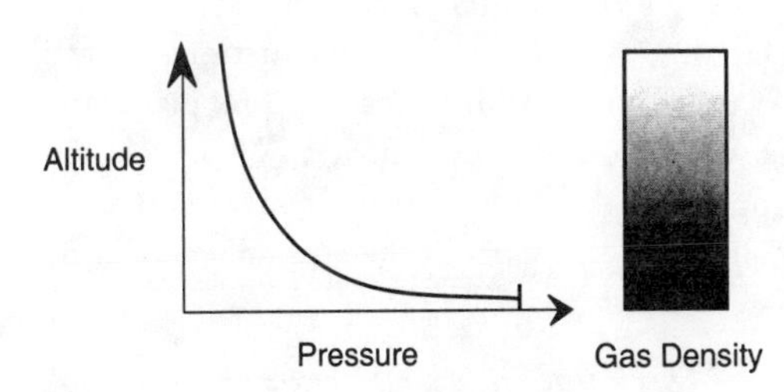

We now know that atmospheric pressure is a function of the mass of air molecules, their velocity, and the frequency with which they impact one another. We also know that the frequency of air molecule collisions is dependent upon the distance between molecules. It is now time to introduce another concept: gas is compressible. We know this because we can force air into tanks to hold propane for gas charcoal grills, to hold oxygen for scuba diving gear, or for a multitude of other purposes. Because gas is compressible, the region of densest gas molecules occurs near the earth's surface where the entire depth and weight of overlying atmosphere forces large numbers of molecules into relatively small space (Figure 7.3). The higher one goes in the atmosphere the less air there is above, therefore, pressure and density decrease. One-half of the mass of the atmosphere exists below an altitude of about 3,500 meters (18,000 ft.). The weight of air is no longer 1.03 kg/cm^2 (14.7 lbs./in^2) as at the earth's surface, but approximately one-half those values, 0.52 kg/cm^2 (7.4 lbs./in^2).

Just because pressure exists does not mean that force is present. Let's attempt an explanation. Suppose you shake-up a bottle of champagne or pump air into an automobile tire. Both actions increase gas pressure. We can continue to increase gas pressure. Yet, this does not mean there is any force affecting the object, for example, the champagne. This is because pressure is uniformly distributed within the object. When you open the bottle of champagne or depress the valve stem on the automobile tire, however, something happens. The champagne spews out of the bottle and air rapidly escapes from the tire to the atmosphere. Force is generating this movement (or push) of mass. But what has happened to change the environment from one of "no force" to one "with force." The secret to the explanation of our example lies in the previous sentence. The environment has changed. By opening the bottle or depressing the valve stem, a substance under high pressure has been exposed to an environment (the atmosphere) of significantly lower pressure. In essence, for force to exist, there must be a difference in pressure. When this occurs objects will be forced in a direction from high pressure to low pressure.

Sir Isaac Newton's Second Law of Motion provides the mathematical derivation of force and is expressed by the equation:

$F = ma$; where F is force, m is mass, and a is the acceleration.

If you recall our previous discussion on weight ($w = mg$) you will recognize that the force and weight equations appear similar.

Figure 7.4 Pressure in Relationship to a Cartesian Coordinate System. P1 and P2 represent atmospheric pressure values at given distances from the X, Y and Z axes

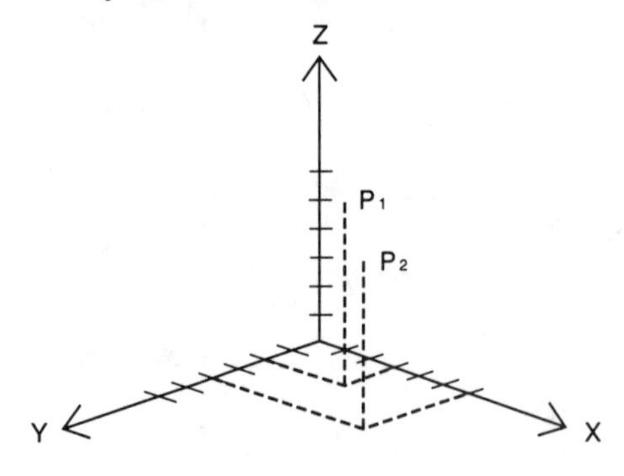

What is different about them? In the weight equation g equals acceleration of gravity. In the force equation a is acceleration. Weight is a force. Your body's mass, as affected by gravity, imparts a force on the earth's surface. Thus, g is acceleration in the vertical; whereas, a may apply in any direction. The weight equation is, in effect, a special application of the force equation, that is: $F = ma$ or $F = mg$.

Pressure Gradient Force

As previously discussed, mass cannot be accelerated (change in speed or direction) without the existence of a difference in pressure on an object. The object of our attention is the combination of gases that comprise the atmosphere and how pressure differences occur within it. Let's simplify our approach to understanding the origin of pressure differences. We already know that there is an inverse relationship between temperature and pressure. These are called thermal-pressure relationships.[5] So we might expect high pressure to exist in the cold, polar areas and low pressure in warmer, equatorial regions; high pressure over cold water and land areas and low pressure over warm water bodies and warm land masses; or high pressure over cold land masses and low pressure over warm bodies of water. There are infinite possibilities for pressure differences to exist within the earth-atmospheric system that result from temperature variations.

Pressure magnitude varies from place to place in horizontal and vertical directions (Figure 7.4). At this moment in time, we will focus on pressure differences in one plane, that is within X and Y coordinates. We choose this approach because it is representative of conditions that exist in the atmosphere near the

[5] *Thermal refers to temperature.*

earth's surface. If the X-axis is oriented east-west and the Y-axis directed north-south, we can hypothesize an inverse temperature-pressure relationship from polar to equatorial latitudes, as demonstrated in Figure 7.5 for the Northern Hemisphere.

The Isobaric Map

Real pressure differences are not as simplistic as those presented in Figure 7.5. As a matter of fact, they can be rather complex. Fortunately, we can readily understand the significance of pressure variations over the earth's surface by employing a technique of analysis we used earlier to make sense of temperature patterns; that is, we can construct an isoline map. In this instance, the isolines are referred to as **isobars** (lines connecting points of equal barometric pressure). Figure 7.6 illustrates sea-level pressure patterns over North America on January 1, 1992.

There is an important concept we should bear in mind when we discuss atmospheric pressure. The concept is this: pressure-relationships exist only in a relative sense. We previously defined the standard sea-level pressure as 1013.2 mb. This might lead the casual reader to interpret any pressure reading above 1013.2 mb as representing high pressure and vice-versa for those below the "standard." This is not true! In Figure 7.6 two high pressure systems (H) and a low pressure system (L) have been identified. Yet, all isobars represent pressure values above standard sea-level pressure.

What, then, does differentiate high pressure from low pressure? Look carefully at Figure 7.6 again. The region of low pressure (L) has a central isobar of 1020 mb; whereas, surrounding isobars indicate greater pressure magnitude. Consequently, the low pressure system is defined as an area of given pressure magnitude, around which pressure magnitude increases in all directions. Now, examine the high pressure systems (H). Note that from the innermost isobars (1032 mb in the west, and 1036 mb in the east), the surrounding pressure magnitudes decrease.

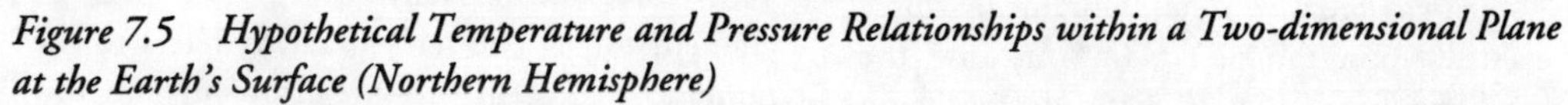

Figure 7.5 Hypothetical Temperature and Pressure Relationships within a Two-dimensional Plane at the Earth's Surface (Northern Hemisphere)

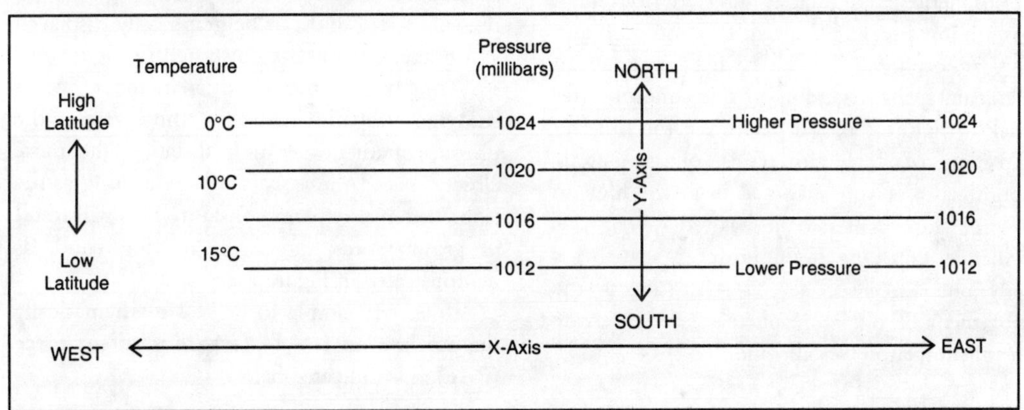

Figure 7.6 Pressure Variation over North America.

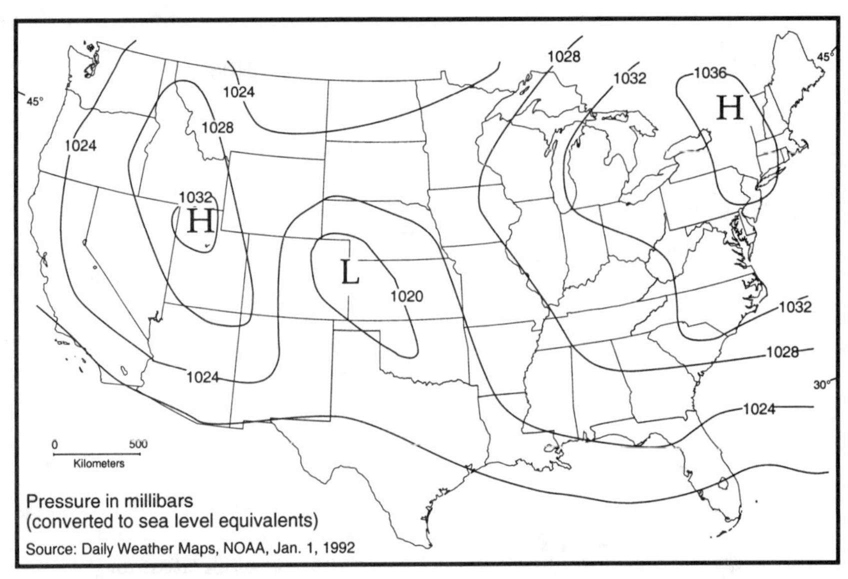

Force

Let's begin by recalling our previous examples of releasing gas from high to low pressure environments (removing the top from a champagne bottle or depressing the stem valve on an automobile tire). These demonstrated that when gas pressure is not contained, a force is exerted upon the object (in our case the atmospheric gases) that is directed from high toward low pressure. Indeed, this also occurs in the atmosphere. To test what is now an elementary understanding of this concept, refer back to Figure 7.6 and determine the primary direction of force in terms of geographic coordinates. If you answered from the high to low pressure centers you have mastered this principle. Please be aware, however, that this is only one force, of several, that affects the movement of our atmosphere's gases to create the phenomenon we call wind.

We now know that difference in atmospheric pressure between areas over the earth's surface creates a force upon air that makes it move. In classical physics we would say that the air is accelerated (changing in speed or direction) when pressure differences are present. This force is formally called the pressure gradient force. The term gradient refers to the numeric change in pressure from high to low pressure areas, and can be graphically displayed as a grade, or slope, as illustrated in Figure 7.7. An important thing to remember when studying isobaric maps is the greater the pressure gradient per unit distance, the more closely spaced the isobars will be as long as the isobar interval remains constant. This will result in greater force and faster moving wind, as demonstrated in Figure 7.8.

It is very simple to express mathematically the acceleration (a) of **pressure gradient force** (PGF) as is indicated below:

Figure 7.7 Pressure Gradient Force and Pressure Gradient

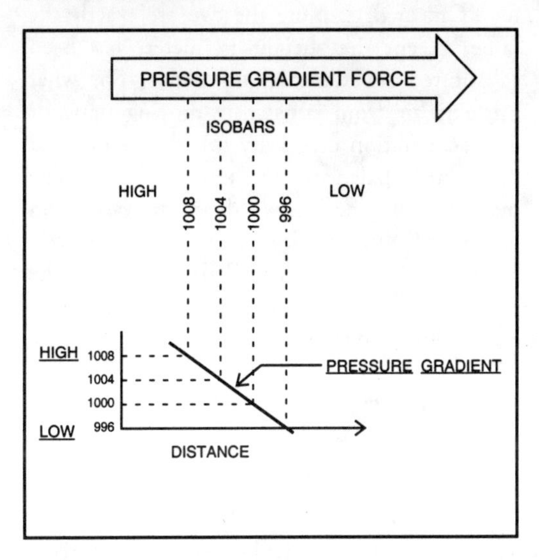

$$\partial_{PGF} = \left(\frac{1}{\rho}\right) \frac{\text{pressure difference}}{\text{distance}}$$

where density (ρ) is expressed in kg/m³, pressure is given in newtons per square meter,[6] distance is defined in meters, and the acceleration (∂_{PGF} is in meters/sec²

Figure 7.8 Contour Interval and Pressure Gradient Relationships

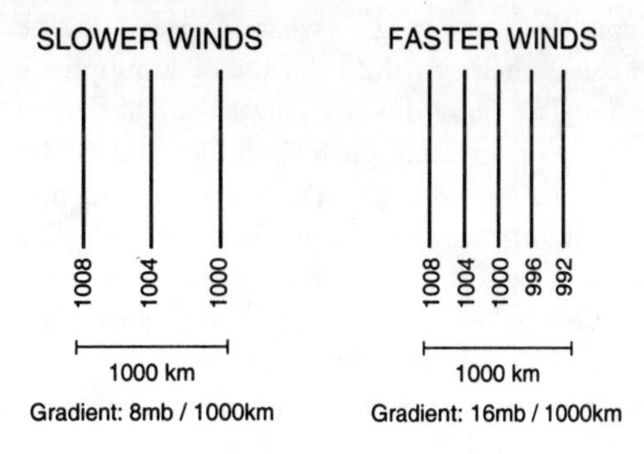

Given an air density of 1.3 kg/m³, let's examine the effect of two different pressure gradients on the acceleration of air. The two gradients are 4 mb/100 km and 8 mb/100 km.

If no other forces were acting upon the air and it only accelerated for one second at a velocity of 0.30 x 10^{-2} m/s² it would travel extremely slow at 0.01 km/hr (.006 mi/hr). Since it continues to accelerate each second at 0.003 m/s², however, its average velocity at the end of one hour would be 60 seconds x 60 minutes x 0.003 m/s² and would equal 10.8 meters/second or 38.8 km/hr (24.1 mi/hr). Doubling the pressure gradient doubles the acceleration. As we shall later see, such velocities do not occur near the earth's surface due to other forces acting upon the air.

$$\partial_{4\ mb/100\ km} = \left(\frac{1}{1.3}\right)\frac{400}{100{,}000} = (0.30)(10)^{-2}\, m/s^2$$

$$\partial_{8\ mb/100\ km} = \left(\frac{1}{1.3}\right)\frac{800}{100{,}000} = (0.61)(10)^{-2}\, m/s^2$$

[6]*Pressure is normally expressed in* **millibars** *(mb); hectopascals (kPa); or newtons per square meter. A Newton per square meter (N/m²) is equal to a pascal (Pa). To convert millibars to pascals, simply multiply the number of millibars by 100, for example, 4 mb equals 400 pascals or 4 hectopascals.*

Hydrostatic Equilibrium

Beginning students of the atmosphere are typically plagued by what appears to be inconsistencies in the behavior of atmospheric gases. Two questions generally arise. The first is: "If the upper atmosphere is colder and colder air is dense, why doesn't the upper air sink downwards toward the earth's surface?" The second is: "If gas always moves from high pressure to low pressure regions, why doesn't air continually lift since the highest pressures are recorded at the earth's surface and decrease markedly as altitude increases?"

Why doesn't cold air aloft sink? We must bear in mind that air pressure at any level in the atmosphere is a function of the density of air molecules, as well as temperature. Referring to the ideal gas law ($P = R\rho T$) and knowing that both temperature and pressure decrease with altitude, while R is a constant, it is clear that air density must also decrease substantially with distance from the earth's surface. Thus, although temperature aloft may be low, the density of upper atmospheric gases is low enough to maintain their buoyancy.

Why doesn't air always ascend? Because pressure gradient force and air movement are always directed from high to low pressure, it would appear that air should always move from the earth's surface (where it is densest) in an upward direction toward where the atmosphere is least dense. In reality, there is a vertical pressure gradient that is in accordance with this principle. The vertical acceleration (a_v) associated with an upward directed force is mathematically defined as:

$$\partial_v = \left(\frac{1}{\rho}\right) \frac{\text{pressure difference}}{\text{height difference (meters)}}$$

Compare the above equation with the previously given equation for the acceleration for PGF, or ∂_{PGF}. Note the two are exactly the same, except that height in meters has been substituted for distance in meters. So, what prevents air from constantly moving upward? The acceleration of gravity (g), that is directed downward, balances this upward force. The majority of the time these upward and downward directed forces are almost exactly balanced. When such a balance exists, the atmosphere is said to be in hydrostatic equilibrium and expressed as:

$$g = \partial v$$

or

$$g = \frac{1}{(\rho)} \frac{\text{pressure difference}}{\text{height difference}}$$

Hydrostatic equilibrium is common. The relationship is rarely out-of-balance, occurring in less than 1 percent of all weather events. The odd exceptions have been known to happen when tornadoes or severe thunderstorms are active. The hydrostatic equation not only permits an evaluation of atmospheric equilibrium; it also predicts change of pressure within specific distances from the earth's surface. Such determinations are valuable for determining wind flow patterns and for calibrating altimeter settings for aircraft pilots. Suppose, for example, that an airplane is traveling a route at an altitude near 3,000 meters (10,000 ft). The pilot cannot keep making observations and calculations to determine the plane's altitude during flight; therefore, he/she flies along a pressure surface that would approximate the 700 mb level for which the altimeter is set. Some altimeters are mechanically very similar to aneroid barometers, but are calibrated to infer altitude as a function of air pressure. Note that the 700 mb level should approximate 3,000 meters above sea-level. Does the hydrostatic equation help in calibrating the altimeter? Assume that surface pressure is 1,000 mb and air density equals 1.02 kg/m^3 and we wish to determine the altitude of the 700 mb level. Then, the hydrostatic equation can be employed as follows:

$$g = \left(\frac{1}{\rho}\right) \frac{\text{pressure difference}}{\text{height difference}}$$

$$9.8\ m/s^2 = \left(\frac{1}{1.02}\right) \frac{30{,}000 \text{ Pascals}}{\text{height difference}}$$

$$\text{height difference} = 3001.2 \text{ meters}$$

Note that the foregoing is an extremely close approximation to the level of flight at which the airplane is expected to cruise. While this is a useful and simplistic formula, it should be obvious that meteorologists around the world are not spending their time determining how many gas molecules occupy a given volume of space, i.e. density. Yet, air density is critical to solving the equation. This apparent problem is easily resolved if we first revert to the use of the ideal gas law, which requires only values for pressure and temperature to derive the density factor.

Other Forces

Pressure gradient force (PGF) provides for the movement of air from areas of high pressure to areas of low pressure. The direction of this acceleration is always at right angles (90°) to the orientation of the isobars, as shown in Figure 7.9. If the earth stood still in space (in other words, was at rest), wind, the horizontal movement of air, would always follow the path determined by the PGF. Earth is not at rest, however. It is continually spinning about its axis. At the equator its velocity of spin exceeds 1,600 km/hr (1,000 mph). This means that air may travel in its accelerated path along a straight line, but it is doing so over a coordinate system on the earth's surface that is constantly changing in position relative to celestial and planetary reference points. This rotational motion has the effect of deflecting the flow of wind from its intended path on the earth's surface and is referred to as the Coriolis Effect.

Figure 7.9 Pressure Gradient Force Is Directed at 90° to the Isobars from the Highs (H) to Lows (L)

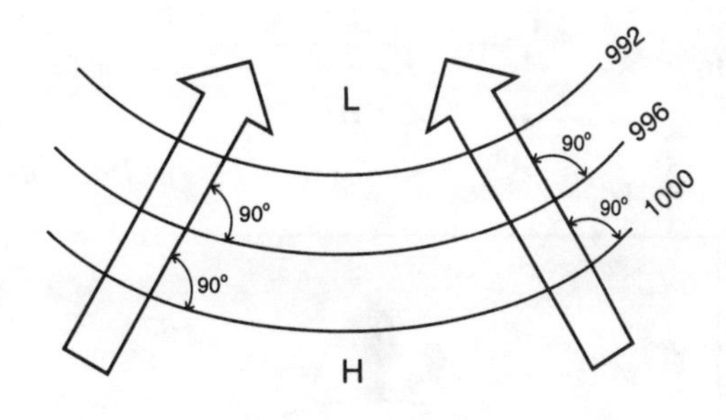

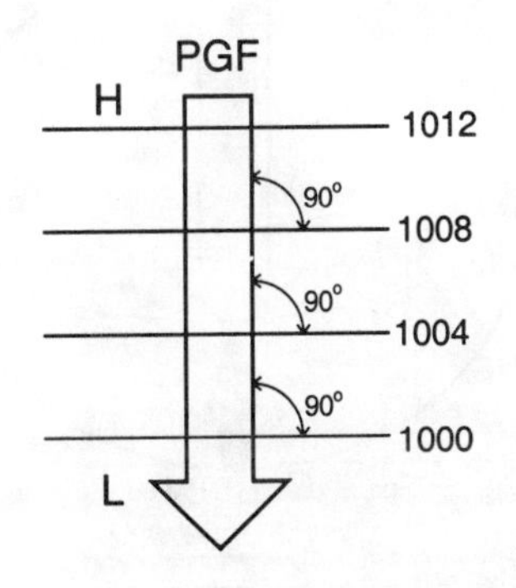

The Coriolis Force is an apparent force that results from earth rotation, consequently it is also referred to as the Coriolis effect. This phenomenon is named after Gaspard Coriolis, a French scientist of the nineteenth century who was the first to mathematically derive its influence on freely moving objects such as air molecules and ocean currents. Coriolis effects are always oriented at a right angle to the wind vector (direction of wind flow). In the Northern Hemisphere this deflection is to the right;

Figure 7.10 Coriolis Deflection. (PGF = pressure gradient force; v = wind vector; and cf = coriolis force.)

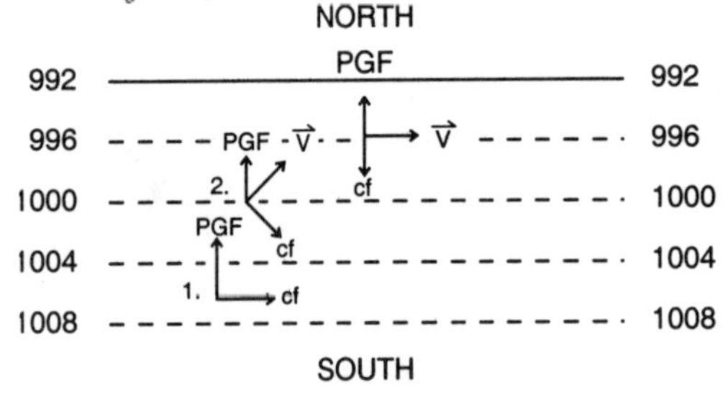

whereas, in the Southern Hemisphere it is to the left. Figure 7.10 illustrates wind direction (and velocity) as a result of PGF and Coriolis.

Beginning at point 1, air is accelerated by PGF in a direction from South to North. As it begins to move its flow is deflected to the right. Note at point 2, PGF remains the same. The wind vector now deviates from a north-south orientation to a NNE direction and cf is still directed at 90° to the right of the wind vector. This process can continue, until the wind vector is deflected 90° from both PGF and cf. When this condition exists, the two forces affecting the direction and magnitude of air flow are said to be balanced. If no other forces affect the wind (for example, friction) and isobars maintain straight and parallel lines, the wind would be called a geostrophic wind, that is, a frictionless wind acted upon by forces, wherein PGF = cf. Figure 7.11 illustrates a simplistic example of the apparent Coriolis force. In Figure 7.11 A, two figures, let's call them Kyle and Cartman,

Figure 7.11 The Principle of Coriolis Deflection

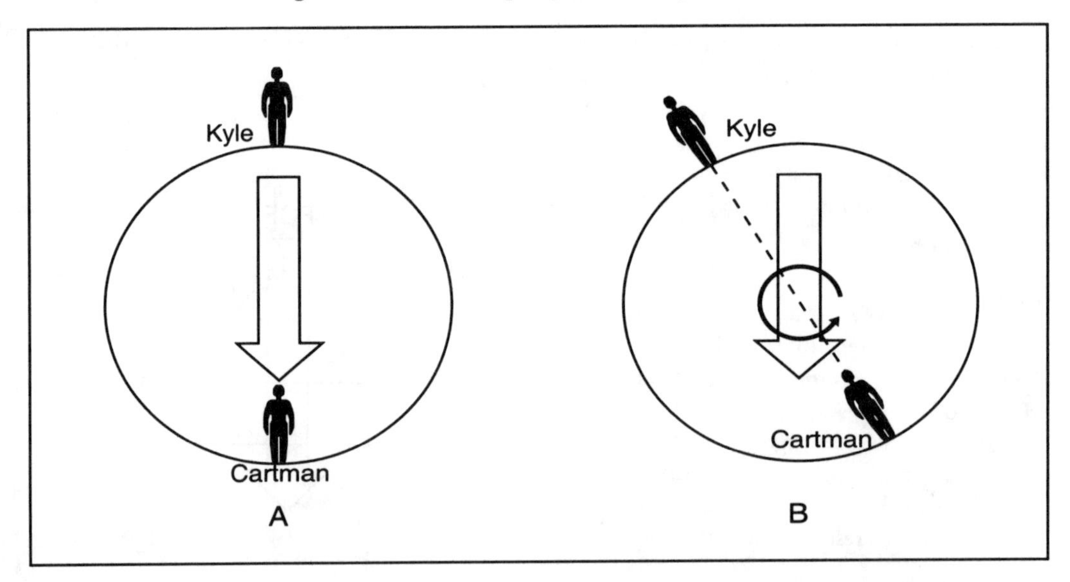

face each other across a horizontal disk that can be rotated like a merry-go-round. Kyle is about to throw some object toward Cartman. The large arrow indicates the direction in which the object is thrown. In Figure 7.11 B, the disk is set in motion as the object is thrown. The object travels in a straight line, but from Kyle's perspective, the object appears to be deflected to the right of Cartman. On the rotating earth, Coriolis deflection has an important influence on winds and ocean currents. Coriolis Force may be expressed mathematically as follows:

Coriolis force (cf) = $2 \text{ m } \Omega \text{ V} \sin \theta$

where m is the object's mass
Ω equals the earth's angular rate of spin (7.29×10^{-5} radian/sec)
V is wind speed
and θ represents the degrees of latitude.

The above equation clearly reveals that cf is dependent upon the speed of earth rotation and has a magnitude that increases with latitude (lowest in equatorial regions and highest in polar realms) and wind speed. Please remember that cf cannot change the speed of a unit of air, but only its direction.

Frictional Force

Thus far, air motion has been presented as if it existed in a frictionless environment. This is basically true for atmospheric conditions above 1. 6 km (about 1.0 mi). Near the earth's surface (the region of air referred to as the boundary layer of the atmosphere), however, friction between the air and the earth's surface is a significant factor in affecting wind flow. It is difficult, if not impossible, to calculate all the impacts of acceleration that friction produces, but two impacts are especially significant. First, friction reduces wind velocity. Second, friction acts in opposition to Coriolis force. In both cases, the possibility of a geostrophic wind with balanced PGF and cf cannot be achieved. Thus, rather than winds blowing in a direction that approximates the isobars, they flow at a distinct angle across them as is described in Figure 7.12. Some rough estimates of frictional influences are as follows: over smooth grassland velocity is reduced 60 to 70 percent and direction is oriented 20 to 25° to the left of the isobar; over large water-bodies values are similar; when dealing with rough land surfaces, deviations can be 45° and speed reduced to about 40 percent of its geostrophic value.

Figure 7.12 Components of Force Affecting the Wind:
A. Geostrophic Wind;
B. Wind Affected by Friction.

(PGF = Pressure Gradient Force,
V = wind vector,
cf = Coriolis force,
and Fr = friction)

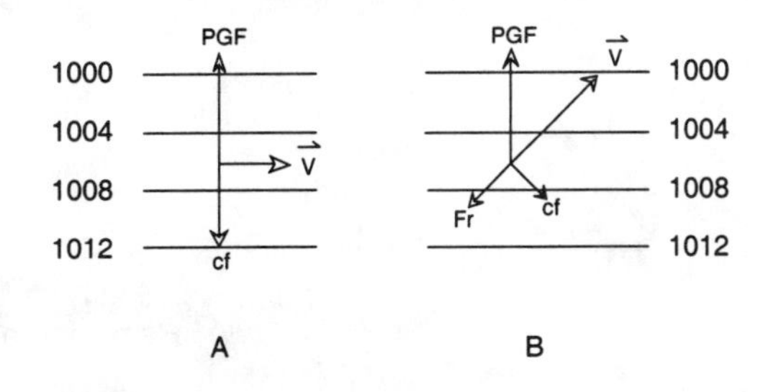

Figure 7.13 Horizontal Perspectives of Windflow at the Surface and Aloft.

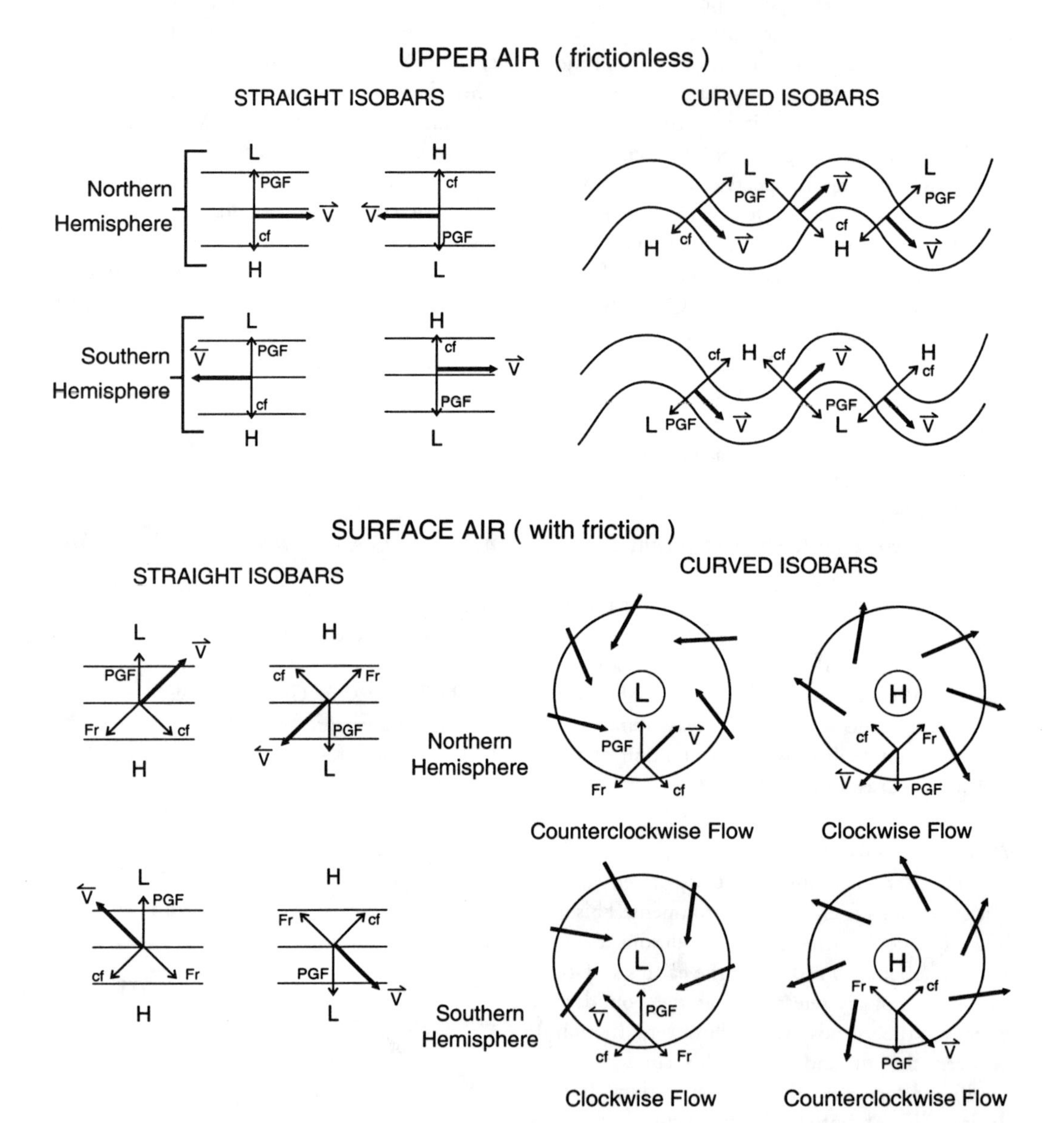

Centripetal and Centrifugal Forces

These are forces that accelerate air into a low pressure system (centripetal) or outward from a high pressure system (centrifugal) and occur when airflow is following curved, as opposed to straight isobars. The explanation of the equations that govern these forces is not difficult, but need not be introduced at this point. Generally, for an object to follow a curved path, some force is necessary to cause it to deviate from a straight line of motion.

Forces and Resultant Winds

Recall that wind velocity and direction are a result of several interacting forces. We can schematically combine these forces into simplistic diagrams that describe wind flow, as is presented in Figure 7.13. As you examine these diagrams, keep in mind the basic principles described in the foregoing pages and the fact that the motions represented are primarily horizontal. Meanwhile, it is important to recognize that horizontal motions do not occur in total isolation from vertical motions. High pressure systems have densely concentrated air molecules relative to low pressure systems. Therefore, as a general rule, air in high pressure regions near the earth's surface tend to subside (settle) toward the surface; whereas, low pressure regions are dominated by a less dense concentration of air molecules and are relatively buoyant. Given these relationships, it is possible to extend our understanding of airflow in surface high and low pressure systems, as illustrated in Figure 7.14.

High pressure systems are associated with descending and diverging air. Viewed from above, the diverging air circulates clockwise in the northern hemisphere and counterclockwise in the southern hemisphere. Low pressure systems are characterized by converging and ascending air. Viewed from above, the converging air circulates counter-clockwise in the northern hemisphere and clockwise in the southern hemisphere.

Temperature, Pressure, Density and Vertical Motion

The atmosphere is free and unconstrained, and it behaves somewhat differently from confined gases. When air is heated the molecules move faster and exert more force. They cannot expand laterally, however, because adjacent molecules also possess higher kinetic energies and are pushing back with equal force. The greater pressure that develops, momentarily upsets the hydrostatic equilibrium. The force exerted by internal air pressure is greater than that of gravity so that air moves upward in response. This upward movement involves higher energy molecules and leaves behind fewer and lower energy molecules and correspondingly results in lower pressure. Unlike a laboratory experiment where heating of confined gases produces higher pressure, when the unconfined atmosphere near the

Figure 7.14 Vertical and Horizontal Motion of Air in Northern Hemisphere High and Low Pressure Systems.

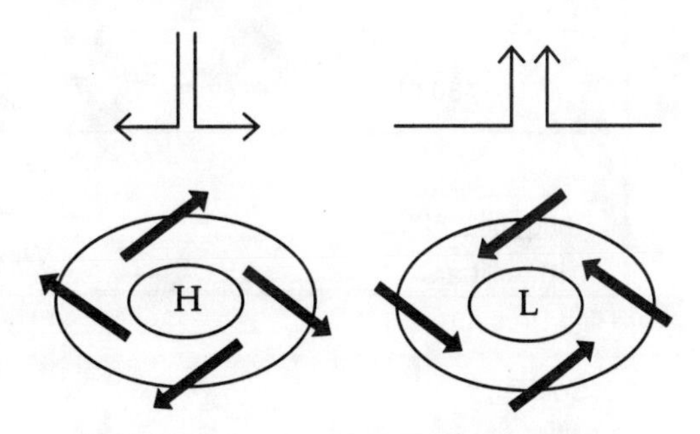

surface is heated, lower pressure is produced. The lower density more than compensates for the higher average kinetic energies of the remaining molecules. High atmospheric temperatures, then, are associated with low pressure. When air is cooled, it contracts, and although the average kinetic energy is lower, the higher molecular density produces higher pressure.

High pressure air is more dense than low pressure air. The vertical rate of pressure change in a high pressure cell is also greater than in a low pressure zone because of density differences. Thus, **isobaric surfaces** (planes that connect points of equal pressures) are closer together in a high pressure area. When high and low pressure areas are adjacent to each other, isobaric surfaces appear as in Figure 7.15. At 1 km, the upper air pressure is actually less over the surface high than it is over the surface low, and the lateral pressure difference increases with altitude. Thermally induced surface pressure differences such as these are frequently overlain by upper air pressure of an opposite type, with surface lows being overlain by upper atmospheric highs and vice versa. The net result is that upper air winds may blow in a direction opposite to that of surface winds, and that upper air winds are stronger.

Localized surface heating produces warm rising air currents known as thermals. As the air ascends, local surface pressure is reduced and air from surrounding higher pressure areas moves in to replace it, thereby setting up a convection cell (Figure 7.16). Ascending air in a convectional cell expands in response to the vertical change in pressure. The work of expansion requires energy and reduces the temperature of the rising air mass. In addition, some molecular kinetic energy is converted into

Figure 7.15 The Effect of Temperature on Isobaric Surfaces

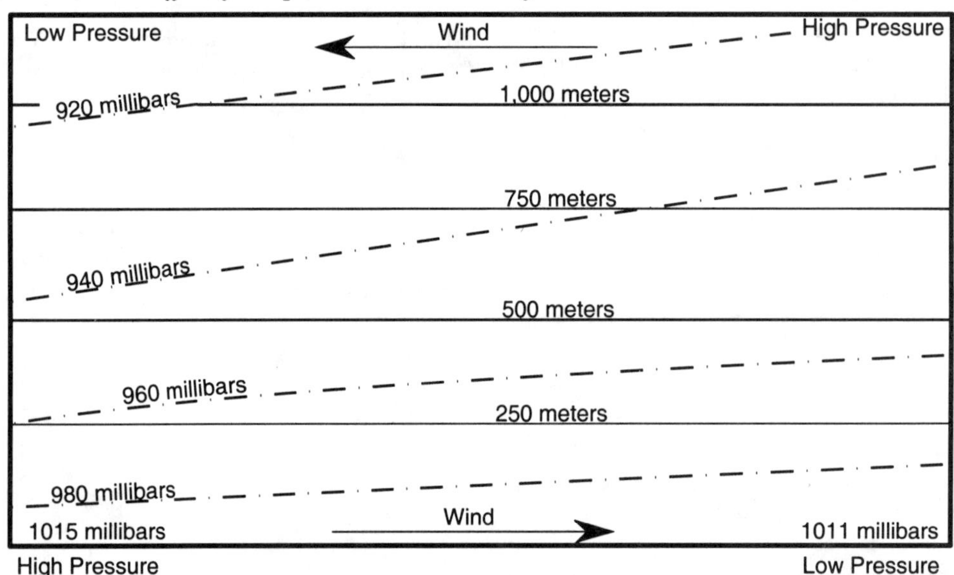

potential form, thereby causing corresponding temperature reduction. This cooling process, called adiabatic cooling, is an internal process entirely. No energy is lost from the lifting air system. If for any reason the air moves downward again, it will be compressed due to the higher pressure, potential energy will be converted into kinetic form, and the air will become warmer.

Warm air is less dense than cool air because it contains fewer molecules per unit volume. Because warm air is lighter, it can be displaced by cooler air and pushed upward. This tendency of warm air to rise is referred to as **buoyancy**. As long as rising air that is undergoing adiabatic cooling is warmer than surrounding air, it will continue to rise. If it is surrounded by warmer air, it will settle downward until its temperature is in equilibrium with surrounding air.

In addition to the cooling of ascending air due to convection, air may also experience vertical motions and temperature changes due to being forced aloft by other mechanisms. These include situations where fast-moving winds overtake slow moving winds, called **speed convergence**; when two large bodies of air, each possessing distinctly different properties, are in contact with another (**frontal lifting**); or when the flow of wind is directed perpendicular to a mountain range (**orographic lifting**). These procedures are more thoroughly discussed in later chapters, but are illustrated in Figure 7.17.

Figure 7.16 Convection: The Transfer of Heat through Mixing

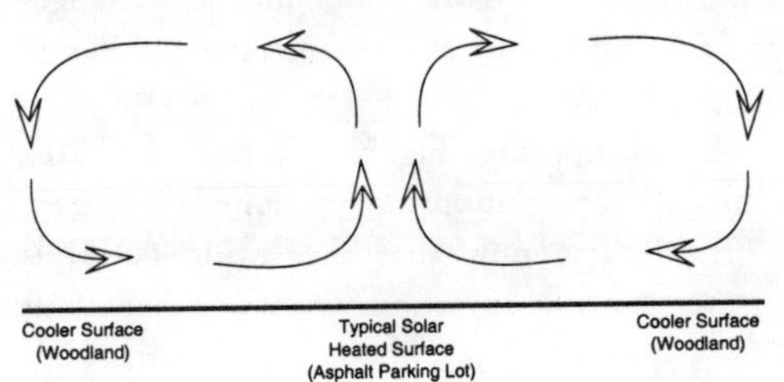

Stable versus Unstable Air

The air lifting processes just described can be considered triggering mechanisms that initiate vertical atmospheric motion to a parcel of air. What happens to the lifted air parcel when the triggering mechanism is no longer present? For example, what vertical motion does air crossing a mountain have when it bypasses the summit? Does it continue to rise? Does it descend? Does it stay at the same altitude and travel horizontally? These questions can be readily answered with an understanding of atmospheric stability and instability.

Air that resists lifting and does not mix vertically is **stable**. Air that spontaneously rises because it is warmer than surrounding air is said to be **unstable**. To determine whether air is stable or unstable, it is necessary to compare the cooling rate of lifting air parcels to that of the surrounding still air.

Earlier in this textbook, we discussed the mean (long term average) atmospheric rate of temperature decrease with altitude (the normal lapse rate). Recall that this was 6.5°C /1,000 m (3.5°F/1,000 ft). In reality, temperature change from the surface upward varies widely seasonally, daily, and at times hourly. Thus, at any given moment it may be 5°C/1,000 m

Figure 7.17 Common Processes that Result in Air Lifting

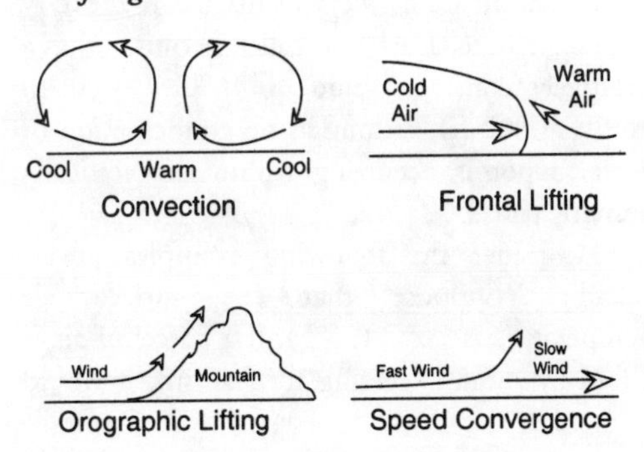

(3°F/1,000 ft.), 15°C/1,000 m (9°F/1,000 ft.) or some other amount of temperature change with altitude. The actual temperature change per unit of vertical distance is referred to as the **environmental lapse rate** (ELR). The ELR is determined from data collected by sensors attached to a weather balloon and subsequently transmitted to a weather station at the earth's surface. For example: assume a weather balloon is released at the surface where the temperature is determined to be 30°C (86°F). Information recorded and transmitted from an altitude of 1,500 m (approximately 5,000 ft) indicates air temperature is 10°C (50°F). Dividing the temperature difference between these two altitudes by a standard unit of distance we obtain the ELR, i.e. (30°C-10°C) divided by 1.5 thousand meters = 13.3°C/1,000 m (similarly, 86°F-50°F divided by 5.0 thousand feet = 7.2°F/1,000 ft).

Air parcels that experience vertical motion within the still atmosphere have a rate of temperature change that is different from the ELR, but more predictable than the ELR. Their temperature changes are referred to as adiabatic. This means the temperature change is related to the parcel experiencing a pressure or volume change and specifically means that the temperature change is not a result of heat being exchanged between the parcel and its surrounding air. There are two adiabatic lapse rates. The first one we will discuss is the Dry Adiabatic Rate (DAR), so called because it has a relatively constant value of 10°C /1,000 m (5.5°F/1,000 ft), as long as no condensation of water vapor is occurring within the vertically moving parcel.

Consider the following examples of air stability. Suppose that the surface air temperature is 30°C (86°F). If a parcel of air is lifted by some triggering mechanism, it would expand (change volume) as a consequence of moving into an environment of lower pressure. Thus, it would cool at the DAR and decrease in temperature by 10°C/1,000 m (5.5°F/1,000 ft), as illustrated in Figure 7.18. If the lifted air were to subside, it would also heat at the same rate it cools during ascent. To determine stability status, compare the DAR cooling to the ELR. Remember ELR is established from **empirical** (observed) data obtained from weather balloons. Suppose that the ELR is 5°C/ 1,000 m (3°F/1,000 ft) and surface air temperature is 30°C (86°F). Then, the air temperature surrounding the lifting parcel would be as presented in Figure 7.19 A.

Now we have set the stage to illustrate the relationship between the ELR and DAR, both of which have been superimposed in Figure 7.19B. Notice that the lifted (and expanding) parcel is cooling much more rapidly than the air surrounding it. This means that when the force that triggered the parcel to lift is removed, the parcel will sink back to the surface. Why? Because it is colder and denser than the air of its surrounding environment. In this case, we refer to the status of the atmosphere as being stable, that is wanting to resist displacement. Stability occurs anytime the ELR is less than the DAR.

The DAR is relatively constant, but the ELR can vary over a wide range. Suppose on a given day, we once again have a surface temperature of 30°C (86°F) and an ELR of 15°C/1,000 m (9°F/1,000 m). The DAR temperature change would be exactly the same as illustrated in Figure 7.18. The ELR changes would be different, as demonstrated in Figure 7.20A.

Superimposing Figures 7.18 and 7.20A results in the relationship presented in Figure 7.20B. Notice now that the lifting parcel is warmer than the surrounding environmental air

Meters °C
2,000 10°
1,000 20°
0 30°
Feet °F
2,000 75°
1,000 80.5°
0 86°

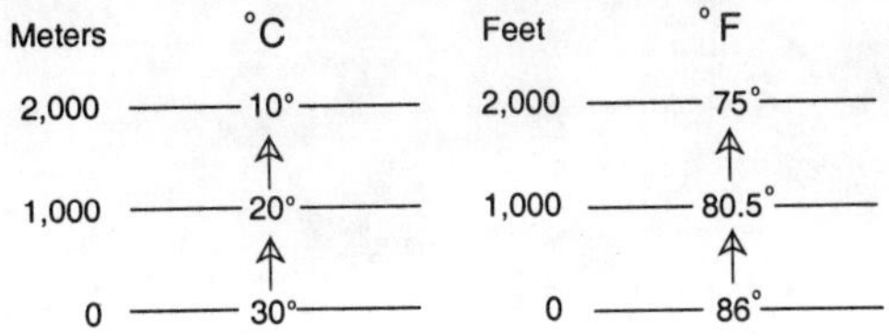

Figure 7.18 Dry Adiabatic Cooling

at all levels. This means the parcel is more buoyant than its surroundings and will continue to ascend of its own accord, like a hot air balloon rises. Atmospheric conditions of this nature are called **unstable.**

Whereas the DAR is considered relatively constant, our second adiabatic rate is more variable and the value assigned to it represents an average. This rate of temperature change is applied to lifting air that is saturated with moisture, is experiencing condensation, and is releasing latent heat that retards the dry adiabatic cooling rate. It is also referred to by several terms. Because we have not yet discussed many of the factors influencing this adiabatic rate, we can only sketch the process. We will use the term **saturated adiabatic rate** (SAR) and define it as 6°C/1,000 m (3.2°F/1,000 ft.). This rate is applicable to lifting air when, in general, it has reached 100 percent relative humidity. Note that if you substituted these values into our examples presented in Figure 7.18B and Figure 7.19B, the atmospheres stability status would not change. As we will see later, this will not always be true.

Conclusion

We have examined the many forces that affect air movement. Yet, these have presented a series of isolated processes. In the following chapter an attempt will be made to integrate these concepts in a discussion of our planet's atmospheric circulation system.

Meters
2,000 20°C
1,000 25°C
0 30°C
A °C
Feet
2,000 80°F
1,000 83°F
0 86°F
°F
Meters
2,000 10 20°C
1,000 20 25°C
B 0 30 30°C
°C
Feet
2,000 75 80°F
1,000 80.5 83°F
0 86 86°F
°F

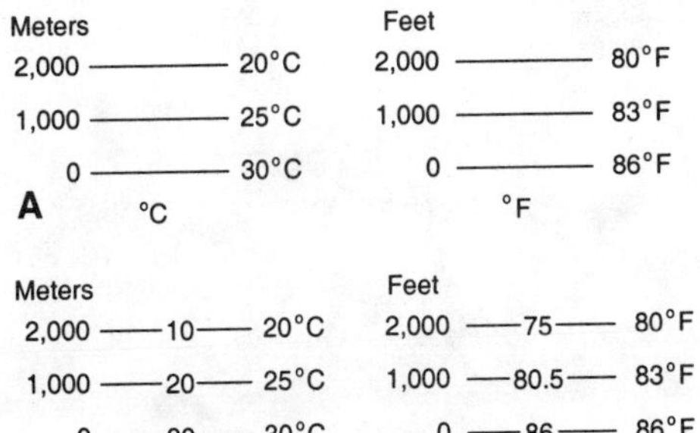

Figure 7.19 A. Example of an ELR Change of 5°C/1,000m (3°F/1,000 ft.) B. Relationship Between ELR and DAR

Meters
2,000 0°C
1,000 15°C
0 30°C
A °C
Feet
2,000 68°F
1,000 77°F
0 86°F
°F
Meters
2,000 10 0°C
1,000 20 15°C
B 0 30 30°C
°C
Feet
2,000 75 68°F
1,000 80.5 77°F
0 86 86°F
°F

Figure 7.20 A. Examples of an ELR Change of 15°C/1000m (9°F/1000 ft.) B. Relationship Between ELR and DAR

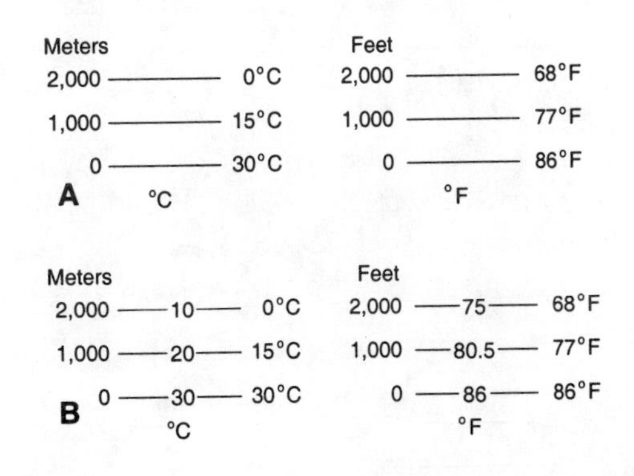

Activity: Atmospheric Pressure

OBJECTIVES

Upon completion of this lab, you will understand: (1) the influence of surface temperature on horizontal pressure patterns; (2) the influence of high and low pressure on air movement; (3) the concept of a pressure gradient; (4) the relationship between atmospheric pressure gradients and wind speed/ direction; and (7) the vertical atmospheric pressure gradient.

INTRODUCTION

Air, like any other substance, occupies space and has weight. The weight of air exerting a force on a surface constitutes air pressure. Air has the ability to become compressed when put under pressure. The air near the surface of the earth is most dense because it is compressed by the weight of the overlying air. Thus, as height above sea level increases, air becomes less dense.

Atmospheric pressure is measured with a barometer. The first practical working barometer was developed in 1643 by Evangellista Torricelli. With minor modifications, his is the type of barometer still in use. This instrument consists of a glass tube about three feet long and sealed at one end. The tube is completely filled with mercury with the open end inverted to rest in a reservoir of mercury. Mercury in the tube drops until the atmospheric pressure bearing on the exposed surface of the mercury reservoir balances the weight of the mercury in the tube. The sealed end of the tube is almost a complete vacuum. The height of a column of mercury at sea level is about 30 inches. Variations in atmospheric pressure are indicated by variations in the length of the column of mercury. Because mercury is very susceptible to expansion and contraction due to changes in temperature, it is necessary, when reading a mercury barometer, to take a temperature reading and compensate for the effect of temperature.

Another type of barometer is an aneroid barometer, which records changes in air pressure without using mercury. An accordion-type box, partially emptied of air, is squeezed or stretched as atmospheric pressure rises or falls. Aircraft altimeters also work on this principle. These instruments indicate altitude on the basis of pressures that prevail at given heights above sea level. Adjustments are made as varying weather conditions affect sea level pressure.

A third type of barometer is a recording barograph. The recording barograph is actually an aneroid barometer consisting of a disk-shaped hollow metal cylinder from which air has been removed to form a vacuum. The cylinder is linked by a series of levers to a pen on a revolving drum. Atmospheric pressure variations act on the evacuating cylinder in the same way that they act upon a column of mercury in a mercurial barometer. These variations are transmitted through the levers to the pen which records them as a line on a calibrated paper on the drum. Today, pressure is usually measured in millibars rather than inches of mercury. As a point of comparison, 30 inches of mercury is the equivalent of 1016 millibars. Standard sea level pressure is 29.92 inches of mercury or 1013.2 millibars.

Changes in the temperature and moisture content of the atmosphere cause changes in barometric pressure. A region which absorbs more of the sun's radiant energy will have a higher temperature than a region where the sun's rays are not as readily absorbed. Since heated air rises, the atmospheric pressure in that heated region of the earth's surface decreases. Like all gases, the higher the temperature of the air, the faster it will rise, and the lower will be the atmospheric pressure exerted upon the surface of the earth. Conversely, as air cools it becomes more dense. As the cool air sinks toward the earth, surface pressure increases because the cooler air is denser .

On a weather map, atmospheric pressure is portrayed through the use of isobars. **Isobars** are lines that connect points of equal (iso) atmospheric pressure (bar). On a weather map, isobars connect to form "cells" or regions of high and low pressure. The intensity of high and low pressure cells is represented on a surface weather map by the spacing of the isobars. The change in pressure shown by the spacing of isobars is called the pressure gradient. The change in pressure Δp per unit distance defines the pressure gradient:

$$PG = \frac{\Delta p}{\text{Distance}}$$

Horizontal movement of air down pressure gradients is what we commonly refer to as "wind." If the isobars are close together, there is a rapid change in pressure within a short distance (a steep pressure gradient) and wind will move across these isobars at a high velocity. If, however, a weak pressure gradient is indicated by isobars drawn far apart on the weather map, the wind will be light.

VERTICAL DISTRIBUTION OF ATMOSPHERIC PRESSURE

The following table gives a number of sets of pressure-altitude data, as defined for the lower portions of the U.S. Standard Atmosphere. Use the data below to plot an accurate, smoothly curving line graph which shows the pressure - altitude profile for a standard atmosphere from mean sea level (MSL) to 30,000 meters (Figure 1). Refer to the completed graph and tabulated data to answer the questions that follow.

Table 1 Typical Pressures at Selected Altitudes

Altitude (meters)	Pressure (Millibars)
sea level	1,013.2
1,000	898.7
2,000	794.9
3,000	701.1
4,000	616.4
5,000	540.2
6,000	471.8
7,000	410.6
8,000	355.9
10,000	264.3
30,000	11.7
60,000	0.2

Figure 1 The Vertical Pressure Gradient

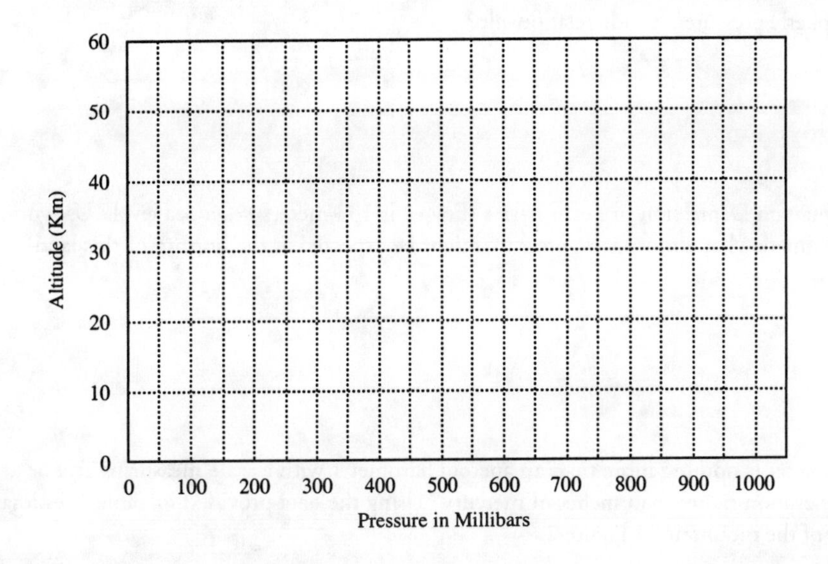

1. How much does the atmospheric pressure decrease for the first 1000 meters increase in elevation?

2. At what altitude would atmospheric pressure become zero, if that rate of pressure decrease found in question 1 did not change with altitude?

3. On your graph, draw a straight line connecting mean sea level pressure and the altitude of hypothetical zero pressure discovered in question 2. Use this line to estimate the difference between actual and hypothetical straight line pressure at 10,000 meters.

4. At higher altitudes the slope of the line is greater. What does this mean in terms of the atmospheric pressure/altitude relationship?

5. The Quattara Depression in north Africa (Egypt) is 133 meters *below* sea level. Use your graph to estimate what atmospheric pressure might be expected at the bottom of this landmark.

6. An altimeter is nothing more than an aneroid barometer with a scale measuring feet or meters of elevation rather than inches of mercury. Using the data provided in Table 1, estimate the height of the mountain in Figure 2.

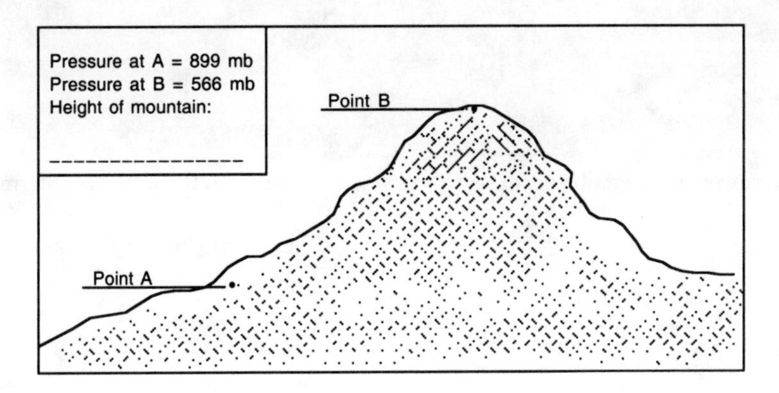

Figure 2 Elevation and Altitude Can Be Estimated from Barometric Pressure

HORIZONTAL DISTRIBUTION OF PRESSURE

1. Examine the world maps of semi-permanent global pressure cells during January and July. What relationship can you see between atmospheric pressure, and land and water distribution? Note that there are important seasonal variations.

2. At the same latitude, pressure varies between land and water areas. What reasons can you suggest to explain the difference?

3. Weather maps and barometric charts plot isobars as sea level equivalent values. Explain why this is necessary.

4. Examine the world maps of semi-permanent global pressure cells during January and July in Chapter 6. Remember, the closer the isobars, the faster the wind. When and where would you expect the greatest wind speed?

When and where would you expect the lowest wind speed?

5. Using the same maps, describe the expected barometric pressure as high, low, or intermediate.

	Location	Season	Barometric Pressure
A.	Siberia	Winter	______________
B.	Aleutians	Winter	______________
C.	Hawaii	Summer	______________
D.	Southern Asia	Summer	______________
E.	Iceland	Winter	______________
F.	Australia	Winter	______________
G.	Charlotte	Summer	______________
H.	Charlotte	Winter	______________

Activity: Air Stability

OBJECTIVES

Upon completion of this exercise you will be able to: (1) describe the main lifting processes in the atmosphere; (2) understand adiabatic temperature change; (3) understand the concepts of stable, neutral, conditionally unstable, and unstable air in terms of adiabatic rates, lapse rates, and lifting condensation level.

INTRODUCTION

Air stability addresses temperature and moisture changes associated with the vertical displacement of units of air (formally called air parcels). Everyone is familiar with the horizontal movement of air (called wind) but not many understand and can visualize the vertical (up or down) motion of air parcels within larger masses of air that have a dominant horizontal flow. Yet it is this vertical air displacement that largely produces the weather changes we experience from day-to-day.

Within the troposphere, air normally becomes progressively cooler as altitude from the earth's surface increases. Recall that this "inverse relationship" (altitude increases while temperature decreases) is associated with the fact that the earth's atmosphere is largely heated from the earth's surface by terrestrial radiation. An example of this process is something you experience on a cold night when standing by a fireplace. The radiant heat is intense in front of the fireplace, but diminishes in intensity as you move toward the opposite side of the room.

GATHERING TEMPERATURE DATA ABOVE THE EARTH'S SURFACE

As we soon shall see, knowing the manner in which temperature changes with distance from the earth's surface is important to understanding vertical (up and down) motions of air. To obtain air temperature above the earth's surface, the meteorologist connects an instrument package to a large helium filled balloon. The instrument package contains sensors of temperature, humidity, and air pressure that are connected to a radio-transmitter. As the balloon ascends, changes in these three weather elements are recorded and transmitted via radio broadcast to a surface-based radio-signal receiver that is attached to a recording device. A typical recording for temperature and pressure might appear as follows:

Table 1. An example of temperature, pressure, and altitude data from a weather balloon.

Pressure (millibars)	Altitude (meters)	Temperature
1,000	0	30.0° C
850	1,500	22.5° C
700	3,000	15.0° C
500	4,000	10.0° C

The data gathered above provide valuable information to the meteorologist who uses it to describe the manner in which temperature changes aloft and how the air in a given area will likely behave. To visualize how these changes occur, it is convenient to graph temperature versus altitude relationships as is shown in Figure 1. (Note: Data are derived from Table 1).

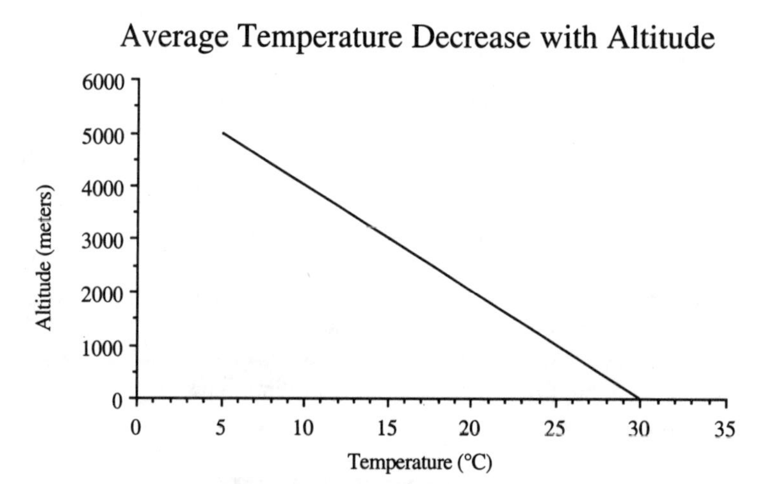

Figure 1. Temperature Plotted Against Altitude

If we refer back to the data from which Figure 1 was derived, we can see that the temperature of the air lapsed (or decreased) from 0 to 5,000 meters by 25° C. Compare this to our previous analogy of walking away from a fireplace, where intensity of absorbed radiation is lower. We can express this relationship as 25° C/5,000 meters. The ELR can differ from place-to-place, from day-to-day and through different depths of the atmosphere; for example, 2° C/1,000 meters, 12° C/4,000 meters, and 6° C/1,500 meters. To make sense of temperature changes throughout varying depths of the atmosphere, the meteorologist converts all temperature lapse rates to equivalent atmospheric depths.

ESTABLISHING THE "EQUIVALENT" ELR

To compare ELR's, their original values are, by convention, converted to an atmosphere depth of 1,000 meters or 1,000 feet (depending on the data-base being utilized). In the previous paragraph, reference was made to an ELR of 25° C/5,000 meters. The easiest way to convert this ELR to an "equivalent 1,000 meter depth" is to divide the numerator by the number of thousand meters in the denominator, that is 25 ÷ 5 = 5, thus establishing a lapse rate of 5° C/1,000 meters which is the "ELR equivalent 1,000 meter depth" of 25° C/5,000 meters. All future references to the "equivalent ELR" will simply be referred to as the ELR.

Problem 1
Convert the following observed temperature-altitude values to "equivalent 1,000 meter ELR's:"

8.1.1.	2° C / 1,000 meters	=	______/1,000 meters
8.1.2.	12° C / 11,000 meters	=	______/1,000 meters
8.1.3.	6° C / 1,500 meters	=	______/1,000 meters
8.1.4.	20° C / 20,000 meters	=	______/1,000 meters
8.1.5.	15° C / 3,000 meters	=	______/1,000 meters

The atmosphere may have several ELR's above the surface. Referring once again to Figure 1, we may conclude that a given atmospheric ELR only holds true for a specific layer of air above the surface. Examining our data base for Figure 1, we can see that air between 0 to 2,000 meters has a lapse rate of 5° C/1,000 meters. The same is true of the layer from 2,000 meters to 5,000 meters. This suggests that the atmosphere between 0 and 5,000 meters has a common source of heating (i.e. from the earth's surface) and is behaving as a uniform layer of air. Different layers of air may behave differently; lapse rates above 5000 meters may be very different. What the foregoing illustrates is that a given ELR is valid only for a layer of the atmosphere that has a relatively uniform temperature change.

UNDERSTANDING ADIABATIC PROCESSES & ADIABATIC LAPSE RATES

Now that we have a basic understanding of how the temperature lapse structure (ELR) of the atmosphere is determined and how to distinguish horizontal layers of air having unique lapse-rate traits, it is possible to examine the behavior of air parcels that possess vertical motions within horizontally stratified air.

If air were to move, at all times, in horizontal flows (which, by the way, is impossible), its temperature characteristics could be directly related to the amount of energy each layer absorbs versus the amount of energy it loses through re-radiation. When air is vertically displaced, however, temperature changes occur within it that are related to processes other than previously described. Before defining those relationships, lets consider a simple example of what happens to a unit of air that is contained within a balloon and lifted from the earth's surface. To begin, it is important to understand that the air is composed of a mixture of gases and that gases behave as a fluid. **Gases exert force or pressure in all directions.** Average pressure at sea level is approximately 1000 millibars, which equates to a force of almost 15 lbs per square inch. If our balloon is filled with air to where its internal pressure is also 15 lbs/in^2, the balloon is in equilibrium with its environment (Figure 2).

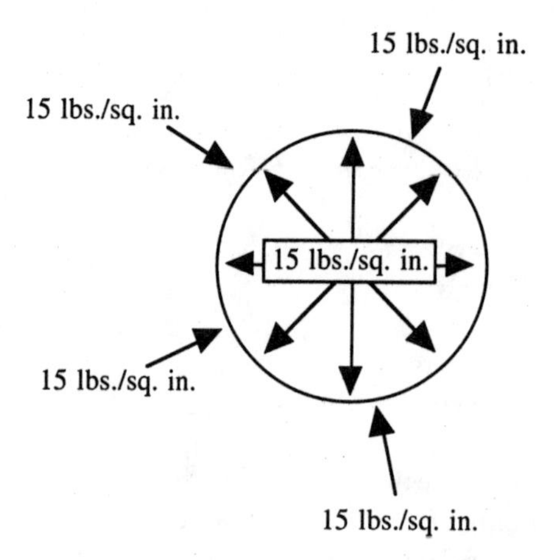

Figure 2. The pressure exerted upon the interior and exterior surfaces are balanced.

Atmospheric pressure decreases with distance from the earth's surface. If the same balloon is lifted to the 500 mb level (pressure about 1/2 of that at the earth's surface), the external pressure drops to 7.5 lbs/ in^2. To regain equilibrium, the air inside the balloon pushes its elastic walls outward (that is, it expands) until the inside wall pressure equals 7.5 lbs/in^2.

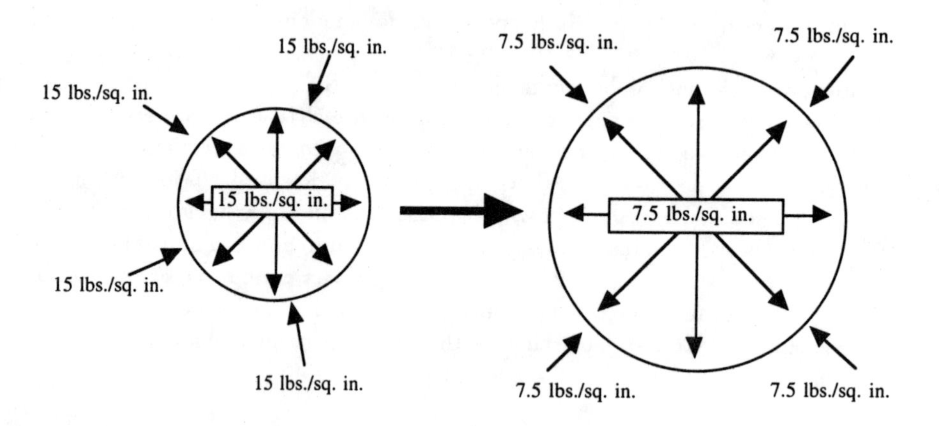

Figure 3. The lifted air expands because it has entered a region of lower pressure.

The significance of the previous example is that air lifting vertically behaves like the air trapped in the balloon; it expands as it moves into the lower pressure environment of the atmosphere above the earth. A classic example of this process occurs along the northwest coast of the United States where westerly winds must be forced upward over the coastal mountain ranges. Air at the surface must be forced upward over the mountain to continue its eastward journey. In doing so, the air expands (Figure 4).

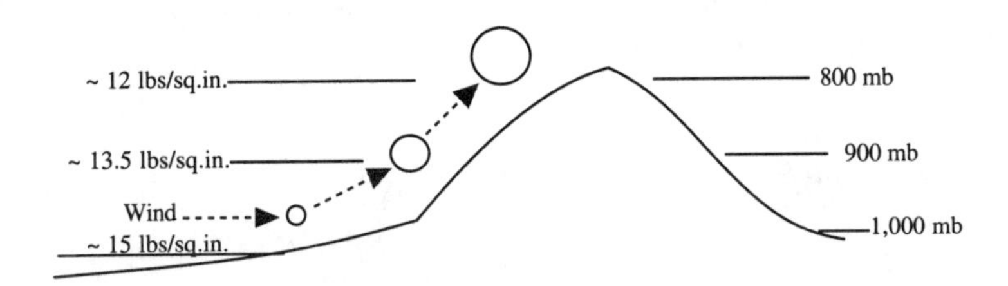

Figure 4. Lifting air expands as it moves to levels of lower pressure.

VERTICAL AIR DISPLACEMENT AND TEMPERATURE

Temperature is a measure of the average kinetic energy available within a given mass. With respect to air at a given volume and pressure, temperature is an indirect measure of the average speed of its molecules.

As a given parcel of air rises, the surrounding gas pressure decreases. As the surrounding pressure decreases, the parcel expands, the air molecules in the parcel slow down, and fewer molecular impacts occur. At sea level, air molecules move at speeds as fast as 500 meters per second, but this is not the speed at which a molecule moves across the room. A molecule does not move in a straight line, but has several billion collisions per second with other molecules. As air rises, fewer collisions occur and the molecules slow down. Therefore, rising, expanding air always cools, and subsiding, compressed air likewise heats (Figure 5).

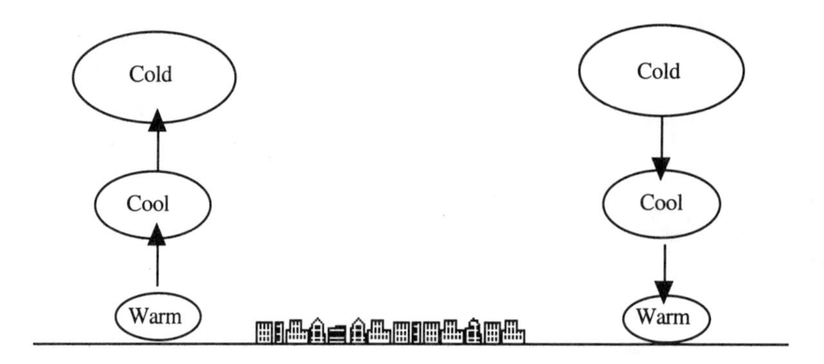

Figure 5. Lifting air cools by expansion. Descending air warms by compression.

WHAT CAUSES AIR TO RISE

Knowledge of the characteristics of lifting air is important because most clouds form as air rises, expands, and cools. The majority of clouds are formed by the following mechanisms.

1. Convection

When air is heated at the earth's surface it expands and becomes lighter than the surrounding air and tends to rise. Convection provides most of the rainfall in the tropics and throughout the humid subtropics during summer.

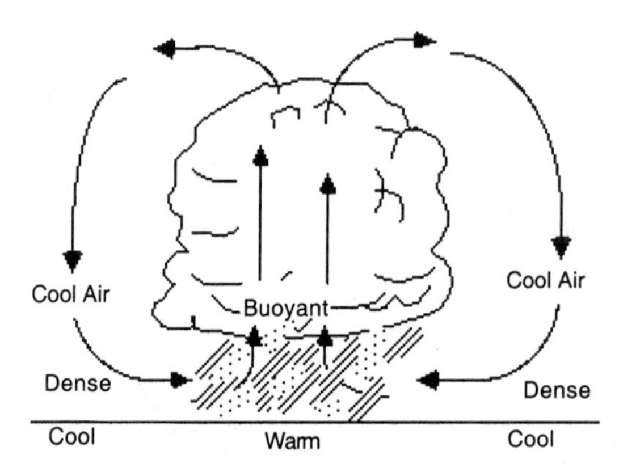

2. Frontal or Cyclonic Lifting

When a cold front moves into a region, the relatively warmer, lighter air is pushed aloft by the leading edge of the advancing mass of relatively colder, denser air. When a warm front moves into an area, the leading edge of the advancing mass of relatively warm, less dense air is pushed up over the mass of stationary, cooler, more dense air.

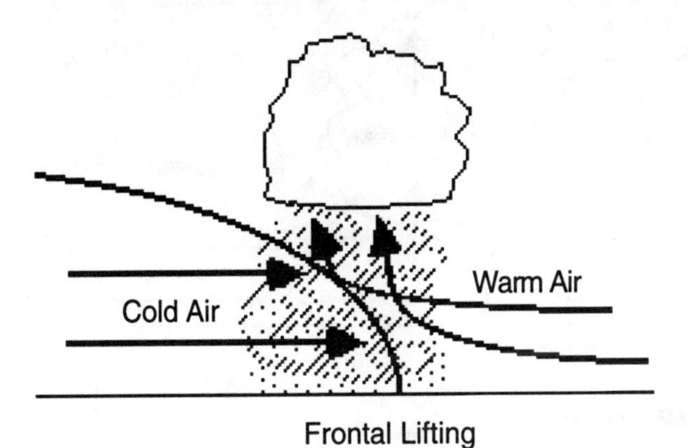

Frontal Lifting

3. Orographic Lifting

When air is pushed up against a physical barrier such as a mountain range, it is forced to rise upward along the slope of the mountains. Orographic lifting is responsible for much of the precipitation occurring on windward mountain slopes.

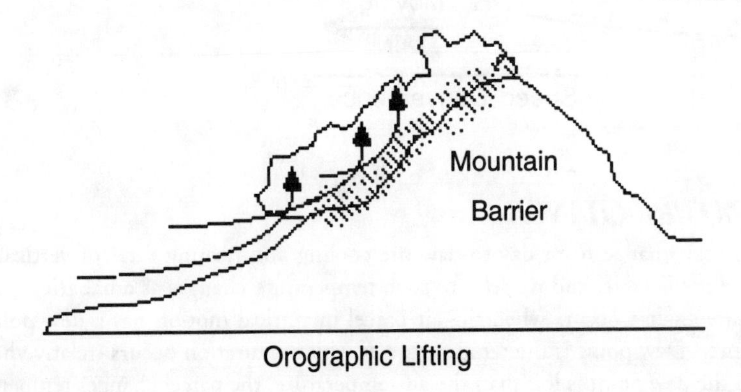

Orographic Lifting

4. Convergence

The confluence of air (convergence) requires that lifting take place.

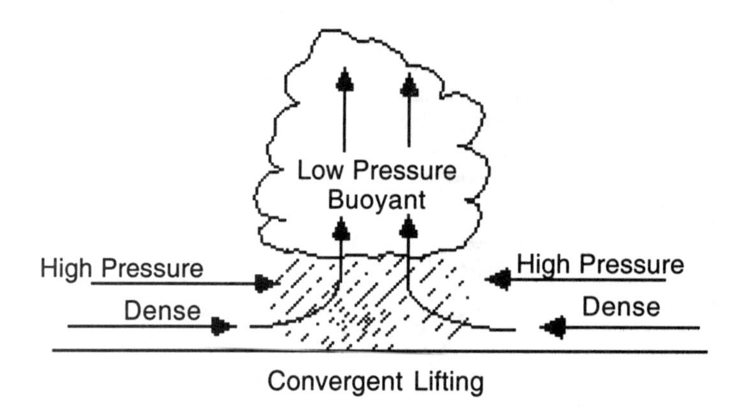

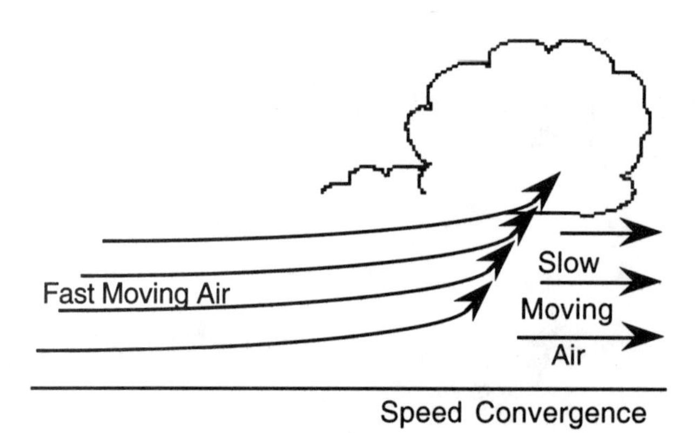

ADIABATIC TEMPERATURE CHANGE

Unlike the ELR which may change from day-to-day, the cooling and heating rates of vertically displaced air parcels are more predictable, and we refer to such temperature changes as **adiabatic**. The simplest adiabatic temperature change occurs when the air parcel in vertical motion has a dew point lower than the air temperature. Dew point is the temperature at which saturation occurs (relative humidity = 100%). As long as the dew point is less than the air temperature, the parcel changes temperature at the **dry adiabatic rate** (DAR) of 10° C/km. If, on the other hand, the parcel has cooled to the dew point and is saturated, the rate of change averages 6° C/km and is referred to as the pseudo-adiabatic lapse rate (PAR). The PAR is also referred to as the **saturated adiabatic rate** (SAR) or the **wet adiabatic rate** (WAR). It is lower than the DAR, because latent-heat is released as water vapor condenses. Cool air can hold less water vapor than warmer air; consequently, cooling forces water vapor to condense.

Problem 2

A. On the graph provided (Figure 6), draw with a solid line the DAR for air having a surface temperature of 35° C.

B. What is the temperature of the lifted air at each of the following altitudes?

(1) 2,000m T = _____ C°

(2) 5,000m T = _____ C°

(3) 10,000m T = _____ C°

C. On the same graph, draw similar lines for surface temperatures of 30°C, 20°C, 10°C, 0°C, -10°C, and -20°C. If you have properly constructed your graph, the DARs should be parallel to one another. This indicates that, regardless of initial temperature altitude level, temperature change is linear (consistent).

D. Using the same surface temperatures in C above, draw dashed-lines to predict the temperature change in vertically moving *saturated-air;* that is, use the PAR.

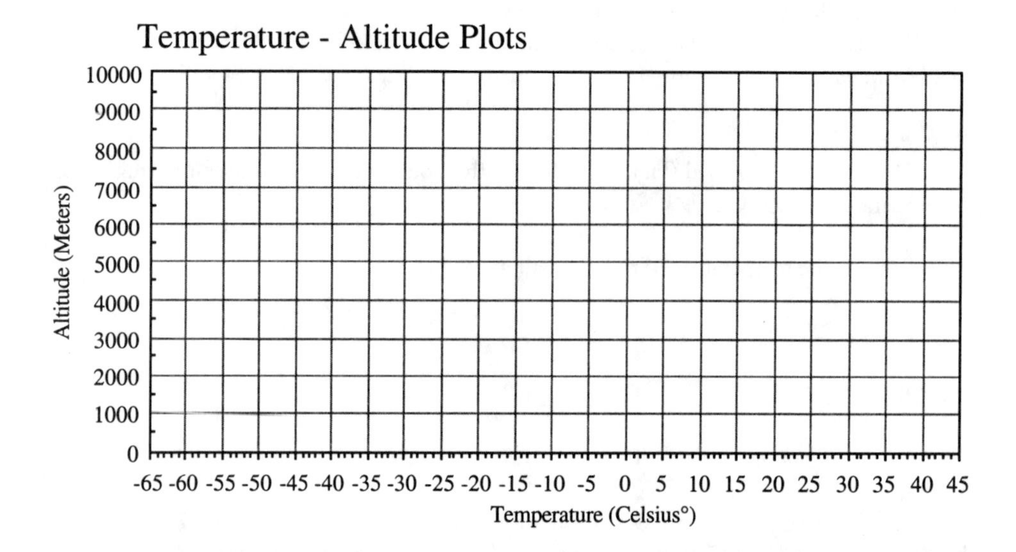

Figure 6. The above graph is provided for plotting the DAR and PAR of air parcels having selected surface temperatures.

In completing problem 2, you have produced a **thermodynamic diagram** that predicts the temperature behavior of vertically moving air. When this diagram is combined with knowledge of the ELR, you will be able to define *atmospheric stability status.*

STABLE ATMOSPHERIC CONDITIONS

Air is considered stable when it resists vertical displacement. Note that it is still possible for it to move vertically, but only when strong forces cause it to do so. For the present we will consider only the classic characteristics of stable air. Consider the scenario in which a parcel of air cools at the DAR (10° C/km). The surface temperature is 26.5° C and the air 5,000 meters aloft is –2.5° C.

Problem 3

A. The ELR of the above air is ________° C/km.

B. Plot this lapse rate on the graph provided in Figure 6.

Note the following:

1. You have plotted the actual temperature structure of the atmosphere.
2. If air is lifted from the surface (whether dry or saturated), it cools at a rate that is greater than

the ELR. Remember that the predicted changes in temperature are determined from the adiabatic lines supplied on your graph.

3. At any level between 0 and 5000 meters, lifted parcels will be colder than the surrounding air. Cold air is denser than warm air. Thus, when the force that caused the air to be vertically displaced is removed, the parcel returns to its original altitudinal position. This represents stable atmospheric conditions.

RULE 1
When air is unsaturated and the DAR is greater than the ELR, the atmosphere is stable.

RULE 2
When air is saturated and the PAR is greater than the ELR, the atmosphere is stable.

Under stable atmospheric conditions, typically there would be clear skies with little or no chance of rain. In such situations, air stagnation and pollution could be problematic since fumes and particulate matter expelled into the air would tend to stagnate near the surface and increase in concentration.

HOW CLOUDS CAN FORM IN STABLE AIR

If stable air is forced to ascend, it will, as all air does, cool adiabatically until it becomes saturated with respect to water vapor (that is, relative humidity equals 100%). This occurs when cooling of the air, lowers its temperature to the dew point temperature. The altitude at which temperature and dew-point temperature become equal (due to lifting) is referred to as the **lifting condensation level** (LCL) and represents the base of cloud formation. The LCL is easily approximated by the following formula:

$$\text{LCL (\_\_\_\_\_\_\_) meters} = \frac{\text{Air Temperature (°C) - Dew Pt. (°C)}}{\text{Dry Adiabatic Lapse Rate}} \text{ X 1,000 meters}$$

Let's apply the foregoing to the following data:

Air Temperature = 30° C
Dew Point Temp. = 10° C

$$\text{LCL} = \frac{30 - 10}{10} \text{ x 1,000} = \text{2,000 meters (2 km)}$$

It is important to remember that once the LCL is reached, further cooling associated with lifting will take place at the PAR. You may question why air would lift and develop clouds within air that is stable and resists lifting. Figure 4 provides one example of how this may occur.

UNSTABLE ATMOSPHERIC CONDITIONS

When vertically displaced air changes temperature at a rate less than the ELR, it is said to be unstable. Consider the following data derived from a weather balloon: Surface air temperature = 30° C; temperature at 4,000 meters = -18C; dew point = 10° C.

RULE 3 When air is unsaturated and the DAR is less than the ELR, the air is unstable.

RULE 4 When air is saturated and the PAR is less than the ELR, the air is unstable.

Problem 4

A. The ELR of the above air = ______° C/km.

B. Plot the DAR, the PAR, and the lapse rate on the graph provided in Figure 7. Use colors and a legend with labels.

C. Develop statements that explain: 1) the relationship between lifted air and surrounding air temperatures in unstable air and 2) why air would have a tendency to keep lifting once the process was initiated.

1) explain:

2) why:

D. Calculate the LCL of this air: ________meters.

E. Calculate the temperature of the lifted parcel at the LCL: _________° C.

Figure 7 Unstable

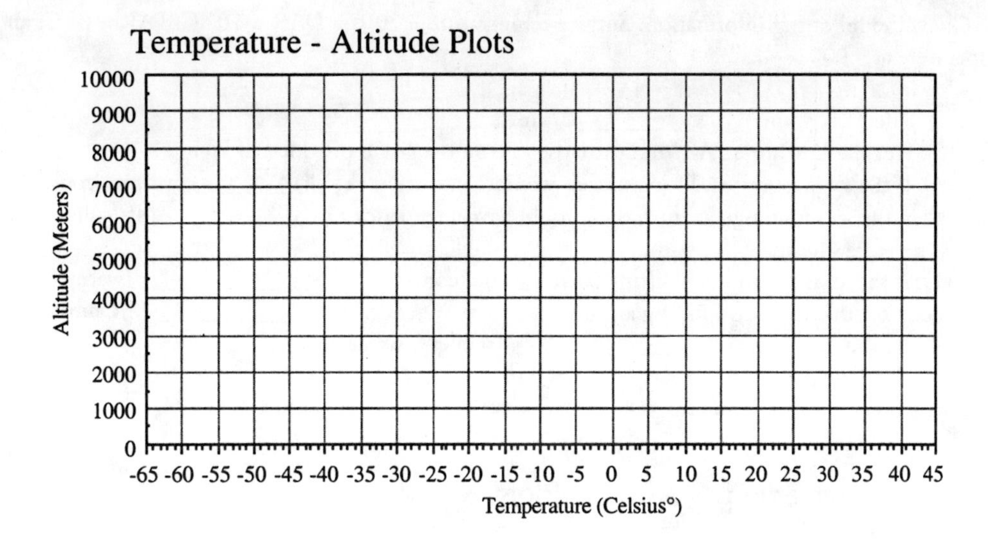

NEUTRAL ATMOSPHERIC CONDITIONS

In the somewhat rare case when the ELR is exactly equal to the DAR, a rising or subsiding unsaturated parcel changes in temperature at the same rate as the air around it. At every altitudinal level, the parcel has the same temperature and density as the surrounding air. Therefore, it neither continues rising or sinking. For saturated air, neutral stability exists when the ELR is equal to the PAR.

RULE 5 When air is unsaturated and the DAR equals the ELR, the atmosphere has neutral stability.

RULE 6 When the air is saturated and the PAR equals the ELR, the atmosphere has neutral stability.

CONDITIONALLY STABLE ATMOSPHERIC CONDITIONS

By definition the atmosphere is conditionally stable (or unstable) when the ELR has a value lying between the DAR and PAR. The best way to understand conditional stability is to graph the ELR and examine its relationship to adiabatic processes.

Problem 5

Given the following information: Surface temperature = 20° C; DAR = 10° C; PAR = 6° C; and temperature at 3,000 = -4° C

A. The ELR of this air = _______° C/km.
B. Plot the DAR, the PAR, and the lapse rate on the graph provided in Figure 8.
C. If the air is saturated, lifting parcels would continually be __________ (warmer/cooler ?) than the surrounding air; thus, they would have a tendency to __________ (lift/subside ?). Therefore, the air is: ______________ (stable/unstable?).
D. If the air is unsaturated, lifting parcels would continually be ____________ (warmer/cooler ?) than the surrounding air; thus, they have a tendency to __________ (lift/subside ?). Therefore, the air is: ______________ (stable/unstable?).

Figure 8 Conditionally Stable

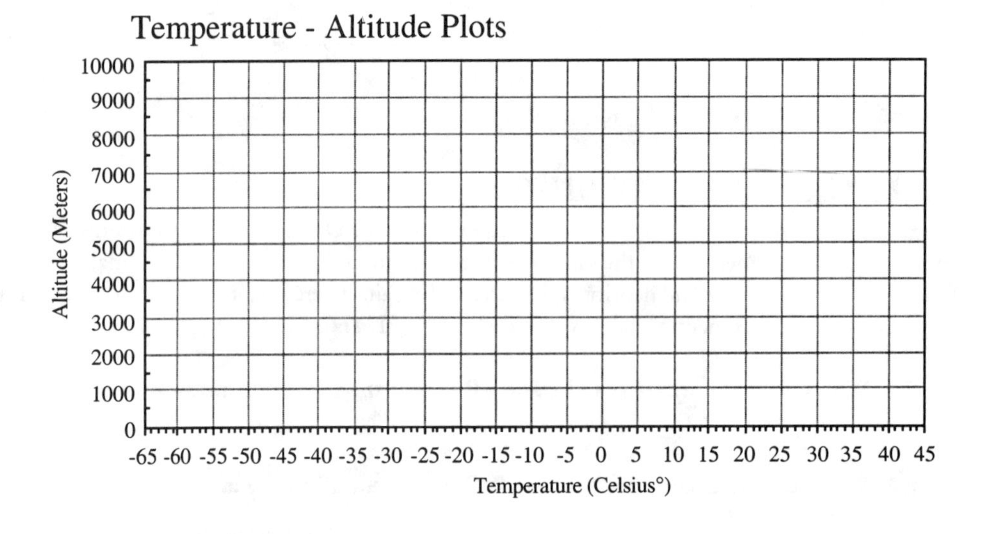

NOTE: The above examples illustrate that, when the ELR lies between the DAR and PAR, the stability status depends on the air's moisture content. Properly used, the term "conditionally stable" air exists according to Rule 7:

RULE 7 Conditional stability occurs when the ELR lies between the DAR and PAR and when moisture-status is unknown.

Problem 6

Given the information provided below, answer the following questions. Remember that the PAR is, in effect, above the LCL.

	A	B	C
Surface Air Temperature:	35° C	30° C	20° C
Dew-Point Temperature:	15° C	25° C	-10° C
Air Temperature Aloft:	19° C @ 4000M	-30° C@ 5000m	-16° C@ 4,500m
1.Lifting Condensation Level (LCL)	2000	______m	______m
2.Environmental Lapse Rate	______° C/km	12 C/km	______° C/km
3.Temperature of lifted air at:			
a. 2000 meters	______° C	______° C	0 ° C
b. 3000 meters	9 ° C	______° C	______° C
c. 4000 meters	______° C	4 ° C	______° C
4.Is the lifted air warmer or colder than the environmental air	______ (warmer/colder/=)	______ (warmer/colder/=)	______ (warmer/colder/=)
5. Compare the ELR with the lapse rates and classify the air as stable, neutral, unstable, or conditionally stable:	stable	______	______
6. Would cloud development be likely to occur without forced lifting?	______	Yes	______

Summary of Rules:

1. When air is unsaturated and the DAR is greater than the ELR, the atmosphere is stable.
2. When air is saturated and the PAR is greater than the ELR, the atmosphere is stable.
3. When air is unsaturated and the DAR is less than the ELR, the air is unstable.
4. When air is saturated and the PAR is less than the ELR, the air is unstable.
5. When air is unsaturated and the DAR equals the ELR, the atmosphere has neutral stability.
6. When the air is saturated and the PAR equals the ELR, the atmosphere has neutral stability.
7. Conditional stability occurs when the ELR lies between the DAR and PAR and when moisture-status is unknown.

Chapter 8

Atmospheric Moisture

Over half of the energy given up to the atmosphere by the earth is transferred through the process of evapotranspiration. Evapotranspiration refers to the combined effects of evaporation from water and soil and transpiration from plants. The process is important not only for the energy it supplies to the atmosphere, but because it sets in motion a chain of events that is necessary to sustain life processes and that plays the dominant role in the evolution of landscapes.

Evapotranspiration is the beginning of the hydrologic cycle (Figure 8.1). The water vapor transferred to the atmosphere condenses, releasing its latent heat, and falls as precipitation, supplying water for plant growth, streamflow, and storage as soil moisture and groundwater. In this section we will examine the atmospheric components of the hydrologic cycle: evapotranspiration, condensation and precipitation.

Evapotranspiration

The temperature of water is a function of the mean kinetic energy of molecules, but at any given temperature not all molecules possess the same kinetic energy. As molecules constantly collide, they are either slowed down or speeded up so that there are always some low energy and some high energy molecules. There are proportionately more high speed molecules in warm water than in cool water.

The forces which hold the molecules of liquids together are analogous to gravity. A tremendous force is required to accelerate a rocket so that it can escape the earth's gravitational field. A water molecule near the surface of the water needs only enough kinetic energy from collisions with other molecules to overcome the force of molecular tension and escape into the atmosphere. High energy molecules are the only ones that can do this so that the average kinetic energy of the remaining molecules is reduced, thereby cooling the water.

The number of high speed molecules in water is proportional to its temperature. Water temperature affects how many water molecules escape from the water. However, molecules of water vapor also move back into the water, and as they become more numerous in the air above the water the probability increases that some of them will move back to the water surface. The exchange of water vapor between a water surface and the air above it is a function of the respective temperatures of the two substances. At some point the density of water molecules in each cubic centimeter of air above the water surface reaches an equilibrium. Under equilibrium conditions, the probability of a water vapor molecule leaving the air and entering the water is the same as the probability of a molecule leaving the water so that the net flow is zero. A steady state exists with both water and air gaining as many molecules as they lose. In the very thin layer of air immediately above

Figure 8.1 Model of the Global Hydrologic Cycle

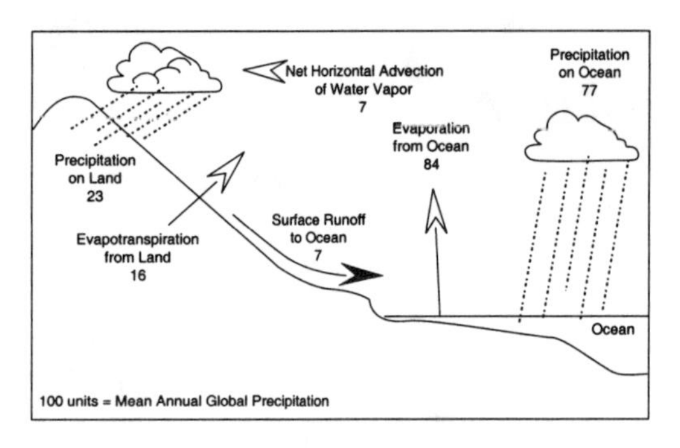

any water or ice surface, this condition of maximum water vapor concentration always exists. The thin air layer is said to be 100 percent saturated.

Water vapor in the air can be measured and expressed in several important ways: 1) mixing ratio — mass of water vapor per unit of dry air which is expressed as grams per kilogram; 2) absolute humidity — mass of water vapor per volume of air expressed as grams per cubic meter; or 3) vapor pressure — molecular pressure expressed as millibars. In discussing evaporation, vapor pressure is the most appropriate measure of humidity to employ. Water vapor molecules exert a pressure proportional to their number, and total atmospheric pressure is the sum of dry air pressure plus vapor pressure. However, air with high vapor pressure is less dense than dry air.

The limited amount of water vapor that air can hold is a function of air temperature. The ratio of actual water vapor present in the air to the maximum it can hold (or actual vapor pressure to saturation vapor pressure) is called relative humidity. Table 8.1 shows the relationship between air temperature and vapor pressure.

Vapor pressure is a function of the number of water vapor molecules and their velocities. We have already seen that high air temperatures produce low densities and low pressures. Warm air has more space that can be occupied by water vapor molecules. Warm air, then, has greater

Table 8.1 Vapor Pressure for Selected Temperatures and Relative Humidities

Temperature		Vapor Pressure in mb				
°C	°F	20%	40%	60%	80%	100%
0	32	1.22	2.44	3.67	4.89	6.11
5	41	1.74	3.49	5.23	6.98	8.72
10	50	2.45	4.91	7.36	9.82	12.27
15	59	3.41	6.82	10.22	13.63	17.04
20	68	4.67	9.35	14.02	18.70	23.37
25	77	6.33	12.87	19.00	25.34	31.67
30	86	8.49	16.97	25.46	33.94	42.43
35	95	11.25	22.50	33.74	44.99	56.24

Source: R. J. List, Smithsonian Meteorological Tables, 6th Revised Edition, 1949. Table 94, "Saturation Vapor Pressures Over Water," Smithsonian Mis. Collections, Vol. 114.

capacity to hold water vapor molecules and higher saturation vapor pressures than cool air. Thus, the vapor pressure in unsaturated air is a function of its temperature and its relative humidity.

For computational purposes, it is often convenient in meteorology to use mixing ratio as an alternate measure of humidity. Figure 8.2 shows the relationship between the maximum water vapor holding potential (i.e., saturation mixing ratio) and air temperature. Table 8.2 compares saturation values between the several common measures of humidity.

If a warm water surface is overlain by saturated air of the same temperature, no evaporation occurs. In fact, the evaporation rate is a function of the vapor pressure gradient (the difference between the vapor pressure immediately above the water surface and the vapor pressure of the air). This is an important observation. Even partially saturated warm air over a cold water surface may result in condensation rather than evaporation.

Consider the following illustration of the preceding statement. Table 8.1 shows that the vapor pressures of 25°C (77°F) air with 60 per cent relative humidity and 5°C (41°F) air with 60 per cent relative humidity are 19.00 and 5.23 mb, respectively. Correspondingly, the saturation vapor pressure (100% relative humidity) of 15°C (59°F) air is 17.04 mb. Thus, if a water surface of 15°C were overlain by air of 25°C and 60 per cent relative humidity, the vapor pressure gradient would be directed from air toward water (from 19.00 mb to 17.04 mb), and no evaporation would occur. On the other hand, if the water were overlain by 5°C air with 60 percent relative humidity, the vapor pressure gradient would be from the water surface to air (17.04 to 5.23) and evaporation would proceed until the vapor pressure reached 8.72 mb, the saturation vapor pressure for 5°C air. Actually,

Figure 8.2 Saturation Mixing Ratios (w)

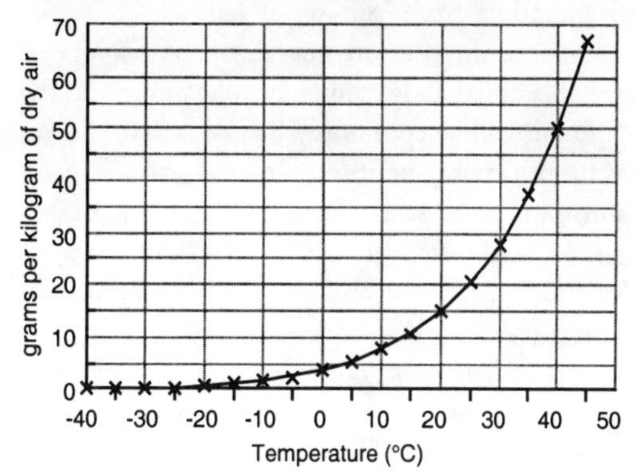

Table 8.2
Saturation Vapor Densities and Pressures

Temperature (°C)	Saturation Mixing Ratio (g/kg)	Maximum Absolute Humidity (g/m3)	Saturation Vapor Pressure (millibars)
45	67.0	65.50	98.86
40	50.0	51.19	73.78
35	37.5	39.63	62.40
30	27.8	30.38	42.43
25	20.5	23.05	31.67
20	15.0	17.30	23.37
15	10.8	12.83	17.04
10	7.8	9.40	12.27
5	5.5	6.80	8.719
0	3.9	4.847	6.108
-5	2.6	3.407	4.215
-10	1.8	2.358	2.863
-15	1.2	1.605	1.912
-20	0.79	1.074	1.254
-25	0.51	0.705	0.807
-30	0.32	0.453	0.509
-35	0.195	0.286	0.314
-40	0.125	0.176	0.189

when water is warmer than the air above it, evaporation from the water surface continues because of the pressure gradient. The evaporated water immediately condenses, however, in the cold air. This commonly happens over steam vents in cold weather. It also occurs over unfrozen polar seas where it is called arctic sea smoke, and over warm lake surfaces in the mid-latitudes when the air above the water surface becomes cold. Fog may be produced in this manner when warm rain falls through cold air.

So far we have seen that air temperature determines the amount of water vapor that the air can hold, and that vapor pressure and water temperature influence the rate of evaporation. There is, however, another important variable affecting evaporation rates. Anyone who has felt the cooling effect of a summer breeze knows that the wind is a factor in evaporation. Without wind, the air layers immediately above an evaporating surface would approach saturation vapor pressure quickly, and the evaporation rate would be reduced. Turbulent surface winds remove these saturated layers, replacing them with dryer air. In effect, the wind reduces the local vapor pressure, encouraging maximum evaporation rates.

The temperature-moisture relationship is rather simple: warm air can hold more water than cold air. This is why air from warm tropical oceans contains more water than the cold polar air over the Arctic ocean. Air with more water vapor and liquid water can trap more of the earths infrared radiation, and the counterradiation of this moist air prevents excessive heat loss by the earth at night. Dry desert air intercepts little heat emanating from the earth's surface at night causing deserts to experience extreme daily temperature ranges.

The rate at which water evaporates is a function of the vapor pressure gradient away from the water surface. This, in turn, is influenced by the water temperature and the temperature and relative humidity of the air. Wind is an important consideration because it constantly reduces the local vapor pressure by mixing the moist surface air with the air above it. The ideal conditions for maximum evaporation include warm water, warm dry air, and wind. The desiccating effect of a warm wind on a hot summer day provides ample proof of the effectiveness of this mechanism.

The temporary storage of water vapor in the atmosphere affects organisms as well as atmospheric processes. The negative feedback effect that high relative humidities have on evaporation rates has a direct effect on human comfort levels. Our sensation of temperature is, in large part, a function of relative humidity. When high vapor pressures retard evaporation rates, the cooling effect of sweating is reduced and we tend to feel uncomfortably warm. Likewise, when the air is dry, evaporation is more rapid, especially when the wind blows. If protected from direct solar radiation, one might not feel uncomfortable despite temperatures in the 38°C (100°F) range.

As previously noted, the amount of water that can enter the atmosphere at any location through the process of evapotranspiration is a function of water temperature, air temperature, wind speed, and the humidity of the air. Even with favorable conditions, however, evapotranspiration does not always proceed at the maximum rate possible. The availability of water for evapotranspiration is a limiting factor. The maximum rate at which evapotranspiration could occur at a location, if unlimited water were available, is called potential evapotranspiration (PE). In hot, dry areas there may be a considerable difference between potential and actual evapotranspiration rates, whereas the two might be nearly the same most of the time in humid areas. Potential evapotranspiration is greatest in areas with the largest inputs of solar radiation. PE declines from the equatorial regions toward the poles and reaches a maximum under the clear skies of the subtropical deserts. When more water is supplied by precipitation than can be evaporated, a temporary "surplus" accumulates and some may be stored in the soil for later withdrawal by plants or evaporation. The bulk of this "surplus" water, however, either runs off or is used to recharge groundwater.

Although several methods exist for calculating potential evapotranspiration, the most widely accepted is the one employed by C. W. Thornthwaite.[1] This empirical method depends upon temperature as the dominant control. Potential evapotranspiration in centimeters can be calculated for climates similar to those in the southeastern United States in the following manner

$$e = 1.6\left(\frac{10t}{I}\right)^a$$

where e = unadjusted PE
t = mean monthly temperature in °C
I = the annual heat index

$$I = \sum i \;\; where \;\; i = \left(\frac{t}{5}\right)^{1.514}$$

and
a is dependent on the value of I

This unadjusted estimate of potential evapotranspiration (e) merely describes a single 12 hour day and must be adjusted to compensate for variations in length of daylight and number of days in each month. By calculating adjusted PE and subtracting from precipitation, a series of positive and negative values, a water budget, can be generated that identifies typical periods of water abundance or scarcity.

Water budgets which detail the exchange of water among atmospheric subsystems may be constructed for geographical areas (Figure 8.3). These budgets provide some insight into water availability by describing effective moisture regimes. Water budgets provide much greater information about moisture stress to plants than precipitation data alone. Annual patterns of moisture deficit are valuable in evaluating agricultural potentials and requirements, especially where irrigation is practiced or contemplated. There are, however, several difficulties encountered in water budget preparation. Perhaps the most critical is the matter of accuracy and sufficiency of the data

Figure 8.3 Water Budget for Charlotte, NC (based on 30 year normal data for 1990)

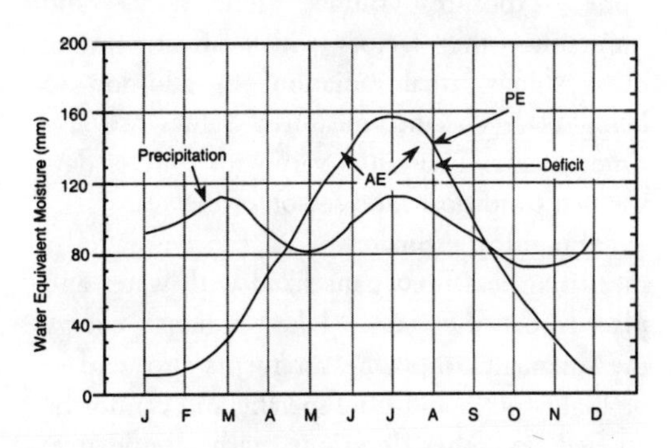

[1]*John R. Mather. Workbook in Applied Climatology. Elmer, New Jersey: C. W. Thornthwaite Associates, Laboratory of Climatology, Publications in Climatology 30 No. 1 (1977).*

needed to construct the budget. In order to compute storage one needs to have accurate data on inputs and outputs, especially precipitation, evaporation, and soil water storage.

Accurate estimates of rates of potential evapotranspiration can be computed theoretically or estimated empirically for specific sites. Theoretical computations are based on two types of approaches, one using wind speed and one using energy. Accurate wind speed data are not available for most locations. In addition, wind is highly variable over time so that there is serious doubt whether average values can produce reliable results.

The second theoretical approach to estimating potential evapotranspiration is based on the energy available for evaporation. There are two difficulties here. First, total incident radiation amounts are required. Second, it is necessary to know how much of this total radiation is used in evapotranspiration and how much is reflected, radiated, conducted to the ground, or used in photosynthesis.

While it is possible to measure these values, it is not practical to do so for very many locations. Average values at one place are not valid for other locations because of variations in slope, exposure, surface reflectivity, and numerous other factors which affect, among other things, total radiation. In addition to being place specific, measured values are also time specific. The values vary with time of day, weather conditions, and season of the year.

Empirical estimates of PE can be made by measuring loss from pans filled with water and placed in the sun. Like radiant energy measurements, these measurements also tend to be highly place and time specific and cannot be applied to other locations with confidence. While water budgets can be produced for specific sites where reliable data are available, it is difficult to compute budgets for large areas with accuracy.

Condensation and Precipitation

When air containing water vapor is cooled, there is some point at which saturation vapor pressure is attained. The temperature at which this occurs is called the dew point, and it depends on the amount of water vapor the air initially contains. By definition, the air has been cooled to a temperature where it reaches saturation vapor pressure of 100% relative humidity, and further cooling ought to produce condensation, the change of state of water from vapor to liquid molecules.

Cooling below the dew point does not necessarily cause condensation in air containing few aerosols. Instead, a condition of supersaturation may exist until the air reaches a temperature of –40°C (–40°F). At that temperature water vapor experiences deposition, changing directly into ice crystals. Sublimation refers to a state change in ice directly to gas, but the term deposition applies only to the change from vapor to ice. Neither deposited or sublimated molecules pass through the liquid stage. (Note: deposition may occur at any temperatures below freezing.)

At warmer temperatures, and if a sufficient number of small aerosols are present in the air, condensation will take place on the surfaces of the particles once saturation vapor pressure is reached. In fact, if certain types of particles, like salt, are present, condensation will begin to occur as soon as approximaely 80% relative humidity is attained. Such particles are referred to as either condensation nuclei or hygroscopic nuclei because of their affinity for water. The affinity of sodium chloride for water is readily apparent to anyone who has ever tried to shake damp salt from a shaker on a humid day.

When water vapor condenses in the atmosphere, it becomes visible because of the light it scatters. Haze, clouds and fog are the most obvious examples of water vapor condensation in the atmosphere, and four basic forms of condensation are dew, frost, fog and clouds.

Dew

At night rapid radiation heat loss occurs at the earth's surface, and objects at the surface grow cold faster than the air. The air near these surfaces may lose heat to the surface by conduction and radiation. As the temperature of the air gets low in the immediate vicinity of the surface, the saturation vapor pressure can be attained (100% relative humidity) and condensation may occur on the surfaces of objects. The resulting water is called dew. Dew is most likely to occur when calm air with high relative humidity is cooled. If air motion is rapid it will, however, cause mixing and prevent the temperature stratification necessary for condensation.

Frost

Frost is the crystalline structure of ice which is most readily observed on flat surfaces like glass and metal. When temperatures drop below freezing, atmospheric water vapor is deposited and delicate ice crystals grow. Frost is not frozen dew. If it were, it would not form the crystalline structure. When sublimation continues for long periods, crystals may grow perpendicular to the surface producing a "fuzzy" look. This condition, called hoar frost, occurs only when saturated air and cold conditions prevail for extended times.

Fog

Weather generates emotional responses in people that range from romanticism to fear. Perhaps no single phenomenon illustrates this better than fog. Poets take delight in describing its strange, silent beauty, airline pilots and ship captains dread it; and duck hunters love it. Fog is a deadly menace to transportation and is directly responsible for hundreds of lives and millions of dollars lost each year in airline, automobile, and ship collisions. Additional millions in lost revenue and overhead costs occur when airports must be closed to regularly scheduled traffic.

Fog may be composed of water droplets or, when the temperature is below freezing, ice crystals may form. The former is called warm fog, the latter is called ice fog. Fog develops near the ground when air with high humidity is cooled below its dew point temperature. Cooling may occur in several ways. Rapid radiation heat loss on a clear night may reduce air temperatures near ground level drastically. As the earth grows cooler the air within several feet of ground level radiates heat to the cool ground as well as to space, thereby lowering its temperature more than the air above it. Widespread ground or radiation fog occurs when the air is humid.

Warm air may be cooled by passing over a cold surface. This is a common type of fog along coasts where cold ocean currents are found. Warm humid air from the ocean condenses as it passes over cool currents and moves inland as fog. The same thing can occur when warm, moist air passes over cold ground in late winter or early spring, and the resulting fog, called advection fog, normally covers large areas.

Sometimes warm air will be cooled when it comes into contact with cold air along a weather front. These fogs are usually confined to the vicinity of the front. Fogs are also common in valleys where denser cold air drains down-slope displacing warmer, lighter air. Condensation occurs as some of the warm air is cooled, and pockets of fog collect in low places.

Clouds

So far our discussion on humidity has dealt with inputs, storage, and changes in the state of water within the atmosphere. The atmosphere also outputs water to the earth as precipitation. The key element in the process is some triggering mechanism to set the air in upward motion. Rising air cools primarily by expansion (adiabatic cooling). When moist rising air is sufficiently cooled condensation occurs to form clouds.

There are four primary mechanisms by which air is subject to lifting. They are illustrated in Figure 8.4 and consist of:

1. Convection. When air at the earth's surface is heated in one area, more than the adjacent air, it becomes more buoyant than the surrounding air and tends to rise. The rising air is replaced by cooler air that moves in from adjacent areas.

2. Frontal Lifting. When dense air (often cold) and less dense air (usually warmer) converge, they meet along a contact zone called a front. The merging of air along the front causes the more buoyant air to lift over the denser air that hugs the ground.

3. Orographic Lifting. When airflow is directed perpendicular to a physical barrier such as a mountain range, it is forced to rise upward along the windward slope of the mountains.

4. Convergence. The flowing together of air is called convergence and results in air ascent. There are two primary forms of convergence. One form occurs when sea-level lows, such as

Figure 8.4 Primary Mechanisms That Cause Air to Lift

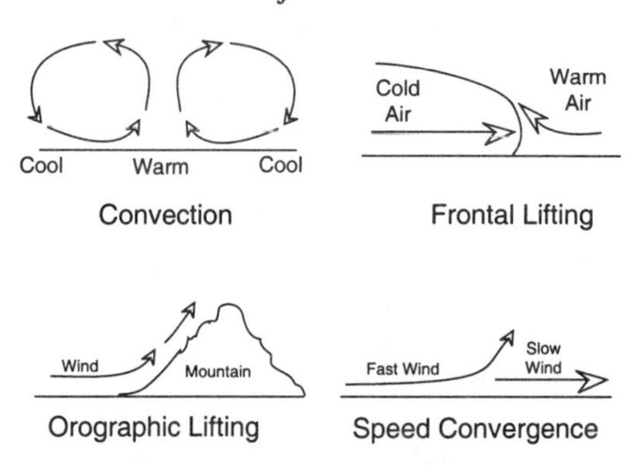

the Intertropical Convergence Zone (ITCZ) lie between regions of high pressure that cause air to merge. Another form is called speed convergence and occurs when high velocity winds flow into a region of low velocity winds.

Precipitation

Precipitation may occur in several forms, most of which are familiar to the average person. The most common forms are rain, snow, hail, and sleet. All precipitation is initially formed as either rain or snow depending upon cloud temperature, availability of nuclei, and rate of cooling. High clouds are composed of ice crystals formed by deposition when very cold temperatures prevail. Liquid water can exist, however, down to temperatures of – 40°C (–40°F). Below that temperature, deposition to ice crystals occurs.

Clouds alone, however, do not precipitation make, as any drought-stricken farmer who hopefully scans the sky can attest. Water droplets and ice crystals in clouds are small enough to remain suspended in the air. How then do raindrops form? It was initially supposed that, when clouds became sufficiently laden with water droplets, collisions between droplets would cause them to grow. As the larger droplets grew they would become too large to be supported by the air and would fall, colliding with other particles and coalescing into larger droplets. This coalescence theory might amply explain the formation of raindrops in warm tropical areas where condensation produces abundant cloud moisture in the form of droplets, but not in the mid-latitudes where water vapor is deposited as ice.

In the mid-latitudes cloud temperatures are low enough that much of the phase change from water vapor occurs as ice crystals. The vapor pressure over ice crystal surfaces is less than the vapor pressure over water droplet surfaces. It is possible for the vapor pressure in the cloud to be greater than the vapor pressure over ice surfaces and less than the vapor pressure over water droplets. Under this circumstance the water droplets evaporate while moisture is deposited on ice crystals causing them to grow. As the crystals get heavy and fall to warmer altitudes, they melt, and the resulting droplet may continue to grow by coalescence. This latter theory of raindrop formation is called the Bergeron theory after the Norwegian scientist who first proposed it. It is important because it provides the theoretical basis for rainfall augmentation.

People have tried to make rain since prehistoric times. The attempts were not confined to witch doctors and medicine men of so-called primitive societies. Even the U.S. Cavalry tried rainmaking in the Great Plains during the 19th century by firing large cannons and exploding dynamite to try to shake moisture from the clouds. All in all the U.S. government spent several tens of thousands of dollars on these attempts which were based on the mistaken belief that rain commonly followed battles.

Starbuck, the rainmaker in the Richard Nash play, "The Rainmaker," explained the mechanics of rainmaking in more sophisticated terms when asked how he could bring rain, "How? Sodium Chloride! Pitch it up high — right up to the clouds. Electrify the cold front. Neutralize the warm front. Barometricize the tropopause. Magnetize occlusions in the sky." Successful rainmaking had to await a more sound theoretical basis, however.

After Bergeron proposed his theory of raindrop formation, scientists hypothesized that rain could be produced by charging super-cooled clouds with dry ice or hygroscopic nuclei like silver iodide. The formation and subsequent growth of ice crystals at the expense of water droplets should significantly increase the probability of rain. Laboratory experiments soon confirmed the theory. The same process is sometimes effective in dissipating fog which is composed of ice crystals and moisture.

No one can yet create rain in the absence of potential rain clouds. Few atmospheric scientists today, however, doubt that the probability of rainfall can be increased by seeding appropriate clouds. There is still considerable debate over the economics of the process as well as concern with other factors.

The economic question is not a simple direct cost-benefit analysis. One has to consider such intangible effects as long-term ecosystem adjustments that might result from increased precipitation, what segments of society must bear the costs, and who receives the benefits. There may be negative effects with legal consequences — one person's drought-breaking rain might be another's disastrous flood — and even such potential legal issues as "rain-robbery." There is, after all, a limited amount of water vapor in the air at one time and if one group seeds clouds and gets more than they would under normal circumstances, it is conceivable that some other area downwind might be deprived of the precipitation it would otherwise receive. Such possible legal consequences as these temper the enthusiasm with which rainmakers proclaim the success of their efforts.

Distribution of Precipitation

Precipitation patterns over the earth reflect the availability of moisture, the temperature of the earth and atmosphere, and the stability of the atmosphere. The bulk of it falls as rain, but snow falls at high altitudes and in high latitudes during winter. Maximum evaporation occurs in the tropical oceans where the warm air can hold much moisture. Accordingly, the equator is an area of heavy rainfall. The polar regions are dry because the cold polar air can hold little moisture (Figure 8.5).

Precipitation is relatively abundant in the mid-latitudes, especially on windward coasts and on the windward sides of mountains. Here the circulation of warm moist air from tropical and subtropical areas provides the source for moisture (Figure 8.6).

Figure 8.5 Precipitation Distribution by Latitude

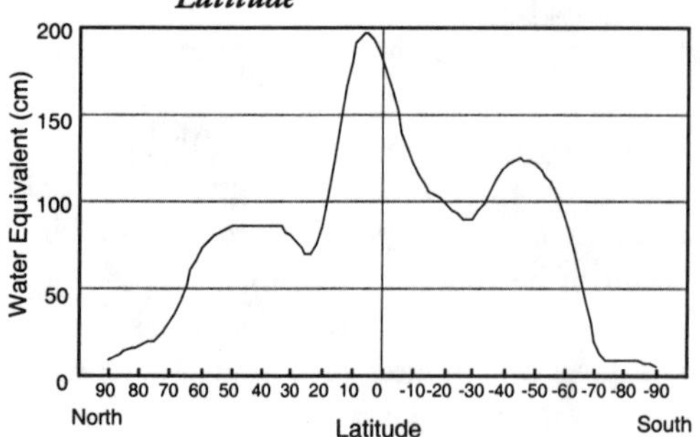

A band of very low precipitation encircles the earth at approximately 25° to 30° north and south of the equator. This is a zone of divergence and descending dry air. In addition, precipitation amounts decline toward the interiors of the largest continents, and amounts are low on the leeward sides of mountains. In the latter case, descending air is warmed and its relative humidity is lowered by the combined effects of heating and the loss of moisture through precipitation on the windward side.

Summary

Moisture continually cycles through the atmosphere. Evapotranspiration supplies moisture to the air. The rate depends on the vapor pressure gradient which is a function of water temperature, air temperature, relative humidity, and wind speed.

Moist air cooled below the actual vapor pressure (100 percent relative humidity) condenses when hygroscopic particles are present. If sufficient condensation occurs, water droplets or ice crystals fall as precipitation. Adiabatic cooling is the main cooling mechanism. Adiabatic cooling takes place when air is lifted orographically, by convection, or through surface convergence along fronts.

The resulting precipitation is concentrated along the equator and along the eastern portion of continents in the mid-latitudes. Heavy precipitation also occurs on windward coasts and on the windward sides of mountains. Sparse precipitation occurs around the earth at 25° to 30° North and South latitude and in continental interiors. The rain shadow effect produces low precipitation along the leeward sides of mountains.

Figure 8.6 Global Distribution of Average Annual Precipitation (cm)

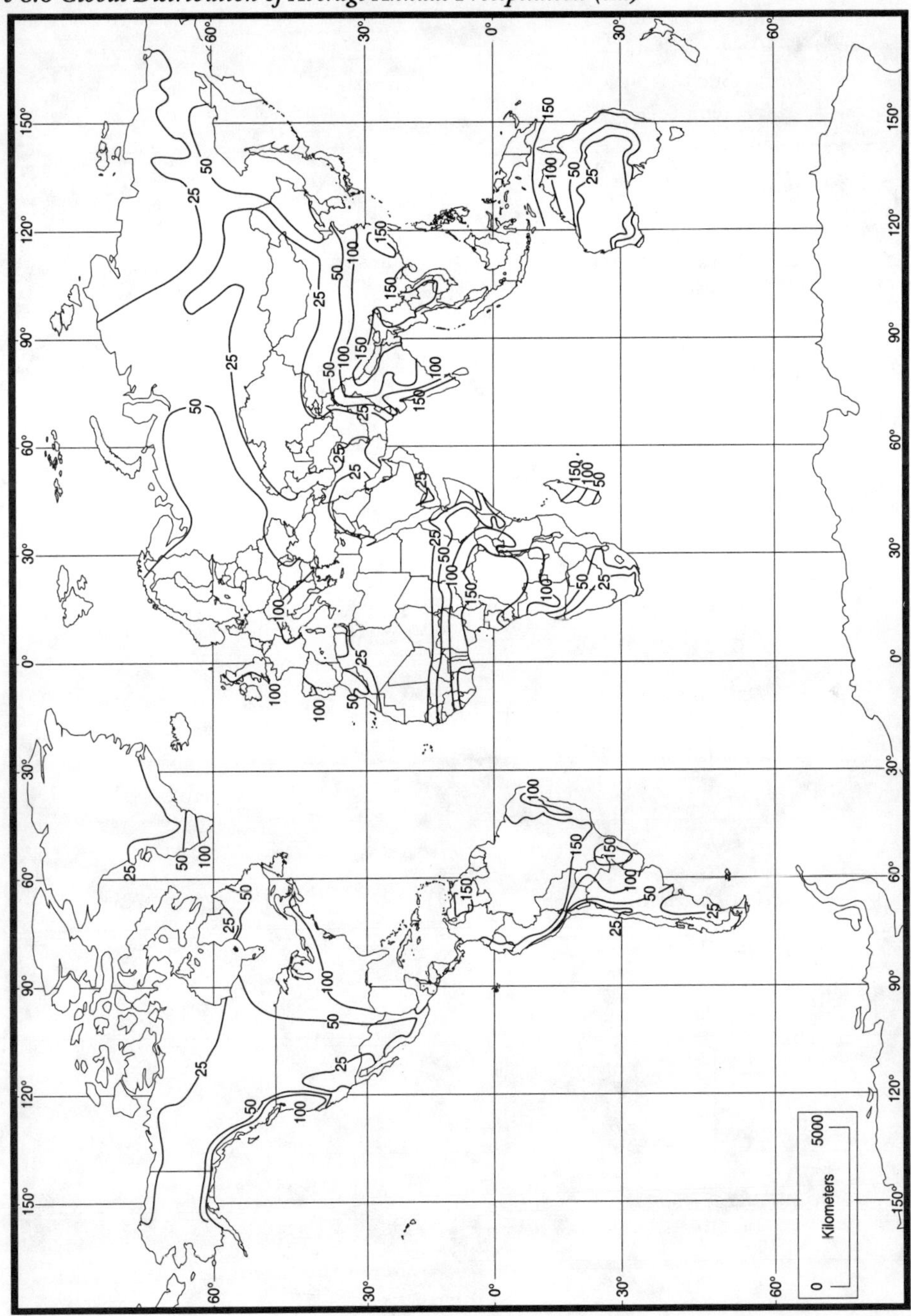

Activity: Moisture

OBJECTIVES

Upon completion of this lab, you will be able to: (1) understand the fundamental relationship between temperature and atmospheric moisture; (2) define saturation, relative humidity, absolute humidity, and dew point; (3) calculate relative humidity; (4) interpret the Temperature-Humidity Index; and (5) determine relative humidity and dew point temperature using a sling psychrometer.

INTRODUCTION

Water in its gaseous state (water vapor) is the most variable gas within the atmosphere. Atmospheric moisture fluctuates markedly from place to place, among different air masses, from season to season, and from time to time through the course of a single day. The water vapor content of the atmosphere is a major controlling factor in precipitation processes and represents the most dynamic moisture exchange within the hydrologic cycle. Its presence is evident in the appearance of types and amounts of clouds and precipitation, such as fog, frost, rain, and snow.

Water vapor also contributes significantly to the heating and cooling of our atmosphere via the storage and release of latent heat. Phase changes (the changes in state between ice, liquid, and water vapor) involve loss or gain of enormous amounts of heat. The evaporation of one gram of water at 0° C (32° F) requires 600 calories which are *stored* in a vaporous moisture phase within the atmosphere. The same 600 calories of heat are *released* when condensation of atmospheric moisture occurs. The next time you observe the formation of clouds or precipitation, you can be aware of the more subtle, yet enormous amount of hidden energy transfer that is taking place.

Temperature determines the atmosphere's capacity to hold moisture at any given time and place. Warm air can hold more water vapor than a similar volume of cool air. Air masses originating in tropical source regions are, therefore, likely to contain far greater amounts of moisture than cooler Arctic air masses, which have much more limited capacities due to their lower temperatures. Air masses may be quite dry, or they may be really charged with moisture so that they contain all the water vapor they can hold given their temperature. This latter condition results in the air reaching **saturation,** or holding all the water vapor that its capacity (temperature) will allow.

RELATIVE HUMIDITY AND DEW POINT

The most commonly used measurement of moisture content in the air is **relative humidty**. Relative humidity (RH) is a ratio between the amount of moisture present and the amount of moisture that could be held if the air were saturated (RH=100%). Relative humidity is usually expressed as a percentage.

$$\text{Re}lative\ \ Humidity = \frac{Amount\ \ of\ water\ vapor\ \ air\ \ is\ \ holding}{Amount\ \ of\ \ watervapor\ \ air\ \ can\ \ hold}$$

Another expression of the moisture content of air is the **dew point**. Dew point is the temperature to which a given parcel of air must be cooled without changes in pressure or moisture content for saturation to occur. Maps produced by the National Weather Service routinely record dew point temperatures to indicate humidity observed at weather stations across the country. For example, a weather station reporting an air temperature of 26.6° C (80° F) and a dew point of 26.6° C (80° F) would be experiencing 100% relative humidity and cloudy to overcast skies or fog. Another weather station might show clear skies, a temperature of 26.6° C (80° F) and a dew point of 15.5° C (60° F). At that location the air would have to cool to the dew point or more water vapor would have to be added to bring the air to saturation. As saturation of an air mass occurs or dew points are reached, the possibility of cloud formation and precipitation increases.

HUMIDITY AND HUMAN COMFORT

Relative humidity also affects human beings' sensitivity to heat. When people sweat in humid heat, little cooling can result because there is little evaporation. Sweating itself doesn't cool you; the evaporation of the sweat does. With a high relative humidity, the air is nearly saturated with water vapor. In this weather, body moisture does not readily evaporate into the air; instead, it collects on the skin as beads of perspiration. The reduced evaporational cooling in hot, moist weather makes most people feel hotter than it actually is. For this reason, people often remark that it's not the heat, it's the humidity.

Home air conditioning systems that depend on refrigeration not only lower the air temperature, but the humidity as well. In its passage through the air conditioning unit, air is cooled, thus increasing its relative humidity to the point of saturation. Condensed water vapor (liquid water) must be removed to produce cooler and dehumidified "conditioned" air that is circulated inside the house.

A different type of air conditioning system, the evaporative cooler, is used in areas of high temperatures and low relative humidities like those of the southwestern United States. In these systems, fans blow hot, dry outside air across a large container of water. Evaporation of the water lowers the temperature of the air above the container and allows cooler air to enter the home. These systems, also known as "swamp coolers" add water vapor to indoor air, and consequently reduce the discomforts associated with dry air.

The discomfort associated with warm temperature and high humidities may be statistically evaluated by calculation of the **Temperature-Humidity Index** (THI). This index is frequently featured on daily weather reports during the summer time and is derived from the following formula:

$$THI = T - [0.55(1 - RH)(T - 14)]$$

Where : T = air temperature in degrees C
RH = relative humidity expressed as a decimal (70% RH would be .70)
The terms 0.55, 1 and 14 are constants in the equation.

THI values in excess of 25 indicate that most people will feel uncomfortable. THI values between 15 and 20 are accepted by most as comfortable. Values below 15 might be interpreted as uncomfortably dry. For illustration, assume that weather observations reveal an air temperature of 30° C (86° F) and a relative humidity of 90 percent. The THI is calculated as follows:

$$THI = 30 - [0.55\ (1 - 0.90)\ (30 - 14)]$$
$$THI = 30 - 0.55\ (0.1)\ (16)$$
$$THI = 30 - 0.55\ (1.6)$$
$$THI = 30 - 0.88$$
$$THI = 29.12$$

At the other extreme, really dry environments with low relative humidities create discomforts such as itchy noses and throats, cracked dry skin and headaches. According to Rand-McNally's Places Rates Almanac, the nation's dampest and driest metro areas are:

Damp	RH(%)	Dry	RH (%)
Galveston,TX	78	Las Vegas, NV	29
Alexandria, LA	77	Phoenix, AZ	36
Houston, TX	77	Tucson, AZ	38
New Orleans, LA	77	El Paso, TX	39
Asheville, NC	77	Albuquerque, NM	43
Biloxi-Gulfport, MS	77	Colorado Springs, CO	49
Corpus Christi, TX	77	Reno, NV	50
Eugene-Springfield, OR	77	Charlotte, NC	69

USING A SLING PSYCHROMETER TO MEASURE RELATIVE HUMIDITY

The amount of moisture in the air can be estimated from rates of evaporation. Evaporation is more rigorous where vapor pressure gradients are steep. See Chapter 8 for more about vapor pressure and moisture gradients. A psychrometer is an instrument designed to estimate humidity by measuring the rate of evaporation. It contains two thermometers: one has a conventional dry bulb-reservoir, and the other uses a damp fabric sleeve or wet bulb to measure the amount of cooling from evaporation. The degree to which the wet bulb temperature is depressed (lower than the dry bulb temperature) is proportional to the vapor pressure gradient. The vapor pressure gradient is a function of how much water vapor the air contains. The greater the wet bulb depression, the drier the air. Meteorological tables can be used to convert wet bulb depression into relative humidity. If a psychrometer is available, try this.

1. Examine the dry bulb and wet bulb thermometers on the sling psychrometer. Record the dry bulb reading, time, date and place of observation.

2. Spin the psychrometer for a couple of minutes to evaporate moisture from the wick attached to wet bulb thermometer. Check the wet bulb temperature, then spin the psychrometer for another 30 seconds or so. If no further temperature decrease of the wet bulb is noted, you are ready to record measurements.

3. Record the dry bulb and wet bulb temperatures (T_d and T_w). Convert Fahrenheit readings to Celsius using the formula: C = 0.55 (F-32)

Dry Bulb ____________° C Wet Bulb ____________° C

4. Record the wet bulb depression ($T_d - T_w$) or the difference between dry bulb and wet bulb readings. Refer to RH tables, tracing dry bulb temperature to wet bulb depression to record RH. Use the Dew Point table in a similar manner to record dew point.

RH ________% DP ______° C

5. Check your results with the lab instructor. Assuming that you've received the lab instructor's blessing, you are now free to go outdoors and make actual observations of temperature and relative humidity.

OBSERVATION OF LOCAL WEATHER CONDITIONS

Now make the following observations outside: one above paved surface and one above a vegetated surface.

1. Record the time, date, and place of observation on the chart below. Observe sky conditions. What is the sky cover in tenths (ex. overcast 10/10; partly cloudy 5/10)? What types of clouds are present? What is the air temperature (record from dry bulb thermometer)?

Place	Time & Date	Sky Cover	Cloud Types*
Paved Surface			
Vegetated Surface			

*Cloud types:
High (> 7 km)cirrus (Ci); cirrostratus (Cs); cirrocumulus (Cc); cumulonimbus (Cb)
Middle (2-7 km)altocumulus (Ac); altostratus (As); cumulonimbus (Cb
Low (< 2 km)stratus (St); stratocumulus (Sc); nimbostratus (Ns); cumulonimbus (Cb

2. Formulate an hypothesis based on observed weather conditions. For example, if it is cloudy will relative humidity be high, or if it is clear, will relative humidity be lower? Assumptions also may also be based on how the air *feels*. Is it muggy and warm or cool and dry? Do you think RH will be higher or lower than the annual average of 69% in Charlotte, NC? How do your observations compare with those taken by a group of students during the summer of 1986? Their observations are summarized below:

	TIME	TEMP	R.H.	Sky Conditions
Paved Surface	11 a.m.	85° F	65%	clear-scattered (0-4/10)
	3 p.m.	93° F	46%	scattered (2-4/10)
Vegetated Surface	11 a.m.	82° F	67%	clear-scattered (04/10)
	3 p.m.	90° F	54%	scattered (2-4/10)

3. Statement of tentative hypothesis:

4. Test your hypothesis by measuring RH following the same procedure as the indoor experiment. Record time and date and wet and dry bulb readings for each place of observation.

	Paved Surface	Vegetated Surface
Time & Date:		
Dry Bulb Reading:		
Wet Bulb Reading:		
Relative Humidity:		
Dew Point:		

5. Accept or reject your hypothesis on the basis of observed measurement. How do your observations compare with your expected values?

6. Summarize the observed relationships between weather (clouds and temperature) and RH.

7. Assuming that temperature and dew point are different, theorize on ways that temperature and dew point could become equal, resulting in saturation of air at the place of observation.

8. Calculate the THI for either of your outdoor observations. What do you conclude about the comfort/discomfort indicated by the resultant THI?

Sources include: Ahrens, C. D. *Meteorology Today* (2nd ed.), St. Paul: West Publishing Company, 1985; Boyer, R. and Savageau *Places Rated Almanac* (2nd ed.), Chicago: Rand-McNally and Co., 1985; Gedzelman, S. D. *The Science and Wonders of the Atmosphere*, New York: John Wiley and Sons, 1970 and Lutgens, F. K. and Tarbuck, E. J. *The Atmosphere*, Englewood Cliffs: Prentice-Hall, 1979.

Dew Point Temperature (1,000 mb)

Dry-bulb temp ⇓ / Wet-bulb depression (T_d-T_w) ⇔

C°	1	2	3	4	5	6	7	8	9	10	11	12	13	14	15	16	17	18	19	20	21	22
-20	-33																					
-18	-28																					
-16	-24																					
-14	-21	-36																				
-12	-18	-28																				
-10	-14	-22																				
-8	-12	-18	-29																			
-6	-10	-14	-22																			
-4	-7	-11	-17	-29																		
-2	-5	-8	-13	-20																		
0	-3	-6	-9	-15	-24																	
2	-1	-3	-6	-11	-17																	
4	1	-1	-4	-7	-11	-19																
6	4	1	-1	-4	-7	-13	-21															
8	6	3	1	-2	-5	-9	-14															
10	8	6	4	1	-2	-5	-9	-14	-28													
12	10	8	6	4	1	-2	-5	-9	-16													
13	11	9	7	5	2	0	-3	-7	-13													
14	12	11	9	6	4	1	-2	-5	-10	-17												
16	14	13	11	9	7	4	1	-1	-6	-10	-17											
17	15	14	12	10	8	6	2	1	-4	-7	-14											
18	16	15	13	11	9	7	4	2	-2	-5	-10	-19										
20	19	17	15	14	12	10	7	4	2	-2	-5	-10	-19									
22	21	19	17	16	14	12	10	8	5	3	-1	-5	-10	-19								
24	23	21	20	18	16	14	12	10	8	6	2	-1	-5	-10	-18							
26	25	23	22	20	18	17	15	13	11	9	6	3	0	-4	-9	-18						
28	27	25	24	22	21	19	17	16	14	11	9	7	4	1	-3	-9	-16					
30	29	27	26	24	23	21	19	18	16	14	12	10	8	5	1	-2	-8	-15				
32	31	29	28	27	25	24	22	21	19	17	15	13	11	8	5	2	-2	-7	-14			
34	33	31	30	29	27	26	24	23	21	20	18	16	14	12	9	6	3	-1	-5	-12	-29	
36	35	33	32	31	29	28	27	25	24	22	20	19	17	15	13	10	7	4	0	-4	-10	
38	37	35	34	33	32	30	29	28	26	25	23	21	19	17	15	13	11	8	5	1	-3	-9
40	39	37	36	35	34	32	31	30	28	27	25	24	22	20	18	16	14	12	9	6	2	-2

Relative Humidity in Percent (1,000 mb)

Dry-bulb temp. ⇓ Wet-bulb depression ($T_d - T_w$) ⇔

C°	1	2	3	4	5	6	7	8	9	10	11	12	13	14	15	16	17	18	19	20	21	22
-20	28																					
-18	40																					
-16	48	0																				
-14	55	11																				
-12	61	23																				
-10	66	33	0																			
-8	71	41	13																			
-6	73	48	20	0																		
-4	77	54	32	11																		
-2	79	58	37	20	1																	
0	81	63	45	28	11																	
2	83	67	51	36	20	6																
4	85	70	56	42	27	14																
6	86	72	59	46	35	22	10	0														
8	87	74	62	51	39	28	17	6														
10	88	76	65	54	43	33	24	13	4													
12	88	78	67	57	48	38	28	19	10	2												
13	89	78	68	59	49	40	30	22	13	5												
14	89	79	69	60	50	41	33	25	16	8	1											
16	90	80	71	62	54	45	37	29	21	14	7	1										
17	90	80	71	63	55	46	39	31	23	17	10	4										
18	91	81	72	64	56	48	40	33	26	19	12	6	0									
20	91	82	74	66	58	51	44	36	30	23	17	11	5	0								
22	92	83	75	68	60	53	46	40	33	27	21	15	10	4	0							
24	92	84	76	69	62	55	49	42	36	30	25	20	14	9	4	0						
26	92	85	77	70	64	57	51	45	39	34	28	23	18	12	9	5						
28	93	86	78	71	65	59	53	47	42	36	31	26	21	16	12	8	4					
30	93	86	79	72	66	61	55	49	44	39	34	29	25	19	16	12	8	4				
32	93	86	80	73	68	62	56	55	46	41	36	32	27	22	19	14	11	8	4			
34	93	86	81	74	69	63	58	52	48	43	38	34	30	26	22	18	14	11	8	5		
36	94	87	81	75	69	64	59	54	50	44	40	36	32	28	24	21	17	13	10	7	4	
38	94	87	82	76	70	66	60	55	51	46	42	38	34	30	26	23	20	16	13	10	7	5
40	94	89	82	76	71	67	61	57	52	48	44	40	36	33	29	25	22	19	16	13	10	7

Activity: The Water Balance

OBJECTIVE

Given monthly values for precipitation and potential evapotranspiration, you will understand the seasonal interrelationship between soil moisture deficit, surplus, and runoff by balancing precipitation, evapotranspiration, and soil moisture storage. You will be able to identify several major surface hydrologic features of North Carolina, South Carolina, and Georgia.

INTRODUCTION

For any given land area of the earth's surface, there is a balance over the years between the amount of water supplied to the area by precipitation (e.g., rain, snow) and the amount leaving the area by evapotranspiration and runoff. Evapotranspiration is the transfer of water vapor into the atmosphere through direct evaporation or transpiration through plant surfaces. Runoff is flow of surface water either as sheet flow (e.g. across pavement) or as channelized flow in streams and rivers. This balance must hold true over a long period of time. If it did not, the area would become progressively wetter or drier. However, a balance does not always hold true over short periods of time such as days, weeks, or months.

It is useful, therefore, to be able to develop and maintain, for any given place, a systematic record of the water balance account. The most widely accepted method for water budget accounting was devised by C. W. Thornthwaite and his associates. Water budgets are presently used to solve a wide variety of environmental problems including flood forecasting, establishing irrigation schedules, understanding plant growth, and managing water resources.

Before using the Thornthwaite method to analyze a local water budget, let's first review the global hydrologic cycle. Refer to Figure 1 and notice the volume of flow between land areas and oceans. It is important to realize that there is a finite quantity of water on the earth and that water is cycled through each reservoir over and over again. Water enters the atmosphere by evaporation from the oceans, lakes, rivers, and land surfaces. Water also reenters the atmosphere by transpiration from plants. In a state such as North Carolina, which has a naturally heavy vegetation cover, transpiration through plants accounts for approximately 80% of the moisture returned to the atmosphere from the ground.

The ascent of moist air results in sufficient adiabatic cooling to cause condensation. If the right conditions exist, precipitation will occur and water will return to the earth's surface. Some of this water reenters the air again by evaporation or transpiration.

Some of the water may enter the area of permanent saturation beneath the earth's surface. The surface of this groundwater is known as the water table. Soil moisture and groundwater, under the influence of gravity, flow beneath the surface and may eventually enter streams, rivers, or oceans. Water that does not soak through the soil enters rivers and oceans by surface runoff.

MATERIALS

*A*tlas or globe, colored pencils, and a calculator

THE GLOBAL WATER BUDGET

The volume of mean annual global precipitation is estimated to be 532,468 km³. Use this parameter to calculate the volume of flow as a percentage between each of the following reservoirs shown in Figure 1.

Mean Annual Volume of Flow	Water Transfer
Evaporation from oceans	_________ %
Precipitation on oceans	_________ %
Horizontal advection of wv to ocean	_________ %
Horizontal advection of wv to land	_________ %
Precipitation on land	_________ %
Evapotranspiration from land	_________ %
Surface runoff to ocean	_________ %

THE LOCAL WATER BUDGET

Before calculating a local water budget, it is necessary to understand the terminology that is involved. Look at the water budget below and notice the letters representing the months of the year across the top of the chart. Next, note the letters down the left-hand side of the chart, beginning at the top.

P is precipitation. This represents the average amount of precipitation for a specific month in millimeters.

AE refers to the actual (or true amount of) evapotranspiration that occurred for the month. Ideally, this should be equal to the potential evapotranspiration but may be less if soil moisture storage is depleted.

PE is potential evapotranspiration. This is the maximum amount of water lost through evaporation and transpiration that could occur under given temperature and precipitation conditions.

P – PE is a calculation of gain through precipitation minus potential losses from potential evapotranspiration. This may be a positive or a negative number. If it is positive, it means there is more than enough moisture income to fulfill the evapotranspiration requirements. If the number is negative, it means there is insufficient precipitation to fulfill the moisture needed. This deficiency will have to be supplied by another source.

ST stands for soil moisture storage, or the amount of moisture that can be held in the soil. This amount can be compared with a savings account. When P – PE results in a negative number, the deficiency can be corrected by withdrawal of moisture from this storage. It is imperative, however, *not* to confuse this soil moisture storage with groundwater or the water table. They are not the same. Also, moisture storage capacity varies widely among thin soils with little water holding capacity and thick soils with larger than average capacity. A limit of 150 mm of water is is typical for many soils. Beyond the storage limit, soils become saturated and unable to hold more water.

Δ ST —The symbol Δ is the Greek symbol delta and is used to indicate change. Δ ST, therefore, depicts change in soil moisture storage from the previous month. This can be a positive or negative number.

The letters "**D**" and "**S**" respectively refer to a deficit of water or a surplus of water. A deficit occurs when soil moisture storage is depleted and requirements for PE cannot be met. A surplus occurs when there was more than enough precipitation to satisfy PE requirements, and soil moisture storage is already

Figure 1 The Global Water Balance

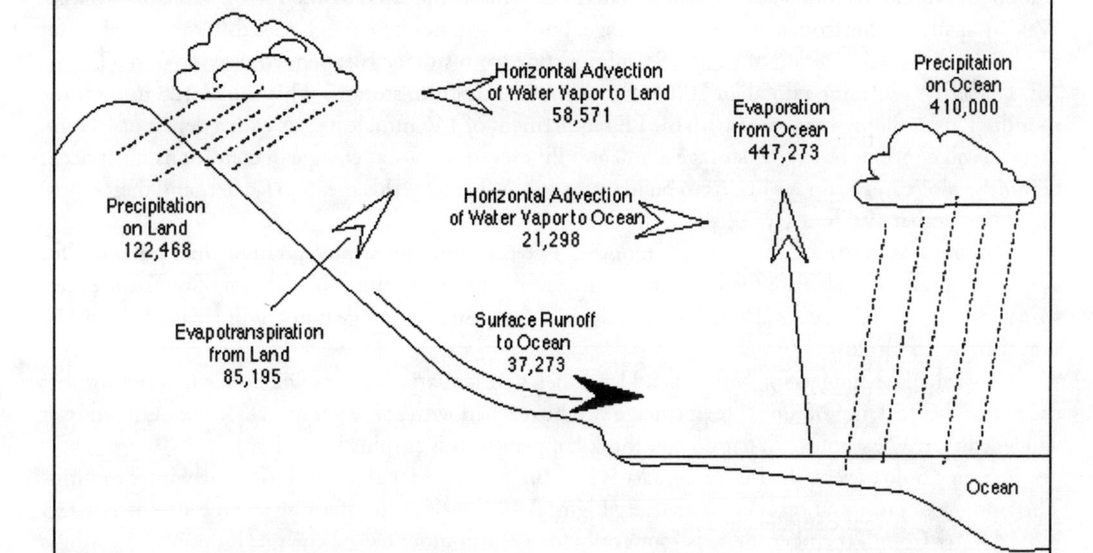

at the maximum amount.

Now, let's calculate a water budget for Charlotte, N.C. Precipitation and potential evapotranspiration amounts have been calculated for you. First calculate all of the P – PE values. Notice that the columns for May and June have been completed and the storage value for April has been included. For problems that you will be solving on your own, a starting place will be provided. Now, let's work through May and June.

The P – PE value for May is -31. This means that all of the precipitation received for that month was evapotranspirated and that there is still a need for 31 mm more. We then examine the storage (ST) for April to see if there is sufficient soil moisture to meet this need. The ST value for April is at a maximum of 150 mm. We can, therefore, borrow the necessary 31 mm from storage to fulfill the potential evapotranspiration requirement. The Δ ST, or change in storage value, then becomes -31 because we took 31 mm away from storage. The amount left in storage becomes 119 mm for May. The AE value is 105 and matches the PE value. This was possible in May because 74 mm available from precipitation and 31 mm available from storage could be combined to fulfill the PE requirements. Notice that there is no deficit or surplus for May. There can be a deficit only when all the water from storage has been removed. There can be a surplus only when soil moisture storage is at the maximum of 150 mm.

June is another potentially dry month because PE exceeded P by 48 mm. Again we are able to borrow this 48 mm from ST. We record this loan with a Δ ST of -48 mm and our balance in ST is now 71 mm. AE = PE since we have again met the required 142 mm. Again there is no deficit or surplus. Complete the column under July on your own.

August appears to be another dry month in which the temperatures are high, but with only a small amount of rainfall to compensate for the elevated PE. This is indicated by the P – PE value of –46 mm. We may again borrow from soil moisture storage, but this will not be enough to satisfy the PE value, for the soil contains only 28 mm of water. So, all moisture from storage is expended, leaving 0 in ST. The AE value is the incoming rainfall of 100 mm plus the 28 mm from storage. This equals 128 mm for the month. This 128 mm, compared with the PE requirement of 146 mm, leaves us with a deficit of 18 mm. Deficits will continue as long as storage is at 0 and PE exceeds P. What change in conditions must occur before the moisture in storage begins to build up again? What does this mean to farmers and their crops? Fill in September's values.

October has, for the first time in 5 months, a greater amount of precipitation than potential for evapotranspiration. Calculate the values for October. P – PE = 10 mm. This 10 mm of moisture enters storage and the Δ ST value is also +10 mm. AE = PE. Because soil storage is not at the maximum of 150 mm there is no surplus.

Now calculate budgets for November, December, and January. January will be the first month that the maximum 150 mm of moisture in storage will be reached with some left over. The amount left over will be entered as a surplus. What do you think happens to this surplus in reality?

As you would expect, the PE for January is low due to cold temperatures during the winter months. The amount of precipitation is comparatively high. P – PE is 81 mm. Because storage already has 133 mm of water from December, there is room only for 17 mm more before the maximum of 150 mm is reached. This determines the Δ ST value of 17 mm and the ST value of 150 mm. The remainder of the excess precipitation appears as a 64 mm surplus.

February, March, and April parallel the kind of situation shown in January. Storage is at a maximum so there will be surpluses of moisture. Now complete Table 1.

Table 1 Water Budget for Charlotte, NC

Storage Maximum =150 mm All values in millimeters

	JAN	FEB	MAR	APR	MAY	JUN	JUL	AUG	SEP	OCT	NOV	DEC
P	89	97	115	86	74	94	116	100	88	68	70	87
PE	8	10	28	65	105	142	159	146	102	58	26	8
P - PE	___	___	___	___	-31	-48	___	___	___	___	___	___
ST	___	___	___	150	119	71	___	___	___	___	___	___
Δ ST	___	___	___	___	-31	-48	___	___	___	___	___	___
AE	___	___	___	___	105	142	___	___	___	___	___	___
D	___	___	___	___	0	0	___	___	___	___	___	___
S	___	___	___	___	0	0	___	___	___	___	___	___

P = Precipitation
PE = Potential Evapotranspiration
P - PE = Precipitation - Potential Evapotranspiration
ST = Soil Moisture Storage
Δ ST = Change in Storage
AE = Actual Evapotranspiration
D = Deficit
S = Surplus

Plot Charlotte's precipitation, potential evapotranspiration, and actual evapotranspiration on the following model. How might soil moisture shortages and surpluses be represented?

___

___

___

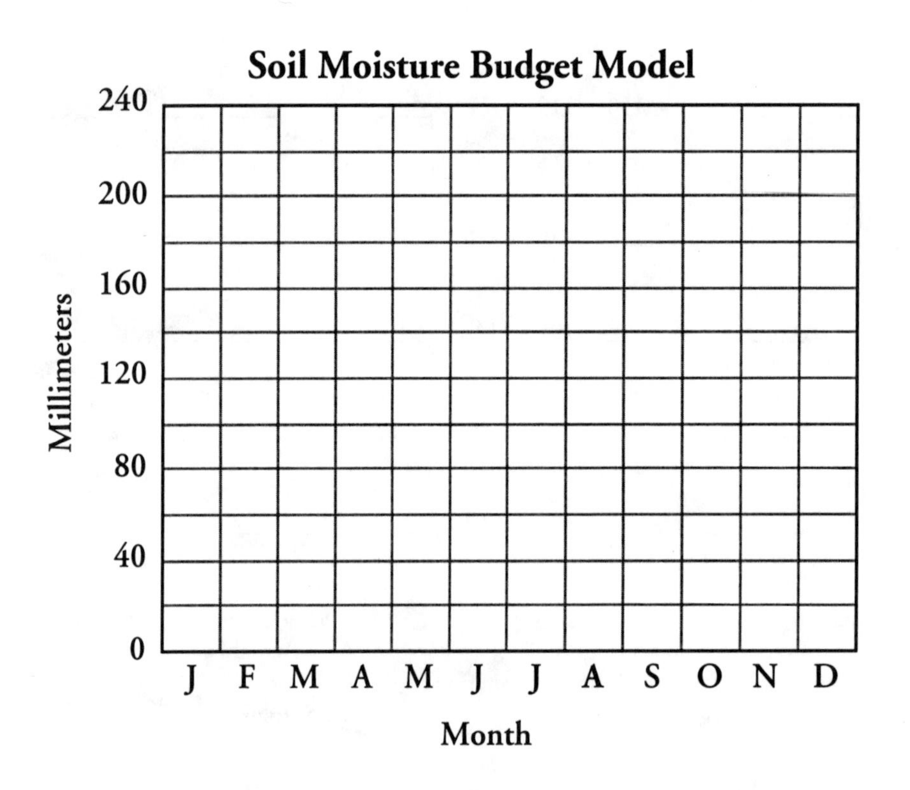

Now see if you can work out the water budget for a location with seasonal aridity.

Table 2 Water Budget for San Francisco
Storage Maximum =150 mm All values in millimeters

	JAN	FEB	MAR	APR	MAY	JUN	JUL	AUG	SEP	OCT	NOV	DEC
P	119	94	77	38	2	1	0.0	0.0	8	3	6	11
PE	33	38	50	58	69	79	81	74	71	66	45	33
P - PE	___	___	___	___	___	___	___	___	___	___	___	___
ST	___	___	150	___	___	___	___	___	___	___	___	___
Δ ST	___	___	___	___	___	___	___	___	___	___	___	___
AE	___	___	___	___	___	___	___	___	___	___	___	___
D	___	___	___	___	___	___	___	___	___	___	___	___
S	___	___	___	___	___	___	___	___	___	___	___	___

P = Precipitation
PE = Potential Evapotranspiration
P - PE = Precipitation - Potential Evapotranspiration
ST = Soil Moisture Storage
AE = Actual Evapotranspiration
D = Deficit
S = Surplus

Plot San Francisco's precipitation, potential evapotranspiration, and actual evapotranspiration on the following model. Identify soil moisture deficits by shading areas where PE > P and soil moisture has been depleted.

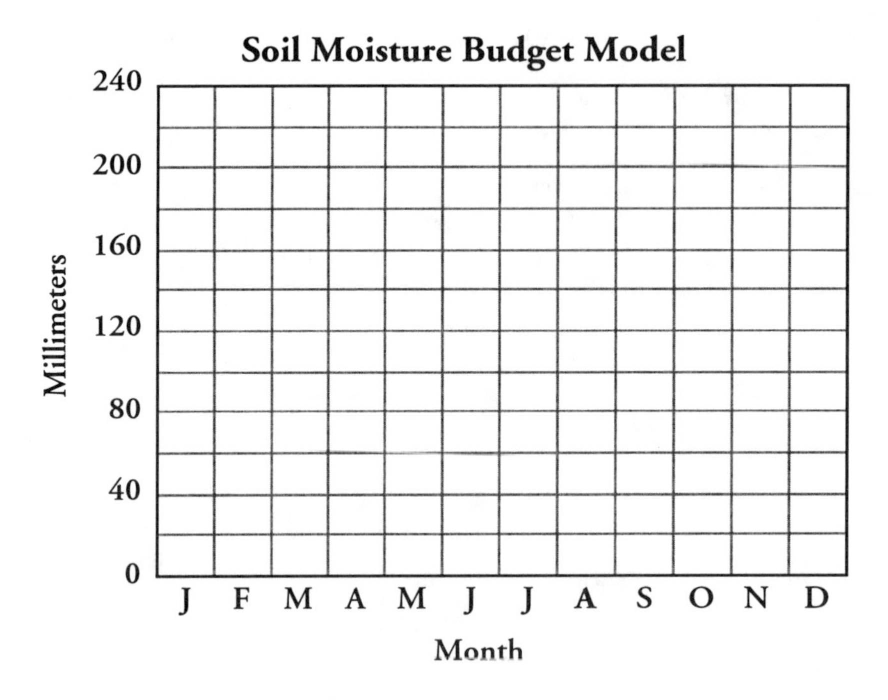

SURFACE RUNOFF

Surpluses from the water budget feed all of the streams, lakes, and reservoirs common throughout the southeastern United States. Water is stored in all of these hydrologic features and represents a valuable resource for industry, agriculture, and residential users. Most of the general population have a relatively shallow awareness of the fresh water resources within their local region. Charlotte for example draws most of its water supply from Mountain Island Lake, yet few would be able to identify the location of that reservoir. To gain a better understanding of the spatial pattern of our local water resources see if you can identify, circle, and label the following hydrologic features (use an atlas to assist you). A template legend has been provided on the stream channel map of North Carolina, South Carolina, and Georgia. Label each of the reservoirs and lakes. Trace the perimeter of each riverine drainage basin and lightly shade with a colored pencil or with a pattern. With labels, or with the use of patterns or colors, or both, identify the location of each feature on the map and in the legend.

Mountain Island Lake	Lake Norman	Cape Fear River	Lake Wylie
Neuse River	High Rock Lake	Catawba River	Pee Dee River
Chowan River	Albemarle Sound	Pamlico Sound	Fontana Lake
Yadkin River	Tar River	Hartwell Reservoir	Lake Sidney Lanier
Santee River	Savannah River		

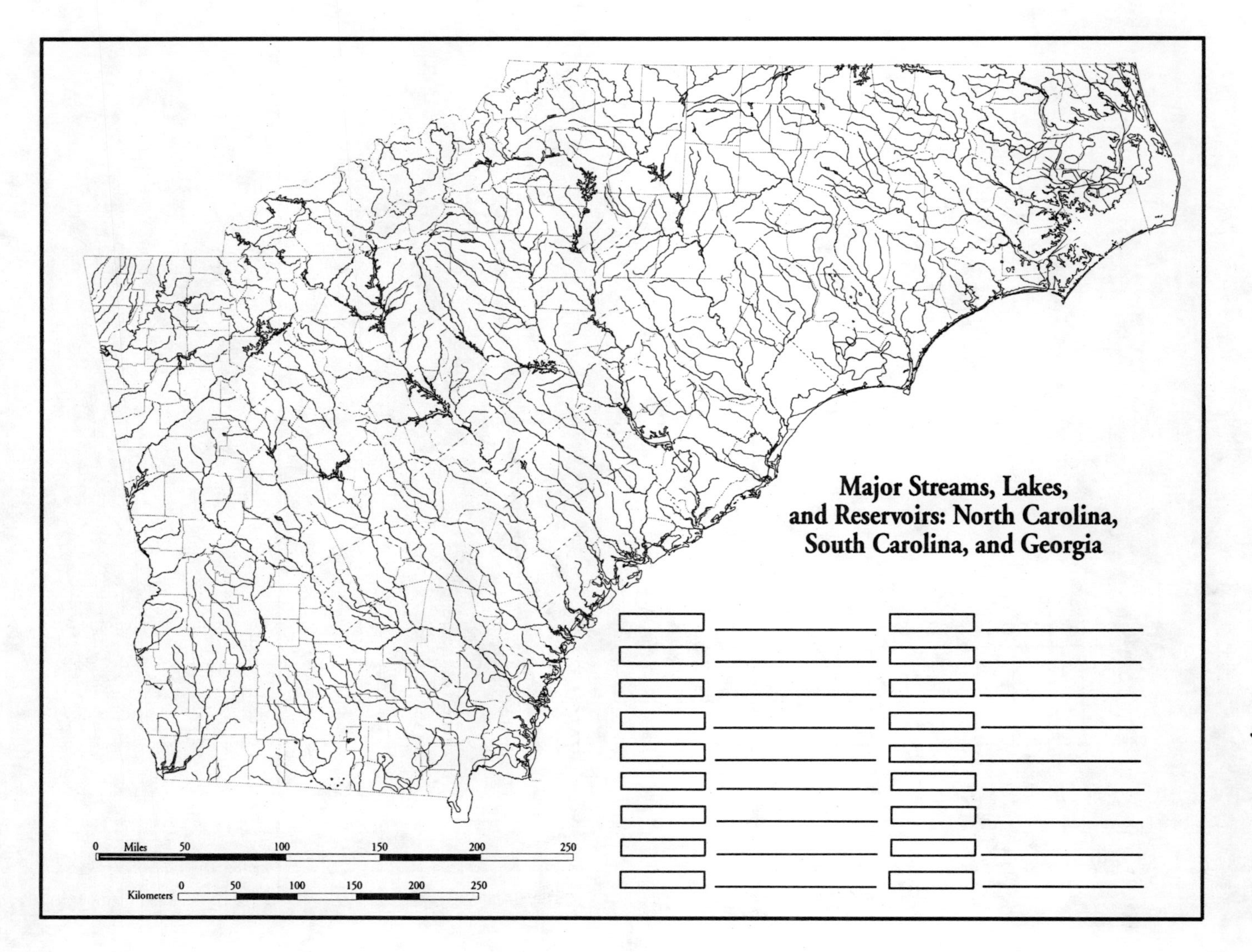
Major Streams, Lakes, and Reservoirs: North Carolina, South Carolina, and Georgia
Miles
0
50
100
150
200
250
Kilometers
0
50
100
150
200
250

Chapter 9

Atmospheric Circulation

Introduction

Even the most casual observer of the atmosphere quickly recognizes that it is **ambient** (constantly in motion and encompassing all substances it contacts). From micro-scale motions that occur within pore spaces located between soil particles to macro-scale motions that accelerate continental-sized bodies of air over thousands of kilometers, the forces that generate airflow are essentially the same, differing only in magnitude.

In the previous chapter, special attention was devoted to explaining temperature-induced pressure changes that set up forces to accelerate air. Let's remember, however, that temperature variation is related to differences in energy budgets: whether a global region is one of surplus or deficit energy. Patterns of air movement cannot be understood fully without referring to all atmospheric parameters.

Numerous theorists have proposed airflow models. Each has been limited by the amount of atmospheric data available to the investigator. It is worth pointing out that meteorology, as a recognized subject meriting scientific inquiry, is a youthful science. The professional legitimacy of this field of study is largely confined to the present century. Thus, it should seem logical that theories related to atmospheric behavior predating the 1900's were based upon meager data resources and were proposed by individuals having occupations that were not focused on weather phenomena.

George Hadley, in 1735, devised a simplistic air circulation scheme that is illustrated in Figure 9.1. Hadley's theoretical model was based upon known facts of the historical period in which he lived. Although today his model is considered naive, we should keep in mind that he did not have the wealth of data resources available today. He did, however, explain the mechanism for **convection** (transfer of heat by a moving body) in the atmosphere. The Hadley circulation model is based upon simple assumptions described in the next paragraph.

Figure 9.1 George Hadley's Air Circulation Model (ca. 1785)

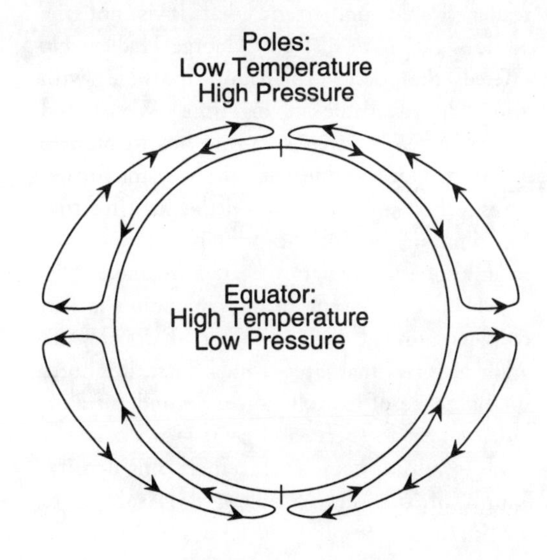

Pressure differences drive winds, and pressure differences arise, in large part, from temperature differences. The latitudinal distribution of temperatures provides hypotheses about resulting pressure and wind patterns. Surplus energy and high temperatures in the equatorial zone ought to produce low pressures, causing air to lift convectively. Because air is affected by gravity, the lifting air must be deflected poleward at some critical altitude aloft. (Note: If this were not the case, atmospheric gases would continue to ascend into outer-space and be lost from the earth-atmosphere system). In the polar regions, low temperatures prevail and ought to produce high surface pressure and subsiding air that should flow toward the equator and complete a closed circulation system. According to Hadley's model it would appear that surface winds ought to blow from polar regions to the equator, and upper air winds ought to blow from the equator to the poles, upper air pressure systems being generally the opposite of surface pressures.

The simple model just described does not fit reality. Before explaining why it doesn't, the reader should understand that it is not the authors' intent to discredit George Hadley. He offered the best explanation possible with knowledge available in his time. We should always remember that someone has to pioneer an effort. Hadley's ideas led the way for further investigations, and as acknowledgement for that contribution, one of the principal components of atmospheric circulation bears his name.

Hadley's concept of air circulation lacked consideration of factors, other than pressure gradient force, that affect winds. First, it ignores the influence of Coriolis force. Second, air aloft radiates heat to space, and, as it grows colder it tends to subside because of its increasing density. Third, air that moves poleward is subject to longitudinal compression. As we noted in Chapter 3, one degree of longitudinal distance at the equator is about 112 km (69 miles), but steadily decreases in length as latitude increases (Figure 9.2). Thus, poleward moving air is forced into smaller space, and molecular density increases.

Mechanics Of Planetary Winds

In the previous chapter we found that the direction and speed of wind are functions of pressure-gradient, Coriolis, and frictional forces. Some clarification and additional perspectives on these forces are in order. First, it is necessary to understand that even calm air at the surface, i.e., having no perceptible horizontal motion, is still traveling at a speed proportional to the rotational speed of the earth. In other words, the

Figure 9.2 Longitudinal Distance Decreases Poleward as a Function of Latitude. To find the number of kilometers in 1° of longitude on the earth's surface, multiple 111.31 by the cosine of the latitude.

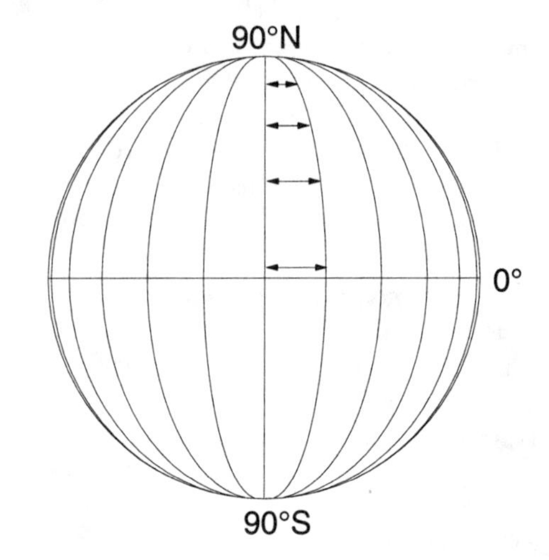

atmosphere always has **momentum**.[1] If the earth could suddenly be stopped, the atmosphere's momentum would keep it moving until an opposing force, such as frictional drag, eventually slowed it to the stopping point. Fortunately, this isn't likely to happen. What is important about this scenario, however, is that air momentum exists, regardless of the presence of PGF, cf, and frictional force. To illustrate this point, refer to Figure 9.3 where the circumferential distances along selected parallels of latitude are given. Because a point along a given parallel must complete a full 360° rotation in approximately 24 hours, the distance travelled is a function of latitude and is equal to the length (circumference) of the parallel. Speed is distance divided by time, thus speed of rotation is fastest at the equator and decreases poleward. This results in a difference in momentum relative to latitude, a factor that can be mathematically expressed as the product of three factors:

$$A_m = mvr$$

[1]*Momentum equals mass times linear velocity.*

where A_m is angular momentum,
m is mass of the air,
v = velocity of the air,
and r = distance of the air from its axis of rotation.

Don't let the above equation confuse you. Instead, let's simplify facts we already know. Assume, for the present, the mass of air (m) at a given latitude is constant. We just discovered in the previous paragraph that its *velocity is inversely related to latitude.* The only remaining component of this equation is the factor r, that is described as radius in Figure 9.3. Combining what we've learned thus far, low latitudes have high A_m per unit mass of air because of high velocities (v) and greater distances from the axis of rotation (r), compared with air at high latitudes.

In case you are not sufficiently confused at this point, we will add yet another factor to this discussion (one that has been addressed before), the Coriolis effect. The Coriolis deflection is heavily dependent upon the latitude and velocity of a freely moving object. Its force of acceleration is zero at the equator and increases

Figure 9.3 Circumference, Radius, and Speed of Rotation along Selected Parallels of Latitude

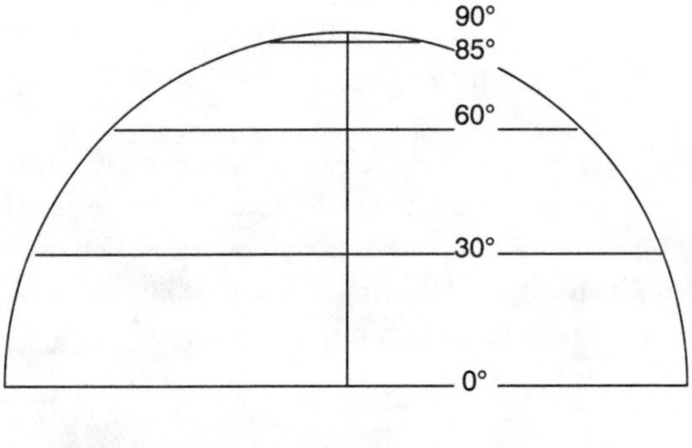

Latitude	Circumference of Selected Parallels km (miles)	Radius km (miles)	Speed of Rotation kph (mph)
90°	0.0 (0.0)	0.0 (0.0)	0.0 (0.0)
85°	3,504 (2,178)	560 (347)	146 (91)
60°	20,084 (12,481)	3,197 (1,986)	837 (520)
30°	34,729 (21,584)	5,528 (3,435)	1,447 (899)
0°	40,064 (24,064)	6,377 (3,963)	1,669 (1,003)

toward higher latitudes, as is graphically summarized in Figure 9.4. Remember that deflection is to the right in the northern hemisphere and to the left in the southern hemisphere. Finally, before returning to circulation patterns, we would like to point out the law of conservation of angular momentum, which states that the angular momentum (A_m) of a unit mass of air at a given time must remain the same at some future time. Therefore, 1 kilogram of air at 85° latitude with a wind speed of 5 meter/sec would have angular momentum of:

A_m = mvr
A_m = (1 kg) (5m/sec) (559.6 km)
$A_m = 5(5.59 \cdot 10^2)$
$A_m = 2.79 \cdot 10^3$

Figure 9.4 Coriolis Force Increases as Latitude Increases. As air moves poleward, Cf increases, but as air moves equatorward, Cf decreases.

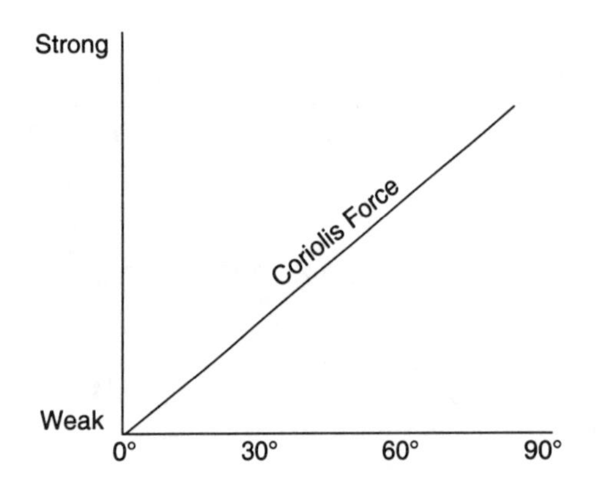

If this air travels equatorward, to 60° latitude, its angular momentum must be the same (according to the Conservation Law), yet r is greater and v must decrease. Therefore at 60° latitude

A_m = mvr
v = $(A_m)/(mr)$
v = $(2.79 \cdot 10^3)/[(1) \cdot (3{,}196.6 \text{ km})]$
v = .87 meters/sec.

Thus, wind speed is reduced to less than one-fifth of its initial value. Equatorward moving winds are slowed down relative to the surface over which they are traveling, and the air is deflected rightward (northern hemisphere) to cause air to have an apparent flow from east to west.

Now, lets reverse the situation and examine air flowing in a poleward direction. At 30° latitude and wind speed 5 mi/sec, A_m equals:

A_m = mvr
A_m = (1) (5 m/sec) (5,527.5 km)
$A_m = 5(5.52 \cdot 10^3) = 27.6 \cdot 10^3$

To conserve its angular momentum while traveling poleward, air velocity must increase. Using our previous formula, v = $(A_m)/(mr)$, and determining velocity of this air at 60° latitude, we get:

v = $(A_m)/(mr)$
v = $(27.6 \cdot 10^3)/[(1)\ (3{,}196.6 \text{ km})]$
v = $(27.6 \cdot 10^3)/(3.19 \cdot 10^3)$
v = 8.6 m/sec

Notice that velocity has increased. We have air with an easterly directed (west to east) momentum that is faster than the region into which it is flowing. This, coupled with increasing Coriolis force, causes westerly winds.

Global Circulation Patterns

George Hadley's concept of two balanced convection cells, one in the northern hemisphere and the other in the southern hemisphere, was gradually improved upon. Today, most meteorology textbooks present a model composed of three circulation cells in each hemisphere. Although future atmospheric data acquisitions may clarify the "unknowns" in this model, it has, in its present state, proven useful to understanding the interlinkages of the atmosphere that produce wind systems. Our presentation of this model will differ from those of most textbooks. Other texts normally present the entire model and then explain its components. This approach will be to explain each component and develop the model in a step-by-step fashion.

Hadley correctly predicted that surplus energy in the equatorial zone would result in continuously warm temperatures, low pressure, and ascending air. In this area weak horizontal pressure gradients (widely spaced isobars) and light winds are common. It is a low pressure environment, referred to as the **Equatorial Trough**, but in wind terminology designated by either the name **Doldrums** or **Inter-Tropical Convergence Zone (ITCZ)**. Daily weather forecasts are uniformly consistent for lowlands: warm, temperature maxima about 30°C (86°F), minimum temperatures in the 20's Celsius (70's Fahrenheit), with development of mid-to-late afternoon huge, cumulous (heaped up) clouds that generate thunderstorms. The clouds release copious amounts of rainfall (well over 250 cm, or 100 in, per year over extensive regions), as well as a tremendous amount of latent heat that increases the vertical acceleration of air and provides energy to drive the circulation cells illustrated in Figure 9.5. This ascending air reaches the tropopause, which in some respects is like a ceiling that restricts further lifting and

Figure 9.5 Air Circulation between the Equator and 30° N and 30° S Latitude

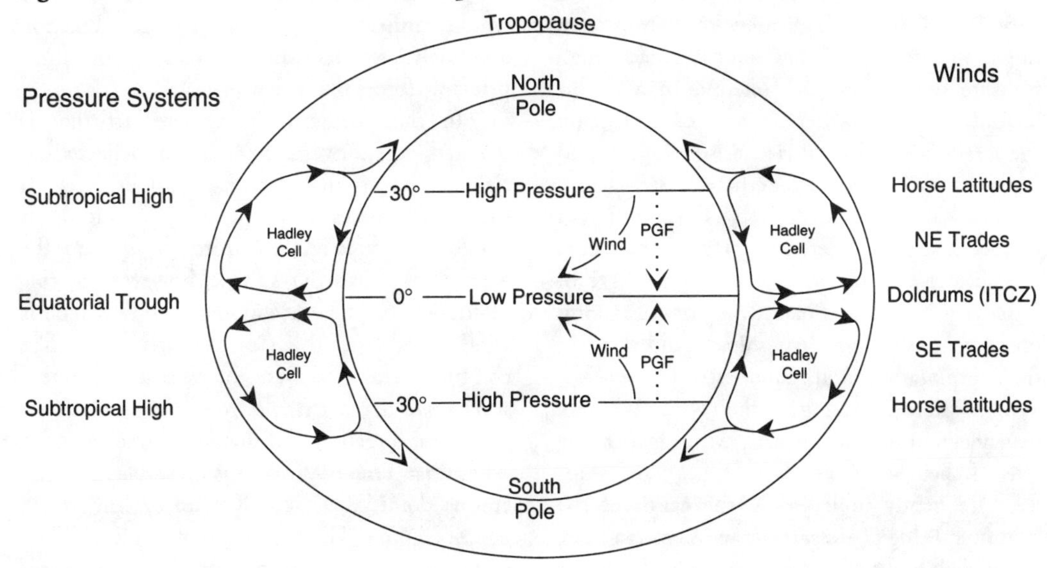

directs the air to move poleward. The poleward upper air flow is then deflected (to the right in the northern hemisphere and to the left in the southern hemisphere) by the Coriolis force to develop westerly winds (directed from west to east). *Note: winds are identified by the direction from which they blow.*

Increased density of the lifted air, as it moves poleward and is compressed, causes a substantial volume of these gases to sink, or settle, in the vicinity of 30° North and South increasing surface pressure. These regions of high pressure are called the **subtropical highs** or **subtropical anticyclones**.

The air that descends in the subtropical highs is compressed by the overlying atmosphere and, at the same time, adiabatically heated. Although surface pressure increases, horizontal pressure gradients are weak and windflow is sluggish. Winds are often **calm** (air having little or no horizontal movement) and of variable direction. Clear skies, relatively high temperatures, and very low precipitation are characteristic. Most global deserts are associated with the subtropical high pressure. It may seem rather strange to find that the subtropical high pressure systems are also referred to as **horse latitudes**. Several writers have tried to explain the rationale for this terminology, and explanations are diverse. Perhaps a satisfactory answer will never be derived. Thus, different textbooks suggest different origins for the name. We suggest the following. During the "Age of Discovery," the Spanish Conquistadores boarded horses upon their ships to aid them in their exploration and conquest of the New World. Remember that their travels were dependent upon the force of wind against the sails of their sea-going ships. Likely, the ships were frequently subjected to the calms of the subtropical high pressure systems where they would drift aimlessly for long periods of time. It is hypothesized that as food and water supplies dwindled, the horses would either be cast overboard to fend for themselves or were butchered to feed the hungry crew. Whether this etymology is valid is a matter of conjecture, but the "horse latitudes" has become a permanent term in the literature of meteorology and climatology.

To refresh your memory of the circulation system that is our current focus of attention, please refer again to Figure 9.5. Notice, in particular, that pressure-gradients must exist between the subtropical high pressure systems and the equatorial trough. The gradient from high pressure to low pressure is directed in the northern hemisphere from about 30°N to 0° latitude, that is from north to south. The southern hemisphere's pressure-gradient is reversed and goes from south to north. Consequently, we should expect winds to be generated in the directions described above, except that the direction of their flow is modified by Coriolis and frictional forces. Indeed, recorded data on the monitoring of wind confirm that their average direction accurately reflects the interaction of these principle forces that drive air.

In the northern hemisphere, air that is directed equatorward by PGF is deflected to right of its original path by Coriolis force. If these two forces were perfectly balanced air would theoretically flow from the east to the west. Near the earth's surface, however, the role of friction must be accounted for. This retards wind speed and the Coriolis deflection. The combined effect of these forces is to generate a surface wind that travels from a northeast to a southwest direction and that is referred to as the **Northeast Trade Wind**. It is important to note, at this point, that like all wind systems to be described, the **NE Trades** do not always blow from the NE; rather, they **prevail** (blow most frequently) from the northeast. Nonetheless, the NE Trades are normally considered the most reliable of all winds, with respect to their direction of flow. They consistently served explorers and traders who, prior to the invention of the steam engine, depended upon the wind to transport them and their goods over sea routes.

The NE Trades have a southern hemisphere counterpart called the **Southeast Trade Wind** or **SE Trades**. This flowing air behaves according to the same forces described in the previous paragraph, except Coriolis deflection is oriented to the left. The parallel processes that operate between the equator and 30°N and 30°S latitude form large convection cells that have been named **Hadley Cells** in honor of George Hadley who pioneered the investigation of planetary wind movements.

It is appropriate now to expand our study of global circulation to include the middle and high latitudes, where patterns of windflow are more complex. To aid your understanding of this air movement, Figure 9.6 illustrates a generalized pattern of air pressure systems aloft, in relation to the temperature-gradient that occurs between the equator and the poles. In warm regions, high kinetic energy and increased air buoyancy cause the mass of gas molecules to be distributed throughout a greater altitudinal distance from the surface than is characteristic of cold regions (Figure 9.6A). At a constant altitude above the surface, shown in Figure 9.6B by the line designated X-Y, horizontal air pressure becomes greater over the warm region (for example, the equator) than in the cold region (in our example, the poles). Thus, an upper air pressure-gradient exists between tropical and polar locations, from high pressure to low pressure. This PGF directs a portion of the lifted tropical air that is not recycled in the Hadley Cell, to move toward higher latitudes and to converge at the poles (Figure 9.7). This air is subject to Coriolis deflection.

A simplified presentation of upper airflow is provided in Figure 9.8A and B. In illustration A, assume that air rises along the equator, generating a high pressure environment as it moves poleward, and that winds are all westerly, except for the region lying above the equatorial realm where a general easterly flow exists. For such a wind pattern to exist, every point along a given parallel of latitude would need to have

Figure 9.6 A Hypothetical Change in Air Pressure Aloft Relative to Latitudinal Location; B Latitudinal Change in Air Pressure Aloft at a Constant Altitude

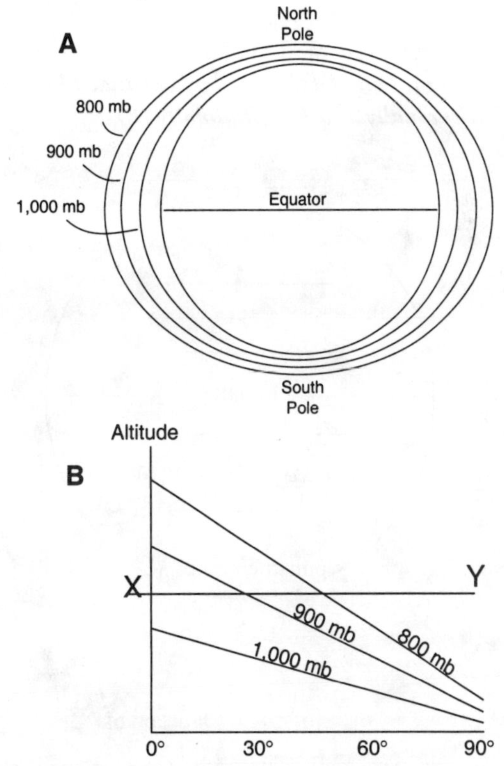

Figure 9.7 Upper Atmospheric Airflow to the Polar Highs and Surface Airflow Emanating from Polar Regions

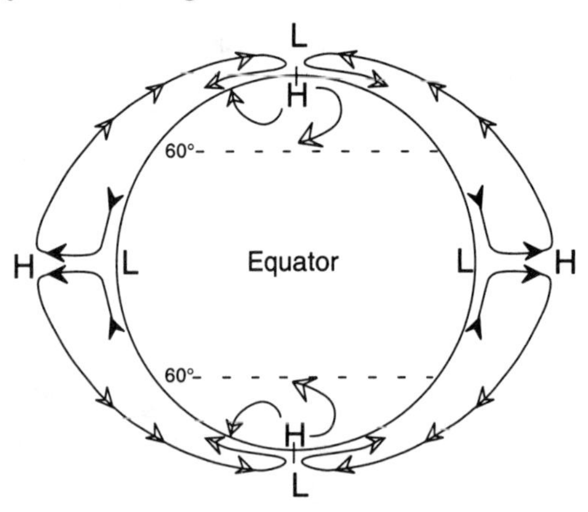

exactly the same temperature, and latitudinal temperature gradient would need to decrease uniformly from the equator to the poles. As we know, however, earth's response to energy receipts is affected by non-uniform surface materials. Thus, the westerly flow undulates, as shown in Figure 9.8B, and provides for an exchange (mixing) of cold and warm air between low and high latitudes. Because the air is well above the surface, however, frictional forces are insignificant, resulting in a flow of air that generally flows from west to east.

Upper atmospheric flow, generated from air lifting in the ITCZ and being directed toward high latitude, eventually converges in the polar regions where it descends to form high pressure

Figure 9.8 Hypothetical Flow of the Upper Atmosphere on an Earth Where Temperature Change is Only Determined by Latitude (A), and on an Earth Where Temperature Varies Both Latitudinally and Longitudinally (B)

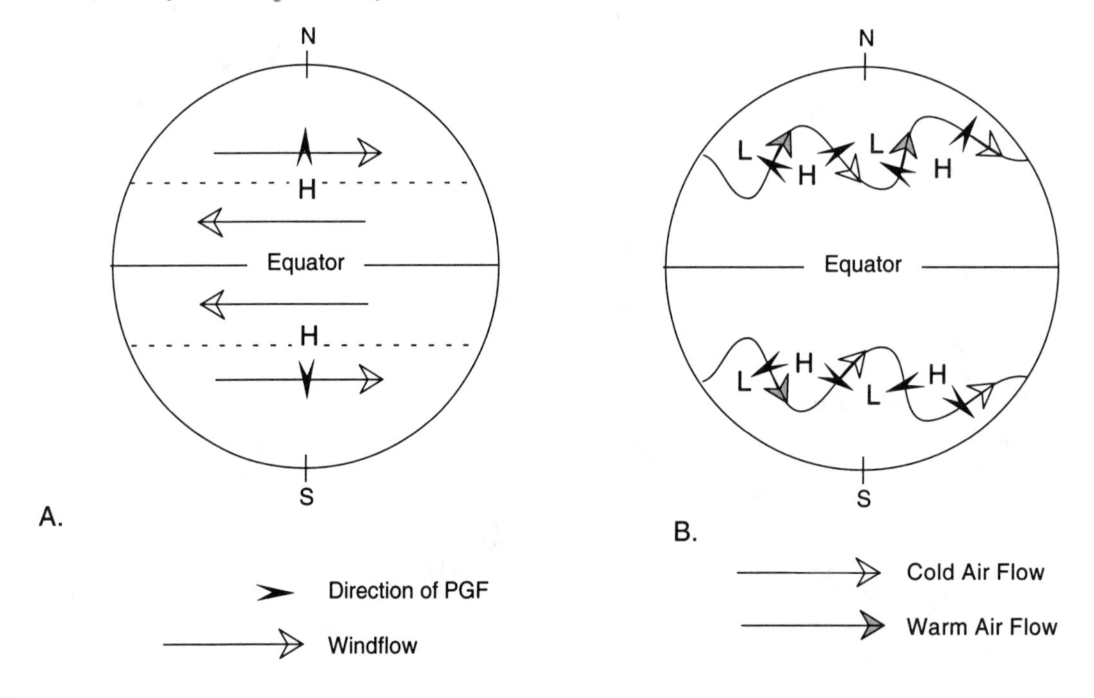

systems that are referred to as the **polar highs** (Figure 9.7). The air descending in the polar highs is relatively dense and cold. During its travel from the tropics to high latitudes, the upper air flow loses heat by radiation to space and is further cooled when it reaches the surface, particularly in winter, by contact with cold, snow covered surfaces. These processes increase air density in the polar regions and cause air to diverge equatorward. The Coriolis deflection (when combined with surface friction) generates an airflow between the poles and 60° latitude, a source of wind from the northeast in the northern hemisphere and from southeast in the southern hemisphere. These winds are referred to as the **polar easterlies.**

At this point, we have discussed surface air circulation within the low latitudes (0–30°) and high latitudes (roughly 60–90°). The final piece of the circulation puzzle is the wind flow pattern of the middle latitudes (approximately 30° to 60°). Recall that the Hadley Cell is characterized by lifting air within the equatorial trough and descending air in the subtropic highs (Figure 9.5). While the Hadley Cell describes that portion of air from the subtropical high that travels equatorward, it does not account for that portion of air that travels poleward. There exists a pressure gradient that decreases from a high at about 30° latitude to a low at approximately 60° latitude. When Coriolis deflection (which increases toward the poles) is combined with the PGF, the general airflow is from a westerly direction, as illustrated in Figure 9.9. These winds are called the **prevailing westerlies,** the most variable of all earth's winds. Although winds most commonly originate from some westerly compass point (i.e. from 180°to 360°), they may over a short span of time come from any direction. This variability in wind direction is associated with the frequent development and

Figure 9.9 Poleward Airflow from the Subtropical Highs to the Subpolar Low

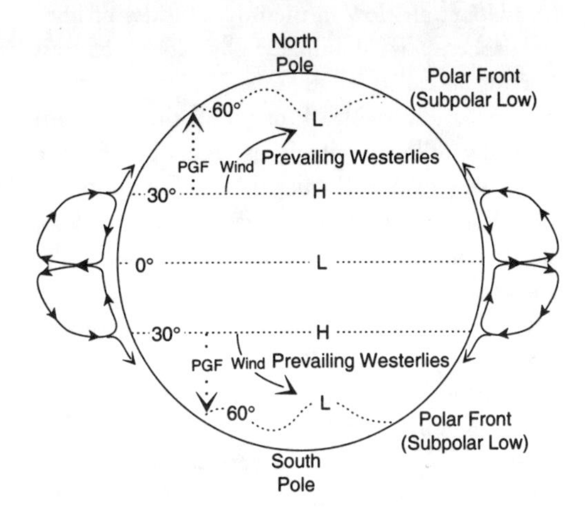

eastward movement of high and low pressure systems that form in these latitudes. The low pressure systems are associated with large-scale storm events that bring the middle latitudes a major supply of their annual precipitation. We will provide a detailed discussion of these storm systems in a forthcoming chapter. What is more pertinent to our present discussion of global circulation is that the poleward moving prevailing westerlies and the equatorward traveling polar easterlies converge along an irregular and discontinuous boundary, known as the **polar front** (Figure 9.10).

The latitudinal position of the polar front is determined by the magnitude of the opposing high pressure systems that cause it to exist, that is the subtropical and polar highs. During the summer when the subtropical highs strengthen and the polar highs weaken, the polar front is found at higher latitudes. The opposite is true for winter. Because air converges along the polar front and higher pressure exists to its north and south, the polar front is a region of relatively lower pressure called the **subpolar low** (Figure 9.9).

Upper air flow in the mid latitudes of both hemispheres is dominated by cores of fast moving air called **jet streams**. The paths of these streams are variable throughout the year but generally follow a circuitous meandering route from west to east (Figure 9.10). The course of the jet stream is important because it provides a steering influence on the path of many storms and thereby provides a mechanism to distribute heat and moisture throughout the mid latitudes.

It is important that we view all these pressure and wind patterns, illustrated in Figure 9.11, as long-term (for example, one year) averages and not as a predictive tool to determine how atmospheric conditions will be

Figure 9.10 Typical Path of the Jet Stream and the Polar Front

Figure 9.11 Average Global Pressure and Wind Patterns at the Surface

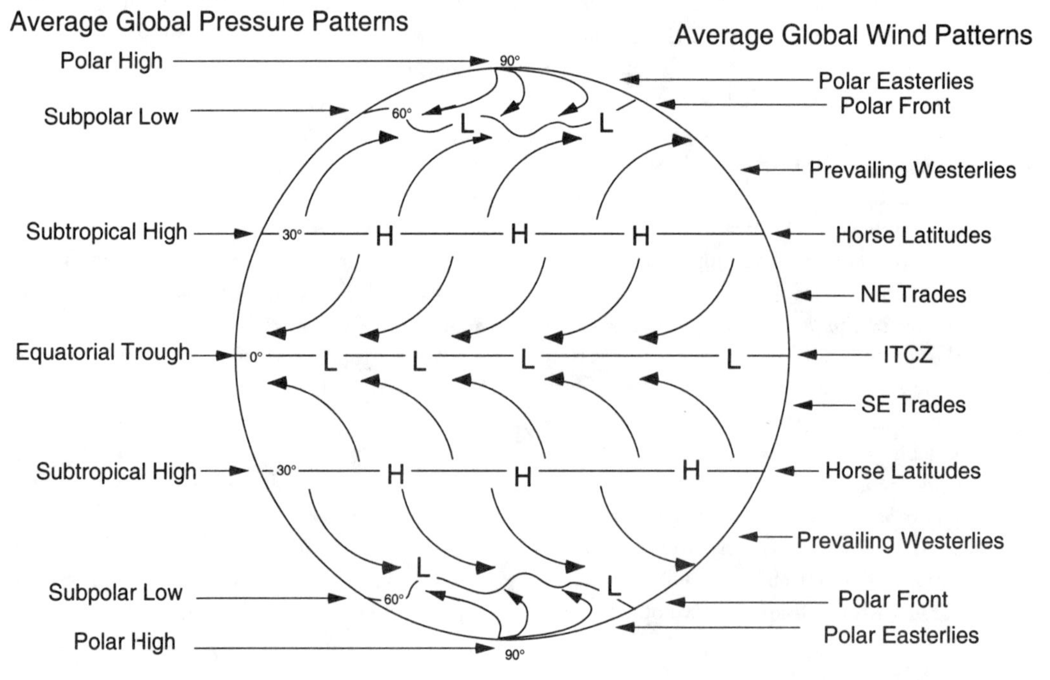

during a given season, day, or hour. Factors other than those discussed in this chapter can alter the basic wind flow patterns we've described.

Variations in Atmospheric Circulation

As an explanation of the overall circulation of the atmosphere, the foregoing model is a useful device. It is of limited value in explaining the actual circulation pattern that might prevail at a particular location at a particular time, however. The problem is with the nature of the model itself. It is designed to illustrate mean conditions whereas actual conditions may vary considerably from the mean. Since the mean circulation is a function of the latitudinal variation in solar radiation, it is not surprising that variations in heat budgets due to surface conditions and seasonal changes produce some striking departures from the mean circulation. These kinds of variations are regular and predictable. Other variations in circulation tend to be more random and have their origin in the random variability of the circulation system itself.

Seasonal Effects

Seasonal variations in net radiation are striking, especially in higher latitudes. Even if we could ignore the effects of surface materials on energy budgets, we would still be able to postulate a pronounced seasonal effect on atmospheric circulation. As the sun's declination changes with season, the locations of the pressure systems and wind belts ought to migrate. Indeed, this does happen. Figures 9.11 and 9.12 show the seasonal pressure systems for January and July. The location of the equatorial low pressure trough migrates between these two periods. The locations of the trade winds, the subtropical high pressure systems, the westerlies, the polar fronts, and the jet streams all experience seasonal variations and are displaced toward the poles in the summer hemisphere.

Not all locations at the same latitude have similar net radiation values. The contrast is most pronounced between land and water surfaces. The ability of water bodies to store vast quantities of heat without great temperature change, and the reduction of sensible heat transfer through evaporation, produce cooler summer temperatures and warmer winter temperatures over oceans than over land surfaces.

Temperature differences are reflected in pressure differences, and high temperatures over large mid-latitude continental areas in the summer produce low pressure centers that disrupt the subtropical high pressure belts. Instead of two continuous zones of high pressure encircling the earth, there are cells of high pressure centered over the mid-latitude oceans. In the winter hemisphere, the subtropical pressure system is more continuous and is intensified by high pressures over landmasses.

Although there is no contiguous subpolar low pressure zone as such, in the northern hemisphere there are quasi-permanent low pressure cells centered over the North Atlantic in the vicinity of Iceland and over the North Pacific near the Aleutian Islands. Those lows intensify strongly during winter because of the relatively warm ocean surface. The low pressure is zonal in the Antarctic because there are no landmasses, other than Antarctica, over which strong high pressure systems develop.

Wind patterns that develop in response to the quasi stationary lows and highs have strong meridional as well as zonal components and also display some striking seasonal variations. The gradient winds of the anticyclonic circulations over the oceans are pronounced in summer and

Figure 9.12 Dominant Planetary Atmospheric Pressure Patterns in January

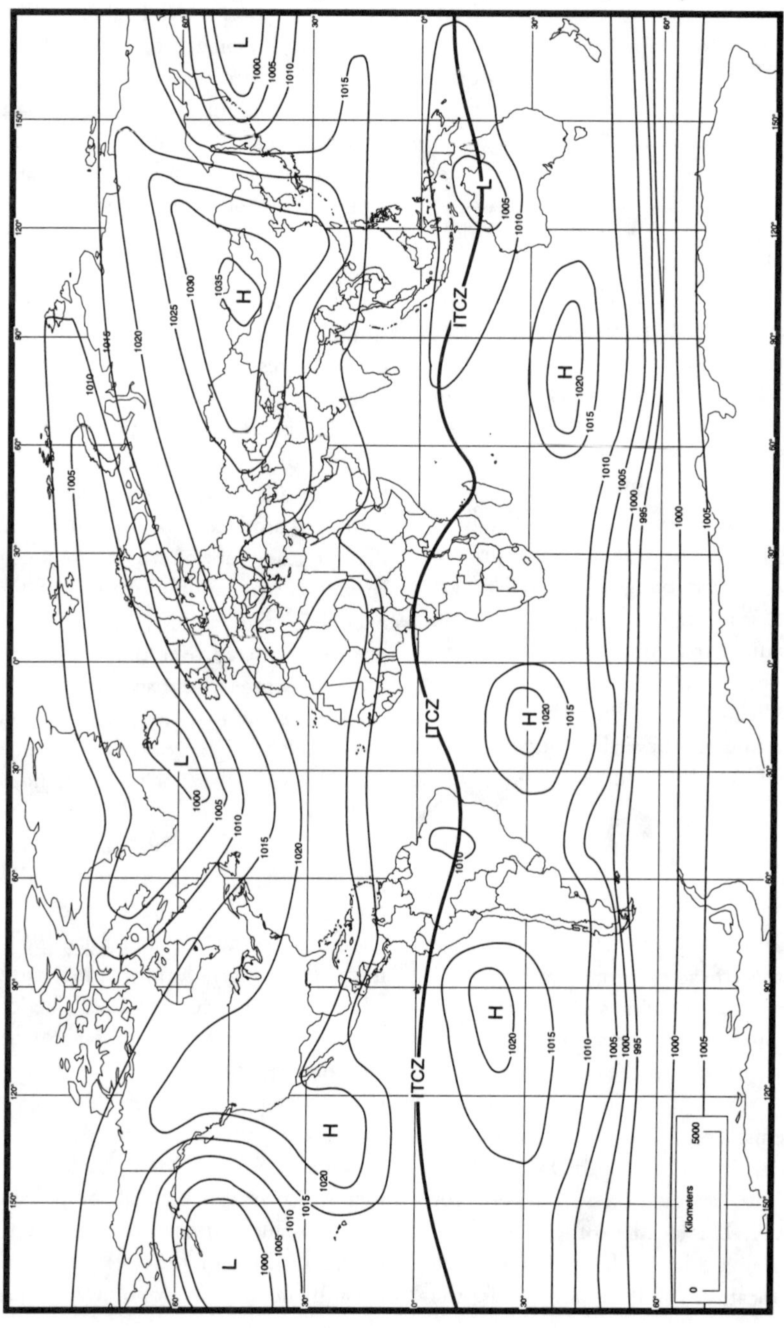

Figure 9.13 Dominant Planetary Atmospheric Pressure Patterns in July

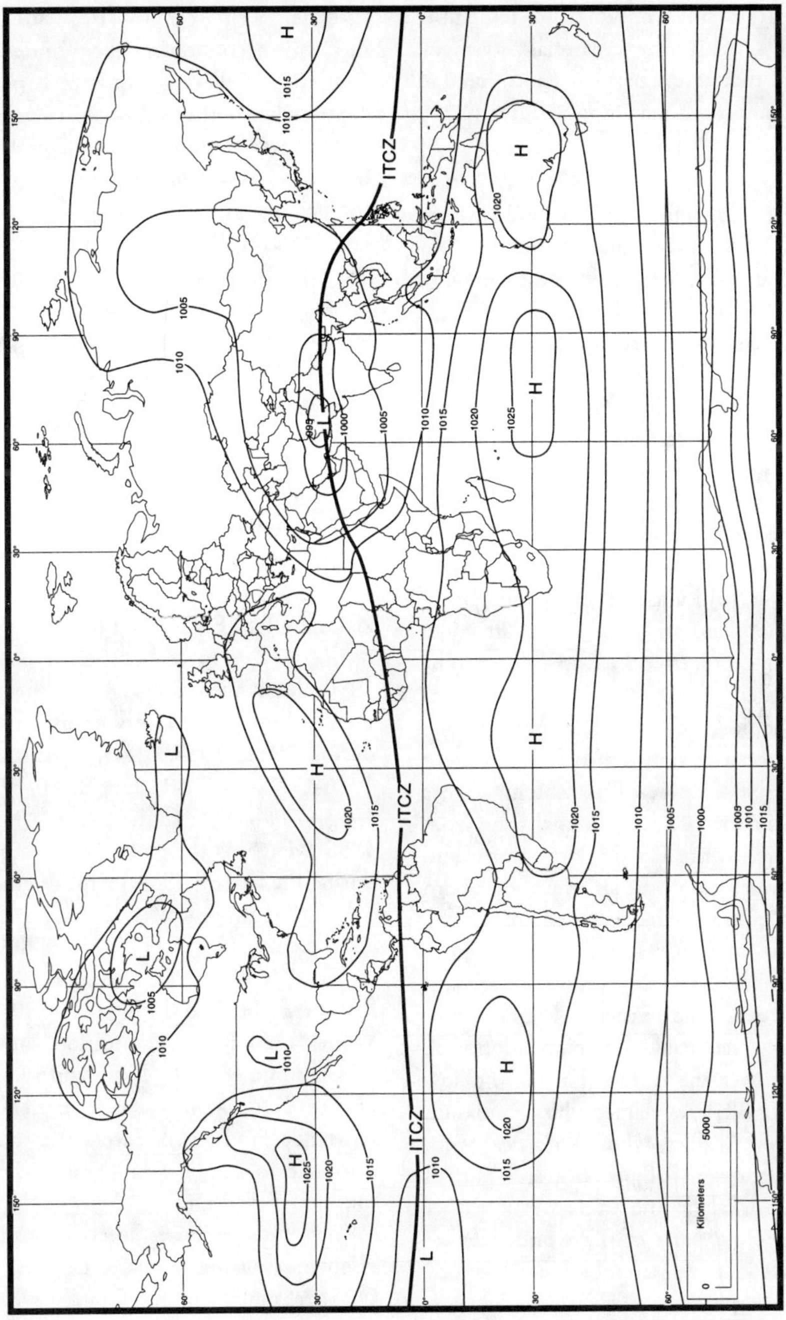

reinforce the mid-latitude westerlies and the tradewinds. It is still important to remember that these systems are seasonal averages, however, and that at any particular location and time, actual conditions may be very different.

ENSO

A contraction for **El Niño/Southern Oscillation, ENSO** events are fundamental but temporary changes in the massive heat engine of the tropical Pacific. This change is the single most prominent influence on the climatic variability of surface pressure and rainfall differences in the tropics and on wintertime circulation in the mid-latitudes.

Jakob Bjerknes was the first to offer, in 1969, the hypothetical explanation for ENSO that has gained widespread acceptance today. As the trades blow west across the Pacific generating the South Equatorial Current, deep water from off the coast of Peru and Ecuador is drawn to the surface in a process called **upwelling**. The cooler upwelled water causes the eastern equatorial Pacific to be unusually cold among low latitude oceans. The warmth of the western Pacific creates a steep sea surface temperature gradient and a corresponding thermal wind from east to west across the Pacific. Now warm and loaded with water vapor, the air over the western Pacific tends to be unstable and rises to join the poleward flow at upper levels of the Hadley circulation. Some of this flow returns to the east to descend over the eastern Pacific, thus completing a zonal equatorial circulation cell: high pressure over the eastern Pacific and low over the west. Bjerknes named this circulation after Sir Gilbert Walker, whose empirical work connected fluctuations of pressure and rainfall over the central Pacific and India with weather conditions in other parts of the world. Walker was aware that in some years the low developed over the eastern Pacific rather than in the west. Bjerknes went on in 1969 to postulate a mechanism connecting the **Walker circulation** and the oscillating equatorial pressures. He suggested that the Walker circulation amplified easterly winds across the equatorial Pacific and increased upwelling off coastal Peru thereby sharpening the contrast of sea surface temperatures (SST) between the eastern and western Pacific. Because the SST contrast between east and west is what causes the Walker circulation in the first place, trends lasting for several years would be expected to result. Any decrease in the equatorial easterlies would likewise weaken upwelling causing the eastern Pacific to become warmer, reducing the SST gradient, and slowing the Walker circulation. These feedback mechanisms permit a never ending series of oscillations in the global heat engine. Causes for perturbations capable of kicking off an ENSO event are unknown but do recur every 4 to 5 years on average, but the 1990s have been unusually prone to El Niño events. Recent El Niño events have occurred during 1951, 1957, 1963, 1965, 1969, 1972, 1976-1977, 1982-1983, 1986-1987, 1991-1993, 1994-1995, and 1997-1998. The 1997-1998 event was the strongest by most measures during the latter half of the twentieth century.

Although much is not known about the concurrent effects that El Niños have on regional weather, seasonal rainfall anomalies have been observed during the 1980s and 1990s. Wetter than normal conditions have occurred during summer in the Basin and Range province of the U.S., along coastal Equador between November and April, across the southern U.S. from Texas to Florida during winter, across southern Brazil and northern Argentina during spring and summer, across the Southern Highlands of East Africa between April and October, and from the Maldives to Sri Lanka between October and December. Dryer than normal conditions have been observed from southern Mexico across the Windward Antilles and Venezuela and the island archipelagoes of southeast Asia between July and October; and the southeastern U.S., south central Australia, southeast Australia, central and northern India, Bangladesh, Nepal, and Pakistan, and central and southern Africa and southern Mozambique during their respective summers.

ENSO events are mechanisms by which large quantities of heat are removed from the tropical Pacific. Initiation, termination, and periodicity of the events are thought to result from an instability in the coupling of two interacting systems: oceanic and atmospheric. The enormous capacity of the equatorial ocean to store heat acts as a flywheel to drive the periodicity of these events beyond mere annual cycles. Although the scientific jury is still debating this issue, the preconditions for triggering an ENSO appear to be as follows: an accumulation of heat over several years to a threshold level; and harmonic phasing of the normal seasonal cycle with interannual cycles and minor perturbations such as the frequent bursts of westerly winds in the western equatorial Pacific.

Changes in the interaction between the tropical Pacific Ocean and the atmosphere have widespread and long term implications. ENSO events have been associated with dramatic climatic extremes. The 1982-1983 ENSO event brought Australia's worst dought this century; drought struck India and Sri Lanka; a rare typhoon slamed into Hawaii; California was deluged with heavy rains and mudslides, and, driving rains flooded the Gulf Coast. The ENSO of 1991-1993 was accompanied by Hurricane Andrew and record losses to southern Florida and the Gulf Coast; Iniki, the strongest hurricane of the century to strike Hawaii; and the worst winter storm of the centruy struck the eastern seaboard of the United States with barometric pressures lower than most hurricanes and 120 kph winds.

The strongest associations between El Niños and weather in the United States are an active storm track across the southeastern states and the lack of frigid arctic air masses across the northern tier of states. Some notable hurricanes have made landfall in Hawaii and in the Southeast during these events, but there is not a scientific consensus as to what if any influence El Niños have on the frequency, intensity, or track of these powerful storms. The atmosphere is very complex and many factors are at work.

Perturbations in the tropical Pacific Ocean have long term effects on the transfer of heat and moisture to other regions of the earth. Although surface ocean currents are important transporters of heat to the mid and high latitudes, a more complex pattern of ocean circulation is emerging that points to the importance of the tropical Pacific as a critical heat source for circulation of the global ocean. Driven chiefly by differences in density, both thermal and haline, a loop pattern called the **great ocean conveyor** circulates water through the world ocean (Figure 9.14).[1] Many of the interactions between ENSO events, weather extremes, and the ocean conveyor are currently being investigated.

The challenge to climatologists and meteorologists to understand and predict the behavior of global systems such as ENSOs continues. The rewards for solving this mystery will be a better understanding of the way the atmosphere-ocean-climate system works and the potential for much more accurate long range weather forecasts.

[1]W. S. Broecker. The Biggest Chill. Natural History Magazine (October 1987): 74-82.

Figure 9.14 The Global Ocean Conveyor

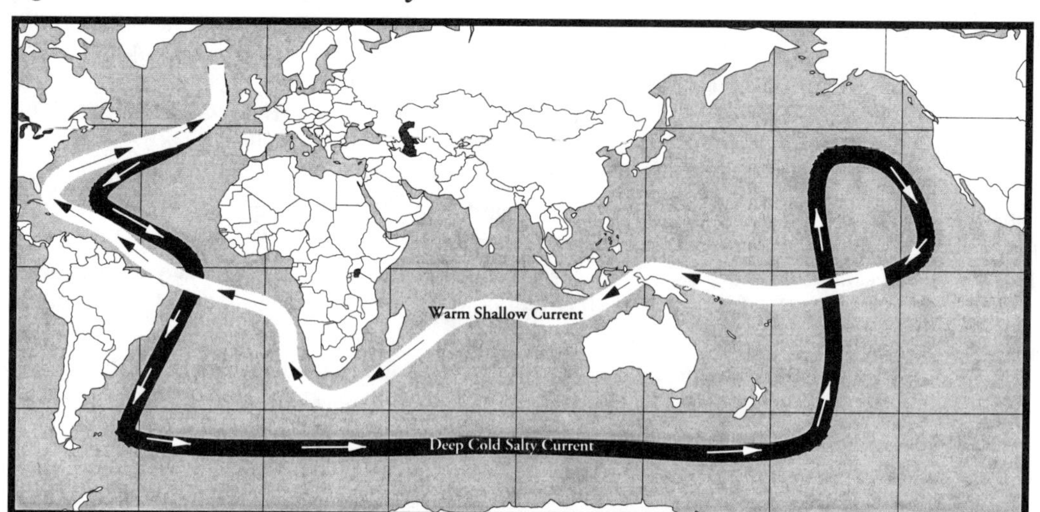

Monsoon Flow

The controls that affect air motion on a global scale also operate at regional levels. One of the best known regional wind systems is the monsoon. Monsoons are winds that essentially reverse their direction of flow between summer and winter. The best developed monsoon circulation pattern is found in south and east Asia.

During the summer, intense heating of the Asian continent produces a thermally induced low-pressure system in south Asia (Figure 9.12). This causes a lower-atmospheric movement of air with a south-westerly flow from the Indian Ocean and the western Pacific toward the continent. This summer monsoon is made up of warm and humid air that produces heavy rainfall as it is forced to rise over mountains and hills.

Radiational cooling of the Asian continent during the winter leads to the formation of a strongly developed high-pressure cell in Siberia (Figure 9.11). The winds blowing from this area are northeasterly across southern Asia. This winter monsoon is dry (originating in the interior of the continent, it has little opportunity to absorb moisture) and brings clear weather to most of the monsoon lands. The term monsoon, meaning "reversing," was introduced by Arab traders who would sail from East Africa for India with the onset of the summer winds. In winter, reverse winds carried them home.

Minor Circulation Systems

Some kinds of atmospheric circulation systems are so localized and last such a short time that they may not even appear on the weather maps. These systems may be important components of local weather, however.

Land-Sea Breezes

Most local circulation systems develop in response to differences in heating and cooling of the air. Along coasts, during the summertime, the land surface heats more rapidly during the day than the water surface. In response, lower pressures develop over the land and air flows

Figure 9.15 Land and Sea Breezes: The Diurnal Wind Shift Along Coastal Area

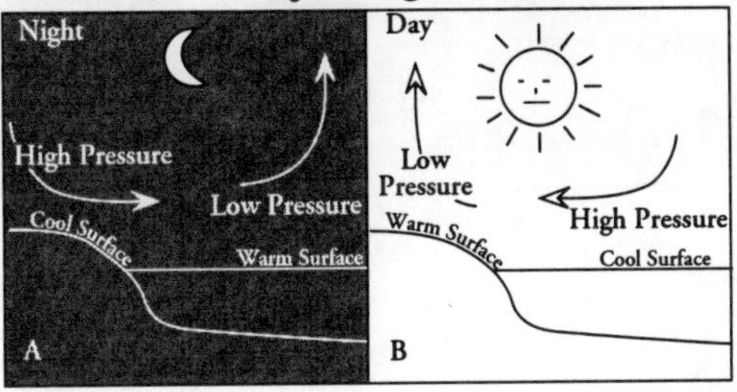

inland from the higher pressure areas over water. A circulation cell develops like that in Figure 9.15, and the onshore wind is called a sea breeze. At night the land surface cools more quickly and develops a higher pressure. The resulting land breeze blows from land to sea. Because of temperature lag the sea breeze is strongest in the afternoon, whereas the land breeze is best developed just before dawn.

Urban Effects

Any condition under which the air over a local area is heated (or cooled) more than the surrounding air will produce an effect similar to the sea breeze. Urban areas absorb more heat than do surrounding areas. The sun strikes building walls at more direct angles, the structural materials store up more heat, and there is more surface area to absorb heat. In addition, heat is added to the air in large quantities over urban areas by the burning of fuel. The net effect is that air temperatures over urban areas are warmer, and lower pressures prevail. Cities, in effect, produce their own convectional circulation cells.

Mountain-Valley Breezes

In mountainous terrain several types of local breezes may develop. Valley breezes commonly occur during the daytime. Mountain tops heat up more than valleys do because of the thinner air (less scattering) and the more direct sun angles on the slopes. If the low pressures which develop over the summits are intense enough, air actually flows up slope (Figure 9.16).

At night, mountaintops radiate heat more quickly and grow colder. The more dense cold air then slips down slope to settle in the valleys. For this reason it is not uncommon to find two frostlines in the spring and fall, one marking the upper limit of the cold air in the valley, the other marking the lower limit of cold conditions in the higher elevations. The intermediate slopes make good sites for orchards because they are less likely to freeze.

The increased radiation that occurs at high elevations affects organisms too. Inexperienced mountain climbers and skiers may suffer loss of body heat (hypothermia) brought on by failure to wear adequate clothing. They lose heat rapidly by radiating it to the cold sky. No cooling sensation accompanies this form of heat loss, as happens with evaporation, and the loss of body heat may go unnoticed.

Figure 9.16 Valley and Mountain Breezes: The Diurnal Wind Shift in Mountainous Regions

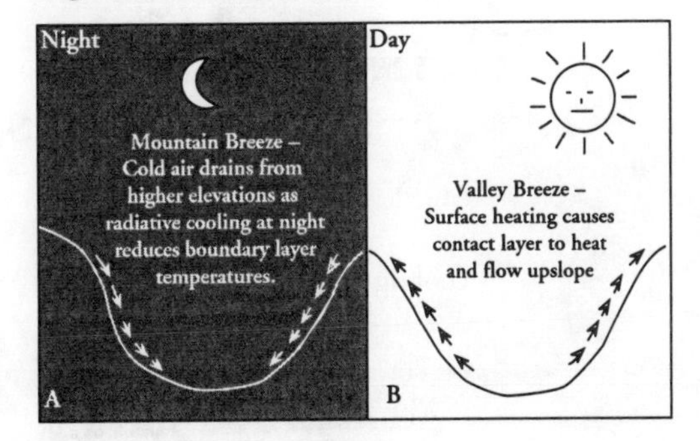

Chinooks

In the wintertime, areas on the eastern side of the Rocky Mountains frequently experience the warming winds called Chinooks. Chinook is a native American word meaning "snow-eater." It is not at all uncommon for these warm winds to melt or sublimate five to six inches of snow in a few hours and raise temperatures as much as 28°C (50°F) above normal. During a 24 hour chinook in Kipp, Montana in 1896, the temperature rose dramatically and melted 75 cm (30 in) of snow. Chinook winds, which may gust to 70 or 80 kph, result from adiabatic warming of the air moving down the mountain slopes. Chinooks are particularly dramatic when they bring an abrupt end to frigid conditions. Temperatures may rise 25°C (45°F) or more in a few hours.

The term Chinook is used in North America, but diverse names are used around the world to identify similar winds: the Föhn winds of Switzerland and southern Germany, the Mistral in France, the Santa Ana in southern California, the Hamsin of Israel, and the Mediterranean Khamsin or Sirocco. It should be noted that the world's highest temperature was recorded during a Khamsin in El Azizia, Libia: 58°C (136°F).

Activity: Weather Maps

OBJECTIVES

Upon completion of this activity, you will be able to: (1) identify and interpret weather map symbols for barometric pressure, cloud cover, dew point, and air temperature; (2) interpret the patterns of temperature, pressure, and moisture on synoptic weather maps.

INTRODUCTION

Collecting and analyzing data from all over the world have given us the ability to predict weather over wide areas and for appreciable periods of time. Benefits of weather forecasting may be small, such as planning a family picnic, or very great, such as issuing major storm warnings.

Two of the most important tools currently used by meteorologists are weather maps and satellite photographs. In this lab we will be concerned with the preparation and the use of a **surface weather map** and its correlation with **satellite photographs** of the same area at approximately the same time. It should be noted that the surface weather map is only one of several different kinds of weather maps used by forecasters.

You are already familiar with the various measurements of temperature, barometric pressure, and relative humidity. Weather vanes are used to measure wind direction. The arrow on the weather vane points *toward* the direction *from which* the wind is blowing. Wind that comes from the southeast is called a southeast wind, and so forth. **Anemometers** are used to measure wind speed. As the wind strikes the cups, the anemometer rotates in proportion to wind speed.

Charts of the directions and velocities of winds aloft give further information on the trend of surface weather, including the location of the important jet stream. The jet stream is a tunnel or river of fast-moving air within the troposphere. The force and the location of the jet stream fluctuate considerably and have marked effects on surface weather phenomena.

A body of air in which temperature and humidity are fairly distinct and which moves over a large area is known as an **air mass.** The characteristics of an air mass depend upon the region of the earth over which it originates. The boundary between two unlike air masses is called a **front**. Air mass boundaries are known as **surface fronts** when they intersect the ground, and as **upper air fronts** when they do not. Important changes in weather, temperature, wind direction, and clouds often occur with the passage of a front. The corridor of energetic activity associated with the leading edge of front is called a **squall line**.

A **warm front** is the boundary between a relatively warm air mass which is moving into an area occupied by colder air. The advancing warm air pushes the cold air ahead of it. Since warm air is less dense than cold air, the warm air mass will climb up and over the cold air mass. This results in gradual cooling of the warm air and the probability of light, steady precipitation. As the warm front passes along the earth's surface, temperatures increase, atmospheric pressure decreases, wind direction and wind speed change, and finally the skies become partly clear.

A **cold front** develops when a cold air mass advances on a warm air mass. As cold air is denser than warm air, the leading edge of the advancing cold air wedges under the warmer air mass, causing the warm air to rise vertically. Convection currents cause the formation of cumulonimbus clouds, and the probability of heavy, intense rain and thunderstorms increases.

When two unlike air masses meet one another and there is little movement of air, the surface between them is called a **stationary front**. Weather associated with a stationary front is similar to that of a warm front. Eventually, one or both of the air masses will move and either a warm front or a cold front will form.

An **occluded front** occurs when a cold front overtakes a warm front and merges with it near ground level. The warm air mass is completely lifted off the ground by the colder underlying air. Because of this, an occluded front is characterized by a combination of weather conditions, warm front weather first, followed by cold front weather as the warm air is forced aloft. Eventually, the warm air mass which was lifted, cools, and one large cold air mass is formed.

WEATHER MAP SYMBOLS

Various weather map symbols are used to represent atmospheric conditions, e.g. cold fronts, warm fronts, stationary fronts, and occluded fronts. Refer to the figures in this activity or search the Internet for "weather map symbols" to find their meaning. One particularly common symbol for weather or weather related information is Wx and your may find it helpful in your search. The American Meteorological Society maintains an informative web site and provides a key for selected Wx map symbols at: http://www.ametsoc.org/dstreme/extras/wxsym2.html. Knowledge of frontal notation is particularly useful because approaching weather systems are commonly described through the use of these symbols. Half circles and triangles are placed along air mass boundaries to indicate the type of front. The side on which the symbols are placed indicates the direction of frontal movement. Surface fronts are drawn in solid black, and fronts aloft are drawn in outline only.

The surface weather map is the compilation of local weather data from many different stations. The relationships between weather phenomena and different localities are shown in order to present a picture of the general movement of weather across the area depicted by the map. The information entered on the map is from every station available and has been gathered at the same hour of the day or night.

Figure 1 deciphers the symbol for a typical weather station as it would appear on a weather map. This is the way weather information for a given station is displayed on weather maps. Refer to the chart available in the lab which explains the weather map symbols. Write a brief explanation for each symbol or number in the space provided.

Figure 1 The Weather Station Model

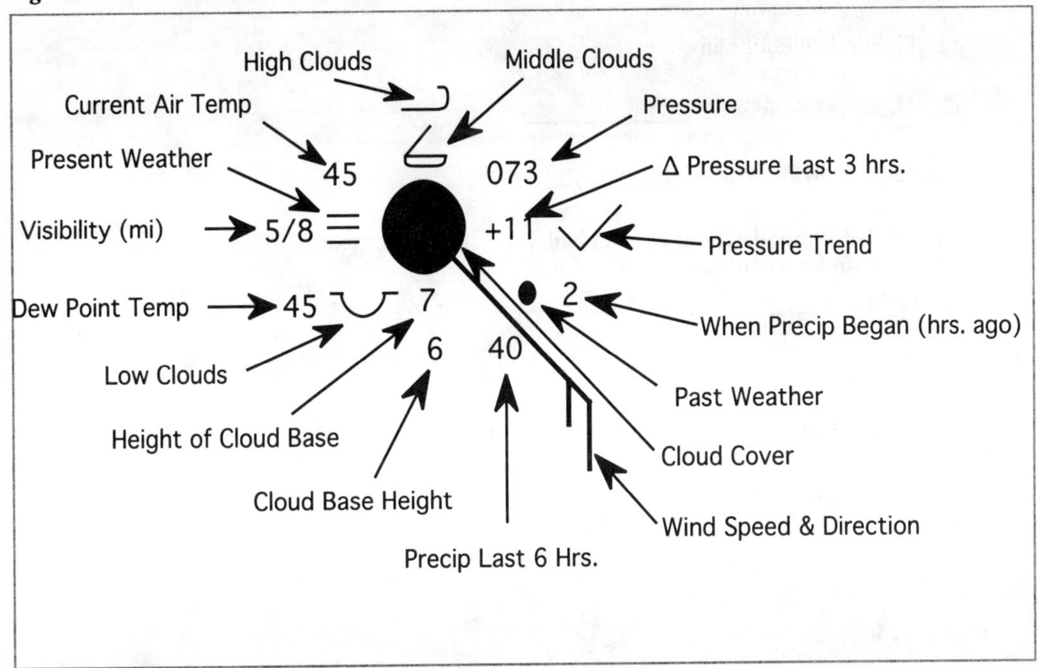

1. Identify the wind direction and speed for the following:

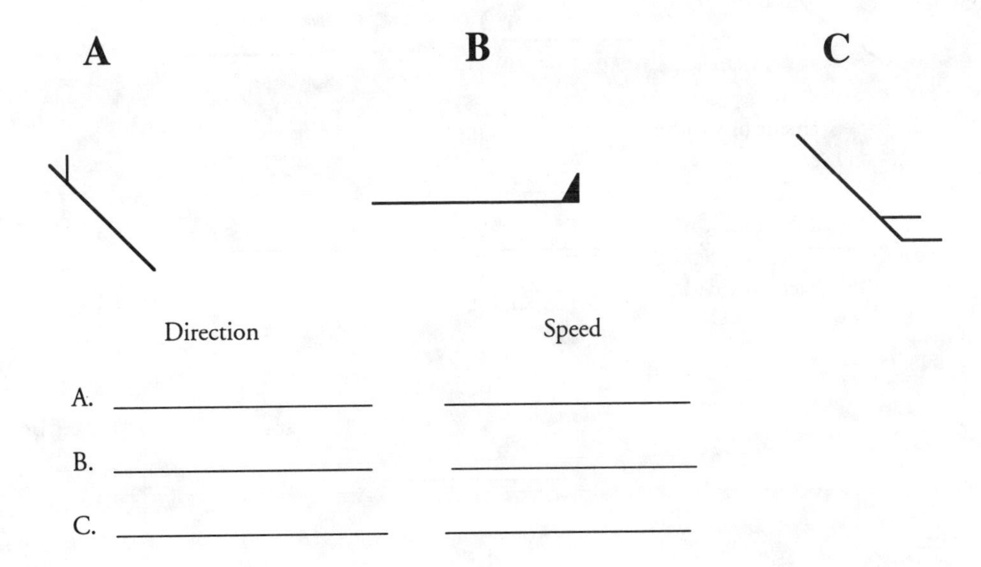

	Direction	Speed
A.	____________________	____________________
B.	____________________	____________________
C.	____________________	____________________

2. What weather symbols indicate the following conditions?

   a. Light continuous rain ___________

   b. Heavy continuous rain ___________

   c. Slight snow shower ___________

   d. Moderate thunderstorm without hail, but with intermittent rain ___________

   e. Light fog ___________

   f. Haze ___________

3. What cloud types are represented by the following symbols?

A **B** **C**

A. ___________ B. ___________ C. ___________

4. Using the following report of weather conditions at Charlotte, North Carolina, construct a weather station symbol complete with the following data. In recording station pressure, the initial 9 or 10 and the decimal are omitted. For example a pressure of 1002.3 would be plotted simply as 023.

Temperature: 46° ; Dew Point : 43° ; Wind: West 15 MPH; Pressure: 1021.2 mb;
Sky: 1/10 covered with cumulus; Weather: Light fog.

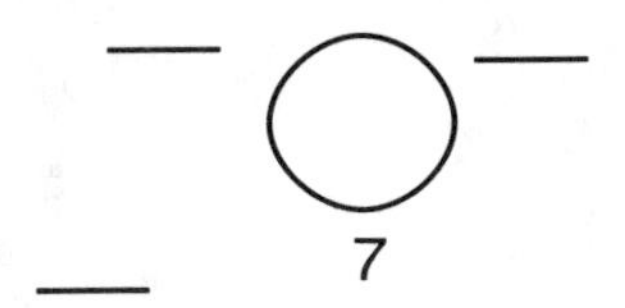

Isobars are lines of equal barometric pressure. Isobars are usually drawn for every 4 millibars of pressure. They are, by convention, 992, 996, 1000, 1004, 1008, and so on.

5. A pressure given on a weather map as 072 is ____________________mb.

698 is _______________mb.   372 is ______________________mb.

6. On the weather station map (Figure 2), construct isobars to identify pressure systems, locate the low pressure center and draw in the position of cold and warm fronts. Each station on the map is coded with the following data:

air temperature and dew point , appear to the left of the station
barometric pressure, appears to the right of the station
wind speed and direction and sky cover.

Sketch in the isobars lightly in pencil first, interpreting their location according to your best judgement. After you have located them satisfactorily, you can go back and darken the lines. The lowest pressure on the map is 992.5 millibars, which you will find at Chicago, Illinois and also Milwaukee, Wisconsin. This pressure would be included within the 992 millibar line, and is probably at or near the low pressure center. Sketch in this 992 millibar line where you think it is located on the map. Next sketch in the 996 millibar line. When you sketch in the 992 and 996 millibar lines correctly, you can begin to see the outlines of a large well-developed low pressure center. Proceed to complete the isobars on your map. Take your time in constructing isobars, it isn't easy, especially the first time.

To complete this map, you need to locate the cold and warm front. Weather fronts are usually defined by the beginning or the end of precipitation, a marked shift in wind direction, by a noticeable change in the temperature, or by a combination of these factors. Examine the map and sketch where you believe the fronts should be located. The location of the fronts in between reporting stations is a matter of interpretation. Use the following summary of conditions associated with each type of front as a guide.

Figure 2 Weather Station Map

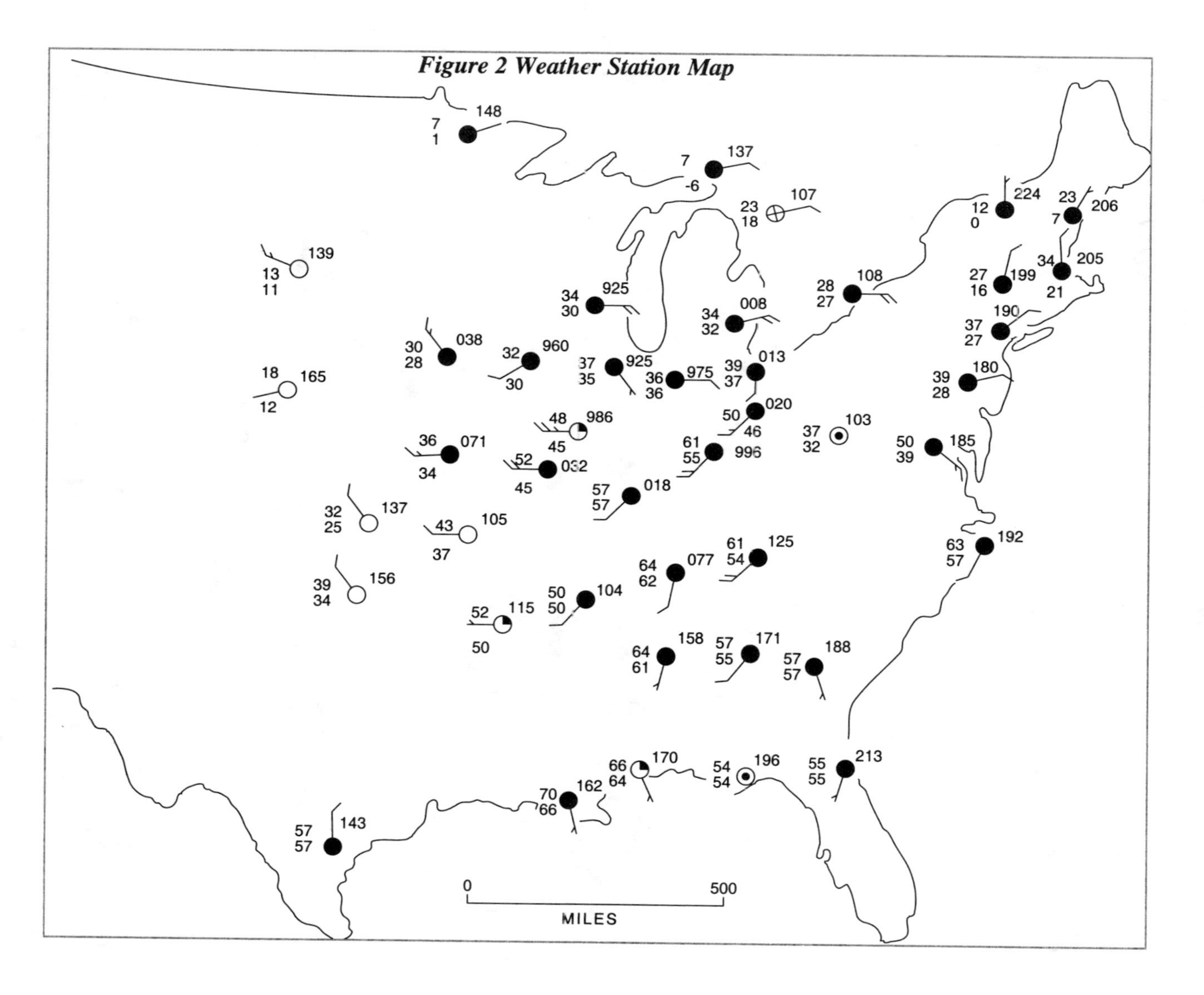

TYPICAL WEATHER CONDITIONS ASSOCIATED WITH FRONTS

Weather Map Symbols for Fronts

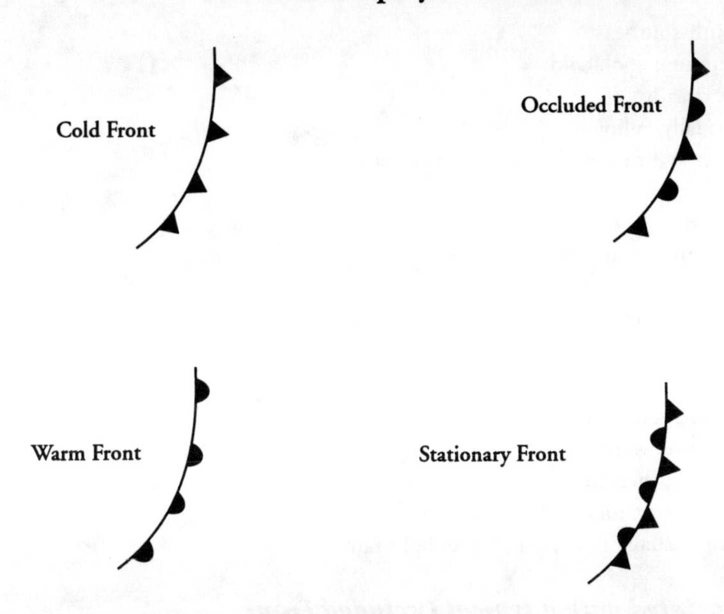

Surface Conditions associated with a typical Cold Front

Before Passing

Winds: south-southwest
Air Temperature: warm
Dew Point: high, remains steady
Pressure: falling steadily
Precipitation: short period of showers

While Passing

Winds: gusty, shifting
Air Temperature: sudden drop
Dew Point: sharp drop
Pressure: sharp rise
Precipitation: heavy showers of rain or snow

After Passing

Winds: west-northwest
Air Temperature: colder
Dew Point: lowering
Pressure: rising steadily
Precipitation: decreasing intensity of showers

Surface Conditions associated with a typical Warm Front

Before Passing

- Winds: south-southeast
- Air Temperature: cool, cold
- Dew Point: steady
- Pressure: usually falling
- Precipitation: light to moderate rain, snow, sleet

While Passing

- Winds: variable
- Air Temperature: steady rise
- Dew Point: slow rise
- Pressure: leveling off
- Precipitation: drizzle

After Passing

- Winds: south-southwest
- Air Temperature: warmer
- Dew Point: rise, then steady
- Pressure: slight rise; followed by fall
- Precipitation: usually none, sometimes light rain

Surface Conditions associated with a typical Occluded Front

Before Passing

- Winds: southeast-south
- Air Temperature: cold-cool
- Dew Point: steady
- Pressure: usually falling
- Precipitation: light, moderate, or heavy

While Passing

- Winds: variable
- Air Temperature: dropping/rising
- Dew Point: usually slight drop
- Pressure: low point
- Precipitation: light, moderate, or heavy

After Passing

- Winds: west-northwest
- Air Temperature: colder
- Dew Point: slight drop
- Pressure: usually rising
- Precipitation: light to moderate

7. Examine the weather map of the contiguous 48 states (Figure 3). The synoptic pattern shown on this map is typical of North American systems during much of the winter season. Use your knowledge of fronts and weather systems to discern current and forecast conditions for selected cities.

a. What type of front is currently located near Washington, D.C? ______________

b. What type of front is currently located near St. Louis? ______________

c. What type of front is currently located near Denver? ______________

d. Describe the current weather conditions in St. Louis. ______________

e. Describe the current weather conditions in Cincinatti. ______________

temperature ________ barometric pressure and trend _______

humidity ________ wind direction ______________

f. Produce a 24 hour forecast for conditions at Kansas City.

temperature ______ barometric pressure and trend ________

humidity ______ wind direction ________

g. Explain the rationale for your forecast.

h. Describe the sequence of changing weather conditions that is likely for Akron during the next 24 hours. Include present and forecast temperature, humidity, barometric pressure, and wind direction.

i. Describe the sequence of changing weather conditions that is likely for Tulsa during the next 24 hours. Include present and forecast temperature, humidity, barometric pressure, and wind direction.

j. Describe the sequence of changing weather conditions that is likely for Indianapolis during the next 24 hours. Include present and forecast temperature, humidity, barometric pressure, and wind direction.

Figure 3 Hypothetical Synoptic Weather Map of the Contiguous United States

Low Clouds	Middle Clouds
Cumulus of fair weather, little vertical development and seemingly flattened	Thin Altostratus (most of cloud layer semi-transparent)
Cumulus of considerable development, generally towering, with or without other Cumulus or Stratocumulus bases all at same level	Thick Altostratus, greater part sufficiently dense to hide sun (or moon), or Nimbostratus
Cumulonimbus with tops lacking clear outlines, but distinctly not cirriform or anvil-shaped; with or without Cumulus, Stratocumulus or Stratus	Thin Altocumulus, mostly semi-transparent; cloud elements not changing much and at a single level
Stratocumulus formed by spreading of Cumulus; Cumulus often present also	Thin Altocumulus in patches; cloud elements continually changing and/or occurring at more than one level
Stratocumulus not formed by spreading out of Cumulus	Thin Altocumulus in bands or in a layer gradually spreading over sky and usually thickening as a whole
Stratus or Fractostratus or both, but no Fractostratus of bad weather	Altocumulus formed by the spreading out of Cumulus
Fractostratus and/or Fractostratus of bad weather	Double-layer Altocumulus, or thick layer of Altocumulus, not increasing; or Altocumulus with Altostratus and/or Nimbostratus
Cumulus and Stratocumulus (not formed by spreading of Cumulus) with bases at different levels	Altocumulus in the form of Cumulus-shaped tufts or Altocumulus with turrets
Cumulonimbus having a clearly fibrous (cirriform) top, often anvil-shaped, with or without Cumulus, Stratocumulus, or Stratus	Altocumulus of a chaotic sky, usually at different levels; patches of dense Cirrus are usually present also

High Clouds	Sky Coverage
Filaments of Cirrus scattered and not increasing	No Clouds
Dense Cirrus in patches or twisted sheaves usually not increasing, sometimes like remains of Cumulonimbus, or towers or tufts	10% or less
Dense Cirrus, often anvil-shaped, derived from or associated with Cumulonimbus	20 - 30%
Cirrus, often hook-shaped, gradually spreading over the sky and usually thickening as a whole	40%
Cirrus and Cirrostratus, often in converging bands, or Cirrostratus alone; generally overspreading and growing denser; the continuous layer not reaching 45° altitude	50%
Cirrus and Cirrostratus, often in converging bands, or Cirrostratus alone; generally overspreading and growing denser; the continuous layer exceeding 45° altitude	60%
Veil of Cirrostratus covering the entire sky	70 - 80%
Cirrostratus not increasing and not covering entire sky	90% or overcast w/ openings
Cirrocumulus alone or Cirrocumulus with some Cirrus or Cirrostratus, but the Cirrocumulus being the main cirriform cloud	Completely overcast
	Sky obscured

Wind Speed	Miles/hr.	Knots
	Calm	Calm
	1 - 2	1 - 2
	3 - 8	3 - 7
	9 - 14	8 - 12
	15 - 20	13 - 17
	21 - 25	18 - 22
	26 - 31	23 - 27
	32 - 37	28 - 32
	38 -43	33 - 37
	44 - 49	38 - 42
	50 - 54	43 - 47
	55 - 60	48 - 52
	61 - 66	53 - 57
	67 - 71	58 - 62
	72 - 77	63 - 67
	78 - 83	68 - 72
	84 - 89	73 - 77

Chapter 10

Storm Systems

Most of the variability in weather and climate results from the migration of intermediate scale systems. **Synoptic**, a word for general view, is often used in meteorology and climatology to show atmospheric data collected simultaneously over large areas. **Meso** refers to smaller, subcontinental-scale, and shorter-lived systems such as thunderstorms and tornadoes.

Air Masses, Fronts, and Cyclonic Storms

The word cyclone conjures up a picture in the minds of many people of a large powerful storm. It is true that, in the tropics, cyclones are dreaded storms because of the devastation they bring, but a different form of cyclonic activity is the pervasive weather maker of the mid latitudes. The **mid latitude cyclone** is tame by comparison with its tropical relative. It is larger but less intense. It and its associated system, the anticyclone, dominate weather and circulation patterns within most of the mid latitudes. It is our goal in this section to explain what cyclones and anticyclones are and how these circulation systems develop, grow, and decay.

Cyclones are centers of low pressure, and **anticyclones** are centers of high pressure. They are transient features. The high and low pressures shown on the television and newspaper weather maps are found in different locations each day. It is apparent that these systems move from west to east along with the mean upper atmospheric flow in the mid latitudes. North America alone may contain as many as 2 or 3 highs and an equal number of lows. Their existence cannot be explained in simple terms by heating and cooling, nor is it readily apparent why they move. We will examine these two questions in more detail.

Air Masses

Because the atmosphere is an open system exchanging heat and moisture with the earth and ocean, the characteristics of the air in the boundary layer depend on the nature of the surface below it. When a large body of air occupies an area for any extended period of time, it gains or loses heat and moisture according to surface conditions. The result is that the air develops similar temperature, moisture and to some extent, horizontal pressure characteristics throughout. The sharp discontinuities (very rapid changes in the temperature and humidity) between unlike bodies of air are useful in classifying distinctive **air masses**.

Figure 10.1 North American Air Mass Source Regions

Air masses develop over large areas of land or ocean where uniformly high or low temperatures prevail. Air masses that come from polar regions are cold while those of tropical areas are warm, and those originating over land are dryer than those that originate over water. In addition, those over water usually display more moderate temperatures than their continental counterparts because of the differential heating and cooling typical of land and water surfaces. Occasionally, air from the extremely cold Arctic and Antarctic areas will invade the mid-latitudes. These latter air masses are exceptionally cold and dry. The major types of air masses are: **continental polar** (cP), **maritime polar** (mP), **continental tropical** (cT), **maritime tropical** (mT), **arctic** (A), **antarctic** (AA) and **equatoric** (E). The primary source regions for these air masses are shown in Figure 10.1.

While air masses possess similar characteristics, they are not identical throughout. Furthermore, once they leave their place of origin and begin to move, they begin to undergo constant modification. Cold air masses start to gain heat whereas warm air masses begin to lose heat. The air mass is modified in three ways: by the ground surfaces it passes over, by mixing and exchanging heat with surrounding air masses, and by gain of solar radiation or loss of longwave radiation to space. This process of modification takes place more rapidly around the margins than near the center for the same reason that ice melts at the edge. The **thermal gradient** is steepest around the edges — that is, for a given distance the temperature differences are greater — so that heat loss or gain is accelerated. Over a period of time (usually a couple of weeks after moving out of the source region) an air mass will undergo gradual modification and lose its distinguishing characteristics.

Air masses that move from warm to cold surfaces lose heat at ground level, producing widespread inversions and very stable conditions. When such an air mass moves over a large metropolitan area the effect is almost like putting a lid over the city. Stability inhibits vertical mixing and restricts circulation. As a result, all of the smoke, fumes, and exhaust emptied into the atmosphere hover near the ground producing a serious hazard to those with respiratory ailments.

Cold air masses, on the other hand, have steep lapse rates. As they move across warm surfaces, the lower layer gains heat from the ground further steepening the lapse rate. Under this condition of instability, any air set in motion vertically is very buoyant and will move upward very rapidly. Large scale convection and turbulent mixing will eventually produce more uniform temperatures and reduce the lapse rate.

Figure 10.2 Typical Stages in the Development of a Mid-latitude Cyclone

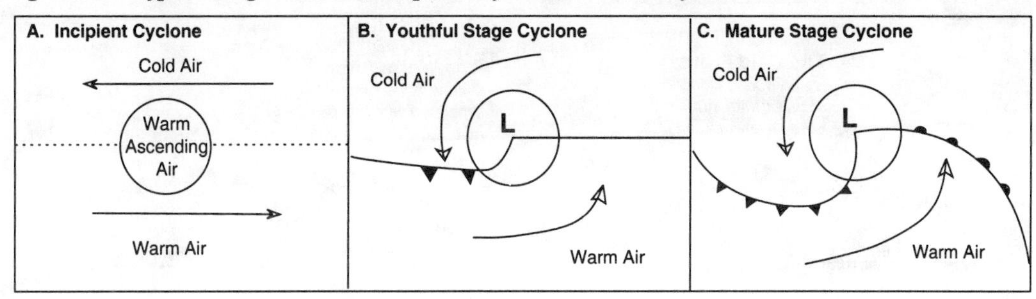

The Development of Fronts and Cyclonic Storms

Low pressure cyclonic storms and fronts most commonly develop along the polar front where polar air contacts tropical air. When the front is aligned east–west, circulation is comprised of easterly winds in the cold air and westerly winds in the warm (Figure 10.2A). A small wave begins to develop and a fall in pressure occurs at the point where the warm air intrudes into the cold air (Figure 10.2B). Once the low develops, circulation across the isobars occurs, and the characteristic cyclonic circulation is evident (Figure 10.2C).

The cold polar air mass eventually becomes detached from the polar front, and the front re-forms behind it. Typical paths followed by cyclones as they move along within the poleward reaches of the westerlies are called **storm tracks** (Figure 10.3). Passage of a mid-latitude cyclone is accompanied by a sequence of somewhat predictable weather conditions commonly associated with fronts.

Cyclonic fronts undergo a characteristic development process. Along the leading edge of the cold air mass, intruding cold air displaces the lighter warm air. This produces a **cold front**, and, if instability exists, widespread precipitation may occur as the warm air rises (Figure 10.4). Cold fronts are plotted on weather maps as lines with small triangles pointing in the direction of the cold air movement. A warm front is found along the leading edge of the warm air mass where it overtakes and overruns the cold air. The warm air does not rise as quickly here as it does along the cold front and generally does not produce as heavy precipitation. Warm fronts are plotted on weather maps as lines with semicircles protruding in the direction of warm air movement.

Figure 10.3 Typical Mid-latitude Cyclone Storm Tracks

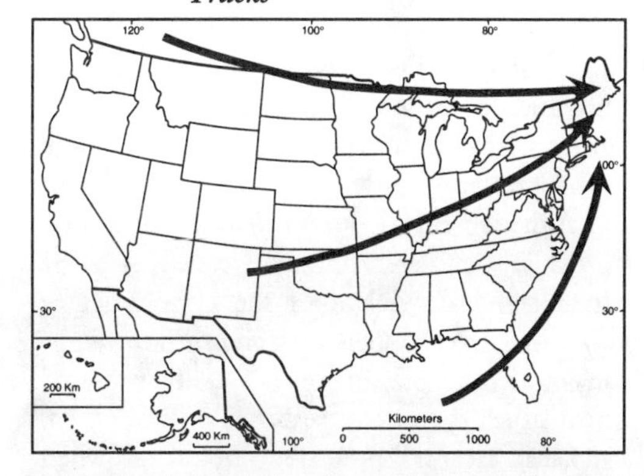

As the cyclone drifts eastward at a speed that typically ranges from 320 km (200 miles) to over 1600 km (1,000 miles) per day (13 kph to 67 kph), the wave deepens. The faster moving cold front overtakes the slower warm front

Figure 10.4 Cross-sectional Profile: A Cold Front and Warm Front

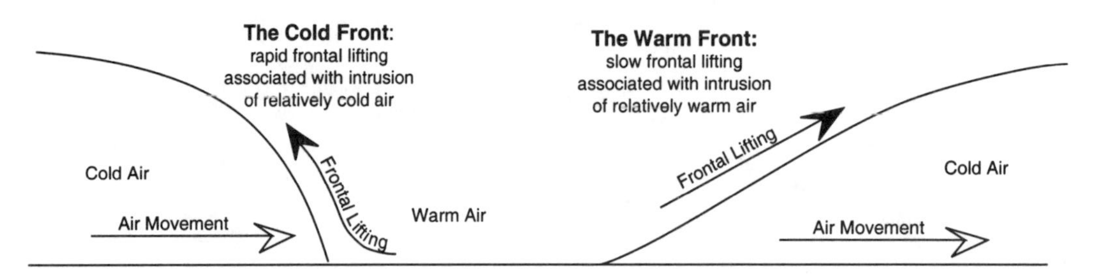

(Figure 10.5) and eventually closes on the warm front to complete the process. This last stage is known as an occlusion. As this cyclone dissipates, new storms typically will have already developed to follow essentially the same path. Mid-latitude cyclonic storms are frequently separated by anticyclones (regions of high surface pressure).

Although mid latitude cyclones, or winter storms as they are sometime called, are the largest and least violent of the cyclones, they may have great turbulence, strong surface winds, and low central pressures. The "century storm" which struck the eastern seaboard on March 13, 1993, had a central pressure of 962.4 mb when it passed Newark, New Jersey. That was lower than the central pressure when hurricane Hugo passed Charleston, Columbia, or Charlotte. This once in a century storm set a record low pressure in Charlotte, NC at 971.2 mb; slightly lower than the 975.2 mb recorded during the passage of Hugo.

Figure 10.5 Cross-sectional Profile: An Occluded Front

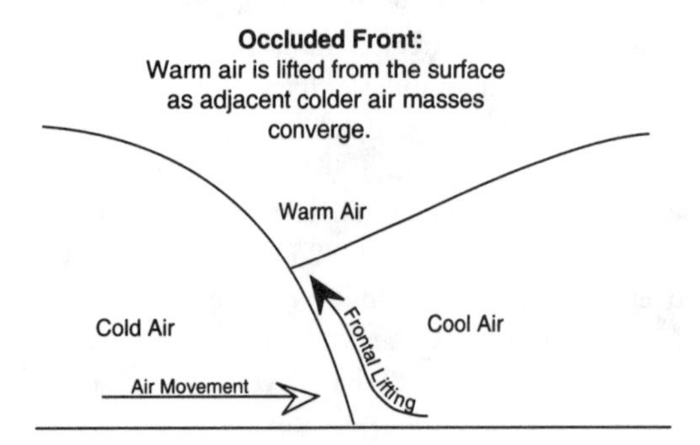

Tropical Cyclones

Tropical regions also experience cyclones. Unlike mid-latitude cyclones, tropical cyclones are not associated with fronts. They differ in other ways as well. They have a distinct seasonal occurrence which varies according to the region, and they are taller and more intense than mid-latitude storms. Northern hemisphere hurricane observations since 1886 suggest that these storms occur most frequently in August, September, and October, with the greatest probability in September.[1] Pressures in the center of a tropical cyclone may be as low as 960 millibars, its winds may reach speeds of over 320 kph, and rainfall may be very intense with some locations receiving as much as 50 cm (20 in.).

Tropical cyclones are known by different names in different parts of the world. In southeast Asia, they are called typhoons; in

[1]*Roger A. Pielke. The Hurricane. (New York:Routledge, 1990) 19.*

Figure 10.6 Selected North Atlantic Hurricane Tracks. Tropical cyclone paths are frequently erratic and difficult to forecast

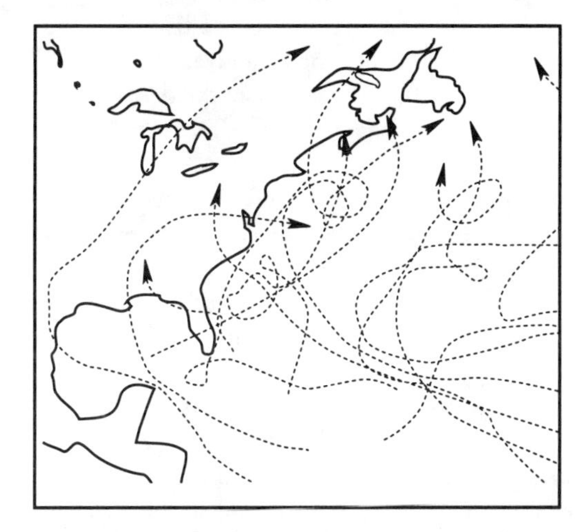

Figure 10.7 Structure of a Tropical Cyclone

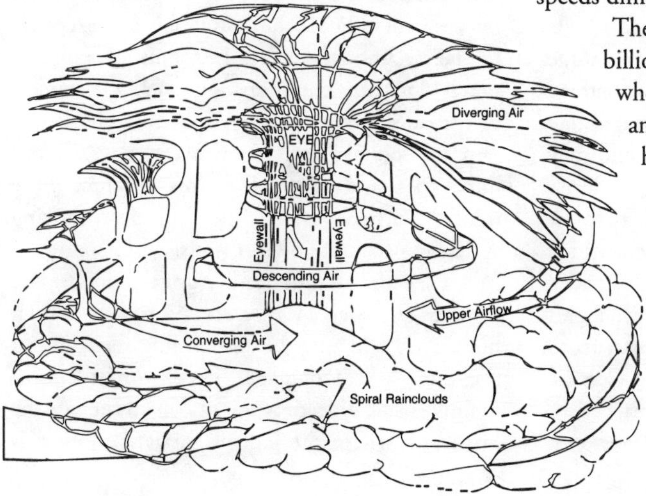

Australia they are called willy-willies in the local vernacular; and, in the U.S. we call them hurricanes. They all originate over the tropical ocean between 5° and 15° north and south of the equator. They cannot form on the equator because coriolis is necessary to oppose the pressure gradient force and impart circular wind motion. Once formed, North Atlantic hurricanes typically move westward, gradually veer poleward, and drift east-northeast, but their stormtracks are notoriously difficult to predict (Figure 10.6). Charts with closely spaced parallels and meridians are used to plot the path of hurricanes. One such chart is included as Appendix D.

The motion of a hurricane is much like a whirlpool in water. Winds spiral inward and a depression forms in the center (Figure 10.7). The central portion of the hurricane, called the eye, is free of clouds and is a zone of subsiding, warm, dry air. Immediately surrounding the eye is a cylinder of very strong winds and rapidly ascending warm moist air. Its location is marked by a wall of clouds, the eyewall, from which intense rains fall. Though the eye itself is calm, the winds surrounding the eye are fierce. Wind speeds diminish outward from the eyewall.

The rapidly ascending air carries literally billions of gallons of water vapor aloft where it condenses, releasing immense amounts of energy to feed the hurricane. The hurricane is an open system and, while over water surfaces, it develops a steady state. It receives a continual input of moisture (and latent heat) and puts out immense quantities of water, dissipating the energy through wind friction. Under these circumstances it is a self-perpetuating system.

Once a hurricane moves over land, it is cut off from its supply of energy and water. Heavy rains continue until the moisture supply is depleted. Without its continual moisture input to supply energy, the intensity of the storm diminishes and it degenerates into a weak cyclonic storm.

Tropical cyclones are the most devastating of the severe storms. It is not uncommon for a typhoon in Southeast Asia to kill thousands of people, more than all the tornadoes that occur in the world for several years. One hurricane striking the Gulf Coast of North America can cause billions of dollars worth of damage. Hurricane Agnes, which struck the Florida coast in 1972, caused $3 billion in damage, but that was dwarfed when Hurricane Andrew did nearly $25 billion in damage to south Florida on August 29, 1992.

There are five major hazards associated with hurricanes: storm surge, strong winds, heavy rainfall, tornadoes, and contaminated water supplies.

By no means is all of the damage and death wrought by tropical cyclones a direct result of strong winds. Flooding of coastal areas and inland stream valleys takes many lives and causes much damage. The greatest flood hazard results from a dome of high water or storm surge beneath the hurricane. The height of the storm surge is inversely proportional to the magnitude of the low pressure at the storm center and may reach more than 20 feet above normal sea levels during severe storms. It is responsible for most of the loss of life from hurricanes. Hugo, which hit Charleston, South Carolina in 1989, had a storm surge of 17 feet and Camille, which hit the Gulf Coast in 1969, had wind gusts up to 220 mph and a storm surge of 23 feet. Camille killed 258 people. Hurricane Andrew reached its greatest intensity just east of the northern Bahamas where the central pressure dropped to 922 mb and maximum sustained winds reached 249 kph (155 mph). The storm surge during Andrew on the northern side of Eleuthera Island exceeded 7 meters (23 feet) and later set a record for southern Florida at 5 meters (16.9 feet). In 1970 a tropical cyclone in India killed an estimated 300,000 people; and again in 1991 more than 150,000 died in southern Asia from disease or drowning in the aftermath of a typhoon.

As wind speed increases, the pressure exerted on stationary objects increases many times. The force exerted on a given area of a vertical wall (F/A) is a function of the square of the wind speed (V^2) times a constant (C_a).

Many structures are not built to withstand the forces produced by wind speeds greater than 150 kph (93 mph). Buildings that are under construction are especially vulnerable. Debris carried by strong winds can become lethal projectiles.

Torrential rainfall may accompany hurricane landfall and contribute to flooding and overwash of causeways and other low elevation areas. Because tropical cyclones and hurricanes are embedded in tropical moist air masses, they hold the potential for excessive rainfall. Although 50 – 100 mm of rain are much more common, a 1979 tropical disturbance poured nearly 500 mm (19 in) of rain on coastal Texas over several days. A few hurricanes and tropical storms produce little rainfall, either because they are fast moving systems or for unknown reasons.

Tornadic winds and funnel clouds are sometimes observed in association with strong hurricanes and may be responsible for some of the most severe damage. These vorticies form mainly in the leading right quadrant of the hurricane where the storms forward motion is

Table 10.1 The Saffir/Simpson Scale

Hurricane Category	Barometric Pressure (mb)	Wind Speed (mph)	Storm Surge (feet)
1	>980	74-95	4-5
2	965-979	96-110	6-8
3	945-964	111-130	9-12
4	920-944	131-155	13-18
5	<920	>155	>18

added to the cyclonic wind speed and where instabilities are most intense.

Often the biggest problems come after the storm. With the destruction of water and sewer systems, water supplies become contaminated. In Southeast Asia, every typhoon that strikes the coast is followed by a cholera epidemic that almost invariably kills more people than were drowned in the storm. The health threat is not limited to developing areas, however. Following the storm that struck Darwin, Australia, on Christmas Day 1974, the government evacuated over half the population of 55,000 largely as a public health measure.

The most commonly used classification system for hurricane intensity is the Saffir/Simpson scale (Table 10.1). This scale provides a convenient method for comparing hurricanes on the basis of typical barometric pressure, wind speed, and storm surge.

Thunderstorms

When strong instabilities and sufficient moist air are present, thunderstorms may occur. They may range in intensity from moderate to violent, and are very common in summer, especially in the Rocky Mountains and the southeastern U.S. Thunderstorms may be convectional in origin, or may also occur in association with frontal conditions. During late

Figure 10.8 Tornadoes Begin with Strong Updrafts inside Cumulonimbus Clouds

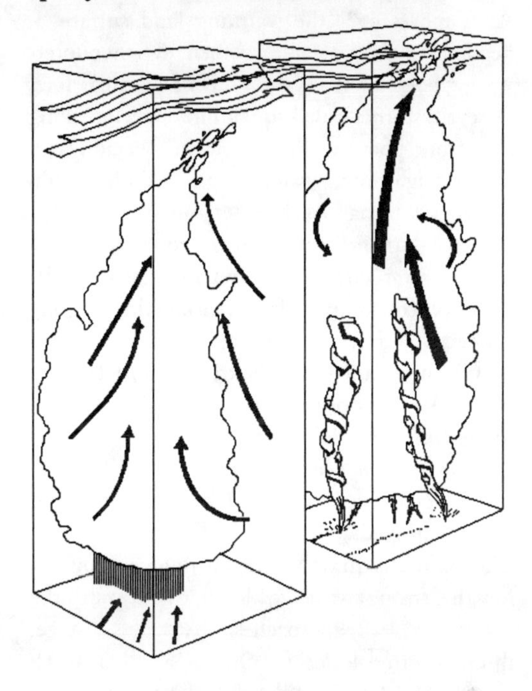

Figure 10.9 The Tornado Season within the United States Based on Frequency of Occurrence 1958–1987[2]

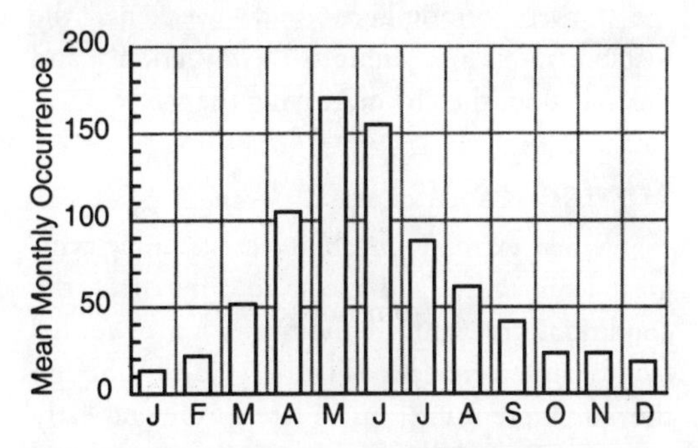

[2]*Edward W. Ferguson, Frederick P. Ostby, and Preston W. Leftwich, Jr. Tornadoes: Slow Start, Fast Finish. Weatherwise 42 (February 1989): 28–35.*

spring and early summer in the southern Great Plains cold polar air moves rapidly southward. As it moves over the warming land surface, it becomes very unstable. When it encounters warm moist air from the Gulf of Mexico, lines of severe storms called squall lines develop along the front. Severe thunderstorms occur with much lightning, hail, high winds, and, frequently, tornadoes.

A thunderstorm tends to develop a characteristic circulation system. Early in its development, up-drafts mark the strong convective activity and rapid cloud growth occurs to altitudes of 7.5 km (25,000 feet) or more. As the storm reaches the mature stage, heavy precipitation falls and down-drafts develop in the central portion of the storm. These down-drafts may be felt as an onrush of cool air immediately prior to the onset of rain. The up-drafts may be very strong, accounting for the strong winds which blow toward the storm as it approaches. At this stage, thunderstorm clouds may reach as high as 18 km (60,000 feet) and develop the characteristic flat-top anvil shape. In the dissipating stage, the up-drafts become weak and most of the air flow is downward. Falling rain cools the surface and the lower atmosphere and weakens the instability. This inhibits convection and precipitation, thereby destroying the storm.

Tornadoes

When extreme instability exists along with great temperature and moisture differences, the conditions are right for the formation of funnel clouds and tornadoes (Figure 10.8). Although they are more numerous in late spring and early summer, tornadoes can occur in any season (Figure 10.9) Tornadoes are not benign storms like the one that delivered Dorothy and Toto unharmed to the Land of Oz. Rather, tornadoes are the most intense storms known. These dark funnels can develop wind speeds of 400 kph (260 mph) even though the vast majority have winds of less than 145 kph. The central vortex has pressures which may be 100 mb lower than the surrounding air so that, as they move over buildings, the sudden pressure differential may cause the buildings to appear to explode. Larger tornadoes may contain miniature vortices (10 meters in diameter) which rotate in or near the wall of the main funnel and are believed to be responsible for some of the more spectacular damage caused by tornadoes.

Tornadoes are confined to only a few locations around the world. The United States has, by far, the highest incidence, but they also occur in Japan, Australia, and other areas. Tornadoes often occur in groups. This was well illustrated in May 1974 when approximately 100 separate tornadoes were spawned from a front as it moved from Oklahoma and Kansas through the Midwest. On the night they struck, Xenia, Ohio, was virtually demolished, and tornadoes even occurred in the Southern Appalachian Mountains, much to the surprise of tornado experts at the time.

A Final Note

In the mid-latitudes, cyclonic activity dominates the circulation patterns. Low pressure storms with counterclockwise winds develop along the polar front and move to the east. Once the storm moves through an area the cyclonic flow is replaced by the counterclockwise winds of the anticyclone.

Much of the rainfall in tropical areas is produced by convectional storms associated with weak systems called easterly waves. Occasionally troughs will develop across the trade winds causing isobars to form a wave whose crest points poleward. The gradient wind parallels the isobars and the narrowing of the isobar spacing causes the winds to converge on the eastern side. Convergence and instability trigger adiabatic cooling which produces rain. On the western side of the wave, winds diverge and subsidence and dryness dominate. The waves move westward with the zonal flow at speeds of 15 to 25 km per hour (10 to 15 mph).

Cyclones also occur in tropical areas where they move from east to west. These violent storms are not associated with fronts and are larger and more intense than the mid-latitude storms. Much of the rain in tropical areas comes from easterly waves which develop in the tradewind zones. Zones of convergence develop in the winds and drift to the west with the circulation system.

Chapter 11

Climates

In a very general sense **weather** is simply the condition of the atmosphere at a particular place at a specific time. It is the state of the atmospheric system as reflected by the variables of temperature, moisture, pressure, density and velocity. To describe the atmospheric state, we measure temperature, precipitation, humidity, pressure, and wind.

Climate differs from weather only in perspective. It is a long-term view of atmospheric behavior rather than an instantaneous one. The concept of climate encompasses the full range of weather conditions that might occur. Since we are generally more interested in knowing what the weather is likely to be rather than what could happen, we generally choose to describe climate in terms of the most likely weather conditions. That explains our preference for "normal" temperature, average rainfall, and other expected conditions.

We have stressed the fact that weather is the instantaneous state of the atmosphere, and that every individual place has its own weather. To be able to describe weather patterns spatially, simultaneous observations of the basic weather variables are required over a wide area. In order to attain the goal of simultaneous observation, the nations of the world have agreed to make weather observations four times each day — at 0000, 0600, 1200, and 1800 hours **Greenwich Mean Time** (GMT). By agreement of all nations, the meridian which passes through the Greenwich Observatory in England was selected as the prime meridian (0°). Greenwich Mean Time is the time as kept by an official clock at the observatory. These four hours are called the **synoptic hours** and plotting the variables on a map yields a synoptic weather chart. The term comes from "synopsis" which means "a general view." In recent years satellite imagery has provided an additional dimension by providing us with pictures of weather conditions from space along with upper atmosphere measurements.

One of the primary reasons for collecting detailed data about the atmosphere is to enable meteorologists to predict the weather. This task used to be accomplished manually, but now computers are used to solve immense numbers of equations to predict the changes that will occur in the atmosphere in the next few hours.

Thorough description of current or predicted weather requires much more information than just temperature, pressure, winds, and moisture content of the air. The orchard owner may need to know whether frost is likely; the airplane pilot needs information on visibility; health officials might want to know if an inversion will develop; transportation officials need advance warning of snow so they can mobilize work crews; and the general public might like to know if thunderstorms are likely to be accompanied by high winds and hail. It is easy to see that different bits of weather information are needed for different purposes, and that complete description of weather involves many variables.

Weather is fickle. It may change suddenly and frequently. Weather may have distinct seasonal characteristics with either temperature or precipitation varying with the time of year. Climate is more stable. It may undergo change, but climatic change occurs slowly and persists for long periods of time.

The same atmospheric variables are used to describe climate as weather. Remember that climate is not just average weather, but that it embodies the full range of weather conditions. The farmer who tills the soil cannot grow his crops in average rainfall. He must depend on the rain that actually falls, and sometimes that is well below average. If he plants his crop so that it emerges the day following the average date of the last killing frost, he runs almost a 50 percent chance of losing it.

The best general description of climate includes information about the most likely conditions (perhaps expressed as mean values) as well as how variable the conditions are. Because of the low probability of occurrence, extreme conditions usually are not included in climate summaries. However, knowledge of rare events is required for some types of planning such as disaster planning and flood control.

Many kinds of climatic data are needed for different purposes. The health planner might need to know wind speed and direction and their variability in order to make decisions regarding possible environmental consequences of urban development. Agriculturally, rainfall and growing season are of utmost importance. Heating fuel companies require information about the intensity and frequency of cold weather so that they can forecast demand and allocate fuel properly. Access to the right kinds of climatic data may play an important role in the success of many endeavors.

The variability of weather has been stressed repeatedly. To the casual observer, it might appear that this variability is entirely random. Such is not the case. Certain weather conditions occur in "runs." For instance, it is much more likely that a sunny day will be followed by another sunny day than might be expected based on just the percentage of sunny days. In common usage people speak of weather runs as "spells.. Thus a farmer might speak of a drought (an extended period of days without rain) as a "dry spell."

From a climatic perspective even longer term runs occur. During the 1920's and early thirties, settlers moved into western Oklahoma, northern Texas, northeastern New Mexico, and southwestern Kansas in large numbers. Rainfall had been abnormally high in this area for several years. The plentiful rain encouraged farmers to plow up the range and plant crops. A few years later, when precipitation returned to normal, the crops dried up, and the soil blew away to create the infamous "dust bowl" of the 1930's. Weather scientists recognize that some of these runs of weather tend to recur on a somewhat regular basis. They refer to them as cycles and spend a great deal of time trying to identify the causes. Weather cycles obviously are an important component of climate, and the concern seems justified.

Factors That Control Weather and Climate

The major control for weather and climate is location with respect to the **zonal wind and pressure systems**. Each of the zones is associated with a set of wind, moisture, and temperature regimes. The time period involved is most often an annual period, but may be longer or shorter. **Zonal flow** is a function of solar radiation and because the general pattern of solar radiation is a function of *latitude*, it is common practice to say that latitude is the dominant climatic control.

The effects of latitude may be either mitigated or magnified by other factors. For instance, those areas in the center of large land masses experience much greater annual temperature ranges and drier air than do those locations near oceans. This effect, referred to as **continentality**, is a major one on the interior climates of large continents located in the middle latitudes of the northern hemisphere. The presence of relatively cold **ocean currents** influences not only downwind air temperatures but also reduces precipitation by increasing the frequency and duration of stable conditions. Warm ocean currents can transport large quantities of heat to locations in the mid and high latitudes.

Landforms, especially mountains, may influence climates. Altitude plays a dual role in weather. First of all, air temperatures decrease with increasing elevation despite the fact that solar radiation is more intense in the thin air. Thus, mountains have cooler weather than lowlands. Secondly, when moisture is present, higher elevations get more precipitation than lower elevations.

Increased precipitation in mountains is especially noticeable on windward slopes. These slopes get more precipitation, more fog, and more cloudy days than leeward slopes do. If the mountains are high enough, leeward slopes may be dry enough to be desert, as in the case of eastern Washington and Oregon. This effect, known as a rain shadow, becomes gradually smaller as one moves away from the mountains.

Locally there are variations in weather and climate caused by topographic conditions and land cover. Where slopes are steep the orientation and slope angle are important. In the northern hemisphere, south facing slopes get more direct solar radiation because of their higher angle of incidence. They warm up quicker, get hotter, and stay warm longer. In high latitudes these slopes lose their snow cover earlier and have higher evaporation rates.

Type and amount of **local vegetative cover** and soil conditions can have some effect on weather. First of all, photosynthesis and transpiration by dense crops can utilize a high percentage of the net short wave radiation during the growing season. This is sufficient to reduce other terms in the heat budget equation significantly. The greatest effect would be on soil heat, and lower soil temperatures result in less convective and radiative heating of the air. Heavily vegetated areas are cooler, all other things being equal.

The site effects that we have briefly surveyed along with the effects of altitude and landforms tend to disrupt what otherwise would be a simple generalized pattern of weather and climate. In the following sections we will ignore the more localized effects while examining the generalized pattern and the factors responsible for it.

Patterns of Weather and Climate

The general climatic controls, latitude, continental location, and location with respect to storm tracks affect large areas in a similar fashion. The pattern of climates they produce is simple and predictable. Superimposed on that pattern are the influences of altitude, landforms, and individual site characteristics that make the actual climate patterns very complex. Rather than trying to examine and explain actual patterns in all their complexity, we will again employ the simplifying concept of a model.

We will begin by reviewing the generalized pattern of circulation and the characteristics within each zone. To further simplify conditions, we will concern ourselves with a simplified continental and oceanic system rather than the more complicated real world and confine our examples to the western hemisphere.

Weather and Climate Zones

Weather and climate patterns are the principal topics in the succeeding pages of this chapter. We will explain the major differentiating characteristics and try to pinpoint the controlling factors in each case. It should be recognized that these are the causal mechanisms of weather, too. Simply, we will concern ourselves less with the variability that characterizes changing weather conditions and concentrate instead on the more general climatic conditions.

One of the oldest climate classifications was proposed by the Greeks. Little was known at that time about the tropical or polar regions, and the Greeks suspected that the tropics were too hot to be inhabited and the polar regions were too cold. Thus, they proposed a three-zone classification with torrid, temperate, and frigid zones. Modern classifications, though much more complex, still recognize comparable zones, although we call them tropical, temperate or mid-latitude, and polar climates.

Tropical Weather and Climate

Tropical climates are differentiated from nontropical climates on the basis of temperature. Of all climates, the tropics, located between the two zones of subtropical high pressure, roughly between 25°N and 25°S latitude, receive the most solar radiation and have the least seasonal variability. It is this lack of seasonal variability, the absence of any cold winter season with freezing weather, that distinguishes the tropical climates from the mid-latitude climates.

Within the tropics the major differentiating climatic characteristic is precipitation. The common conception of tropical weather is one of a monotonous succession of days with high temperatures, thundershowers at 4 p.m., oppressive humidity, and little wind to provide relief. Actual weather conditions in most tropical areas depend on the location and season of year. Although the rainfall may be triggered by synoptic disturbances or convective activity, the annual amount and its seasonal distribution depend primarily upon the zonal circulation — specifically the ITC zone, the easterly tradewinds, and the subtropical anticyclones of the horse latitudes.

The atmosphere along the ITCZ is essentially unstable, and convectional activity is easily triggered by surface heating or by atmospheric disturbances. Along the ITCZ where temperatures are uniformly warm year round, rain is abundant and conditions differ from those described above only by virtue of the fact that it rains neither every day, nor at the same time. It does rain often, however. Seldom does a place go more than a week or 10 days without rain, and it may rain considerably more frequently than that.

At the other extreme are the areas dominated by the dry, subsiding air of the subtropical anticyclones. Clear skies, low humidity, and infrequent rain are the rule.

Between the subtropical anticyclones and the ITCZ lie the tradewind zones, which carry the dry air that subsides in the anticyclones back toward the equator. The tradewinds lose heat to the cooler ocean and take up moisture through evaporation. The cooler air in the lower 600 m (2,000 feet) produces a strong inversion layer which tends to be very stable, especially on the eastern margins of the oceans where cold ocean currents are located. Thus, over the oceans the tradewind zones are dry, despite the fact that the inversion layer may be almost a solid mass of clouds. Along windward coasts with elevations sufficient to induce orographic lifting, the tradewinds may bring abundant rain, but convective rainfall over land masses is virtually nonexistent in these zones because of the extreme stability that confines cloud activity to the lower boundary layer.

The tradewind inversion may be disrupted by synoptic disturbances, however, and low-level convergence may give rise to widespread rainfall. Easterly waves, tropical depressions, and hurricanes are common occurrences especially during late summer and early fall.

In South Asia, monsoonal flows alter the basic pattern so that prevailing winds which blow inland in summer bring heavy rainfall. There is no inversion because these winds originate over the equatorial ocean, and, as they move into the continent, the heated surface initiates strong convective activity that produces widespread rainfall. Orographic effects in mountainous areas help to produce some of the heaviest and most intense rainfall regularly recorded, except for tropical storms. During the winter monsoon the winds that blow from the northeast originate over land and contain little moisture so that dry conditions prevail over most of South Asia.

Synoptic Conditions in Central and South America during January and July

One of the best ways to gain an understanding of the weather regimes that characterize climates and to gain some insight into climatic patterns is to examine synoptic conditions for a real example. Central and South America offer a good opportunity to study tropical climates.

During January much of the Amazon is under the influence of the ITCZ and receives abundant precipitation on a regular basis. Areas to the north and south of the ITCZ are under the influence of the tradewinds. The east coast of Central America and some of the Caribbean Islands with mountainous backbones receive heavy orographic rainfall from the tradewinds. However, places on the west coast, such as Acapulco, experience dry conditions and sunny skies. Interior locations in Venezuela are dry, also, because of the tradewind dominance. Northwestern Mexico and coastal areas of Peru and central and northern Chile are typically cloud free and dry because of the stability associated with the eastern sides of the Pacific anticyclones. Normally an anticyclonic cell, developing off the coast of Baja, California, is reinforced during this season by the high pressure cell over North America, producing exceptionally dry conditions and clear skies over the Sonoran Desert.

Synoptic conditions in July are almost a mirror image of those in January. The ITCZ moves into a position over northern South America and brings rain to Venezuela, the Caribbean Islands, Central America, and southern and central Mexico. Even the Sonoran Desert experiences a rainy season of sorts as the low that develops over the Southwest helps weaken the dominance of the anticyclonic circulation and pulls moist air into the area. Subsequent convection produces scattered thunderstorm activity which may be intense, but highly localized and of short duration. The Central Amazon is still under the influence of the ITCZ during July so that it has an unbroken year-round rainy season. But to the south the dry, stable tradewinds have replaced the ITCZ, and the Brazilian Northeast slips into a winter-long period of drought and high temperatures. Even though there is no cold season it is still appropriate to speak of winter because June, July, and August are the Southern Hemisphere winter season. Although the sun angle is lower than in summer, the absence of clouds allows more solar radiation to be absorbed by the surface.

It is clear from the foregoing discussion of tropical weather that three distinct climatic regimes can be identified: 1) always wet; 2) always dry; and, 3) wet during high sun and dry during the low sun period. Furthermore, the latter regime may be caused by one of two distinct mechanisms – the migration of the ITCZ or monsoonal circulation. Based on these regimes we can recognize four types of tropical climates. They are: 1) the **Rainy Tropics** situated along the equator and extending poleward as far as the Tropics of Cancer and Capricorn in mountainous areas and along eastern coasts where onshore tradewinds and orographic effects combine to produce year-round rain; 2) **Tropical Deserts** which are found in the interior and along the west coasts of continents in the vicinity of 25-30° north and south latitude where the subsiding air of the anticyclones causes dry, stable conditions; 3) **Wet-Dry Tropics**, often called Savanna after the widespread grasslands found there, which are alternately dominated by the ITCZ that brings rain in summer and dry winter tradewinds; and 4) Tropical Monsoon throughout much of South Asia.

Mid-Latitude Weather and Climate

Outside the tropics, temperature is the major differentiating characteristic of most climates, and precipitation is relegated to a secondary role. It is, therefore, seasonality which is the major control of weather and climate in the mid-latitudes. Seasonal temperature differences tend to increase toward the poles, but location with respect to the ocean is a factor of considerable importance. Continental interiors have much higher annual temperature ranges than do coastal locations. Short term temperature variations may be significant, especially during the winter season, as outbreaks of cold polar air penetrate into mid-latitude locations.

Precipitation varies spatially and seasonally in the mid-latitudes. The supply of moisture and the presence of appropriate lifting mechanisms are the major controlling factors. In general, precipitation decreases toward continental interiors as a function of moisture availability. Strong seasonal patterns occur principally in west coast areas where the stability of summertime anticyclonic circulation produces dry summers as far poleward as 35-40°. During winter, however, the stability weakens due to the development of the continental highs, and as the westerly winds shift toward the equator, they bring moisture into these coastal areas.

Orographic effects are noticeable in all major mountain systems in the mid-latitudes where the western slopes receive abundant rain and the lee slopes are considerably drier. Both frontal and convective activity are important precipitation mechanisms in mid-latitude areas. Cold fronts associated with the passage of cyclonic storms are responsible for most wintertime precipitation. Even much of the late spring early summer rainfall that is triggered by convection is associated with fronts. Warm ground surfaces modify cold arctic air to create steep lapse rates and instability.

Seasonality, continentality, the zonal flow (especially the subtropical anticyclones and westerlies), and the migrating cyclones and anticyclones are the major weather and climate controls of the mid-latitudes.

Although the mid-latitudes have a greater variety of climates than the tropics, the climatic pattern is not very complicated. There is a distinct latitudinal boundary due to temperature and a less obvious longitudinal zonation in continental interiors attributable to precipitation. On the North American and European land masses it is difficult for moisture to penetrate into the continental interior so that both have deserts and extensive semiarid climates called **steppes**. Seasonally dry conditions also prevail along western coastal areas for 30° to 40° latitude because of the strong summertime stability that exists on the eastern sides of the subtropical highs. During the winter, however, the westerlies bring rain to these dry summer subtropical climates. These mild, dry summer climates are often referred to as **Mediterranean** climates because that is the prevailing climate around the Mediterranean Sea.

West coast areas poleward of 40° have mild, rainy climates called **West Coast Marine** climates. They are under the domination of the westerlies all year long, and the marine air masses supply ample moisture and have a moderating effect on temperatures in winter and summer.

Along the east coasts in the mid-latitudes where moisture is more abundant the climates are differentiated by temperature. They range from the mild **Subtropical** climates in such places as Argentina and the Southeastern U.S. to the cold **Sub-arctic** climates of Canada, Alaska, and the USSR. There are no sub-arctic climates in the southern hemisphere because there are no significant land masses in those latitudes. Between the humid subtropical with its long, hot summers and short, mild winters and the sub-arctic with its short, cool summers and long, cold winters, lies the **Humid Continental**, a climate with warm to hot summers and rather harsh winters. Much of the wintertime precipitation in the humid continental areas falls as snow. In none of these humid climates is there any distinct seasonality of precipitation.

Polar Weather and Climates

Although there are seasonal changes in the polar regions, polar weather and climates, like those of the tropics, are somewhat monotonous. Polar regions essentially lie inside the Arctic and Antarctic Circles, and the long polar night and the low angle of illumination are the controls of weather and climate. Similar cold conditions are found in high mountainous areas where temperatures are low because of the altitude.

Only two climates exist in polar zones: **Tundra**, which has a short growing season in summer, and **Icecap**, or frost climates, which are frozen year-round. Because of the low moisture content of cold air, neither climate receives much precipitation (which occurs mostly as snow), but the cold temperatures, frozen soil, and poor drainage combine to cause wet-swampy conditions in the low-lying tundra areas.

We have derived a genetic classification of climates that illustrates the descriptive characteristics of the climate and the factors responsible for them. The pattern reoccurs on all the land masses because of the effect of the major climatic controls.

The Köppen System of Climate Classification

Numerous systems for classifying climates have been devised. One particularly suited for identifying large-scale regional climates was designed by a German botanist, **Wladimir Köppen** (1846–1940). The underlying assumption of the Köppen system is that specific forms of natural vegetation are associated with unique thermal and moisture regimes. As a result, plant associations are reflections of climatic regions — areas in which homogeneous sets of climatic conditions are found. The temperature and moisture conditions that exist at the recognized boundaries of plant associations, in turn, identify the limits (or basic requirements) for each climatic type. Although Köppen's approach is not without its shortcomings, it uses easily obtained climatic data, is readily comprehensible, and offers a useful global perspective for the study of climatic types.

Köppen's classification system is structured about an alphabetical code (using capital and lowercase letters) that is used for generalizing climatic data (Figure 11.1). At the most general level, five climatic realms are recognized. Proceeding from the equator to the poles and identified by capital letters, they are: the A climates, or winterless realm of the humid tropics; the B climates which are moisture deficient; the C, or humid and mild winter climates of the lower middle latitudes; the D, or humid and severe winter realm found in upper middle latitudes of the Northern Hemisphere; and the E, polar or summerless climatic types of high latitudes. Each of these climatic realms is subdivided into climatic regions that possess special characteristics, including seasonal variation in temperature and/or precipitation.

Environmental Energy

Climatologists have produced many useful modified or simplified variants of the original Köppen classification to serve various purposes, but few can illustrate the significant interaction between climate and other earth surface processes better than the **Schmudde Model** of world climatic environments.[1] This conceptual model draws upon what we have already learned about the global distribution of moisture and heat to explain the fundamental earth surface processes. Unlike the Köppen model which was conceived to explain global patterns of vegetation, the Schmudde Model uses the concept of energy to link climatic regimes with many hydrologic, fluvial, geomorphic, edaphic, and biological systems. Energy is defined as the capacity to produce an effect. Three fundamental expressions of energy are identified as climatically driven elements of earth surface change. They are biological energy, chemical energy, and erosional energy.

Biological energy can be defined as net primary productivity or the rate at which plants capture and store chemical energy as biomass by photosynthesis minus the rate of energy loss from respiration. Normally this rate is measured in units of energy such as kilocalories per square meter per year. Among generic environmental settings tropical and subtropical estuaries, swamps, marshes, tropical rainforests, and even some temperate forests have the greatest average annual net primary productivity while temperate grasslands, arctic and alpine tundra, and deserts have the least. For biological energy to be high warmth and moisture are required at the same time. Either cold or aridity are unfavorable. Some types of biological activity are interrupted

1Harper, Robert A. and Theodore H. Schmudde. Between Two Worlds: An Introduction to Geography, 3rd Edition, Dubuque, IA: Kendall/Hunt, 1984; and Carter, Douglas B., Theodore H. Schmudde, and David M. Sharpe. The Interface as a Working Environment: A Purpose for Physical Geography, Technical Paper No. 7 Association of American Geographers, Commission on College Geography, Washington, D.C., 1972.

Major Climatic Types in the Modified Köppen System

Letter Code
lst 2nd 3rd Definition

A .... Coolest t ≥ 18°C*
  f .... Precipitation in driest month at least 6 cm
  m .... Precipitation in driest month less than 6 cm but equal to or greater than 10 - P/25*
  w .... Precipitation in driest month less than 10 - P/25
B .... 70% or more of annual precipitation falls in warmer six months and P < 2T + 28*
OR
70% or more of annual precipitation falls in cooler six months and P < 2T
OR
Neither half of year with more than 70% of annual precipitation and P < 2T + 14
  W ....P less than 1/2 upper limit of applicable requirement for B
  S .... P less than upper limit for B but more than 1/2 that amount
    h..................... T > 18°C
    k .... T < 18°C
C ....Warmest t > 10°C and coldest t > 0°C and< 18°C
  s .... Precipitation in the driest month of the warmer 6 months less than 4 cm and less than 1/3 the amount in the wettest month of the cooler 6 months
  w .... Precipitation in the driest month of the cooler 6 months less than 1/10 of amount in the wettest month of the warmer 6 months
  f.................... Precipitation not meeting conditions of either s or w
    a ....Warmest t ≥ 22°C
    b .... All four warmest ts ≥ 10°C and warmest t < 22°C
    c .... One to three ts ≥ 10°C and the warmest t < 22°C
D .... Warmest t > 10°C and coldest t ≤ 0°C
  s .... Same as under C
  w ....Same as under C
  f .... Same as under C
    a ....Same as under C
    b ....Same as under C
    c ....Same as under C
    d ....Coldest t < -38°C (d is then used instead of a, b, or c)
E .... Warmest t < 10°C
  T ....Warmest t > 0°C and < 10°C
  F ....Warmest t ≤ 0°C

*T is the mean annual temperature in °C; t is the mean monthly temperature in °C; and P is mean annual precipitation in centimeters.

Sources: Köppen, W., 1900; Köppen, W., and R. Geiger.,1930, Thornthwaite, C. W., 1933; Trewartha, G. T., 1943; and Critchfield, Howard J., 1974; Strahler and Strahler, 1987.

when air temperatures fall below freezing. Many other processes are virtually terminated when soil moisture freezes and plants experience physiological drought because root systems cannot extract water from the soil.

Chemical energy is defined as the rate of chemical decomposition. Many of these processes also depend upon availability of heat and moisture. Various processes such as leaching, illuviation, and precipitation of iron and aluminum from upper to lower layers within the soil profile are aqueous processes accelerated by heat. These processes, collectively known as chemical weathering, may be sustained at lower rates during periods of substantial aridity or cold. In contrast with biological activity, chemical activity may operate with small quantities of water and heat.

Erosional energy is the rate of mechanical movement. The three agents of this activity involve the physical transportation of materials by moving ice, wind, or water. On a global scale moving water is the dominant agent of erosion. Even within desert environments, the ability of flash floods, driven by infrequent but intense thunderstorms, to transport sediment exceeds that of **aeolian** (wind based) erosion. In many environments the rate of work done by moving water is keyed to the magnitude and timing of soil moisture surpluses and consequent runoff regimes. In most cases erosion takes place as overland flow or stream flow during and after rainstorm events, but in climates where water remains frozen during the winter, the spring thaw may flush water accumulated over several weeks or months into streams and rivers with highly accelerated rates of erosion and sediment transport. Moving water also washes fine particles downward through the soil in a process called eluviation.

Each of these types of activity will vary in importance depending on the annual sequence of heat, moisture, and seasonality (Figure 11.1). Many tropical environments such as the tropical rainforest with abundant heat and precipitation might be identified as year-round high energy environments. Conversely, other areas might be classified as year-round low energy environments because they suffer from perpetual cold, such as atop Greenland's ice covered surface, or nearly perpetual aridity, such as within the hard deserts. There are also seasonal environments that experience a definite closed season where all three types of activity virtually cease. Many of the wet and dry climates such as the dry margins of the tropical savanna, the mediterranean, and the continental climates are examples of seasonally closed environments. In most cases when monthly temperatures drop below freezing for four months or more, the station may be considered seasonally closed by cold. Madison, Wisconsin with four months closed represents one of the mildest examples, but Moscow with five months closed and many Siberian stations with seven months closed are more typical.

Environments like the southeastern United States do not experience a true closed season. Although biological activity may be slow during winter, soil moisture is typically recharged and runoff increases so that erosion reaches peak levels. These special types of environments with no closed season are called seasonal energy rhythm environments. Other examples include Europe, much of the pacific west coast, parts of Chile, parts of Argentina, the southeastern segment of Australia, much of southern China, Korea, and Japan.

Summary

The goals of this chapter have been to introduce the concept of climate and to examine the general patterns of weather and climate and factors responsible for the pattern.

Weather has been defined as the state of the atmospheric system at a particular location and at a specific time. Climate, on the other hand, is the collection of all states that characterize atmospheric behavior at a place.

Weather and climate are controlled by two sets of variables: a general set, regional in effect (latitude, continental location, and location relative to major storm tracks or ocean currents); and a more specific set, local in effect (altitude, landforms, slope aspect, and surface conditions).

The regional pattern of climates produced by the general set of variables is a genetic classification of climatic types. In the Köppen System it includes fourteen basic types: tropical rainforest, tropical savanna, tropical monsoon, arid, semi-arid, humid subtropical, mediterranean, west coast marine, subtropical monsoon, humid continental hot summer, humid continental warm summer, subarctic, tundra, and icecap. The Schmudde Model identifies six climate types based on levels of biological, chemical, and mechanical energy. They are **Year-round High Environmental Energy, Seasonal Environmental Energy Rhythm, Seasonal Environmental Energy with a closed cold season, Seasonal Environmental Energy with a closed dry season, Year-round Low Environmental Energy (cold), and Year-round Low Environmental Energy (dry)**. Each of these six climate types have common characteristics with respect to annual heat and moisture budgets. A general pattern of hydrologic, fluvial, geomorphic, edaphic, and biologic processes is shared within each climate region.

Figure 11.1 Global Climates

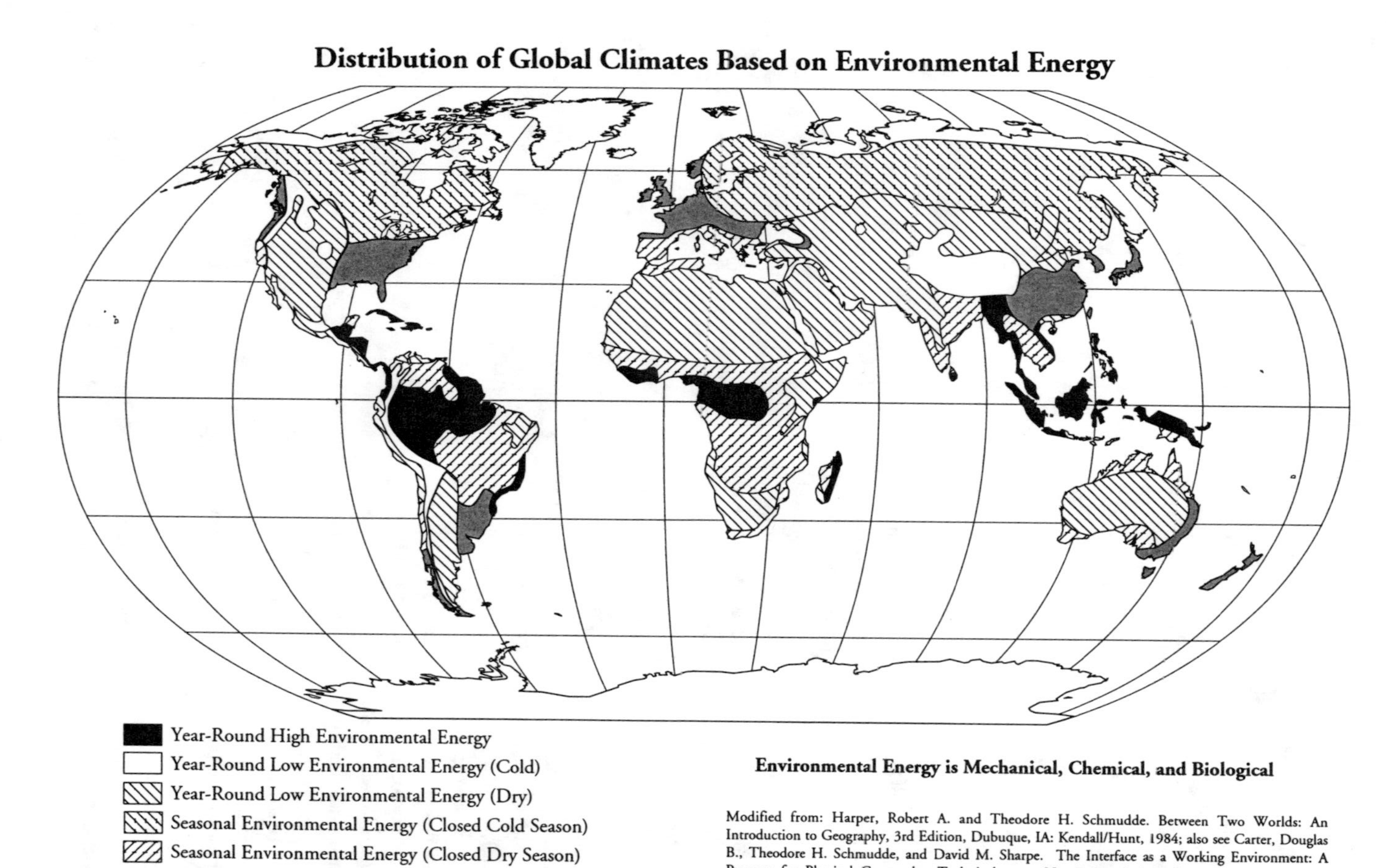

Modified from: Harper, Robert A. and Theodore H. Schmudde. Between Two Worlds: An Introduction to Geography, 3rd Edition, Dubuque, IA: Kendall/Hunt, 1984; also see Carter, Douglas B., Theodore H. Schmudde, and David M. Sharpe. The Interface as a Working Environment: A Purpose for Physical Geography, Technical paper No. 7 Association of American Geographers, Commission on College Geography, Washington, D.C., 1972. Robinson Projection, 1994

Chapter 12

Climate Regions

Year-round High Environmental Energy Climate

The world's high energy climate is humid and tropical. Although the word humid means that a location receives enough precipitation to produce surplus moisture at the surface during at least one season of the year, this climate typically produces surplus moisture throughout the year. Within several climate classification systems including the Köppen system, this climate is called the **tropical rainforest**. The word tropical means that the climate regions are located between the Tropic of Cancer (23.5°N) and the Tropic of Capricorn (23.5°S).

This climate is dominated by equatorial and tropical air masses and generally occupies areas relatively close to the equator (Figure. 12.1). Throughout this area the sun is located directly overhead during part of the year, annual insolation rates are high, warm temperatures prevail, and normal air flow is equatorward and directed toward the west. These are the Northeast and Southeast Trades, occurring respectively in the Northern and Southern Hemispheres.

This climate is dominant in lowland areas that lie within 10° latitude of the equator but extend poleward on windward tropical coasts. The conditions are met best in the Amazon Basin of South America, eastern coasts of Central America, parts of the Congo Basin of Africa, eastern Madagascar, Indonesia, and in the Philippines.

Because of proximity to the equator, Year-round High Energy climates are among the world's leading recipients of insolation. High energy availability and an almost uniform daylight period throughout the year provide for high temperatures and minimal seasonal temperature variation. Each month averages close to 27°C (80°F), and the annual range in monthly temperature can be as small as 2° to 3°C. The diurnal range (difference between daytime high and nighttime low temperature) of 8° to 10°C is considerably greater than the monthly variation. High temperature readings in the tropical rainforest are much lower than for many middle-latitude locations, and rarely exceed 38°C (100°F). There is almost no seasonal temperature change, and the diurnal temperature cycle is repetitious .

Wind and pressure patterns of the tropical rainforest region are largely a response to seasonal shifts of the Intertropical Convergence Zone (ITCZ). As this low pressure area migrates north and south across the equator, it defines a zone toward which the trade winds converge. The high moisture content of the Trade Winds charges the lower atmosphere with energy to produce large convectional storms with

Figure 12.1 Year-Round High Environmental Energy

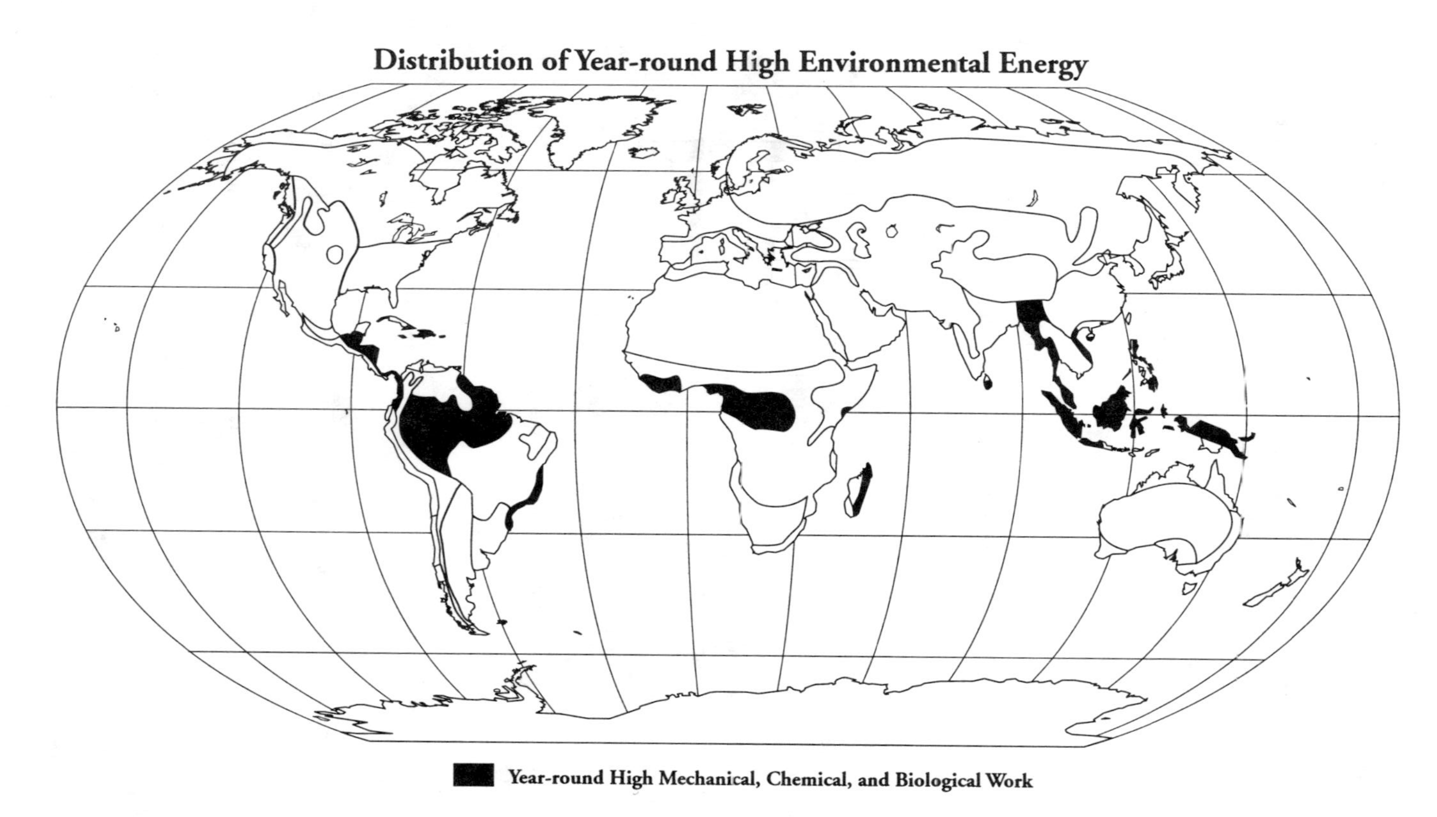

Modified from: Harper, Robert A. and Theodore H. Schmudde. Between Two Worlds: An Introduction to Geography, 3rd Edition, Dubuque, IA: Kendall/Hunt, 1984; also see Carter, Douglas B., Theodore H. Schmudde, and David M. Sharpe. The Interface as a Working Environment: A Purpose for Physical Geography, Technical paper No. 7 Association of American Geographers, Commission on College Geography, Washington, D.C., 1972. Robinson Projection, 1994

abundant lightning and rainfall. Afternoon thunderstorms develop from towering cumulonimbus clouds, and intense rainfall is common. By late afternoon the rain normally ends and skies clear. Annual rainfall in excess of 2,000 mm (80 in.) is characteristic of this climatic region. Figure 12.2 is a climatic graph for Kisangani in Zaire. No month is truly dry, and all months receive at least 6 cm (2.4 in.) of rain. Although droughts can occur, a surplus of water normally is received each year.

Moisture and heat are abundant throughout the year, and consequently streamflow with all attendant fluvial processes help to maintain a high level of mechanical energy during the year. The rivers that drain these regions carry huge quantities of runoff and sediment to the sea. It is not a coincidence that the world's largest river, the Amazon, drains a portion of this climatic region.

With such abundant heat and rainfall, minerals are dissolved and carried downward through the soil. Iron and aluminum in particular, are leached from the upper layers of the soil and deposited in horizons deep beneath the surface. Because of the accelerated rate of chemical activity, soil-forming processes produce layers within the soil that are quite deep and may extend hundreds of meters beneath the surface.

Biological activity within this climate exceeds that in other climatic regions. With abundant heat and water, the stresses on an organism to survive and procreate are minimal, and therefore, a larger number of species populate this biome than any other. Access to light is frequently the limiting factor for many plants. The broadleaf evergreen trees of this climate include species such as mahogany, rosewood, teak, kapok and ebony; the resinous rubber tree; commercial nut trees; and tropical fruit trees (banana, pineapple, mango and papaya). The bulk of biomass within this region is stored in massive tree trunks, branches, and root systems. The broad leaves are very efficient at intercepting sunlight, and competition for light is great with the tallest trees, lianas, and epiphytes that intercept the bulk of sunlight. Only about one percent of direct sunlight reaches the forest floor, and ambient light intensities at ground are only 5 percent of the intensity at the upper canopy. Thus, tree canopies are stratified by species tolerance to sunlight and by plant height (Figure 12.3). Three strata of forest vegetation can be identified: the tallest trees are widely spaced and characterized by intertwined crowns attaining

Figure 12.2 Climate Graph of a Year-Round High Environmental Energy Station

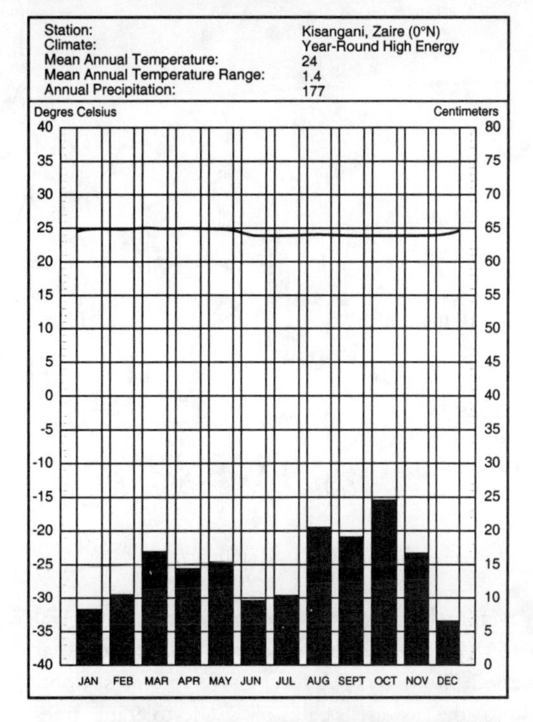

Figure 12.3 Selva or Tropical Rainforest Vegetation

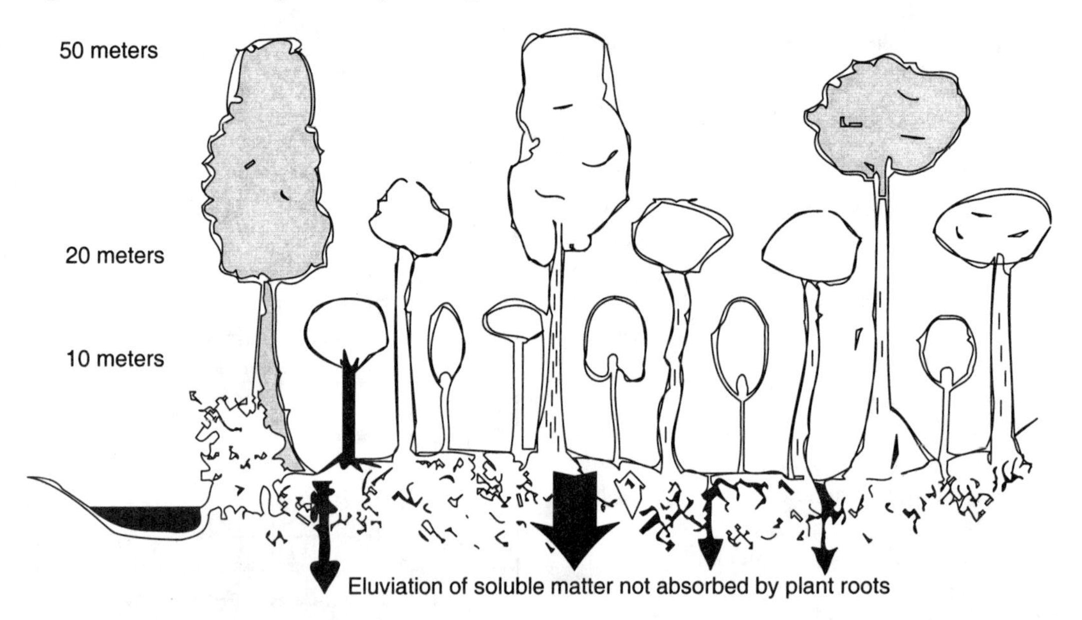

average heights of 50 to 55 meters (150-170 feet); an intermediate understory is developed at about 20 meters (60 feet) above the forest floor; and the lowest strata of shade-tolerant trees occupy a canopy that is around 8 to 10 meters (25-30 feet). Tall trees called **emergents** extend above the general canopy and smaller trees intercept insolation that penetrates canopy openings.

The floristic composition of the rainforest is astounding. There literally are thousands of plant species. Palms, ferns, bamboo, and related plants grow to immense size and may form, along with young trees, a growth layer at the lowest canopy. At ground level are found herbs, ferns, and seedlings

Trees of the rainforest may lose their leaves, flower, and fruit on a continuous basis. High temperature and humidity encourage rapid decomposition, however, so that there is little or no litter on the surface.

Animals in the rainforest are well adapted to the predominating canopy habitat. Insects, birds, and arboreal animals such as monkeys, opossums, squirrels, and sloths are numerous. Because of the scarcity of vegetation at low levels there are relatively fewer ground-dwelling grazing or browsing animals. Perhaps more than any other trait, it is the diversity of flora and fauna that is the most impressive aspect of the tropical rainforest.

Yet, in some areas the original forest has been removed and replanted to single species. This has been done with the cacao and rubber trees and with banana and pineapple plants, all of which are grown on large commercial farms. Some effort also has been made toward the plantation growing of valuable cabinet woods, but on a very limited scale. Environmental conservationists have expressed serious concern about the future of the tropical rainforests.

Although the original extent of tropical forests is thought to have exceeded 15 million square kilometers (10% of the earth's surface), today approximately 100,000 square kilometers (2 percent of tropical forest) are being destroyed each year. Tropical forests provide habitat for half of the world's species, and loss of habitat invariably means species extinction.

Currently, tropical forests provide the livelihood for 200 million people, but the cornucopia of potential fruits, crops, and medicines they represent is priceless. Today 25 percent of U.S. prescription drugs are derived from tropical origins. The rosy periwinkle, a pretty little flower from Madagascar, has been used in Jamaica and elsewhere in the treatment of diabetes. Although none of the 75 alkaloids in its leaves cures diabetes, two of them, vinblastine and vincristine, are used now to fight childhood leukemia and some other cancers. The dart poison used by Amazonian Indians, **curare**, (another alkaloid) interferes with the chemical messages sent between nerve cells and muscle cells. It is now used in surgery as a muscle relaxant. Chemists cannot produce a synthetic drug with all the properties of the original so they harvest the wild lianas from which the poison is made. In the past 30 years drug companies in the wealthy countries have devoted only a tiny fraction of their research budgets to tropical plants. **Ethobotany**, the study of tribal people and their vivid pharmacopoeias, suggests that tropical forests may contain a wealth of undiscovered medicines. With half the world's plant species found in tropical rainforests, probably half the medicinal plants are there as well.

Most tropical forests are cleared for agricultural or logging operations. In attempts to develop tropical areas empty of non-indigenous populations, Brazil built a network of Trans-Amazonian highways and advertised for settlers with the slogan "A land without men for men without land." Rondonia, with only 10,000 people in 1960, saw explosive growth rates and, by 1985. contained more than 1 million people. One quarter of tropical forest loss results from harvesting of timber. The tropical hardwoods such as teak, ebony, and mahogany generate substantial revenues for third world governments, but harvesting the wood at non-sustainable rates can be shortsighted from global and local perspectives. According to the World Bank, only 10 of the 33 exporting countries will still be doing so in 10 years. Ghana and Ivory Coast have banned export of 14 tropical woods in an effort to slow depletion rates.

The Tropical Forest Action Plan, sponsored and advocated jointly by the World Bank, the Food and Agriculture Organization, the United Nations, and the World Resources Institute calls for a halt to destruction of the rainforest. Key elements of the plan include: reforestation programs, construction of water catchments, reducing fuelwood shortages, implementing agro-forestry systems, and delimiting and managing national parks.

Tropical forests are critical reservoirs of carbon, and the effects of massive clearing operations on global atmospheric carbon stores are not fully understood. There are good theoretical reasons to expect atmospheric carbon, in the form of carbon dioxide, to increase as burning releases biogenic carbon to the air.

Seasonal Environmental Energy Rhythm

The middle latitudes are characterized by marked seasonal temperature contrasts. Unlike the tropics, the middle latitudes are areas where polar and tropical air masses meet. Thus, in addition to the cyclical pattern of heating and cooling, the frontal conditions common throughout much of this realm provide for unsettled weather and surplus precipitation in at least one season of the year. In these mild winter representatives of the humid middle latitude climates, temperatures are usually not low enough to support a cold season snow cover. Although snow may occasionally fall, it is normally of short duration and is not retained on the ground for any significant length of time. There are two primary subdivisions of this climate type: the humid subtropics and the west coast marine (Figure 12.4).

Humid Subtropics

The humid subtropics represent that part of the seasonal rhythm climates in which no dry season occurs and in which summers are long and hot. These areas are located on the east side of continents, roughly between latitudes 25° and 35° N and S and occur in the eastern United States, southeastern Europe, eastern China, Taiwan, southern Japan, northeastern Argentina, southeast Brazil, Uruguay, eastern South Africa, and eastern Australia.

The temperature range for a typical humid subtropical station reflects the seasonal contrasts in insolation, as well as the southeastern continental location (Figure 12.5). The normal monthly range at Charlotte, NC, for example, is about 21° C (36°F).

Within this climatic region precipitation exceeds evapotranspiration and numerous permanent streams drain surplus water, erode, and transport sediment toward the ocean. Rainfall is evenly distributed throughout the year, but higher rates of evapotranspiration during the summer mean that mechanical energy as measured by streamflow is at a maximum during the winter and spring. Mechanical effects from the formation and movement of ice are minimal.

Biological activity peaks during the long summers and is dominated by broadleaf deciduous trees with some needle-leaf evergreens. The principal species vary, but in North America the oak and hickory trees characteristic in the south give way to beech, maple, and birch in New England and southern Ontario. In some places large areas of pine and other conifers may be found. Conifers commonly occur in highland areas and along some western coasts in the mid-latitudes. In the United States the southeastern coastal plain and the west coast of California, Oregon, and Washington support predominately coniferous forests.

This forest has several distinct habitat layers. Litter production is greater in the deciduous forest than in tropical forests. The greater leaf fall and slower decomposition are responsible for a layer of litter that contains numerous fungi, insect larvae, and other decomposers. Above this, but only a few centimeters above the ground, a rich variety of herbs, low shrubs, and seedlings may be present, especially in spring and early summer. In the southeastern United States dogwood, red bud, buckeye and other shrubs, and sapling trees may form still another stratum. The large crowns of the dominant trees, which may occur in nearly pure stands, form an almost closed canopy; however, light intensity underneath this canopy may be two to three times the level found in tropical rainforests.

The layers of this humid subtropical forest provide a diversity of habitats that is occupied by a variety of animals, birds, and insects. Species from each of these categories occupy every habitat. There are ground-dwelling birds such as grouse and quail in addition to numerous species of songbirds. There are squirrels in the canopy, and fox, raccoons, rabbits, skunks, and groundhogs at ground level. The inhabitants

Figure 12.4 Seasonal Rhythm Environmental Energy

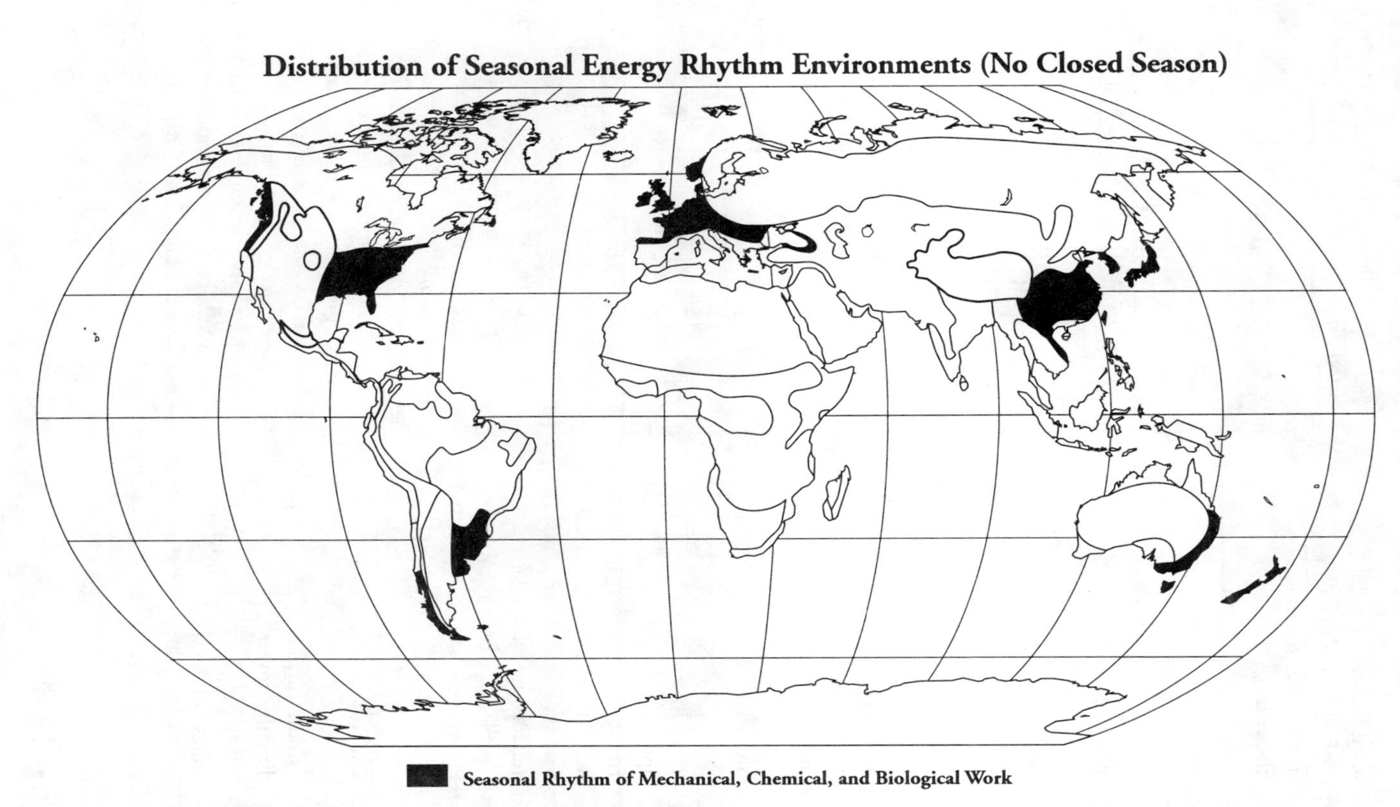

Modified from: Harper, Robert A. and Theodore H. Schmudde. Between Two Worlds: An Introduction to Geography, 3rd Edition, Dubuque, IA: Kendall/Hunt, 1984; also see Carter, Douglas B., Theodore H. Schmudde, and David M. Sharpe. The Interface as a Working Environment: A Purpose for Physical Geography, Technical paper No. 7 Association of American Geographers, Commission on College Geography, Washington, D.C., 1972. Robinson Projection, 1994

Figure 12.5 Climate Graph of a Seasonal Rhythm Environmental Energy Station

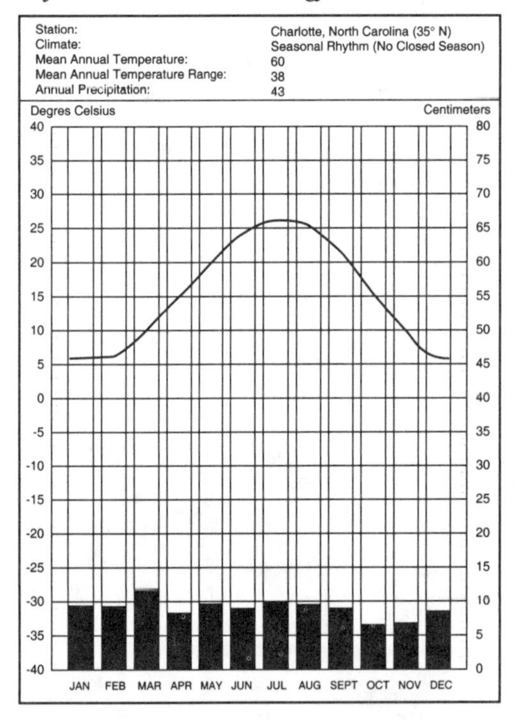

Figure 12.6 Climate Graph of a Seasonal Rhythm Environmental Energy Station

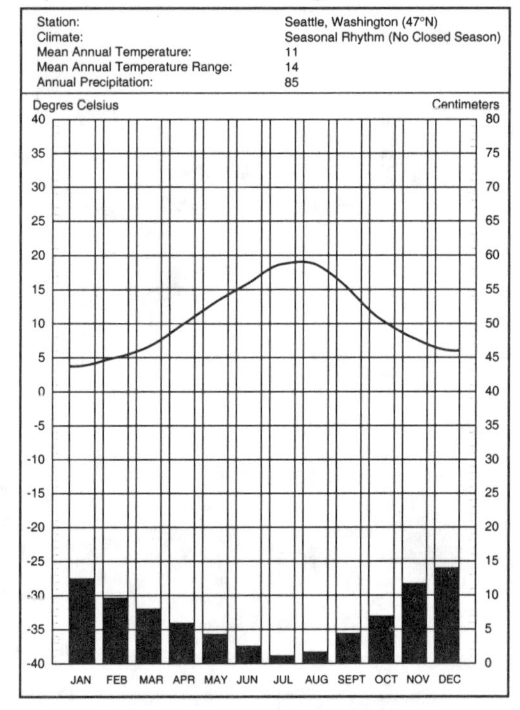

include browsers such as deer, hunters such as wildcats and raptors, and scavengers such as crows and opposums. Despite this diversity, the number of species present is considerably less than in the tropical rainforest.

Marine West Coast Climates

The Marine West Coast Climate is located generally on the west coast of continents between 40° and 60° north latitude including the west coast of North America from California to Alaska; the British Isles and northwest Europe; southern Chile; and southeastern Australia and New Zealand.

The marine west coast climate has adequate precipitation throughout the year. No month is considered dry, but there is a distinct concentration of rainfall during the winter season (Figure 12.6). Seattle, Washington is a good example of a marine west coast climate.

Streamflow and fluvial processes reach their maximum during the winter and spring after the abundant rains and low evapotranspiration demands of winter. Chemical activity peaks in spring when moisture and heat are available. Biological processes are most active during the early summer, but continue at a slower rate throughout the other seasons.

Because this climate does not have a closed season, many magnificent and valuable varieties of needle-leaf evergreen trees flourish. The Douglas fir, prized for home construction and furniture, can reach heights of 70 meters. Other valuable species include **red cedar**, **Sitka spruce**, and **hemlock**.

Seasonal Environmental Energy (Closed Dry Season)

Climates where precipitation is insufficient to maintain adequate soil moisture throughout the year are seasonal climates with a closed dry season (Figure 12.7). Without adequate rainfall many stream channels dry up and the mechanical transportation of water borne sediment is largely postponed until seasonal change reinstates rainfall and streamflow. Soluble minerals and salts are temporarily allowed to accumulate on the surface and the transport of soluble materials downward through the soil is punctuated during the dry season. Native plants are adapted through a variety of methods to endure the dry conditions with minimal damage. Native wildlife also employ a variety of methods, including migration, to survive the closed season. There are three climate types within this climate group: Mediterranean, Subtropical Monsoon, and Tropical Monsoon.

Mediterranean Climates

Mediterranean climates are located on the west side of continents between latitudes 30° and 45° N and S. The largest region is located along the margins of the Mediterranean Sea, hence the name. Other areas with a mediterranean climate include central and coastal California, central Chile, the tip of South Africa, and southern Australia.

Environmental energy is virtually closed during the summer because of the lack of moisture. Rainfall during these months is meager and during the heat of summer, rates of evapotranspiration easily exhaust soil moisture storage (Figure 12.8). Fluvial and aqueous processes virtually stop. Biological activity is adapted to the seasonal aridity.

These environments with their hot, dry summers, nearly always contain woodlands composed of broad-leaf evergreen shrubs and scrubby trees. Woodlands in this climate are called **sclerophyll forests**. The plants have xerophytic properties such as thick, leathery leaves with waxy or hairy coverings to retard evapotranspiration. Some have extensive laterally developed root systems to capture rainwater from a larger area, while others have deep tap roots that may reach 8 meters (28 feet) down in search of the water table. Many of the species are dormant in the summer, and susceptible to fire. In southern California the scrub woodland is called **chaparral**. Nearly every summer one or more major fires occurs in the California chaparral. Burned-over areas are very susceptible to erosion during the suceeding rainy winter and mudflows are a considerable hazard in newly burned lands. Eventually grasses stabilize the surface and if the area is subjected to repeated fires, grass may be perpetuated as the dominant vegetation. Otherwise, the area reverts to scrub-woodland or sclerophyll forest.

Animals of these woodland areas are limited by available habitat, but in North America bears, elk, deer, cougar, and various rodents, reptiles, and birds are representative species. In the Mediterranean basin wild animal life is not very abundant, but at one time hyenas, jackals, apes, wild goats and sheep, and even tigers, were numerous. The natural vegetation is called **maquis** in Mediterranean Europe.

Tropical and Subtropical Monsoon

The subtropical monsoon climate regions are located in northern India; southwestern China; a narrow zone along the northeast coast of Australia; parts of Mexico, Paraguay, and eastern Bolivia; and the plateaus of tropical and subtropical Africa and South America. This

Figure 12.7 Seasonal Environmental Energy (Closed Dry Season)

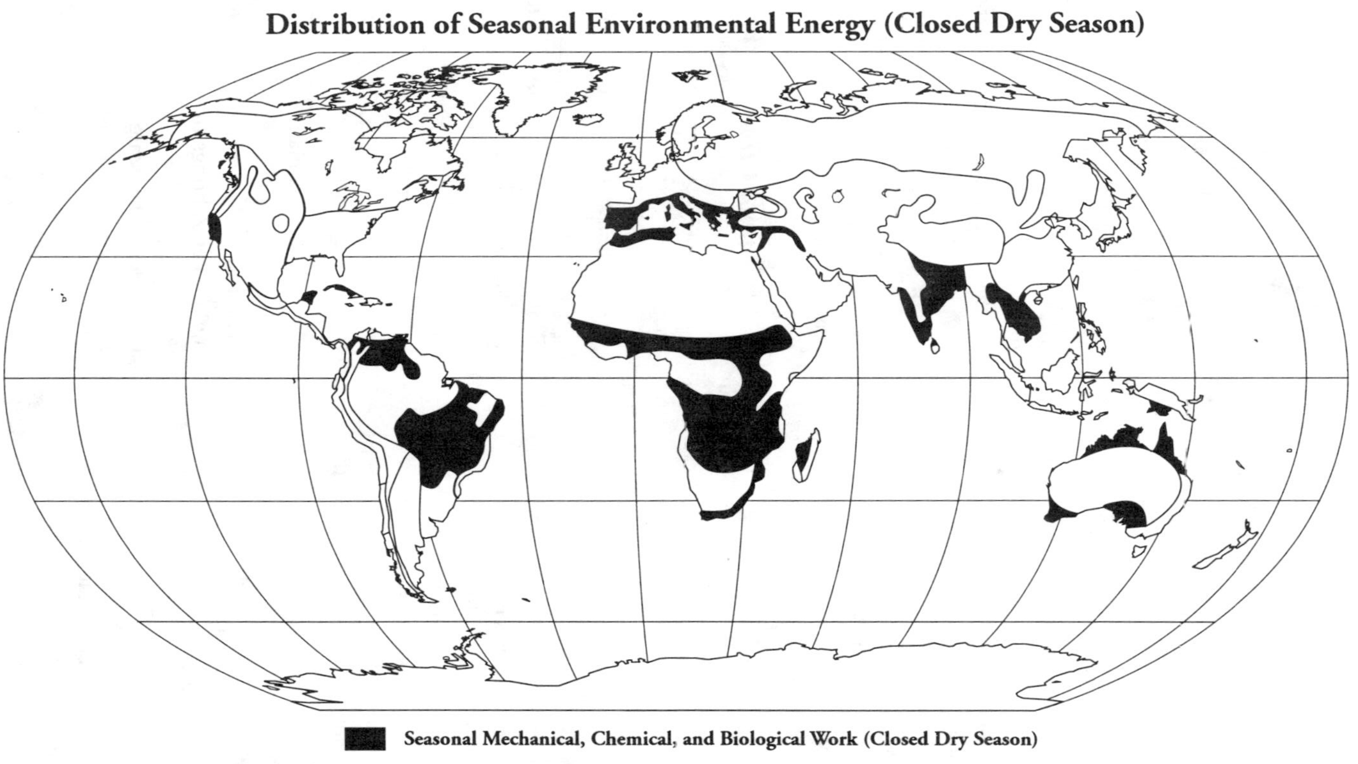

Modified from: Harper, Robert A. and Theodore H. Schmudde. Between Two Worlds: An Introduction to Geography, 3rd Edition, Dubuque, IA: Kendall/Hunt, 1984; also see Carter, Douglas B., Theodore H. Schmudde, and David M. Sharpe. The Interface as a Working Environment: A Purpose for Physical Geography, Technical paper No. 7 Association of American Geographers, Commission on College Geography, Washington, D.C., 1972. Robinson Projection, 1994

Figure 12.8 Climate Graph of a Seasonal Energy (Closed Dry Season) Station

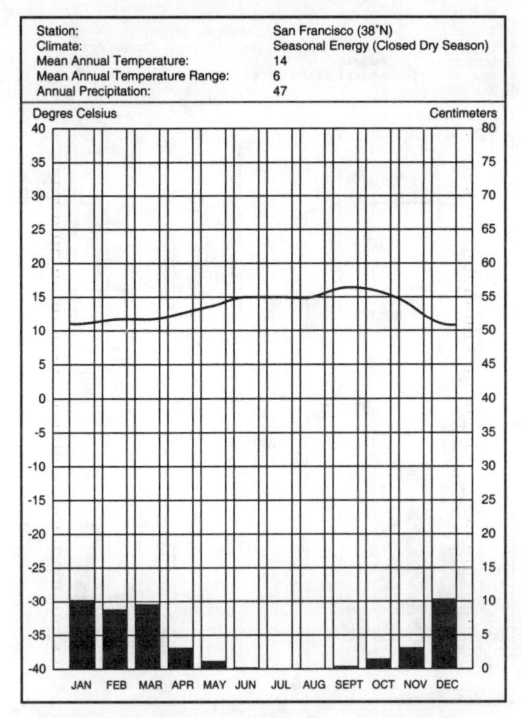

climate type has meager rainfall during the winter and heavy rainfall during the summer. The tropical monsoon climate is found along the west coasts of India and Burma, the Indochina Peninsula, the Philippines, the western Guinea Coast of Africa, northeastern South America, and along windward coasts of the Caribbean. The subtropical monsoon is cooler and has a greater range of temperature than the tropical monsoon climate.

The high sun or summer season permits all three types of environmental energy, mechanical, chemical, and biological to reach their maximum level of activity. With aridity and cooler temperatures, the closed season occurs during the low sun or winter season.

Many of the trees are deciduous and are not so tall as those of the rainforest, forming a discontinuous canopy less than 30 meters (100 feet) high. Below this is a thick evergreen layer of trees and shrubs. The number of species per unit area is less than in the rainforest. Leaves of the trees in the seasonal forest are smaller and the trees may contain thorns or spines.

If the wet monsoon is late and accompanied by meager rains, planting time and irrigation schedules are disrupted, threatening higher food costs, famine, or mass starvation. When the wet monsoon arrives early and with intense rainfall, flooding, drowned crops, and massive landslides often occur. Mechanical energy peaks along with the massive acceleration in fluvial activity associated with the onset of the wet monsoon.

Tropical Savanna Climate

The tropical savanna climate is located between 5° and 25° North and South. As the temperature and precipitation regimes of Miami, Florida and Darwin, Australia reveal (Figures 12.9 and 12.10), no large temperature contrasts are found, and yet there is a hint of seasonal temperature variation. The most distinguishing feature of this seasonally closed climate is a well defined dry period that coincides with the time of year when the sun is lowest in the sky. The length of the dry period varies from about two months on the equatorial margins to six months on the poleward margins.

Rainfall in the savanna is determined by the seasonal migration of the subtropical high pressure and the intertropical convergence zone (ITCZ). The ITCZ brings the rainy season and a supply of water that often exceeds 1,000 mm (40 in). This is the season when biological activity flourishes. The savanna vegetation becomes lush and green. Streams are swollen and large expanses of lowland are flooded. This

Figure 12.9 Climate Graph of a Seasonally Closed (Dry) Station

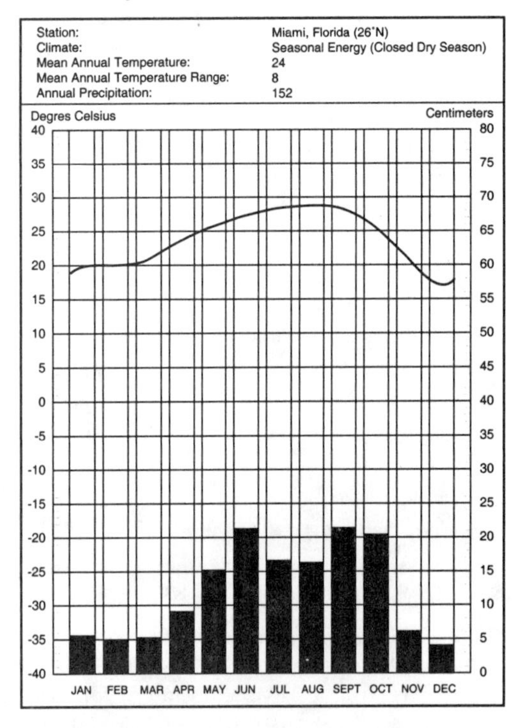

Figure 12.10 Climate Graph of a Seasonally Closed (Dry) Station

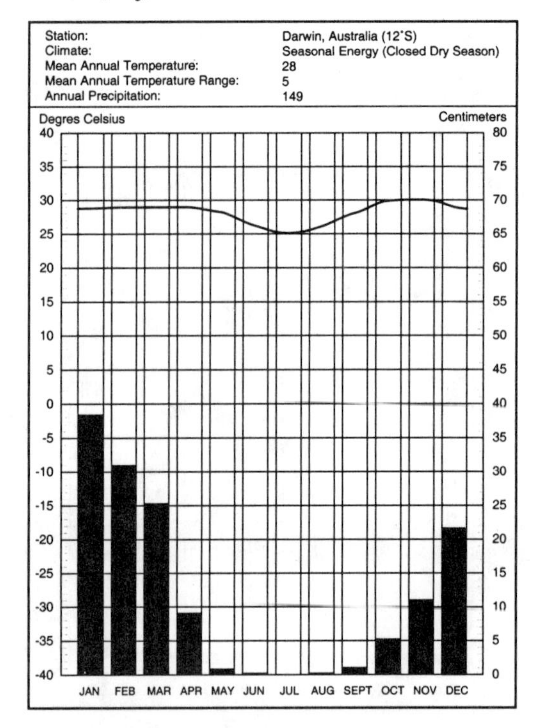

is the season of maximum environmental work, fluvial, chemical, and biological.

The rains end for this climate region when the seasonal migration of the planetary wind systems replace the convergence zone with trade winds and the descending air of the subtropical high pressure. The dry period is more like that of the deserts and may be considered a case of seasonal aridity sufficient to virtually close off most of the fluvial, chemical, and biological activity.

The Savanna climate is named for the vegetation type found there. They are dominated by grasses and sedges, but they may also contain numerous scattered trees and shrubs which give them a two-storied character.

Generally, savanna grasses are tall, 1 to 4 meters (3 - 12 feet), and trees are short, 5 to 10 meters, (15 - 30 feet), often flat-topped, and mostly deciduous. Because of the numerous trees present in savanna areas, most ecologists consider the grasslands to be a fire-climax vegetation perpetuated by repeated burning during the dry season. Burning kills the young seedling trees, but does not harm the grass roots which are safely underground.

Seasonal Environmental Energy (Closed Cold Season)

In the vernacular of the Köppen climate classification system this is the **continental** climate type. Continental refers to the thermal

Figure 12.11 Seasonal Environmental Energy (Closed Cold Season)

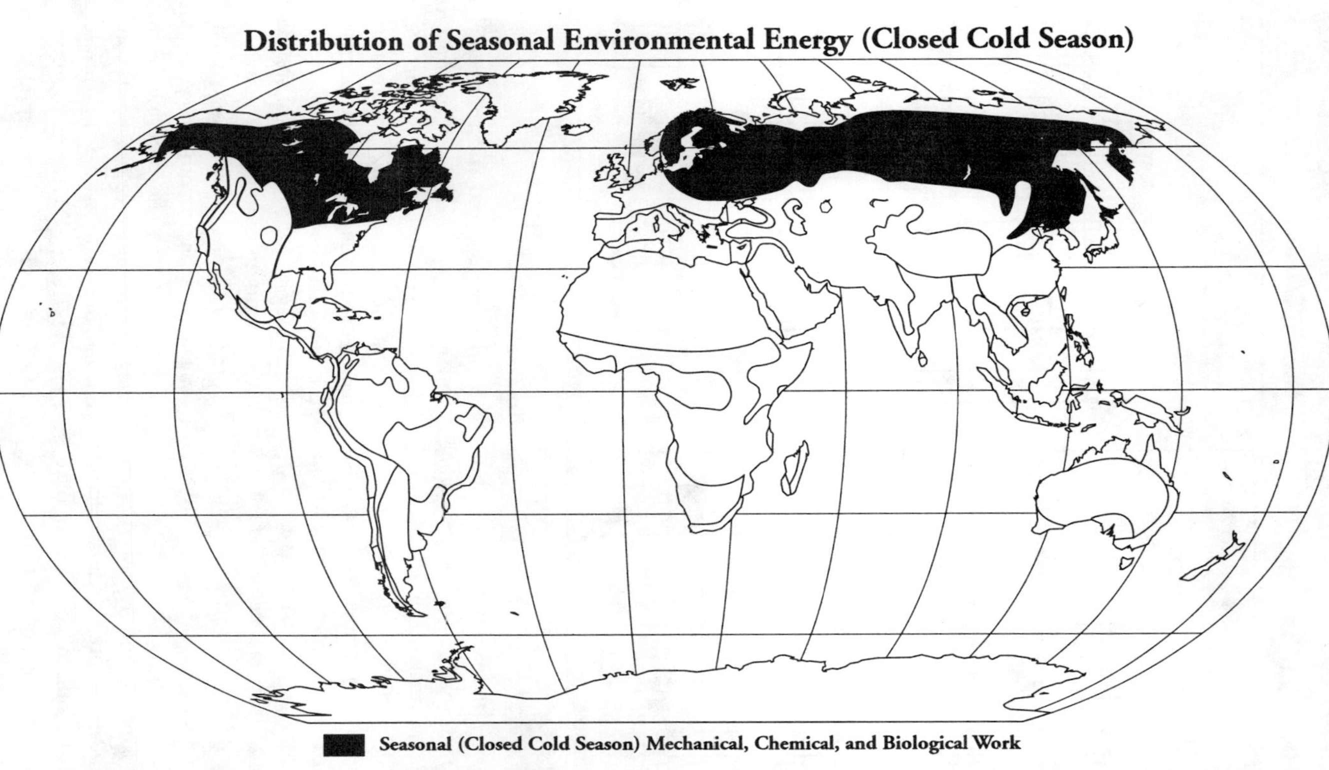

Modified from: Harper, Robert A. and Theodore H. Schmudde. Between Two Worlds: An Introduction to Geography, 3rd Edition, Dubuque, IA: Kendall/Hunt, 1984; also see Carter, Douglas B., Theodore H. Schmudde, and David M. Sharpe. The Interface as a Working Environment: A Purpose for Physical Geography, Technical paper No. 7 Association of American Geographers, Commission on College Geography, Washington, D.C., 1972. Robinson Projection, 1994

Figure 12.12 Climate Graph of a Seasonally Closed (Cold) Station

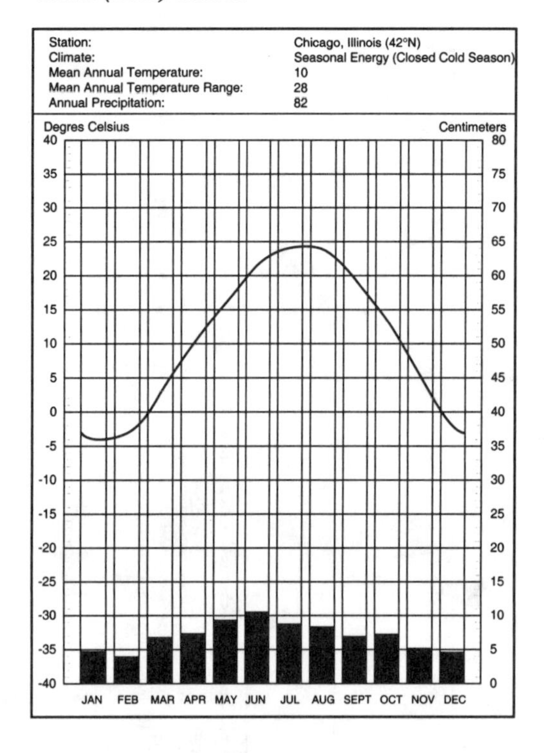

Figure 12.13 Climate Graph of a Seasonally Closed (Cold) Station

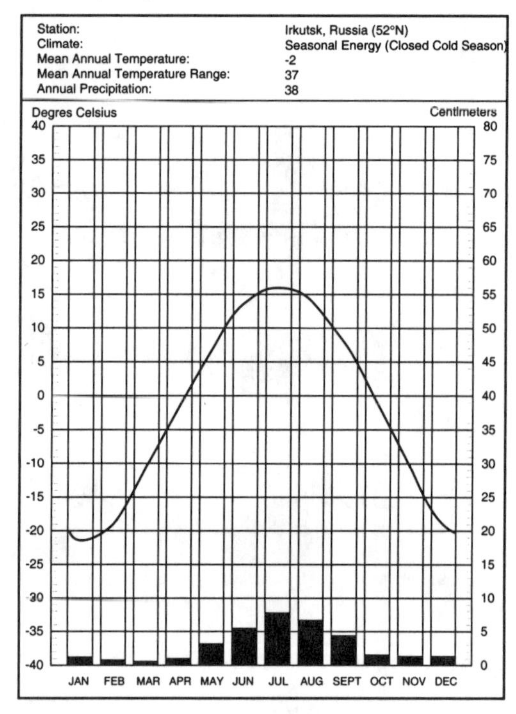

characteristic of large landmasses to become hot during summer and cold during winter. The range in temperatures makes this climate type quite interesting. Verkhoyansk in eastern Asia has recorded summer high temperatures of 36°C (97°F) and during the same year a winter low of -72°C (-98°F). Although most stations have far less severe winters, they are cold enough to sustain a snow cover for long periods during winter. These continental climates occur only in the northern hemisphere and stretch across much of North America and Eurasia (Figure 12.11). Three subdivisions of this climatic realm are recognized: the humid continental (hot-summer) region, the humid continental (warm-summer) region, and the continental subarctic region. Chicago, a moderate example located on the southern boundary of this region, and Irkutsk located in the northeastern Siberia represent extreme examples of continentality (Figures 12.12 and 12.13)

Biological and chemical processes reach their maximum during the summer when warm temperatures and water are available to drive aqueous and photosynthetic cycles. Mechanical processes are maintained during the summer as a result of fluvial processes. Frost heaving occurs during the early winter as surface and soil moisture freeze. Mechanical processes reach their maximum during the spring when ice and snowmelt released from from frozen streams rush downstream scouring stream channels and often inundating floodplains.

Vegetation in this climatic region ranges from broad-leaf deciduous forest along the southern boundary to pure stands of needle-leaf evergreens. In the northern reaches of this climate in both North America and Eurasia there is a coniferous forest that practically encircles the earth. Although precipitation is low, the cold temperatures and short summers reduce the evapotranspiration rate so that the soil is moist and bogs are numerous. This forest, known as the **boreal forest**, the **taiga**, or the great northern coniferous forest, consists of relatively few species. In North America white spruce, black spruce, and tamarack are the dominants, although white pine, jackpine, and fir are also common. Some deciduous trees, most notably aspen, cottonwood, birch, and alder, are present in most areas. When disturbed by fire or lumbering activities, open areas may be quickly invaded by aspen and cottonwood. In time the forest reverts to its coniferous character. In Eurasia, Scots pine, Norway spruce, larch, and fir are the dominant conifers, whereas birch and aspen are the most common broad-leaf species. Larch and tamarack are deciduous needle-leaf species adapted to a long closed season where water is frozen and unavailable to plants.

The appearance of the boreal forest is markedly different within its southernmost and northernmost fringes. In the south where it merges with the deciduous forest of the mid-latitudes, not only are there more deciduous species, but the trees are taller. The cold temperatures, shorter growing season, and shallow soil restrict the growth of trees in the northern section, and forest there is more open and the trees much smaller.

The variety of habitats in the seasonally closed (cold) climate is less than in forests of the mid-latitudes and equatorial regions. Beaver, moose, bear, wolves, and deer are present year-round, but caribou migrate into the area during winter to escape the harsh winters of the tundra, the adjacent climate region to the north.

Year-round Low Environmental Energy (Dry)

Wherever the average potential evapotranspiration exceeds precipitation, surface water is scarce and the rate of mechanical, chemical, and biological processes is reduced. The common names for these climatic environments are deserts and steppes. These dry conditions result from various climatic controls operating either singly or in concert to diminish the chances of precipitation. They include: proximity to the semipermanent subtropical highs, rainshadow effects on the leeward sides of mountain ridges, the stabilizing effect of cold ocean currents on air masses over coastal regions, and great distances from sources of moisture (Figure 12.14).

Deserts

Deserts are characterized by water scarcity caused by low precipitation and high potential evaporation rates. Most deserts receive less than 25 cm (10 in) of precipitation annually, although somewhat greater annual totals may occur where potential evapotranspiration rates are high (Figures 12.15 and 12.16). Within the humid subtropics it is possible to receive that much rain during a single severe weather event. Parts of southern Mississippi, for example, received more that 100 cm (40 in) of rainfall from Hurricane Georges in 1998. Precipitation in the desert can also be intense, but storm

Figure 12.14 Year-round Low Environmental Energy (Dry)

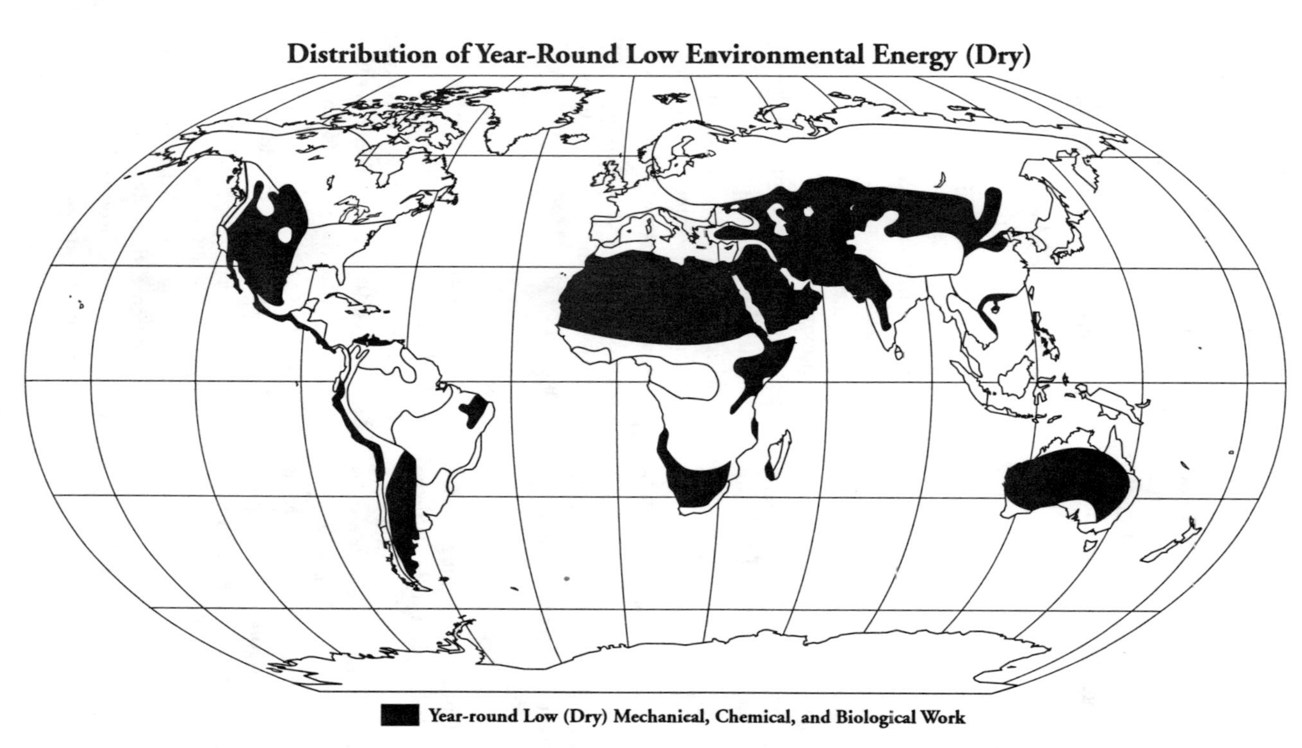

Modified from: Harper, Robert A. and Theodore H. Schmudde. Between Two Worlds: An Introduction to Geography, 3rd Edition, Dubuque, IA: Kendall/Hunt, 1984; also see Carter, Douglas B., Theodore H. Schmudde, and David M. Sharpe. The Interface as a Working Environment: A Purpose for Physical Geography, Technical paper No. 7 Association of American Geographers, Commission on College Geography, Washington, D.C., 1972. Robinson Projection, 1994

Figure 12.15 Climate Graph of a Year-round Low Energy (Dry) Station

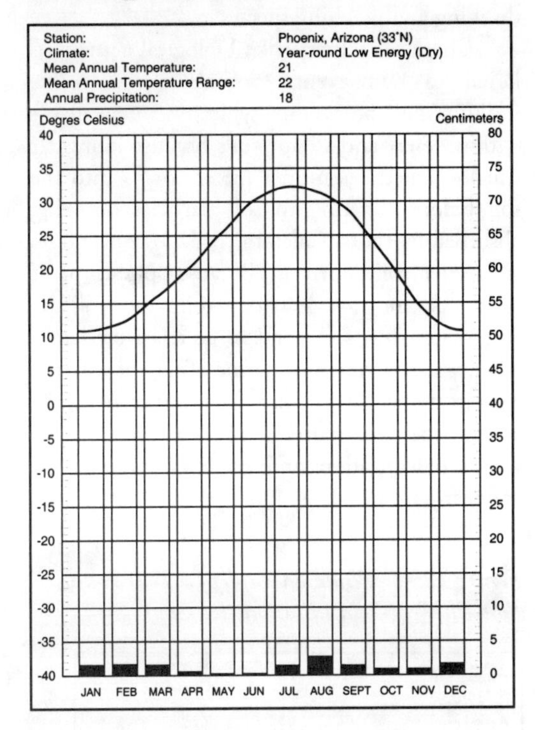

Figure 12.16 Climate Graph of a Year-round Low Energy (Dry) Station

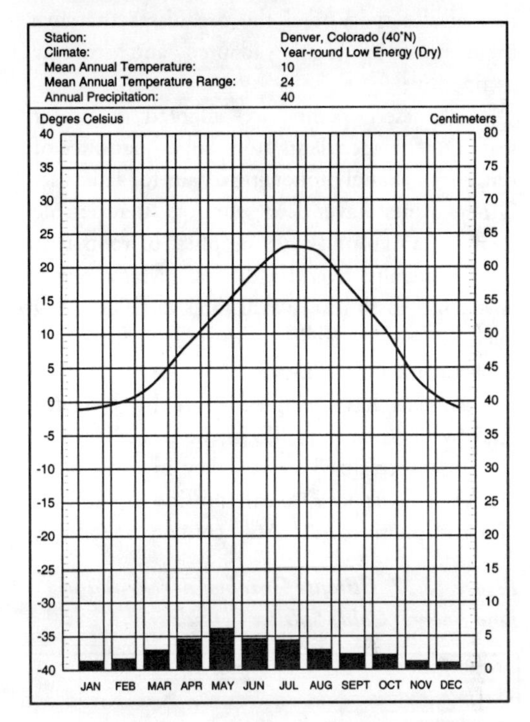

events are short-lived and infrequent. Much of the rainfall may occur beneath passing thunderstorms that quickly fill previously dry stream channels. Most of the mechanical work in deserts is accomplished by fluvial processes during these infrequent flash flood events. As water courses through the stream channels, sediment is literally flushed downstream. Although permanent streams do not originate from within desert environments, a few large streams, such as the Nile River and the Colorado River can traverse deserts. The erosional effects of such streams on the desert landscape can be profound as anyone who has viewed the Grand Canyon can atest.

Secondarily, dry conditions facilitate transport of fine particles saltating along the desert floor or held in suspension by strong winds and turbulence near the ground. Abrasion from theses particles is responsible for carving many of the spectacular geomorphic features that epitomize Monument Valley and other reaches of the desert in the American southwest. The mechanical work of abrasion, erosion, and deposition by wind is more pervasive, but greater work is accomplished by streams and their fluvial processes however punctuated by rainless periods.

As with mechanical activity, biological activity peaks after rain events, and then declines toward dormancy until the next rain. The desert is a harsh environment capable of limited biological activity, and the organisms that live there must be highly adapted and strongly competitive.

All desert plants are adapted to water scarcity. The succulents store large quantities of water immediately following a rain for later use. Other plants have deep taproots that enable them to tap groundwater supplies, or roots that spread laterally over large areas. Many plants have small thick leaves with thick cuticles. They may even shed their leaves during short periods of dry weather dormancy.

Animals have adapted more by behavior than by physiology to life in the desert. Except for a few specialized species they include many of the same species found outside of deserts. North American deserts contain antelope, coyotes, lizards, snakes, scorpions, rodents, cougar, and numerous birds, both ground dwelling and perching species.

Chemical activity, like biological activity, is keyed to rain events. Soluble elements like sodium and potassium typically are available within desert and steppe soils because rainwater usually penetrates only a meter or so into the soil before adhering to soil particles or being absorbed by roots. Calcium tends to concentrate in deeper soil horizons. Because soluble plant nutrients are not leached from the soil by abundant rains, soils tend to be relatively fertile when water is available. Some of the most fertile soils in the world, mollisols, developed naturally within the wet margins of this climate type where precipitation and potential evaporation are nearly balanced.

Figure 12.17 Climate Graph of a Year-round Low Energy (Cold) Station (Tundra)

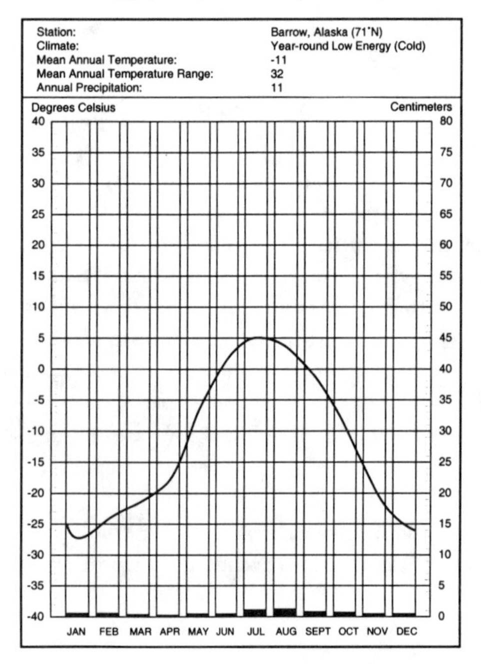

Figure 12.18 Climate Graph of a Year-round Low Energy (Cold) Station (Ice Cap)

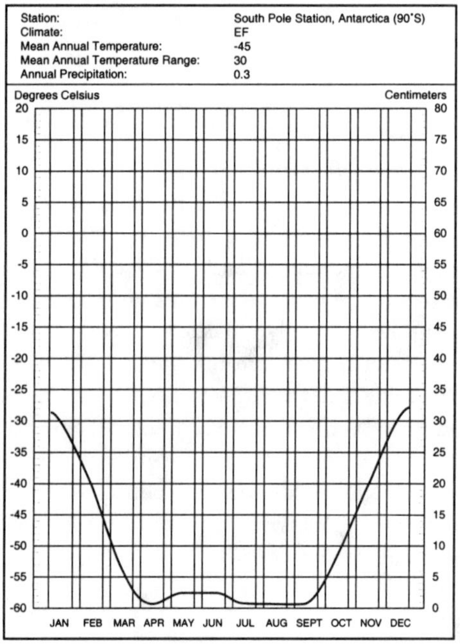

Station: South Pole Station, Antarctica (90°S)
Climate: EF
Mean Annual Temperature: -45
Mean Annual Temperature Range: 30
Annual Precipitation: 0.3
Degrees Celsius
Centimeters
20 15 10 5 0 -5 -10 -15 -20 -25 -30 -35 -40 -45 -50 -55 -60
80 75 70 65 60 55 50 45 40 35 30 25 20 15 10 5 0
JAN FEB MAR APR MAY JUN JUL AUG SEPT OCT NOV DEC

Year-round Low Environmental Energy (Cold)

There are two climate types with cold monthly temperatures throughout the year: tundra, which has a short growing season in summer and ice cap, or frost climates, which are frozen year-round (Figures 12.17 and 12.18). Because of the low moisture content of cold air, neither variant receives much precipitation (which occurs mostly as snow), but the cold temperatures, frozen soil, and poor drainage combine to cause wet, swampy conditions in the low-lying tundra areas.

Beyond the boreal forest in the northern hemisphere lies a vast area underlain by **permafrost** (Figures 12.19). Only the upper few feet thaw for a brief period in summer to create a water-soaked soil that provides inadequate anchorage for all but the most shallow-rooted plants. The **tundra** is neither forest, woodland, nor grassland. It contains no trees, except for a few gnarled, stunted willows or birches that look either like diminutive trees or grotesque caricatures of real trees. The dominant plants are lichens, sedges, grasses, mosses, heather, and various shrubs and flowering herbs and forbs. All grow close to the ground and commonly form a thick, damp mat over the surface.

Many of the tundra animals and birds migrate to avoid the cold winters. The ones that remain have various adaptations to protect them from the cold. All of the tundra habitats are ground level or below. No perching birds or climbing animals are present. The stress of the closed season is extreme. Neither are there many cold-blooded terrestrial species. Survival is difficult for them because of the low temperatures throughout the year. The seasonal abundance of some populations is in contrast to the restricted number of species present in this harsh environment. Swarms of mosquitos and other insects and large herds of caribou might give the misimpression that the tundra is a productive area. Such is not the case. Although herds may be large, the number of herds is small.

Biological activity is limited. Most of the mechanical energy exerted within this region is by moving ice during brief freeze-thaw cycles. Chemical activity lacks heat and progresses more slowly than in other regions.

Figure 12.19 Year-round Low Environmental Energy (Cold)

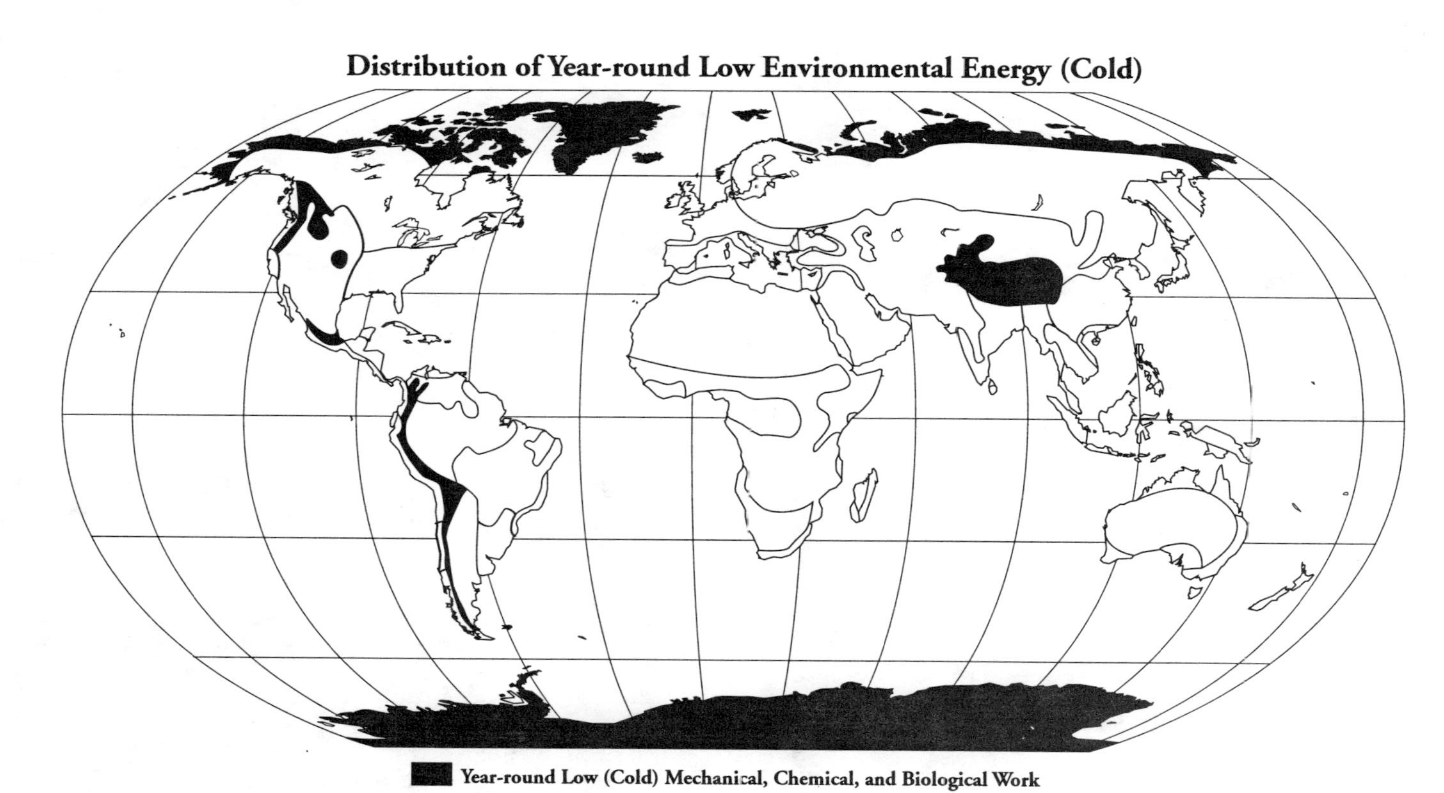

Modified from: Harper, Robert A. and Theodore H. Schmudde. Between Two Worlds: An Introduction to Geography, 3rd Edition, Dubuque, IA: Kendall/Hunt, 1984; also see Carter, Douglas B., Theodore H. Schmudde, and David M. Sharpe. The Interface as a Working Environment: A Purpose for Physical Geography, Technical paper No. 7 Association of American Geographers, Commission on College Geography, Washington, D.C., 1972. Robinson Projection, 1994

Activity: Climate Regions

OBJECTIVES

The purpose of this exercise is twofold: 1) to allow you to examine characteristics of global climate patterns and, 2) to introduce you to a widely used grouping algorithm of the world climates.

INTRODUCTION

In the broadest sense, climatology is the study of the long-term behavior of the atmosphere. Recorded observations of temperature and precipitation over periods of several years yield statistical averages or trends that smooth out the short-term fluctuations of weather. Such trends can be thought of as the composite of weather conditions, or climate, for a given place over a long period of time.

Understanding this extended view of atmospheric behavior holds many useful implications. Architectural design, heating and cooling costs, styles and types of clothing, and agricultural practices and potentials are just a few expressions of human decisions based on an awareness and understanding of climate. Natural environments clearly reflect adaptations to climate through inseparable relationships among the atmosphere, biosphere and lithosphere. Geographic distributions of natural vegetation, wildlife and, more generally, the world's soils, all are indicative of geographic patterns of climate.

Regional climatology focuses on identification and explanation of the spatial distribution of climate types. Identification and classification of climate types are based on the recognition of significant characteristics of climates. Explanation of climate patterns involves application of concepts such as the earth-sun relationship, latitudinal variations in insolation receipts, land and water heating contrasts, precipitation processes and atmospheric circulation systems.

MATERIALS

Atlas or globe, and a calculator

CLIMATE CLASSIFICATION

Classification is an important procedural component of the scientific method. To be meaningful, observational data must be organized for comparison and interpretation. Systematically grouped data can be analyzed to reveal causal or statistical relationships, test hypotheses, evaluate theory, and ultimately extend knowledge. Without systematic grouping of data, the work of data collection and observation and the excitement of data analysis would be wasted, lost in an intractable heap of meaningless statistics.

Classification is especially problematic where large and varied data sets, like those of the world's climates, are involved. For example, Cherripunji, India, receives an average of 1,143 cm of precipitation annually. Halfway around the world, Arica, Chile, recorded no more than a trace (less than 3 mm) of precipitation during each of twenty-two consecutive years. Extremes in temperature are pronounced, too. Dallol, Ethiopia, has an annual average of 34° C while Plateau Station, Antarctica, experiences a 57°C annual average. That is a global range of 92° C in annual averages! Working out a classification scheme for such a varied data set is a necessary, if formidable, task. Fortunately, climatologists have already worked up such classification schemes, allowing us more time to study characteristics and relationships among the world's climates.

The most widely employed systematic grouping of world climates was developed by Wladimir W. Köppen at the University of Graz, Austria in 1918. His classification, with some modification, forms the basis for the world-scale climate maps which appear in many atlases and textbooks. The Russian-born Köppen, like many European geographers of his time, was interested in relationships between climates and other features of the world. His climate classification scheme evolved from an interest in botany as he attempted to develop relationships between climate and vegetation. Consequently, his criteria for climate classes were strongly keyed to the heat and moisture requirements of various vegetation associations. His criteria included: 1) average annual temperatures; 2) average annual precipitation totals; 3) average annual temperature range; 4) seasonal distributions of precipitation; and 5) mathematical formulae relating precipitation and evaporation in arid and semi-arid climates. Using data recorded from a number of places, Köppen generated numerically defined classes for regional climate grouping which, with modification, yielded alphabetically keyed categories (Table 11.1). The result is a comparably based system that neatly arranges similarities and differences among characteristics of world climates. Köppen's son-in-law, the German meteorologist Alfred Wegener, relied heavily on paleo-climatic data to espouse the earliest well-developed theory of continental drift, a precursor of the modern theory of plate tectonics which has revolutionized the field of geology.

Table 11.1 Climatic Types in the Köppen Classification
Adapted from: Howard J. Critchfield, General Climatology, 4th Ed. Englewood Cliffs: Prentice-Hall, 1983

Af	Tropic rain forest. Hot; rainy all seasons
Am	Tropical monsoon. Hot; seasonally excessive rainfall
Aw	Tropical savanna. Hot; seasonally dry (usually winter)
BSh	Tropical steppe. Semiarid; hot
BSk	Mid-latitude steppe. Semiarid; cool or cold
BWh	Tropical desert. Arid; hot
BWk	Mid-latitude desert. Arid; cool or cold
Cfa	Humid subtropical. Mild winter; moist all seasons; long hot summer
Cfb,c	Marine west coast. Mild winter; moist all seasons; warm/cool summer
Csa	Interior Mediterranean. Mild winter; dry summer; hot summer
Csb	Coastal Mediterranean. Mild winter; dry, short, warm summer
Cw a/b	Subtropical monsoon. Mild winter; dry winter; hot summer
Dfa	Humid continental. Severe winter; moist all seasons, long hot summer
Dfb	Humid continental. Severe winter; moist all seasons, short warm summer
Dfc	Subarctic. Severe winter; moist all seasons; short cool summer
Dfd	Subarctic. Extremely cold winter; moist all seasons; short summer
Dwa	Humid continental. Severe winter; dry winter; long hot summer
Dwb	Humid continental. Severe winter; dry winter; warm summer
Dwc	Subarctic. Severe winter; dry winter; short cool summer
Dwd	Subarctic. Extremely cold winter; dry winter; short cool summer
ET	Tundra. Very short summer
EM	Polar Marine. Very short summer; cold winter
EF	Ice Cap. Perpetual ice and snow
H	Undifferentiated highland climates

PROCEDURE

The Köppen climate classification summary (p. 229) can be used in conjunction with climate station data to classify and analyse regional climates. Before you begin, compare the qualitative features of the climates shown here with the quantitative features of the Köppen system. Temperature and precipitation data for Penang, Malaysia, a place with a Tropical Rainforest climate, provide a simple example for climate classification.

Station: Penang, Malaysia Climate Type: Af

	JAN	FEB	MAR	APR	MAY	JUN	JUL	AUG	SEP	OCT	NOV	DEC	YEAR
T	27	28	28	28	28	28	28	27	27	27	27	27	28
R	9.4	7.9	14.2	18.8	27.2	19.3	18.8	29.2	39.9	42.9	30.2	14.7	273.3

First, examine the distribution and values of monthly average temperatures and precipitation. The year-round high temperatures and the small range of temperatures (only 1 degree separates the high of 28° C during February through July from a low of 27° C during the other months) suggest a nearly equatorial location. The total 273.3 cm of precipitation is relatively evenly distributed, with even the driest months having nearly 8 cm of precipitation.

If we refer to the quantitative summary we can rule out the possibilities of the Mesothermal C, Microthermal D, and Arctic E climate types simply on the basis of temperature. Since the average annual temperature in all months is greater than 18° C, Penang must be in a tropical A climate. With over 270 cm of precipitation, Penang is clearly not in an arid or even semi-arid B climate group. Imagine how hot a place would have to be for 273 cm of precipitation to be lost almost entirely to evaporation! Penang's climate, then, fits neatly into the criteria for the Af climate group.

Suppose that Penang's coolest month were 17° C, instead of 27° C. It could no longer be included in the A group but would meet the limits for inclusion into the C group. Suppose now that we return to the original "A" temperature data but reduce the precipitation of the driest month to less than 6 cm. It could no longer be classed as an Af type and would have to be considered as either an Am or an Aw type. The appropriate precipitation suffix can be determined by using the definitions which appear in the A climate section of the classification summary.

Climate classification is a systematic process. If a climate station's temperature and precipitation data fail to meet the criteria for inclusion into one group, the climate must belong to another category. Such an approach works well for climate groups based on temperature criteria, as are the A, C, D and E climates. The B climates, however, are defined on the basis of how greatly potential evapotranspiration, a function of temperature, exceeds precipitation and requires more attention and explanation.

CLASSIFICATION OF B CLIMATE GROUPS

Köppen's semi-arid BS ("S" from the Russian *Steppe* for a short grass vegetation association) and the arid BW ("W" from the German *Wuste* for wasteland) climates are found where the demands of temperature exceed the supply of moisture by precipitation. Annual temperatures and precipitation totals, as well as their seasonal distributions, must be considered. Annual precipitation totals of 60 cm or less, combined with high temperatures, are suggestive of B climates. Notable exceptions to this general figure occur where average temperatures are low, as with D and E climates. Less than 30 cm of precipitation, especially in tropical areas, is usually indicative of BW or desert environments. Relying solely on precipitation totals for classification is risky.

Khartoum, in the Sudan, located in the Sahara of Northern Africa, is a typical example of the BWh climate type.

Station: Khartoum Climate Type: BWh

	JAN	FEB	MAR	APR	MAY	JUN	JUL	AUG	SEP	OCT	NOV	DEC	YEAR
T	23	25	28	32	33	33	32	31	32	32	28	25	29
P	0	0	0	0	0.3	1.0	5.1	7.4	1.8	0.5	0	0	16.0

Over 80 percent of the annual 16 cm is clearly concentrated in the summer months, a time of high temperatures and high moisture demands. Using the appropriate formula:

$$AP < 2T + 28$$

where AP = annual precipitation (16 cm) and T = mean annual temperature (29° C), we can verify a B climate classification:

$$16 < 2\,(29) + 28$$

$$16 < 86$$

Second, we apply the appropriate formula to determine whether this is a BW or BS climate:

$$AP < 2T + 14$$

Substituting the values for Khartoum:

$$16 < 29 + 14$$
$$16 < 43$$

Since 16 is less than 43, Khartoum's climate is a BW, or desert type.

The B group also carries a temperature suffix based on the annual average. If the average annual temperature is 18° C or above, an h is used; if the average annual temperature is below 18° C, a k is used. It is worth noting here that the C and D groups also require a temperature suffix for complete classification. When you encounter them, add the appropriate suffix based not on annual temperature, but on the mean temperature of the warmest month.

CLASSIFICATION OF MESOTHERMAL C CLIMATES

One more example is presented here to help you get started.

Station: Climate Type:

	JAN	FEB	MAR	APR	MAY	JUN	JUL	AUG	SEP	OCT	NOV	DEC	YEAR
T	17	16	14	12	10	8	8	8	9	11	13	15	12
P	6.6	7.4	14.0	23.6	39.9	43.7	41.1	34.8	21.3	14.0	12.7	11.2	270.3

First examine general characteristics of the data for the station above. The distribution of monthly average temperatures (cooler in July than in January) reveals that this place is located in the southern hemisphere. Bear this in mind when the seasonality of precipitation is considered. The annual average precipitation of 270.3 cm precludes any possibility of this station being classed as an arid or semi-arid B climate. Temperatures are far too cool for classification as an A climate. Having eliminated the possibility of A and B climates, see if the C, D or E classes are appropriate. Temperatures are too warm for either the D or E climates, but the parameters for the C climate are met.

Now we will determine which precipitation suffix to use in describing this C group climate. We can determine that neither the "s" (dry summer) nor the"w" (dry winter) suffix apply. The "f" suffix, which denotes relatively evenly distributed precipitation, is appropriate.

So far, we have found this to be a Cf climate. The C type must also carry a temperature suffix based on the mean temperature of the warmest month. In this example, January averages 17° C, which is too cool for the "a" suffix. The "b" suffix does apply. This station is therefore classified as a Cfb, or Marine West Coast climate. Refer to the map in your textbook to see where this climate occurs on a regional scale.

Now see if you can correctly classify each of the following climate stations.

CLIMATIC STATION DATA

Station: 1 Climate Type: ______________________

	JAN	FEB	MAR	APR	MAY	JUN	JUL	AUG	SEP	OCT	NOV	DEC	YEAR
T	22	21	21	19	16	14	14	14	19	22	22	22	19
P	14	10	8	4	1	0	0	0	1.0	2	8	13	61

Station: 2 Climate Type: ______________________

	JAN	FEB	MAR	APR	MAY	JUN	JUL	AUG	SEP	OCT	NOV	DEC	YEAR
T	23	22	22	21	19	17	17	18	21	23	24	23	21
P	28	27	22	10	3	2	1.0	1	1	4	14	28	140

Station: 3 Climate Type: ______________________

	JAN	FEB	MAR	APR	MAY	JUN	JUL	AUG	SEP	OCT	NOV	DEC	YEAR
T	-5	-6	-6	-4	-1	2	5	6	3	0	-3	-4	-1
P	4	4	3	2	1	2	2	3	6	6	4	3	40

Station: 4 Climate Type: ______________________

	JAN	FEB	MAR	APR	MAY	JUN	JUL	AUG	SEP	OCT	NOV	DEC	YEAR
T	25	25	25	26	27	27	28	28	28	27	26	26	26
P	3	2	3	3	11	10	7	9	10	19	8	3	89

Station: 5 Climate Type: ______________________

	JAN	FEB	MAR	APR	MAY	JUN	JUL	AUG	SEP	OCT	NOV	DEC	YEAR
T	-14	-10	-3	4	9	14	19	21	20	9	-1	-10	5
P	1.0	1	2	3	5	7	8	12	11	5	3	2	60

Station: 6 Climate Type: ______________________

	JAN	FEB	MAR	APR	MAY	JUN	JUL	AUG	SEP	OCT	NOV	DEC	YEAR
T	14	17	23	30	34	37	35	33	32	26	20	15	26
P	1.0	1.0	1.0	1.0	0	1.0	3	3	1.0	0	0	0	12

Station: 7 Climate Type: ______________________

	JAN	FEB	MAR	APR	MAY	JUN	JUL	AUG	SEP	OCT	NOV	DEC	YEAR
T	17	16	14	12	10	8	8	8	9	11	13	15	12
P	7	7	14	24	40	44	41	35	21	14	13	11	270

Station: 8 Climate Type: ______________________

	JAN	FEB	MAR	APR	MAY	JUN	JUL	AUG	SEP	OCT	NOV	DEC	YEAR
T	25	27	28	29	28	26	25	25	26	26	26	26	26
P	1.0	1.0	2	8	24	89	76	39	21	26	12	3	301

Station: 9 Climate Type: ____________________

	JAN	FEB	MAR	APR	MAY	JUN	JUL	AUG	SEP	OCT	NOV	DEC	YEAR
T	-13	-13	-5	6	12	18	21	20	15	7	-2	-9	5
P	3	3	4	5	6	8	6	5	3	3	2	2	50

Station: 10 Climate Type: ____________________

	JAN	FEB	MAR	APR	MAY	JUN	JUL	AUG	SEP	OCT	NOV	DEC	YEAR
T	7	8	11	14	18	22	25	24	21	17	12	8	15
P	8	7	7	7	6	4	2	3	6	13	11	10	84

Station: 11 Climate Type ____________________:

	JAN	FEB	MAR	APR	MAY	JUN	JUL	AUG	SEP	OCT	NOV	DEC	YEAR
T	13	13	14	16	17	19	21	21	21	18	16	14	17
P	4	5	2	1	1.0	0	0	0	0	1.0	1	3	18

Station: 12 Climate Type: ____________________

	JAN	FEB	MAR	APR	MAY	JUN	JUL	AUG	SEP	OCT	NOV	DEC	YEAR
T	9	9	8	6	4	2	2	2	4	6	7	8	6
P	7	6	6	7	7	5	5	5	4	4	5	7	68

Activity: Climatic Controls

OBJECTIVES

In this exercise, you will: 1) learn to recognize significant features of several climate types by analyzing monthly temperature and precipitation data; 2) relate the effect of latitude, elevation, prevailing wind, and maritime or continental location to global climate patterns; and 3) associate soil and vegetation patterns with global climate zones.

INTRODUCTION

Areas near the equator receive more of the direct rays of the sun for longer periods of time. As one moves toward the poles, the amount and duration of the sun's heat energy decrease. There is also a relationship between the specific heat of substances (e.g., water and land) and its effect on air temperature. Because water surfaces heat and cool more slowly than land surfaces, the climates of coastal areas are strongly influenced by their proximity to water. These and other factors, such as global air circulation and water availability control the climate of an area.

In this exercise, you will examine the various climate types found throughout the world, as classified by the modified Köppen system. The eight stations that you will analyze have been selected because they are representative of the more common climate types. Complete the questions in your manual for each of the stations. The annual temperature range is calculated by subtracting the temperature of the coldest month from the temperature of the warmest month. Next, determine the Köppen climate type. In classifying each climate, check first to see if the station is a moisture deficit (B) climate. When you have figured out the climate type, you will be able to answer the questions on seasonality of precipitation and whether it is a wet or dry climate.

First, you will examine the A type climates of stations 1 and 2. Station 1, Iquitos, Peru, is located in the Amazon River Basin. Station 2, Cochin, India, is located along the southwestern coast of India. A climate types are located in the equatorial belt, generally between 20° North and 20° South latitude.

Station 1: Iquitos, Peru (4° S 73° W)

	JAN	FEB	MAR	APR	MAY	JUN	JUL	AUG	SEP	OCT	NOV	DEC	YEAR
T	26	25	24	25	25	23	23	24	24	25	26	26	25
P	25.1	25.9	30.5	16.8	24.6	18.3	16.5	12.2	22.1	18.5	21.1	28.7	261.6

1) Annual temperature range:____________________________

2) Station locatio: coastal or interior?____________________________

3) Is there a definite seasonality to precipitation?____________________

4) Kö ppen Climate Type:____________________

Station 2: Cochin, India (10° N 76° E)

	JAN	FEB	MAR	APR	MAY	JUN	JUL	AUG	SEP	OCT	NOV	DEC	YEAR
T	27	28	28	29	28	27	26	26	26	26	27	27	27
P	2.0	2.0	6.9	9.4	29.5	67.6	63.5	23.6	32.5	33.0	17.0	5.1	292.1

1) Annual temperature range: ____________________

2) Station locatio: coastal or interior?____________________

3) Is there a definite seasonality to precipitation?____________________

4) Köppen Climate Classification:____________________

The tropical rainforest (or Af climate) is found closest to the equator where there is the least variation in temperature between the seasons. This is also where the equatorial winds are converging, which helps to account for the equal and abundant distribution of precipitation throughout the year. On either side of the Af climate, we often find the Aw or tropical savanna climate. This can be considered a transition zone from the Af to the B or C climates occupying more area than any of the A climates. Am or tropical monsoon climates are restricted to coastal areas. Monsoons are seasonal phenomena that bring strong winds and either heavy rainfall or rainless periods for a few months of the year. The warm, tropical air currents associated with the monsoons gather moisture and momentum as they travel across the open water. This air is then forced to rise orographically as it approaches barriers such as the Brazilian Highlands and the Himalayas. The instability of this air results in abundant rainfall along the coastal areas. During the alternate season, winds blow from the continental interior and the dry monsoon prevails.

Next, let us examine some of the B climate types. Remember that the B climates are subdivided into Deserts (BW) and Steppes (BS). Stations 3 (Salt Lake City, Utah) and 4 (Timbuktu, Mali) are examples of B climates.

Station 3: Salt Lake City, Utah (41° N 112° W)

	JAN	FEB	MAR	APR	MAY	JUN	JUL	AUG	SEP	OCT	NOV	DEC	YEAR
T	-2	1	5	10	15	19	25	24	18	12	3	0	11
P	3.4	3.0	4.0	4.5	3.6	2.5	1.5	2.2	1.3	2.9	3.3	3.1	35.3

1) Annual temperature range: ____________________

2) Station location: coastal or interior?____________________

3) Is there a definite seasonality to precipitation?______________

4) Köppen Climate Classification:______________________

Station 4: Timbuktu, Mali (17° N 3° W)

	JAN	FEB	MAR	APR	MAY	JUN	JUL	AUG	SEP	OCT	NOV	DEC	YEAR
T	22	23	28	32	34	34	33	31	32	31	32	22	29
P	0.0	0.0	0.0	0.1	0.2	0.8	2.7	2.6	1.3	1.1	0.0	0.0	8.8

1) Annual temperature range:______________________

2) Station location, coastal or interior?______________

3) Is there a definite seasonality to precipitation?______________

4) Köppen Climate Classification: ______________

The Steppe climate is not nearly as arid as the true desert areas, and is often considered the transitional zone between the desert and other climate types. There are two types of deserts that are associated with latitudinal location. First, there are low latitude or tropical deserts. These are generally located between 15 and 30 degrees of latitude North or South of the equator. These are often thought of as hot deserts. The Sahara in northern Africa is at approximately the same latitude as the deserts of Mexico and the southwestern United States. The Kalahari, in southwestern Africa, and the Great Australian Desert, which covers almost all of Australia, are located within 15 and 30 degrees South latitude. These, too, are referred to as low latitude, or tropical deserts.

The other type of desert is referred to as a mid-latitude or subtropical desert or cold desert in which at least one month is colder than 0° C. These colder deserts are located between 40 and 50 degrees North or South latitude.

Of all the major groupings, the C and D regions contain the greatest range of climatic conditions, landscapes, and cultural environments. Instead of the Greek word "temperate" it might be more appropriate to label these areas as "temperamental." These are the areas of cyclones and frontal action, of droughts and thunderstorms, of snowfall and palm trees. Natural vegetation ranges from cypress swamps to boreal forests and from coastal marsh grasses to broad prairies. The diversity of the C and D climates is the key to unity and the basis of classification! The other climate groupings, A, B, and E have a high degree of uniformity whereas the C and D areas are characterized by their variability.

Station 5: Charleston, SC (33° N 80° W)

	JAN	FEB	MAR	APR	MAY	JUN	JUL	AUG	SEP	OCT	NOV	DEC	YEAR
T	11	11	14	20	23	26	29	29	26	20	14	10	19
P	7.6	8.1	8.6	5.8	8.4	13.7	15.2	16.5	14.0	9.7	5.6	7.6	121.4

1) Annual temperature range:________________________________

2) Station location: coastal or interior?________________________

3) Is there a definite seasonality to precipitation?______________

4) Köppen Climate Classification:____________________________

Station 6: London, England (51° N 0° W)

	JAN	FEB	MAR	APR	MAY	JUN	JUL	AUG	SEP	OCT	NOV	DEC	YEAR
T	5	5	6	8	13	16	17	17	14	11	7	5	10
P	4.8	4.1	4.1	4.1	4.6	5.1	5.6	5.6	5.1	6.6	5.8	5.8	61.2

1) Annual temperature range:________________________________

2) Station location: coastal or interior?________________________

3) Is there a definite seasonality to precipitation?______________

4) Köppen Climate Classification:____________________________

Station 5 is a Cfa or humid mesothermal climate. This climate type dominates the southeastern portions of North and South America, and the southeastern portion of China, Northern Burma, and India. The Csa type is called a Mediterranean climate. These climates are found primarily around the Mediterranean Sea and also along the western coast of all climates at middle latitudes. The Csb is also a Mediterranean climate.

One of the most important crops of France is its wine grapes. French wines are considered among the world's finest. Notice that France is mainly within the Csa climate type. What area in the United States is also known for its vineyards? It should not be surprising that California's San Joaquin Valley shares the same climate type as France.

If you were only to examine the latitude of Station 6 (London, England: 51° N), you might think that it had a fairly cold climate. The milder conditions experienced there result from the effect of water, especially major ocean currents, on the weather. The North Atlantic Current transports warm water from the tropics along the eastern coast of North America and over to Northern Europe. This current is responsible for much of the precipitation in our area, a phenomenon very important in the early development of our country. The same current warms the waters of the Grand Banks, one of the most productive fisheries in the world. Finally, the North Atlantic Current warms western and northern coastal Europe. It keeps the harbors of northern Europe, such as Murmansk, open during the winter. How does this help to explain why we associate fog with London?

Station 7 is a D climate. As you work through the problem, think of three physical factors that account for this station's climate. Station 8 is an E climate.

Station 7: Moosonee, Ontario, Canada (51° N 81° W)

	JAN	FEB	MAR	APR	MAY	JUN	JUL	AUG	SEP	OCT	NOV	DEC	YEAR
T	-20	-18	-12	-3	5	11	17	14	7	-5	-11	-18	-2.7
P	3.3	3.0	3.8	3.8	4.8	7.1	8.9	8.1	6.1	5.3	4.8	4.3	63.5

1) Annual temperature range:__________________________________

2) Station location: coastal or interior?___________________________

3) Is there a definite seasonality to precipitation?____________________

4) Köppen Climate Classification:______________________________

Station 8: Barrow, Alaska (71° N 156° W)

	JAN	FEB	MAR	APR	MAY	JUN	JUL	AUG	SEP	OCT	NOV	DEC	YEAR
T	-27	-24	-22	-18	-7	2	5	4	-1	-9	-19	-25	-11
P	0.5	0.5	0.3	0.2	0.3	1.0	2.0	2.3	1.5	1.3	0.5	0.5	10.9

1) Annual temperature range:______________________________

2) Station location: coastal or interior?________________________

3) Is there a definite seasonality to precipitation?_________________

4) Köppen Climate Classification:______________________________

The first and most obvious factor influencing the climate at Station 7 is its latitudinal location. Generally, as distance from the equator increases, the colder the climate becomes. Second, this station's inland location prevents it from being affected by warm ocean currents. Compare the location of this station with that of Station 6, which is at approximately the same latitude. Third, stations in this area are influenced by the prevailing northwestern winds which bring in cold arctic air.

Examine the climate maps in your textbook and note the lack of D and E climates in the southern hemisphere. Most of the landmass of the world is located in the northern hemisphere, and much of that is located north of 50°. Very little land is located south of 50° latitude, and that portion of South America that is south of 50° is influenced by warmer ocean currents.

The coldest regions in the world are classified as E type climates and are located within the Arctic and Antarctic circles. There are two types of E climates. The first is the icecap, where ice is actually on the Earth's surface year-round. The average monthly temperature at these locations never goes above freezing. There are only a few areas that are covered by ice. These are the North and South Poles and Greenland. The very northernmost reaches of North America, Europe, and Asia are referred to as tundra, or ET climate type. These areas are mostly within the Arctic circle. There is very little vegetation in these areas due largely to the fact that the water within the soil is frozen throughout the year, sometimes to a depth of more than 300 m. Although the upper layer of soil thaws during the brief tundra summer, water in the deeper soil remains frozen throughout the year. This soil moisture condition is known as permafrost. In the Dfc areas during the summer, the top layer, or active layer, melts and the surface temperature is warm enough for plant growth.

Just as there is a change of temperature with increased elevation, there is a marked change in climate types from base to summit in the mountainous regions of the earth. Because of the large climatic variation within relatively limited regions, these mountainous areas are labeled simply as highlands on climate maps. Climatic change is best illustrated by the changing vegetation types. Travelling from base to summit in any of the highland regions, you would pass through a succession of climate zones marked by vegetation communities typically found at lower elevations in the higher latitudes.

QUESTIONS FOR REVIEW

A. How can you account for the fact that Station 4 at 17° N is considerably warmer in summer than Station 1 located at 4° S?

____

____

____

B. For each of the following stations calculate the annual range. Using temperature and precipitation data (range, mean, totals), see if you can identify the general location for each station.

	JAN	FEB	MAR	APR	MAY	JUN	JUL	AUG	SEP	OCT	NOV	DEC	YEAR
T	17	18	17	16	14	13	12	12	12	13	14	16	15
P	10.3	8.8	8.8	15.3	14.4	18.7	14.7	19.3	12.9	13.9	11.7	16.7	165.5

Temperature range ______________________________

Northern or southern hemisphere______________________________

Continental or maritime ______________________________

If maritime, east coast, west coast, or insular______________________________

Station:______________________________

	JAN	FEB	MAR	APR	MAY	JUN	JUL	AUG	SEP	OCT	NOV	DEC	YEAR
T	-47	-43	-30	-14	3	13	16	11	3	-14	-36	-44	-15
P	0.7	0.5	0.5	0.4	0.5	2.5	3.3	3.0	1.3	1.1	1.0	0.7	15.5

Temperature range ______________________________

Northern or southern hemisphere______________________

Continental or maritime ___________________________

If maritime, east coast, west coast, or insular_______________

Station:_____________________________________

C. Account for the difference in the annual variation in temperature (range) for Stations 5 (Cfa) and 6 (Cfb).

____

____

D. What three factors account for the climate at Station 7?

1)____

2) ____

3)____

E. How do the Earth/Sun relationships and seasonality affect the annual range of temperatures as you go from the equator to the polar regions?

____

Excluding the arid climates, what appears to happen to precipitation as you go from the equator to the poles?

____

Account for this: ____

PART 2: CLIMATES, VEGETATION, AND SOILS

Fill in the spaces below relating climate type, vegetation, and soils.

CLIMATE NAME	*KÖPPEN SYMBOL*	*VEGETATION*	*SOIL*
Ice Cap	____________	____________	____________
Tundra	ET	Mosses, Lichens, and Herbaceous Plants	____________
Subarctic	____________	____________	____________
Humid Continental	Dfa, Dfb	____________	Mainly Spodisols
Marine West Coast	____________	Needleleaf evergreen (coniferous) and needleleaf deciduous	N. hemisphere: Spodisols; S. hemisphere Inceptisols
Humid Subtropical	____________	____________	Mainly Ultisols
		____________	Spodisols, Alfisols, Mollisols
Mediterranean	Csa	____________	Aridisols

CLIMATE NAME	***KÖPPEN SYMBOL***	***VEGETATION***	***SOIL***
Desert	______________	Mixed grass and herbaceous plants and Xerophytic vegetation	______________
Steppe	______________	Grass and other herbaceous plants	______________
Tropical Savanna	______________	______________	______________
Tropical Rainforest	______________	______________	______________
Tropical Monsoon	______________	______________	______________

Chapter 13

Climate Change ?

Could the patterns of climate regions change? Certainly. Evidence of climate change in the geologic past is unmistakable (Figure 13.1). There is ample evidence that the pattern of climate we recognize today has not always existed. The evidence takes many forms – tropical animal and plant fossils unearthed in polar locations; relics of ancient humid climates preserved in the aridity of modern deserts; and evidence that glaciers which are now restricted to high mountains and polar regions once existed as far south as the Ohio River in the United States and in such unlikely places as South Africa.

What could cause such radical changes? When we examine the trend of temperature over the past million years, clearly we can expect climate change in the future. Could the rate of climate change make important changes in the next century, in the next 50 years? Maybe. Could human activities trigger such a change? Again, maybe. The measurements made over the last half of the 20th century indicate that temperatures have been rising in many locations. The interrelationships between climate and land surface form, vegetation, soils, water features, agriculture, forestry, electricity generation, transportation, and other human activities are so numerous and interlinked that many atmospheric scientists, meteorologists, geographers, economists, and other decision makers have become concerned about the prospects of global climate change. Whether any change is occurring, what the direction and rate of change might be, and what consequences such changes might hold for human activities has been widely examined during the past 20 years.

Although there are valid theoretical reasons to expect global warming from increases in greenhouse gases, the short answer is: we simply have a lot to learn about the atmosphere and many years of research ahead before we will have clear answers to such difficult questions. To illustrate the difficulty in examination of climate

Figure 13.1 Geologically Recent Changes in the Global Temperature

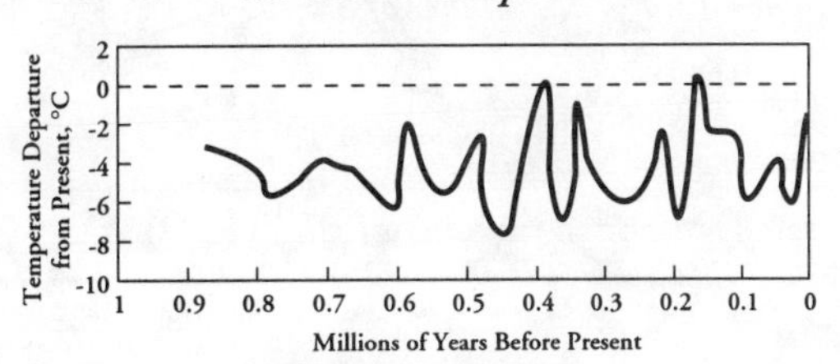

change please consider Figures 13.2 and 13.3. One depicts the global temperature trend during the past 25,000 years, and the other describes the average annual temperature in the southeastern United States since 1900 as illustrated by observations at Charlotte. First, note the large variations in each data set. Also note that the trend for the last 20,000 years seems to have been warming with a slight cooling during recent millennia. In the Charlotte data set, note that since 1900 there appears to have been warming, then cooling then a slight increase during recent years. The nature of climate is change, and small compounded changes in the rates of heating (or cooling) could have important implications within the next century, but these are difficult questions that will require the continuing efforts of many new, clever, and curious atmospheric scientists.

A number of possible causes are capable of altering the earth's heat budget and they fall within three categories: 1) changes in the solar constant that result from changes in solar radiation, 2) changes in incoming shortwave radiation attributable to changes in the earth's albedo, and 3) changes in outgoing longwave radiation.

Changes in the solar constant result from two very separate causes: changes in the earth's orientation to the sun and sunspots. Slight variations in the earth-sun geometry are responsible for most common paleoclimatic shifts. The frequency of sunspot occurrence,

Figure 13.2 Temperature Trend over the Past 25,000 years

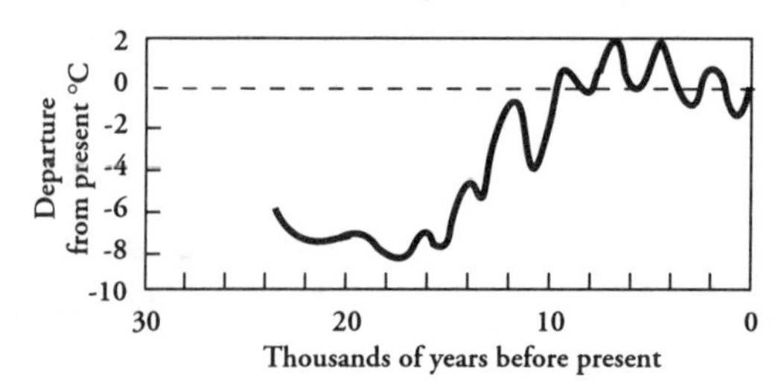

Figure 13.3 Average Annual Temperature (F) in Charlotte, NC 1900-1992

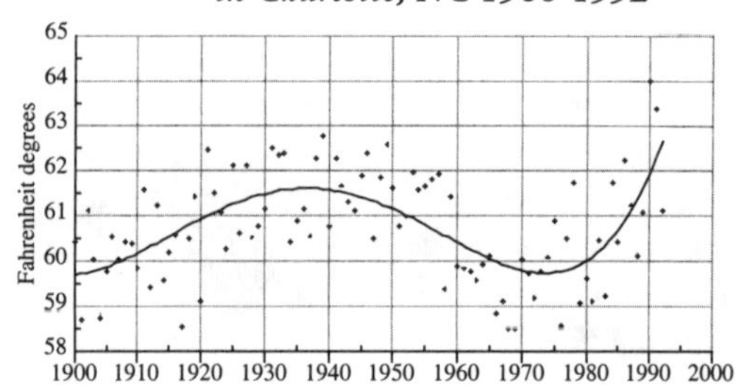

however may be responsible for more recent climatic perturbations. Sunspots are dark areas on the surface of the sun and because they are visible from earth astronomers have been keeping count since 1650 (Figure 13.4). You might think that solar radiation would diminish during years when these dark patches are more numerous, but just the opposite occurs. Sunspots are accompanied by bright patches called faculae that more than compensate for the

Figure 13.4 The Number of Sunspots

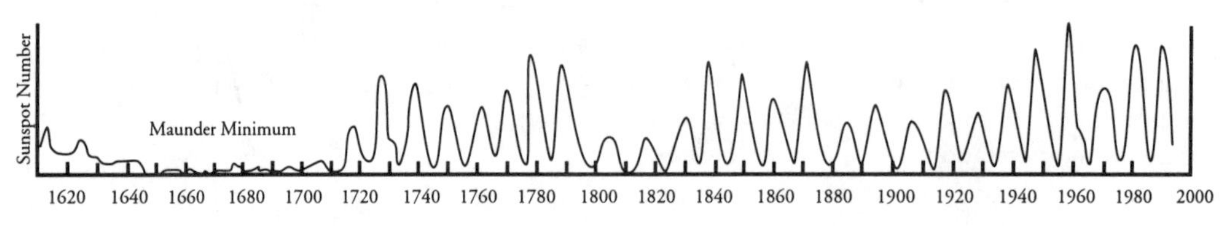

dark patches and result in an increase in solar radiation of about 0.1% during high sunspot activity. Furthermore, the sun has been getting steadily brighter since the Maunder minimum, a period of low sunspot activity that coincided with an unusually cold period known as the Little Ice Age during the 1600s (Figure 13.5). The coincidence is unmistakable, but the magnitude of change in the solar constant from sunspot activity is not sufficient to explain the nearly 1.0°F increase in global temperature during this century. Some as yet unknown positive feedback mechanism could be magnifying the effect, or the explanation might rest with slight changes in the rate of terrestrial radiation.

Increased concentrations of greenhouse gases such as carbon dioxide, methane, nitrous oxide, and halocarbons are theoretically capable of forcing changes in terrestrial radiation that would result in higher global temperatures. Atmospheric chemists are in agreement that levels of carbon dioxide have increased (from about 280 to 356 ppmv) since the 1850 (Figure 13.6). The increased levels of carbon dioxide are of great interest to atmospheric scientists because the ability of carbon dioxide to absorb terrestrial radiation is well known, and because the trend of temperature and carbon dioxide have tended to covary through the recent geological past. Unlike sunspots or changes in the earth-sun geometry, many greenhouse gases like carbon dioxide are by-products of human activities. International attention in 1998 was focused on carbon emissions at a conference in Kyoto, Japan where many nations agreed to explore ways to limit the rate of growth in atmospheric carbon dioxide.

There is ample evidence that increases in aerosol concentrations can reduce incoming shortwave radiation on a global scale. Volcanic

Figure 13.5 The Little Ice Age

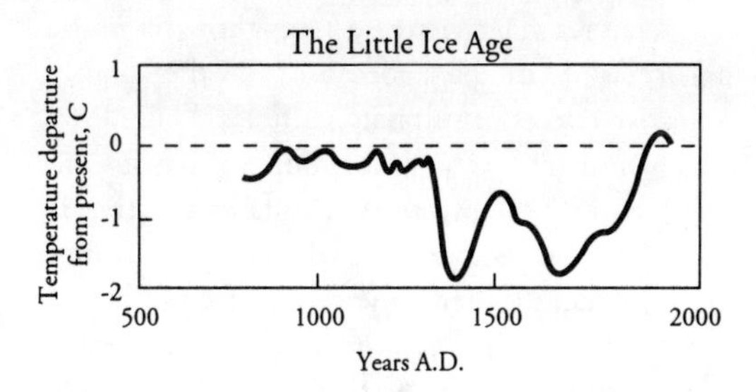

eruptions may eject enough particulate matter into the stratosphere to intercept insolation and reduce global temperatures. Those from Mt. Pinatubo in June of 1991 are credited with reducing global temperatures by 0.3° to 0.5°C in 1992. Most atmospheric scientists feel that anthropogenic particulates, largely as dust from agricultural activities and fine particles from sulfur dioxide emissions and carbon from biomass burning, are also capable of altering the radiation balance. There are direct and indirect effects from the presence of these particulates. Fine particles, particularly sulphate aerosols, are effective in scattering solar radiation back to space. Sulphate particulates are so effective in reducing insolation that daily high temperatures are thought to be depressed slightly in areas

Figure 13.6 Estimates of Global Carbon Dioxide Levels as Measured at Mauna Loa

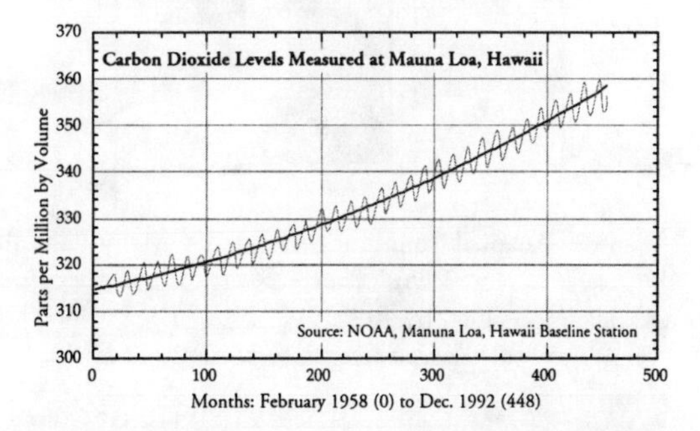

where particulates are abundant. Attempts to isolate empirical evidence that average daily temperatures have increased are thought to be unsuccessful in part because of the highly reflective effects of sulphate particles. Daily lows which usually occur just prior to dawn, are unaffected by the presence of sulfates and trends of daily low temperatures do show warming more frequently than time series of daily highs. Indirect effects include the role of particles as condensation nuclei. As these particles provide opportunities for water vapor condensation and deposition, the chances for increased cloudiness is enhanced. Increased cloudiness, especially high clouds, intercepts and reflects sunlight, and thereby further attenuates insolation. In areas with high concentrations of sulfates, the range between daily highs and lows is decreasing slightly. These areas are regions of industrial activity; the eastern United States, Europe, the Middle East, and East Asia.

In summary, there are three factors that determine the energy available for weather and climate: solar inputs, greenhouse gases, and aerosols (Figure 13.7). The cooling effect of aerosols and stratospheric ozone are generally offset by a warmer sun and increased levels of ground level ozone (smog). Increaes in greenhouse gases carry the risk of increasing global temperatures in the lower atmosphere. If that should happen it would probably not be manifested as a uniform small change in all

Figure 13.7 The Direction and Magnitude of Variables in the Radiation Budget

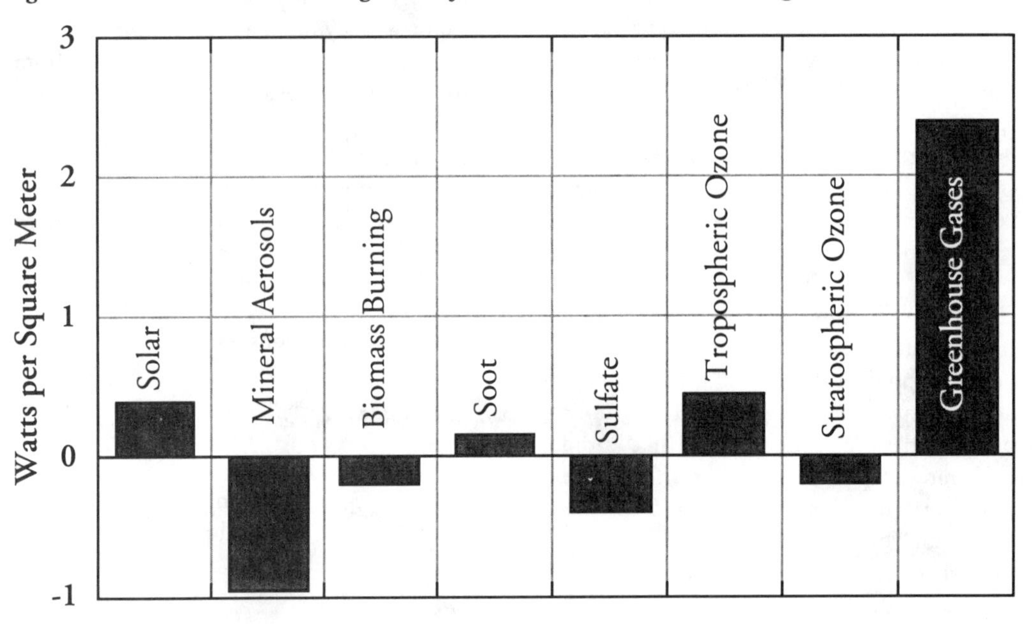

Sources: Adapted from Houghton, J. T.; Meira Filho, L. G.; Callander, B. A.; Harris, N.; Kattenberg, A.; and Maskell, K., eds. 1996, *Climate Change 1995: The Science of Climate Change.* Cambridge: Cambridge University Press; and Balling, Jr., Robert C., 2000, The Geographer's Niche in the Greenhouse Millennium, *Annals of the Association of American Geographers*, 90 (1) 114-122.

Figure 13.8 Regional Climate Deviations Inferred from an Ancient Warm Period

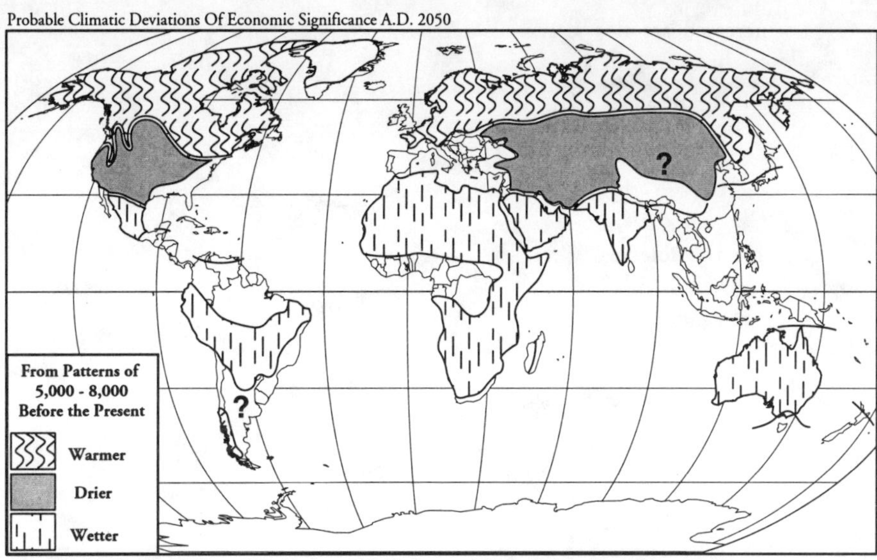

Source: Modified from Karl W. Butzer, 1980. Adaption to Global Environmental Change, *Professional Geographer*, pp 269-278.

locations, but rather as warmer conditions in some places, cooler conditions in others, wetter conditions here, or drier conditions there. Note in Figure 13.2 that the period from 4,500 years to 8,000 years before the present was warmer than today for the earth as a whole. The spatial pattern of climatic conditions during that period as it can be discerned from the geological record suggests what residents of a warmer world might face by the mid-twenty-first century (Figure 13.8). Before we are able to resolve all our questions about the future of climate regions and the global climate, a better understanding of these elements (especially the carbon cycle), the rates of change, and the feedback mechanisms that may exist between these elements will be needed.

Within the previous chapters, the authors have attempted to illustrate the intimate relationship existing between people, climate, and the land. World history is rich with evidence of human-environmental interactions: harnessing naturally occurring geothermal, water, fossil fuel, and nuclear power; domestication and hybridization of indigenous plants and animals; and altering climate through heating and air conditioning. Human ingenuity in resource-use has fostered development of sophisticated civilizations with complex economic and social structures, and of technological advances that have provided us with countless amenities (not the least among them being the automobile). Development, however, has not been without cost. Environmental and urban deterioration, loss of habitat, atmospheric pollution and potential climatic change are but a few of the challenges that await the thoughtful and curious minds of this new century.

Chapter 14

The Hydrosphere

The physical components of the hydrosphere include the rivers, oceans, glaciers, water in the ground, and atmospheric water. Each of these components can be treated as a system with its own set of inputs, outputs, and responses. The unity of the system is confirmed by observation when we see rain fall and water flow over the surface to streams which, in turn, eventually discharge it into oceans so that it can evaporate again and be dropped as precipitation. This cycling of water through the atmosphere, lithosphere, and biosphere helps to provide unity among these components of the earth system just as the flow of energy does (Figure 14.1).

It is customary to subdivide the hydrosphere into the subsystems enumerated in the previous paragraph. This is a simplifying convenience that we will use. Before proceeding to discussions of those subsystems, however, we need to examine some of the physical and chemical properties of water that are essential to understanding how water behaves and the functional roles that water plays in natural systems of all kinds.

Water Properties

Water is, without doubt, one of the most unique and interesting substances known. Its many properties give it a wide range of functions in nature and make it useful to human beings for many purposes. The versatility of water is derived from its molecular structure, and, although a thorough discussion of the molecular structure is neither a necessary nor appropriate topic for this book, a limited introduction seems to be necessary for an adequate understanding of water's role in earth surface systems.

Molecular Structure and Phase Changes

Water is composed of hydrogen and oxygen, two very simple elements. Two hydrogen atoms are bound to one oxygen atom by chemical bonds to produce one water molecule (H_2O). The bonding is such that the two hydrogen atoms, which carry a slight positive charge, are separated by an angle of 105° (Figure 14.2). The water molecule that is produced has a slightly negative charge in the vicinity of the oxygen atom and a slightly positive charge near the hydrogen atoms.

In the vapor state water molecules behave like any other gas. There are no molecular forces binding the molecules together, so that each molecule moves freely and independently of all others. In the liquid state molecular bonds that form between the positively and negatively charged ends of the water molecules hold the molecules together in a loose chains (Figure 14.3). These forces are not strong enough to prevent molecules from moving, and freely sharing their molecular partners. Ice, in contrast to the fluid phases, has a rigid crystalline

Figure 14.2 The Water Molecule

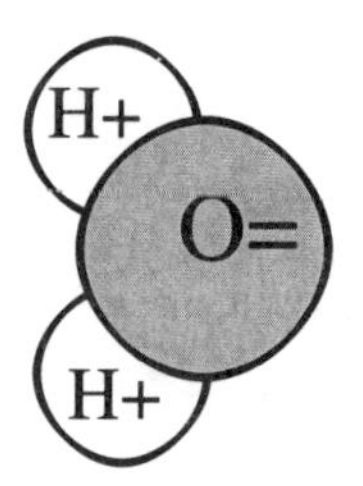

Figure 14.1 Flows and Reservoirs within the Hydrosphere

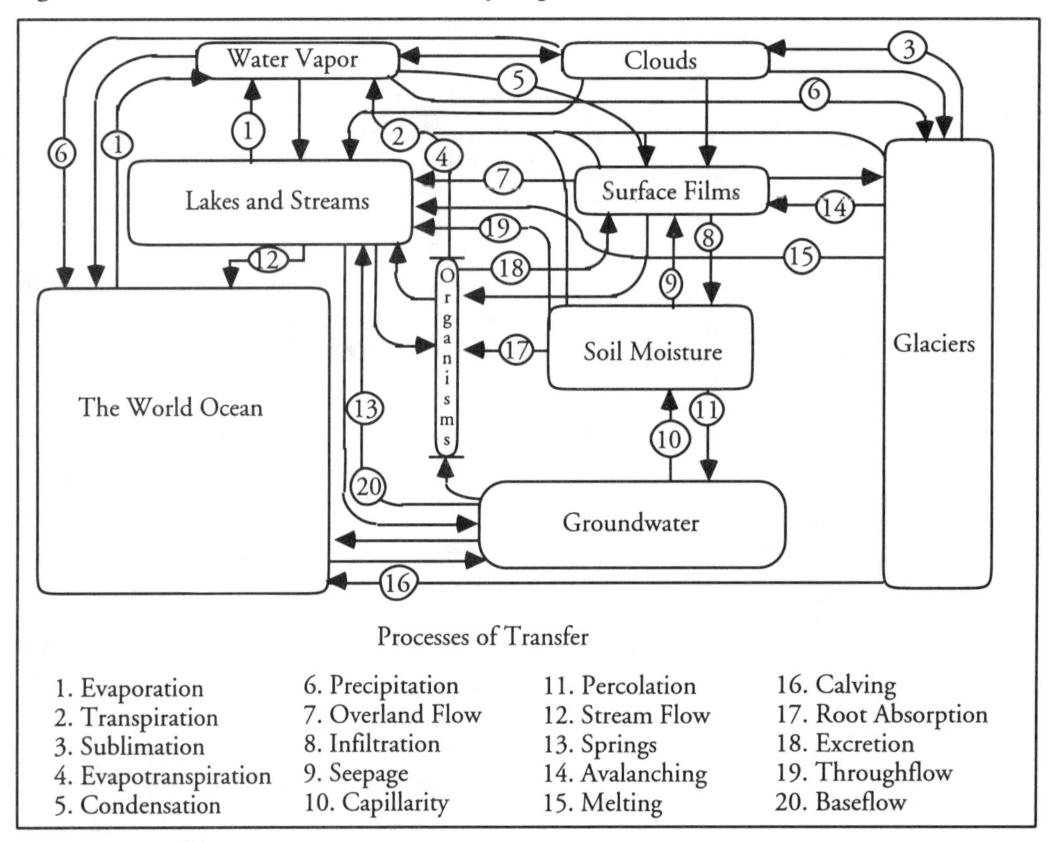

Figure 14.3 A Chain of Water Molecules

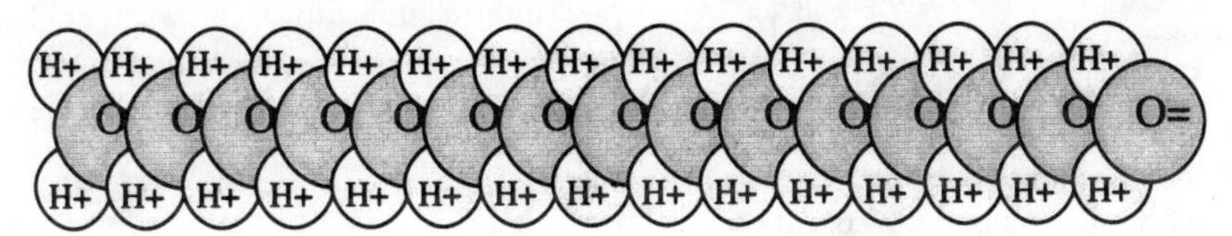

structure the same as other solids. In crystals molecules are arranged in a specific manner similar to soldiers in a military formation.

Energy is required to break the strong bonds between water molecules to change ice into water. The energy required, called **latent heat of fusion**, is 80 calories per gram for ice that is 0°C (32°F). If the ice is colder than 0°C, then more heat is necessary to increase the vibrations of the molecules enough to break the hydrogen bonds.

Much more energy is required to overcome the bonds between molecules in liquid water (540 calories per gram). The heat required to change water into a gas is called **latent heat of vaporization**. Heat of vaporization is "stored" in the water vapor, but is released during condensation, which is so important in transferring energy from the earth to the atmosphere. It is the high heat of vaporization and heat of fusion that give water its high boiling temperature of 100°C (212°F) and high freezing temperature of 0° (32°F), thereby allowing water to exist in all three phases at earth surface temperatures and pressures.

Specific Heat

When heat is added to a substance it may cause the molecules to move around faster, thereby increasing the average kinetic energy and the temperature. In fluids molecules may also rotate or spin, a motion which does not affect temperature. Therefore, how much of a temperature increase a substance will have when heated depends on how much of the energy is stored in rotational motion of molecules as opposed to translational or vibrational motions.

A great deal of heat is required to raise the temperature of a gram of water one degree Celsius. Thus water has a high specific heat, which allows large water bodies to gain or lose much heat without experiencing much of a temperature change.

The storage of heat by water is important for reasons other than its effect on climate. Water is the dominant component in the environment of aquatic organisms, and its heat storage properties reduce the range of environmental temperatures that these organisms must bear, especially in areas outside the tropics. They are subjected to neither the extremes nor the rates of change in environmental temperatures that terrestrial creatures must tolerate.

Pressure and Density

In our discussion of the atmosphere we spoke of the relationship between pressure and density in gases. We said that in gases pressure and density are directly related and depend on temperature. Such is not necessarily the case with liquids. Water is essentially an incompressible fluid, and pressure and density are not significantly related. Although density varies with temperature, pressure does not. Pressure does increase with depth. Water is such a heavy substance that its pressure is measured in atmospheres. It takes only about 10 vertical meters (32 feet) of water to exert as much pressure as one atmosphere, so the pressure of water increases by one atmosphere every 10 meters, reaching a maximum of about 1100 atmospheres in the deepest ocean trenches.

In the formation of ice, water molecules arrange themselves so that a maximum number of bonds are formed between the negatively and positively charged ends of the water molecules. The resulting structure is a rather airy one that includes considerable unoccupied space. When ice melts some of the molecular bonds are broken and the roomy structure begins to break down, allowing molecules to move closer together, thereby increasing the density.

The density of water continues to increase until the temperature reaches 4°C. Above 4°C additional heating increases the kinetic energies of water molecules and produces expansion and corresponding density decreases.

Water bodies freeze from the top down because of the unusual density changes that take place as water is cooled. As surface water cools, the more dense water settles to the bottom, displacing lighter warm water that rises to the surface. Once all the water cools to 4°C, further cooling decreases the density of surface water which will then continue to cool until the surface freezes. As additional freezing must occur from the surface downward, further cooling depends on heat loss that must occur through the ice by conduction. The thicker the ice, the slower the rate of heat loss. As ice forms under the surface, it occupies more space than the water, and the resulting increase in pressure helps lower the freezing temperature. Both processes act as negative feedbacks which slow the rate of ice accumulation. This is one reason why lakes seldom freeze solid unless they are shallow or are located in very cold climates. The interrelationship between the physical hydrosystem and biological systems is apparent, for those lakes that freeze solid contain no fish populations and simplified ecosystems.

Viscosity

Viscosity is a measure of a fluid's internal resistance to flow. It is a function of the molecular structure of a substance, and the viscosities of different fluids are not necessarily proportional to their relative densities. For example, motor oil is more viscous than water yet it is clearly less dense because it floats on water. The viscosity of liquids such as water, does vary with density, but only as a function of temperature. Viscosity of water also varies with pressure to some extent.

The viscosity of water is great enough that movement through water is retarded significantly by friction. Mobile marine organisms such as fish, whales, and seals have evolved efficient streamlined shapes for moving through water with minimal energy, and the forms of ships and submarines are streamlined for the same reason. Yet the viscosity of water is small enough that water flows easily and quickly in response to pressure and gravity and may easily attain velocities in excess of a meter per second (3 feet/s) in streams.

Ionization

Ionization is one of the chemical properties of water that is extremely important in understanding the linkages between the hydrosphere and other earth subsystems. A basic understanding of ionization is essential to explanations of weathering and erosion, soil formation and plant and animal growth.

Ions are electrically charged atoms or molecules. Normally, an atom has enough negative electrons to exactly balance the positive charge in its nucleus. If for any reason it gains or loses an electron, the equilibrium is disturbed and the atom takes on a correspondingly negative or positive charge.

When the molecular bonds are broken in individual water molecules, one of the hydrogen atoms moves away, leaving its electron attached to the remaining hydrogen atom and the oxygen atom.

The positively charged hydrogen atom (H^+, called a ***cation***) is attracted to substances with a negative charge. Solutions with high concentrations of cations are acid. The hydroxyl ion (OH^-) has a negative charge, due to the extra electron, and is attracted and bound to positively charged substances. These are the ions of bases such as lye.

Many compounds are split by water into two compounds, one of which may then combine readily with hydrogen or an hydroxyl ion. The heat released in the process can be used to trigger chemical reactions to create other substances. An example of a simple ionic reaction is the process of carbonation in which carbon dioxide gas (CO_2) combines with the hydroxyl ion (OH) to produce carbonic acid (HCO_3^-) and release heat. The chemical formula is:

$$CO_2 + H_2O \Leftrightarrow H^+ + HCO_3^- + heat$$

This is the reaction that occurs when carbon dioxide gas dissolves in water or other liquids. The process is important in weathering and erosion. The arrows indicate that the process is reversible, providing that heat is available for the reaction. In fact, the heat required is small enough that it is readily available at normal environmental temperatures, resulting in spontaneous release of carbon dioxide gas from carbonated beverages at room temperatures. This is one of the important chemical processes referred to in the previous chapter that performs environmental work within those climates that are not closed.

Solvent Power and Gas Diffusion

Water is sometimes referred to as the universal solvent. A wide variety of earth materials are solvent in water. Some, such as salts and atmospheric gases, are readily dissolved, while others, such as silica, dissolve slowly. The solvency of water makes it an efficient agent in weathering and erosion as well as an effective medium for the growth of organisms.

There are two types of solvent action in water. One type depends on hydrogen bonding between the hydrogen atoms in water molecules and oxygen or hydrogen atoms in the solute. Sugars, organic acids, nitrates, phosphates and other nutrients involved in life processes are dissolved this way. The second type of solvent action is polar bonding which depends on the separation of negative and positive charges between the oxygen and hydrogen atoms in water molecules. This allows the electrical charges to interact with charged particles of various salts, most notably sodium chloride. The potassium salts necessary for plant growth are dissolved in this way.

Solution, ionization, and other chemical state changes that involve water consume or release energy in the same manner as physical state changes do. However, the amounts of energy involved in work and released as heat are not as large as those involved in changing water from solid and liquid states into the gaseous state. Even though chemical changes involving water transfer some energy among earth subsystems the bulk of the energy transferred by water is in the form of latent heat of fusion and vaporization. Since we have already examined the flow of energy between the earth surface and the atmosphere and their associated energy budgets in some detail previously, we will turn our attention at this point to flows of water and water budgets.

The Hydrologic Cycle and the World Water Inventory

The cycling of water is the unifying aspect of the hydrologic system, and by tracing the flows of water, we can identify the subsystems and establish their relationship to each other. The flows are shown diagrammatically in Fig 14.1. Let's trace the flow of water through the hydrologic cycle, beginning with the input of precipitation.

When precipitation occurs as rain over a vegetated area the rain is intercepted by the leaves and is prevented from striking exposed soil with great force. The intercepted water is temporarily stored as surface films and some then drips to the ground or runs down the trunk or stem of the plant as stemflow. In either case, the effects are to reduce the velocity (and, therefore, the force) with which large drops strike the ground and to reduce the effective rate, or intensity, of heavy rainfall. The intensity of very light rainfall may be increased. In addition, some of the intercepted moisture absorbs heat from the plant and may evaporate if the pressure gradient is high enough and the air is unsaturated. It is possible that some of the rain may evaporate as it falls, a process called **virga**, not uncommon in deserts.

Once the water strikes the ground it begins to interact with the lithosphere. It may be temporarily stored in small surface depressions. The ability of the ground to store water in this manner is limited, especially in areas where the ground surface is sloping, and if much rain falls it soon begins to run downslope. Simultaneously, if the soil is not already **saturated** (full of water) some of the water **infiltrates** into the soil by moving through pore spaces and other openings. A number of factors influence the percentage of rainfall that infiltrates and the percentage that runs off.

If enough water infiltrates to saturate the soil, gravity pulls some of the water downward through the soil (a process called **percolation**). This water may collect in a permanently saturated layer called the **groundwater** zone (the upper surface of this zone is called the groundwater table, or **water table** for short). Water remaining in the soil may be withdrawn by plants and transpired or may evaporate directly from the soil surface. Some of it may even flow slowly downhill through soil as **throughflow**, to be discharged into streams or lakes or to emerge as springs. Groundwater may flow even more slowly and may be discharged into springs, lakes and streams. The time lag from rainfall until discharge in groundwater flow is much greater, but provides a steady and less erratic flow.

Most of the water that reaches streams eventually runs to the oceans where it may be output from the hydrologic system as we have defined it (and input to the atmospheric system) to begin the cycle again.

Only one subsystem is missing in our description and that is glaciers. In very cold climates precipitation falls as snow and may be locked up as ice in glaciers for long periods of time. As snowfall after snowfall accumulates, the pressure from the overlying weight compress snow into glacial ice. Glaciers contain the bulk of the world's fresh water.

As this brief description of the hydrologic cycle indicates, water may be temporarily stored, to be withdrawn from the cycle at various stages. The length of time involved may vary from a few days for soil and atmospheric water, to years for glaciers and groundwater, and to millennia for the oceans. In systems terminology such reservoirs of matter or energy are called **sinks**. These sinks are the subsystems that constitute the elements of the larger hydrologic system. Let's look first at the world supply of water, how it is allocated to the subsystems, and magnitudes of the exchanges among them.

World Water Inventory

The bulk of the world's water (97.2%) is in the oceans. This should not be surprising given the fact that oceans cover over 70 percent of the surface of the earth, and typically reach depths of 4 to 6 km (2.4 to 3.6 miles). Except for the 105,000 km^3 of water in saline lakes, the rest of the world's water is fresh, and most of it is tied up in glaciers (2.14% of the world total) and groundwater (0.61%). Stream storage and atmospheric storage are both very small, less than 0.01 percent. Soil moisture is even less (0.005%). Each reservoir in Figure 14.1 is scaled to represent the rank-order size of the sink it represents.

Figure 14.4 Percentage of Water Flows between Continents and Oceans

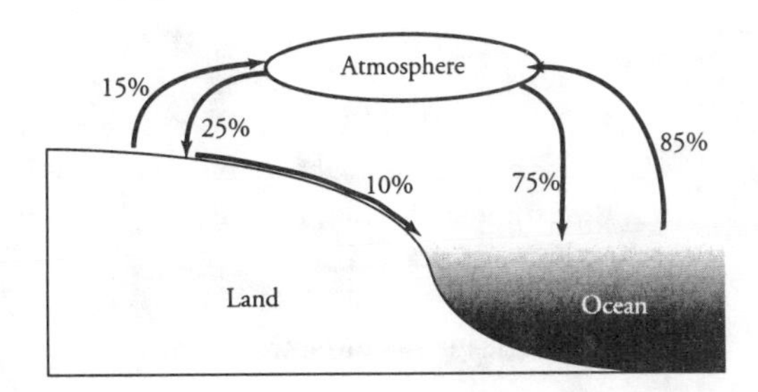

The importance of the atmosphere in the hydrologic cycle is apparent. All of the cyclic water moves through the atmosphere (some 516,000 km^3 per year), and approximately 80 percent of the precipitation falls directly into the ocean. About two-thirds of the 104,000 km^3 that falls on land surfaces evaporates before it can be discharged into the sea. A disproportionate amount of water falls on the land area compared with the ocean area (Figure 14.4). The ocean surface is uniformly and continuously wet, but land masses are not, therefore the average moisture content of air advected to land exceeds that of air advected from land toward the ocean. The result is that continental areas receive more precipitation than they evapotranspirate and the surplus moisture is returned in river channels to the sea.

Nearly half of the 29,500 km^3 discharged into the ocean by rivers is discharged by the 10 biggest streams, and one stream alone, the Amazon, discharges nearly 20 percent of the total.

The world water inventory is a crude device useful only for showing approximate values. They are, at best, educated guesses derived from theoretical calculations and inadequate samples. Despite its shortcomings, however, the inventory does provide us with a reasonable idea of the relative magnitudes of the values. A more important question is how the rainfall and evaporation are distributed over the earth.

Water Budgets

Introduction

The terms "water budget" and "water balance" are virtually synonymous to many hydrologists and water resource managers who use water budgets primarily to calculate soil moisture balance and runoff.

Water budgets are calculated using the basic budget equation: I = O ± ΔS. Precipitation is the major input, and evapotranspiration and runoff are the major outputs. If there is an excess of input over evapotranspiration the excess may be stored temporarily in the soil or it may be output directly as runoff. Total stored moisture is carried forward to the next budget period as a balance and is available for withdrawal at some later period, either as evapotranspiration or to supple stream flow.

Water balances can be applied at any scale and over any time period. Even though the general assumptions are the same in all cases, the specific results vary depending on how generalized and how accurate the input data are. The degree of generalization in input data depends, in turn, on spatial and temporal considerations.

The size of the area for which budgets are computed may vary from whole continents to a specific local site, and the precipitation and temperature data used in the computations may be annual, monthly, or daily averages. The differences in results are attributable to the spatial and temporal variability in precipitation and temperature data. Even if the question of data accuracy is ignored, daily budgets for small areas produce only close approximations of the actual input distributions which supply the moisture and the driving force for evaporation.

The Evapotranspiration Term in Water Budgets

Measurements of evapotranspiration are not normally available and thus, potential evapotranspiration must be estimated from theoretical assumptions or from evaporation pan observations. Many formulas exist for estimating potential evapotranspiration, but we have confined our interest to energy-based estimates such as the Thornthwaite Method (see Chapter 8). The Thornthwaite Method has the advantage of using weather data that are collected from many locations. It is an empirical formula that requires two measures: 1) the average air temperature of each month, and 2) the number of hours of daylight for each month. The latter term varies monthly depending on latitude and the number of days in the month.

In applying the water budget equation we will assume that actual evapotranspiration proceeds at the potential evapotranspiration rate when sufficient moisture is available from input and storage. When input and storage combined are less than the calculated potential evapotranspiration, actual evaporation is less than potential and is limited to the total amount of moisture available from storage plus whatever falls as precipitation. No evapotranspiration is assumed to occur when average monthly temperatures are below freezing.

Continental Water Budgets

When the budget formula is used in annual calculations that employ average values it is simplified even further because the average change in moisture stored in the soil is zero. The equation then becomes:

$$P = E + R$$

$$\text{or} \quad R = P - E$$

where P is average annual precipitation
E is the estimated average annual evapotranspiration
and R is runoff.

South America is the rainiest continent with

1350 mm of precipitation, and it has the largest surplus, 490 mm. The effect of deserts on continental water budgets is evident in Africa and Australia. Africa receives 670 mm of precipitation and has a surplus of 160 mm and Australia gets 470 mm and has a surplus of only 60 mm. The interaction between the atmosphere and hydrosphere is apparent once again.

Soil Moisture Balance and Local Water Budgets

Not all of the water that falls as precipitation evaporates directly or runs off. Some of it infiltrates into porous soils. Once in the soil some water remains there to be used by plants or later evaporated, some flows downslope through the soil as throughflow, and some may percolate downward until it reaches the groundwater table. Water that flows downslope as throughflow eventually emerges as springs or is discharged directly into streams and lakes. On the other hand, water that reaches the water table enters the groundwater system, and it too, may eventually emerge as stream flow. Both represent forms of runoff that must be lumped with direct surface runoff in budget calculations. It is the remaining water, that which is stored in the soil and is withdrawn in the process of evapotranspiration, that makes up the storage term in water budget computations.

Water that enters the soil occupies pore spaces within the soil. Pore space, or **porosity,** is a measure of the total amount of open space contained within a volume of solid or unconsolidated rock fragments. It is expressed as a percentage of void space to the total volume of mass:

$$P = \frac{v}{V} \bullet 100$$

where P is porosity
v is total void space
and V is volume

If pores are large, as they are in sand, water moves freely with minimum resistance and the soil is described as **permeable**. The degree to which liquids can migrate through inter-connected pore spaces is called **permeability** or **hydraulic conductivity.** Materials with small pore spaces such as clay restrict water movement. Yet, the total amount of pore space in clay is very high and the potential water storage of clay soils is higher than for coarser soils. Generally speaking, permeability is directly related to pore size.

Water does not completely fill the pore spaces in soil. After a rain, excess soil moisture, called **gravity water**, quickly drains away as throughflow or percolates through the soil. Once gravity water is gone (a process that may take a couple of days following extended rains), the soil is said to be at **field capacity** (Figure 14.5). The amount of water a soil can hold at field capacity is its maximum storage capacity, the term employed in water budget calculations. Maximum storage capacity depends on soil texture, organic content and soil depth. **Available water** is the difference between field capacity and the **wilting point**, which is the amount of moisture that remains in the soil after plants have withdrawn all they can. Over a period of time available water is transpired by plants or evaporated directly from the soil.

Although most moisture is lost by transpiration from plants, water may evaporate directly from the soil surface. Water is raised to the soil surface by capillary action. At one time soil scientists felt that frequent cultivation preserved water by destroying the capillary networks. We now know that this is not true

Figure 14.5 Soil Moisture

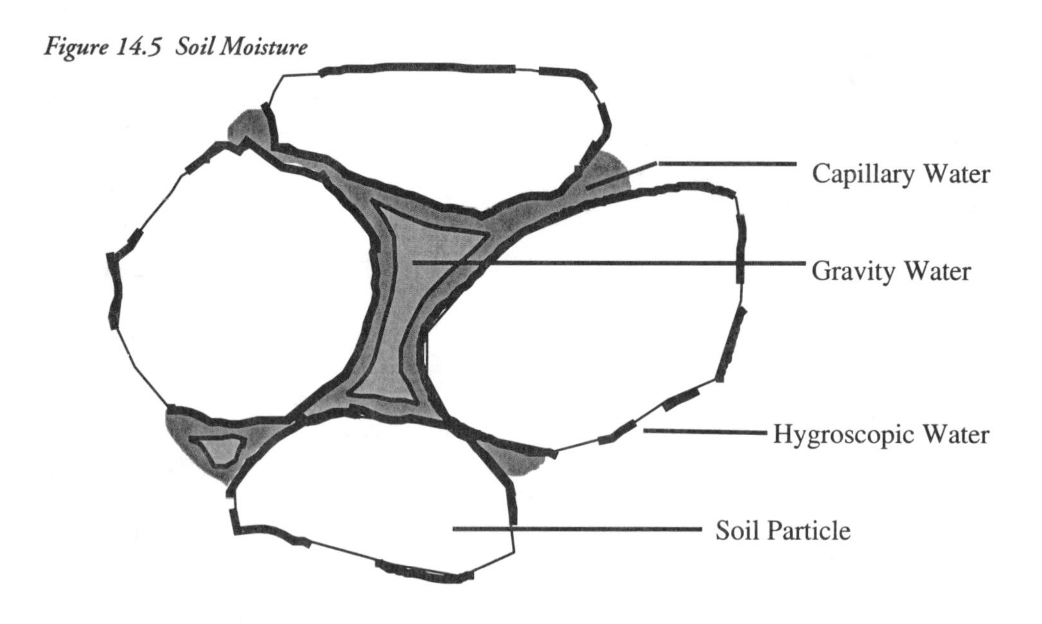

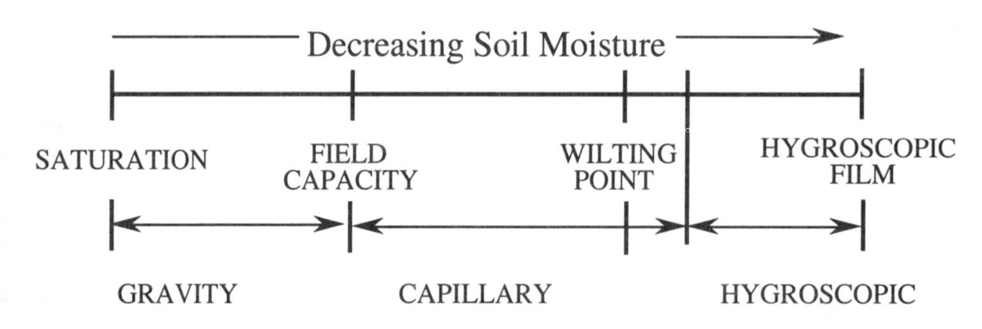

and that cultivation has the effect of reducing infiltration once rain begins by allowing fine particles to clog the soil pores.

Some of the water in soil is attracted by soil colloids and held very tightly by them. This water, called hygroscopic water, is not available to plants. Finely textured soils hold more water this way than sandy soils. It is this water that is left in the soil after plants have withdrawn all they can. Even after the wilting point has been reached, hygroscopic water may account for as much as 30 percent of the weight for clay soils.

Soil Moisture Deficit and Drought

In the application of water balance models normally it is assumed that no water deficit exists until soil moisture storage drops to zero. Drought conditions prevail once water storage drops to zero, the permanent wilting point.

Obviously, this definition fits the concept of drought better than a definition based entirely on rainfall. Drought refers to the availability of water for plant growth and this depends not only on how much rainfall occurs, but also on evapotranspiration rates.

In fact, it might be argued that a water deficit exists (but not necessarily a drought) whenever rainfall drops below potential evapotranspiration for a given time period such as a month. It is known that plant growth is slowed and a plant can be under water stress before the permanent wilting occurs. This happens whenever PE exceeds the rate at which water can be made available, and this is to some extent a function of soil moisture.

The time frame for water budget calculations has a major impact on the amount and duration of surpluses and deficits. Although we have used average monthly values, daily computations would reveal even more variability and some months would have deficits and surpluses.

Application of Water Budgets

There are three model assumptions that restrict the utility of water balance models such as the ones used in this chapter. First, no surplus occurs unless the soil is a field capacity; second, no storage occurs unless precipitation exceeds PE; and finally, when temperatures are below freezing, PE is assumed to be zero. All three assumptions are unrealistic. Most desert soils are unsaturated, yet runoff sufficient to cause widespread flash flooding can occur in deserts following heavy rain. Similarly, some soil storage can occur, even in hot, dry climates, with moderate amounts of rainfall.

Despite their shortcomings soil moisture budgets have been applied to a variety of water management problems, including estimating irrigation intervals and amounts, forest fire forecasting, forecasting crop yields, and climatic classification. The accuracy and value of the method are not independent of the time period used. For irrigation studies and forest fire forecasting, daily periods are most appropriate, whereas climate classification requires average values. Daily computations produce some problems, such as how to deal with runoff. To assume that all surplus water will drain away in 24 hours (especially that which percolates and that which moves as throughflow) is not a valid assumption. Many of these problems are minimal, however, when monthly or annual time periods are employed.

Conserving Soil Moisture

Soil moisture limits agricultural productivity in many areas. The effect is obvious in dry areas, but even in some humid locations, sandy soils with limited water retention can create drought conditions. In areas like the Great Plains where rainfall is low and evaporation is high, soil moisture must be conserved in order to grow crops. Summer fallow is the principal dryland farming technique employed in the wheat growing areas of the western United States and Canada. Land is set aside and not cultivated for one growing season to allow it to accumulate soil moisture. The stored water plus the growing season precipitation are frequently enough to produce a good crop. Fallow storage can be increased by preventing the growth of weeds with their high transpiration rates.

Leaf mulching is often employed by the home gardener as an effective moisture conserving technique. A layer of leaf litter may reduce surface evaporation to a tiny fraction of the rate in uncovered soil, and has the desirable effect of adding much-needed organic matter to the soil.

Other moisture conservation practices are still widely used in many locations. Contour plowing, strip cropping, crop rotation, terracing, subsoiling, and water spreading are all used to some degree. The first four serve double duty by not only increasing infiltration and storage, but also by reducing runoff and erosion. Subsoiling breaks up heavy clay that retards percolation and thereby increases the soil storage capacity. Water spreading is one of the more unique approaches to moisture conservation. Water is transferred from steeply sloped areas with heavy runoff to flat areas where it can infiltrate and increase soil moisture. In some arid land experiments, yields of alfalfa have been increased by as much as 100 percent in years with below average rainfall, and by as much as 29 percent during wet years. The technique seems particularly appropriate for pastures and ranges.

Soil Drainage

Too much water can be as disastrous for farmers as too little. For centuries farmers have been draining wet lands to improve yields. The system employed may be as simple as a series of open ditches or as complicated as a specially designed system of underground tiles. European drainage systems were adapted for use in colonial America. One of the more common techniques was to dig a ditch, throw brush and/or rocks into it, and fill it with sod and soil. The rocks were open enough to allow for drainage, and once the branches decayed, the open passageways they left promoted effective drainage. In the mid-nineteenth century, the invention of the porous seamless drainage tile revolutionized the concept of drainage, and opened up vast areas of fertile soil to agriculture.

Permafrost

In polar and sub-polar locations in both hemispheres temperatures are low enough and summers cool enough that the soil is permanently frozen. Permafrost exists in large unbroken patterns around both polar areas. Beyond this is a zone of discontinuous permafrost with large patches separated by thawed ground. The discontinuous permafrost stretches as far south as lower Hudson Bay in North America and Lake Baikal in Russia.

Permafrost areas thaw near the surface in the summertime, with the depth of thawing dependent on the intensity of solar radiation and length of summer. If depth of winter freezing exceeds summer thawing, the thickness of permafrost layers will increase on an annual basis. The ultimate thickness is determined by the depth where the geothermal gradient and the mean annual temperature for a location intersect. Internal temperatures (geothermal temperatures) normally increase at a rate of about 1°C per 30 meters (1.8°F per 100 ft). The depth at which heat flows from the interior offsets heat loss to the atmospheric environment and determines the depth of the ice.

Permafrost environments are nightmares for engineers and environmentalists. From an engineering perspective, sewage and garbage disposal, water systems, and building and road construction which are routine in most areas become major problems in permafrost areas. Water and sewage pipes freeze solid in the winter unless heated. In rural and suburban areas sewage must be collected throughout the winter and disposed of in spring. Houses must be specially insulated underneath to prevent extreme heat loss and subsequent melting of the permafrost. Ground compacted in the winter conducts heat better and stores more heat in summer, promoting melting of the permafrost and disruption of vegetation. Disturbed vegetation may take years to recover, if it ever does. Differential freezing and thawing of the surface produces warped rail lines and crooked roads. In some respects, efforts to cope with the challenges of permafrost frequently result in large environmental impacts.

Summary

The purpose of this chapter has been to introduce the hydrosphere, examine some of the physical and chemical properties of water that help to explain its behavior and its role in earth systems, and to review the water budget and its uses.

The polar nature of the water molecule, which carries a slight positive charge in the vicinity of the two hydrogen atoms and a small negative charge in the vicinity of the oxygen atom, is the key to many of its chemical and physical properties. When water freezes, for example, the structure of the water molecule dictates the arrangement of the molecules in forming crystals, and the open arrangement gives ice a density that is considerably less than water. Water has its maximum density at 4°C and either further cooling or warming will decrease density. Other properties of water that are related to its molecular structure are viscosity, solvency, specific heat, and its ability to ionize many substances.

The hydrologic cycle illustrates a true open system in which matter and energy are interchanged among the subsystems and with the environment. The atmospheric subsystem was discussed in earlier chapters and the other hydrosystems, groundwater, lakes and streams, glaciers, and the ocean, will be considered in detail in succeeding chapters.

Water budget computations focus attention on the inputs, outputs, and storage of water and energy. Although water budgets can be computed for any of the hydrosystems, most people associate the concept most strongly with the soil moisture balance and its associated outputs of evapotranspiration and runoff. Soil moisture balance is determined by subtracting evapotranspiration from the water supply available from precipitation. Any excess of precipitation over evapotranspiration is used to bring soil storage up to its maximum, then any surplus is treated as runoff. If precipitation is less than potential evapotranspiration the difference is supplied from the soil moisture storage, which is then reduced by that amount. Water budgets are used for such diverse purposes as deciding when to irrigate crops, classifying climates, and determining forest fire hazards.

Soil moisture deficits can be alleviated by manipulating the input sides of the budget equation through such techniques as artificial rainfall augmentation or irrigation, or by restricting output by reducing evapotranspiration.

Low evapotranspiration and cold temperatures in polar areas cause the soil to be saturated with water and permanently frozen below the surface. The depth of **permafrost** and the depth of the summer surface thaw are functions of the intensity and duration of solar radiation.

In the next chapter we will discuss the major hydrosystems of continental areas: glaciers, groundwater, lakes, and streams. We will discuss their structure, but particular stress will be placed on processes.

Chapter 15

Glaciers, Groundwater, and Lakes

Introduction

In previous chapters we have seen that when precipitation exceeds potential evapotranspiration for a given budget period, the bulk of the surplus was treated as runoff — the exception being when soil moisture storage is below maximum so that some of the excess is diverted to soil storage and withdrawn during some later period of low rainfall to supply the needs for evapotranspiration.

Not all surplus water runs off immediately. Some of the surplus may be stored temporarily in glaciers, in lakes and swamps, or as ground water before being discharged into a stream system. Because of their ability to store mass (water, ice, and dissolved and suspended sediments) and energy which are used to modify the earth's surface, glaciers, groundwater and lakes, as well as streams, must be considered as individual hydrosystems.

Over 75 percent of the world's fresh water is tied up in glaciers, and most of the remainder is held in groundwater systems. Only small amounts are stored in lakes and streams. Cyclic water presents a different view. Only about 4,000 km^3 cycle through glaciers and groundwater systems annually, whereas streams discharge several times that amount. It is apparent, then that only a small amount of the water carried by streams is supplied by glaciers and groundwater. The rest is derived primarily from overland flow and throughflow following periods of precipitation. Nevertheless, because glaciers, groundwater, and lakes act as buffers for stream flow by storing and slowly releasing water to streams, we will examine these first.

Glaciers

Glaciers are accumulations of snow and ice that form in locations where, on the average, annual snowfall exceeds snow melt. Modern glaciers are confined to high altitudes and high latitude locations where cold temperatures prevail, but they still cover an estimated 10 percent of the earth's land area. Past glaciers have been much more extensive, however, and have extended as far south in North America as the Ohio and Missouri rivers. At their maximum extent during the **Quarternary** (the geological period covering the last 2 or 3 million years) glaciers may have covered as much as one third of the earth's land area.

Glaciers are by no means static features and the extent of glaciation has fluctuated throughout the Quarternary. There have been periods of rapid advance, called **glacial periods**, during which glaciers covered large land areas, and there were periods of retreat, called **interglacials**, when most of the ice melted away. Glacial expansions have a tremendous impact on climatic conditions and the world's water budget, yet little is known about the mechanisms that control glacial responses, despite significant advances that have been made in glaciology in the past several decades. The paucity of our glacial knowledge should not be surprising considering where glaciers are located and the fact that the first scientific paper on glaciation was not published until 1840.

In the short treatment that follows we will introduce the concept of glacial systems. Special emphasis will be placed on mass and energy budgets and how these are related to environmental conditions, glacial motion, and the ability of glaciers to perform work.

Mass and Energy Budgets

Glaciers are open systems that have significant temporal and spatial variations in inputs and outputs of mass and energy. Gains and losses of mass and energy are cyclic. Mass is added and heat is lost during winter, and mass is lost and heat gained during the summer. The process of melting and evaporation is called **ablation**, and summer is the ablation season.

Gains and losses of mass and energy are no more uniform spatially than they are temporally, and glaciers have distinct zones of accumulation and ablation. Snow accumulates and is transformed into ice in the **zone of accumulation**, and melting and evaporation are the dominant processes in the **zone of ablation**.

We can use the basic budget equation to study mass and energy balances in glaciers. If it is assumed that glaciers are equilibrium systems, then on the average, accumulation must equal ablation. It is apparent, however, that glaciers are not steady state systems for there is abundant evidence that their sizes may vary significantly over periods of just a few decades. Careful monitoring has revealed considerable variability between rates of annual accumulation and ablation. At the end of the twentieth century, more glaciers are shrinking than are expanding.

Changes in mass storage may be computed by measuring snow depth and density and the amount of ablation at a number of sample locations on the glacier being studied. The resulting change in mass can be computed by the mass budget equation:

$Ct = At \pm \Delta S$ or

$\Delta S = Ct - At$

where Ct is the total annual accumulation

At is the total annual ablation

and ΔS is the change in mass storage.

Often mass changes are monitored continuously over a mass balance year and the results are plotted. From such a plot it is not only possible to determine annual changes in storage, but seasonal changes as well. The mass balance year begins when accumulation exceeds ablation and continues through one complete winter accumulation and summer ablation cycle.

Several classifications of glaciers exist but this discussion uses two simplified systems — one based on ice temperature and one based on size and location. A distinction is made between temperate, subpolar, and polar glaciers. **Temperate glaciers** have internal temperatures of ≤0°C during winter and >0°C during summer, therefore they produce meltwater only in summer. **Subpolar glaciers** have <0°C temperatures in winter and 0°C in summer. **Polar glaciers** have internal temperatures <0°C year round. A distinction is also made between glaciers on the basis of ice thickness and areal coverage. Continuous ice sheets which move outward in all directions are commonly called continental glaciers. Glaciers confined to mountain valleys are much smaller and move down slope following the valley floor. These are commonly called **valley**, **alpine**, or **mountain glaciers**.

Mass budgets are directly related to energy budgets in two ways. First of all, mass represents potential energy and the ability to do work. Secondly, the mass balance is partially controlled by the amount and distribution of heat available for melting and evaporation. Although energy balance studies are difficult and expensive, they are essential to an understanding of the dynamics of mass exchanges.

The basic energy budget equation for glaciers is:

$$Rs\downarrow + Rl\downarrow + H\downarrow + C + F = Rs\uparrow + Rl\uparrow + H\uparrow + M + Sub \pm \Delta S$$

where the direction of the arrows represents incoming or outgoing heat and
Rs is shortwave radiation
Rl is longwave radiation
H is sensible heat transfer
C is heat released by condensation
F is heat of fusion released by freezing
M is heat used in melting
Sub is heat consumed in sublimation
and ΔS is change in heat storage.

By far the greatest source of heat for glaciers is net radiation (incoming Rs plus incoming Rl less outgoing Rs plus outgoing Rl). Net radiation varies with albedo (incoming RS/outgoing Rs), latitude, and exposure, reaching its highest values in summer on dirty snow for temperate valley glaciers with southerly exposures. The utilization of heat varies with glacial locations. Polar glaciers lose the bulk of their heat by sublimation, whereas temperate glaciers lose more of their heat by melting and sensible heat loss in summer.

The elevated mass of ice in a glacier represents potential energy that is converted into kinetic energy when the glacier moves. By virtue of their motion and weight, glaciers are capable of displacing soil and rock fragments, thereby performing mechanical work. The potential to do mechanical work is proportional to the mass of the glacier and its kinetic energy as reflected in its mean velocity. As a glacier moves from one elevation to another, the work is equivalent to the product of its density, volume, acceleration of gravity, and the difference in elevation.

$$W = dVg\Delta e$$

where W is work
d is density per unit volume
V is volume
g is the acceleration of gravity and
Δe is the difference in elevation (e1 – e2).

Glacier Motion

Glaciers respond to flows of energy and matter by changing their density and volume, which in turn are translated into motion. It is important to recognize that motion occurs even if the glacier size remains stable because some of the snow and ice must be transported from the accumulation zone to the ablation zone.

Glacier ice velocities typically vary from 3 to 300 m/yr (10 to 1000 ft/yr), but values are as high as 6 km/yr (3.75 miles/yr) have been measured. These rapidly moving glaciers are called **surging glaciers**. Only a small percentage of all glaciers surge, but surging glaciers may be of any size or type. Although no general explanation for surging exists, it is probably related to the slope and the thickness of the ice in the accumulation zone. Normally, if ice accumulates rapidly, the glacier responds by increasing its ice flow or by transmitting a compressional wave-form downslope. For some unknown reason surging glaciers do not respond immediately to increased input. Instead, there is a lag during which increased mass storage causes a buildup of stress which is then released as a surge.

Environmental Linkages to Glaciers

The linkages between glaciers and their environments is a dual affair. Glaciers influence and are influenced by climate, landforms, and other hydrosystems — especially the oceans. Glacial development may be initiated by an increase in winter precipitation, a decrease in summer ablation, or some combination of the two processes, either of which could result from relatively minor global climatic changes. There is no general agreement among glaciologists as to the cause of such changes, and there are numerous theories. Certainly, changes in the earth's heat budget are thought to have triggered such changes in the past. These changes include fluctuations in the solar constant that could reduce temperatures: changing concentrations of atmospheric dust from volcanic eruptions that could decrease temperatures, and increased precipitation by supplying condensation nuclei, and reductions in CO_2 concentrations that would impair the natural greenhouse effect. It is clear that a change in glacial area could become a factor in climate change. Atmospheric heat budgets could be severely altered by the relative amounts of solar radiation used to melt and evaporate ice and by altered surface albedos.

The linkage between glaciers and the oceans is even more obvious for the water tied up in glacial ice is withdrawn from ocean storage, thereby reducing sea level. During the maximum extent of Pleistocene glaciation, the amount of water in glaciers was about 2.5 times greater than the present amount, and sea level was an estimated 100 to 130 meters (300 to 400 feet) lower than today. The sheer weight of all that ice depressed continental masses, whereas the ocean basins had to adjust to the corresponding loss of pressure. The combined effects of all of these processes were to expose many shallow areas that are currently underwater. Consequently, some continents were connected by land bridges which became major migration routes for the spread of plant and animal species. It was by such a land bridge, connecting the Eurasian and North American land masses at what is now the Bering Strait, that the first Americans came to inhabit North America some 40,000 years ago.

The glacial oscillations of the Quanternery Period were not only times of great upheaval climatically and hydrologically, but during this time many species of plants and animals became extinct, new species arose, and existing species spread to new areas. Biological responses to climatic changes were much more rapid than the glacial responses. Glacial response to climatic change is normally slow because of the lag time necessary for ice to accumulate and be transformed into glaciers. As much as 15,000 years might be required for a major glacial advance to begin. Termination of glacial periods would be more abrupt, requiring perhaps as little as 5,000 years to melt the accumulated ice. Biological responses would occur much more quickly. The potential impact of glaciation can be imagined by supposing that civilization had risen to its present level at the time sea level was 120 meters lower than it is now. It is generally believed that the melting of the ice may have taken only 5,000 years or less, producing a sea level rise of some 2.4 meters per century (8 ft/century). Imagine what chaos such an event would create today in seaports and coastal areas. As an example consider what is happening to Venice today. The actual impact may have been only slightly less traumatic for those living during the Pleistocene because some archaeological sites have been found submerged in as much as 50 meters of water.

Debris Budgets of Glaciers

The ability of glaciers to perform mechanical work makes them significant agents of erosion. The kinetic energy of glaciers, coupled with their high viscosity and weight, enable them to erode materials that may be transported considerable distances before being deposited. Glaciers erode by two major processes — **abrasion** and **plucking**. Plucking occurs when glacial ice freezes around loose or jointed surface materials that are then plucked away when the ice moves. Loose rocks suspended in glacial ice abrade rock surfaces in much the same manner that sandpaper abrades wood. The amount of abrasion that occurs is not dependent on ice thickness because ice will yield under the stress of its own weight at a depth of about 22 meters (70 ft.). Because of the nature of abrasion and plucking processes, the total amount of glacial erosion may depend to a large extent on how weathered and eroded the preglacial landscape was.

Materials plucked by glaciers and materials which fall onto glaciers from valley walls are delivered by the glacial ice to the ablation zone where melting and evaporation cause the material to be deposited. Debris transported by glaciers is referred to by the general term drift and may be identified as **superglacial**, **englacial**, or **basal drift**, depending on whether it is transported upon, within, or at the base of the glacier. There are some distinct relationships between the modes of transportation and the properties of the drift and the landforms they produce upon melting.

In this section we have tried to stress some of the aspects of glaciers that are important to an understanding of the linkages to other hydrosystems and to the environment. The potential energy that glaciers possess and the fact that glaciers covered as much as 33 percent of the earth's land surface in recent geological history has made them significant agents in shaping modern day landscapes.

Groundwater

Introduction

It was implied in our water balance discussions that, for most locations, the demand for water for evapotranspiration is exceeded by the supply from precipitation during some season of the year, and that a part of this surplus infiltrates. When soil is saturated some of the water percolates through passageways in the soil and eventually collects within a permanently saturated zone of the crust called the groundwater zone. The input of water into a groundwater system is called recharge. Groundwater systems may also be recharged by water percolating through stream beds.

If the pore spaces through which water moves are small, as they are in soil, then sediment and bacteria are filtered out, so that ground water is potable without further treatment. People who live in rural areas still depend largely upon wells and springs fed by groundwater for their supply of drinking water. In addition, groundwater is also used for irrigation and for industrial purposes.

The top of the saturated zone is called the groundwater table or just **water table**, and wherever it intersects the surface there is a natural discharge of ground water into springs, lakes, and streams. The continual (but not necessarily constant) recharge and discharge of groundwater systems suggests that they are open systems that are in quasi-equilibrium. In fact, if we ignore minor seasonal variations in the change in storage (the water table in most places usually rises in winter and falls in summer) we might consider natural groundwater systems to be steady state systems.

In areas where use of groundwater has significantly increased discharge, we frequently find that the equilibrium has been disturbed and systems must adjust by some combination of decreased storage (with lowered water tables) and decreased natural discharge. In this section we will examine the nature of groundwater systems and discuss some of the problems associated with groundwater use.

The Nature of Groundwater Systems

Water-bearing substances in the groundwater zone are called **aquifers**. The water in aquifers occupies openings or pores in rocks or unconsolidated materials that compose the aquifer. Aquifers vary in quality, depending on their ability to store and transmit water and upon the quality of water which they yield.

The storage capacity of an aquifer depends upon the amount of open space within the substance, measured as a percentage of total volume. This concept, defined as porosity, is best stated in the following relationship:

$$P = \frac{v}{V} \bullet 100$$

where p is porosity
v is total void space within the aquifer
V is the total volume of the aquifer.

Porosity alone is not very important. Unless groundwater flows freely through an aquifer it will not yield water readily. The ease with which a liquid can be transmitted through a medium depends on the amount of resistance that it encounters, and that is a function of the viscosity of the liquid and the sizes of the openings through which the liquid must flow. If we make the simplifying assumption that groundwater viscosity is essentially uniform, then the transmission of water depends primarily upon the sizes and connectivity of the pores within aquifers, and we can associate with the aquifer a value, called **intrinsic permeability** or **hydraulic conductivity**. Rocks such as limestone, in which the pores that are essentially channelways, have the highest permeabilities, whereas fine-pored substances such as clays have the lowest permeabilities.

Many substances have very low abilities to store or transmit water and, therefore, do not make good aquifers. Such substances, which include unfractured rocks such as granite and slate, are called **aquifuges**. Often a aquifer will be confined between two impermeable layers. As we will see later, confined aquifers behave differently than unconfined aquifers.

The Behavior of Groundwater Systems

The amount and quality of water available from groundwater systems depend on the nature of the aquifer and the balance between recharge and discharge. Changes in mass balance in groundwater systems can be calculated from the basic mass balance equation. Because water is basically an incompressible fluid, there must be continuity of water flow — that is, input must equal output plus or minus the change in storage.

$$R = Q \pm \Delta S$$

where R is recharge
Q is discharge
ΔS is change in storage.

Groundwater storage varies with the balance between recharge and discharge, and the water table fluctuates accordingly. In natural systems recharge tends to exceed discharge during wet periods and during cool weather, whereas discharge exceeds recharge during dry periods when recharge is reduced.

Water in the groundwater zone exerts pressure in proportion to its weight just as it does in water bodies. This pressure, called **hydraulic head**, is a measure of the potential energy in the groundwater system. Groundwater discharge depends, in part, on head pressure. Therefore, as the water table rises the increased head pressure acts as a positive feedback to increase discharge. If recharge were to remain high, then a new equilibrium state, with greater storage and higher discharges would be achieved. Conversely, reduced recharge or increased discharges, through such activities as pumping groundwater for irrigation or mining, will cause the water table to fall and pressures to drop, with a concomitant reduction in discharge.

A complete description of flow in an aquifer must take into account the permeability and the length of the flow path. The ratio of head loss to path length is the **hydraulic gradient** of the system and is analogous to the gradient of a stream. Thus, all other things being equal, sloping aquifers have higher discharges than horizontal aquifers.

Due to changing climates, geological uplift, or other factors some aquifers have been permanently isolated from input and no longer receive significant recharge. The Nubian sandstone in North Africa and the Ogallala Formation in the High Plains of Texas and Oklahoma are good examples. Natural discharge and pumpage steadily reduce the storage in isolated aquifers. The water level in the Ogallala formation has dropped as much as 600 meters (1900 feet) or more, since the first wells were drilled in the High Plains. This amounts to 25 to 30 percent of the estimated original content and 50 to 60 percent of the recoverable water in the aquifer. Increased pumping costs act as a negative feedback to slow consumption in such cases.

Confined aquifers are a special case. In an unconfined aquifer the location of the water table is the same as the hydraulic head (called the **piezometric surface**). In a confined aquifer the location of the water table is restricted by the overlying bed so that it is well below the piezometric surface. The water level in a well drilled into a confined aquifer will rise to the level of the piezometric surface. If that surface is above ground, the well will flow freely. Confined systems of this sort are called **artesian aquifers**. Artesian systems are often found in coastal plains and other areas underlain by sloping rock layers of varying permeabilities.

Well Flow

When water is withdrawn from wells by pumping, the piezometric head in the immediate vicinity of the well is lowered (Fig 8.6). The amount of drawdown is greatest near the well so that the piezometric surface is depressed into a shape similar to an inverted cone. The magnitude of the depression cone depends on the pumping rate and the rate of flow of the aquifer. The higher the pumping rate, the bigger the cone, and the less permeable the aquifer, the bigger the cone.

Figure 15.1 Groundwater Depression Cone

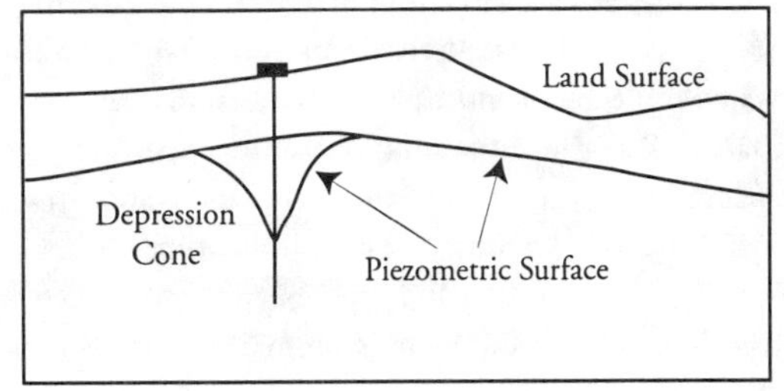

Excessive drawdown in an aquifer can lead to several possible groundwater related problems. Some kinds of aquifers are not able to support the weight of overlying materials after the water has been withdrawn, leading to collapse of the aquifer and surface subsidence. A notable example is found in the Houston area where ground water occurs in clay lenses within the soil at shallow depth. Widespread and rapid population growth placed increased demand on ground water supplies in the Houston area, especially in the vicinity of Nassau Bay. Surface subsidence of as much as 2 to 3 meters (6 to 10 feet) occurred with some areas that had been a couple of meters above sea level and subsided to a couple of meters below sea level. Other notable areas of subsidence occur in Santa Clara, California and Mexico City. The best remedy for subsidence problems appears to be reduced pumping rates.

Salt water encroachment is another major problem caused by drawdown. Saltwater encroachment refers to any situation where saline water enters a freshwater aquifer, but the problem is most widespread and has received the most attention in coastal areas. Saline water can be generally defined as water in which dissolved solids of sodium, calcium, magnesium, potassium and other salts exceed 1,000 parts per million. Although drawdown is a major cause of coastal salt water intrusion, it can occur for other reasons. Dredging and deepening coastal waterways can contribute to encroachment problems, as can surface drainage. In the Miami area drainage of the Everglades has lowered the water table by as much as 2 meters (6 feet) in places. As the hydraulic head dropped, the reduction in pressure allowed sea water to intrude coastal aquifers, thus indicating rather clearly the direct linkage between surface drainage and ground water systems.

The seriousness of salt water intrusion problems has prompted a number of kinds of solutions, including: 1) artificial recharge; 2) the erection of barriers in the form of pressure ridges or systems designed to reduce permeability of the aquifer to sea water; and 3) a line of wells near the coast to intercept the sea water and pump it out as fast as it flows in. The best solution, to this and other groundwater problems, however, lies in proper management of groundwater resources.

Water Quality

The potential use of groundwater may depend as much upon quality as it does upon yield. Water quality is a function of the amounts and types of dissolved and suspended matter in water. The major substances contained in groundwater are listed in Table 15.1, and their principal sources are indicated. It is clear from Table 15.1 that many of the substances are picked up as water percolates through the soil or as water flows through aquifers. Water that contains large quantities of dissolved solids is hard water. The removal of the dissolved substances, called softening, is a multimillion dollar business in the United States.

In fissured aquifers (aquifers in which water flows along joints and cracks or other large openings) there may be little or no filtering of suspended organic or inorganic materials. Water pumped from such aquifers may be contaminated with pathogenic organisms that are filtered out routinely as water flows through fine-pored aquifers. Pathogens are living organisms capable of causing disease. Because of the potential health problems associated with groundwater use, many states have laws that require regular testing of well water that is used for domestic consumption in suburban and rural areas.

Table 15.1 Groundwater Quality

Major Dissolved Solids	Major Sources	Abundance
Chloride	Sea salt trapped in rocks; solution of halite	Very common
Carbonate and bicarbonate	CO2 in air and soil; solution of carbonate rocks	Very common
Sulfate	Sedimentary rocks; precipitation	Very common
Sodium	Sea salt trapped in rocks; weathering of halite and feldspars	Very common
Magnesium	Solution of halite, dolomite, limestone	Very common
Calcium	Solution of calcite, dolomite, gypsum; weathering of metamorphic and igneous rock	Very common
Silica	Weathering of silacate minerals	Very common
Nitrate	Fertilizer; natural organic sources	Common
Acids	Decaying vegetation; coal mine drainage	Common
Iron	Weathering of iron minerals	Common
Potassium	Weathering of igneous and metamorphic rock	Common
Gases		
Oxygen	All are absorbed by rain as it falls through the atmosphere and from vegetative surfaces, and by infiltrating soil water.	Varies with location and water temperature; greatest concentrations in coldest waters
Suspended Sediments		
Inorganic (silt, clay)	Derived primarily from percolating runoff	Rarely >500ppm and usually <5ppm
Organic (bacteria)	Poor sanitation, contamination of wells, lack of adequate filtration	Rare-to-common most common in developing areas

Many groundwater quality problems are anthropogenic, and to a large extent are related to our needs to dispose of wastes of all types. Groundwater supplies are often contaminated by poorly designed or improperly functioning septic systems. In a properly functioning septic system sewage flows into a large buried tank where the wastes are partially digested by bacteria. The liquid product of digestive processes is introduced into the soil over a large area by a series of lateral lines made of porous tile that diverge from the tank. In a permeable soil the percolating liquids are filtered of any bacteria or other solids.

Septic systems may contaminate groundwater supplies if the system is not large enough to handle all the waste it receives, if the percolation capacity of the soil is low, or if the system is located in close proximity to groundwater supplies and proper filtering does not take place.

Leachates from sanitary landfills are also a common source of ground water contamination. Decomposition of buried debris releases gases, acids, and other substances that may be dissolved by percolating water and carried downward to groundwater systems. If there are wells in the immediate vicinity, the problem may be very severe. A gradient may develop from the landfill to the well so that contaminated water can flow directly into the well. Most states now have laws restricting the location of landfills to areas with soils of low permeability and away from groundwater supplies. They also require that new landfills be encapsulated in impermeable liners.

As the effectiveness of treatment facilities is upgraded in attempts to clean up stream pollution, the problem of disposing of the end product, sludge, becomes more serious. One widely advocated solution is land treatment, which consists of spraying or spreading the sludge over agricultural lands as a fertilizer. Critics of land application cite the danger of groundwater contamination as a major danger. Sludge may contain pathogens and toxic by-products of industrial activities such as heavy metals and other persistent pollutants.

Groundwater Management

Management of groundwater resources (where any attempts at management have been practiced) has frequently proceeded independently of other water resources. Groundwater has been treated as a free good, available to anyone willing to invest the necessary capital to extract the water for use. Such a philosophy has tended to encourage waste and exploitation. Naturally though, the major problem that groundwater management must attack is excessive use.

Proper management of rechargeable groundwater resources is directed toward development of a steady state system in which annual input balances natural discharge plus pumpage and where water quality is preserved. The concept is referred to as safe yield, and although various authors have provided slightly different definitions, the concept is basically the same in each case.

Theoretically, safe yield can be accomplished in most aquifers by decreasing pumpage or by increasing recharge. The former is the most feasible management strategy, overall. As the groundwater table is lowered through pumpage, the increased cost of pumping acts as a deterrent to the drilling of additional wells. The number and spacing of wells can also be limited by legislation.

There have been several attempts to increase recharge of aquifers. If recharge areas can be identified, then watershed management techniques designed to reduce runoff and increase infiltration can result in increased recharge. Water spreading is a more positive approach to recharge. Spreading is somewhat like irrigation except that rapid percolation is encouraged rather than holding the water in the root zone. Local rises in the water table are often observed in heavily irrigated areas, and the concept of the spreading can probably be attributed to this observed effect.

Although in its original use spreading referred to the flooding of recharge surfaces, the term spreading is now applied to any recharge technique. Nearly all spreading techniques can be grouped into four basic categories: flooding, basin spreading, the furrow method, and the use of pits and wells. In the flooding method water is slowly passed over surfaces in a thin sheet. Undisturbed vegetated surfaces with gentle slopes work best for flooding. If the surface is too irregular or land is too expensive, basin techniques may be used. In this system water is held in shallow basins formed by low levees or terraces. The system works well in coarse alluvial soils although, if the water is turbid, deposition of silt and clay may eventually clog the soil pores. The furrow method uses shallow flat-bottomed furrows or ditches to distribute the water over the infiltration area. Effectiveness is related to the density of the furrow system. Wells and pits have been used with mixed success. The major problems in well recharge are clogging of pore spaces by silt and some danger of contaminating groundwater supplies with surface pollutants.

Well recharge has been attempted in the High Plains area of Texas. The flat surface contains many shallow depressions where water collects following rain. Water evaporates rapidly from the depressions, the surfaces of which are covered with silt that retards infiltration. Wells were drilled from the depressions in the surface to the Ogallala aquifer so that surface waters would drain directly into the aquifer. Results have been mixed with little recharge taking place in some wells and substantial recharge in others. In either case, over a period of time the pores become clogged with silt, and water has to be pumped from the wells periodically to flush them.

The safe yield concept involves more than just sustained yield. The broadest definitions encompass a steady state concept that goes beyond a simple amount of water. For instance, safe yield management techniques should involve safeguards to protect water quality. Groundwater quality can be impaired by agricultural seepage, leachates from sanitary landfills, and, in some cases, organic pollutants and pathogens which flow into fissured aquifers.

Lakes and Ponds

Lakes and ponds are an important link in the surface water system. Much runoff is temporarily stored in lakes and ponds providing a negative feedback that reduces maximum stream discharge and increases minimum stream discharge. In addition, the bulk of water-based activities involve or are oriented toward lakes: recreation, transportation, irrigation and water supply, power generation and, to some extent, food production.

Water and Mass Budgets in Lakes

If we view lakes as simple input-output systems we can identify a number of mass inputs other than water: sediment, salts, gases, nutrients and, from time to time, various kinds of organisms. We will examine each of these inputs and see how they affect lakes and water systems in general.

All lakes are open systems. Insofar as water is concerned, most natural lakes in humid environments are steady state systems in which output balances input so that the lake level remains constant over time. In some natural lakes, lake levels and discharge fluctuate regularly around an average equilibrium value because there is some variation in input and output. We can still view such lakes as systems in dynamic equilibrium. Ponds that depend on runoff alone for their inputs of water are good examples of transient response systems because they fill up in rainy weather and frequently dry up during extended dry periods.

The water that flows into a lake brings with it suspended sediment and various materials in solution. Unlike the water that flows out or evaporates, these materials frequently remain behind. As water flowing into a lake slows down, it loses the energy necessary to transport sediment and tends to deposit it near the end of the lake where the stream enters. Over a period of time the accumulation of sediment may fill the lake, gradually converting it into marsh.

The interaction of water, sediment, plants and animals can be observed best in small man-made lakes and ponds. As the upstream area is filled in with sediment, plants gain a foothold in the shallow water. The growth and death of plants adds organic matter, and the accumulation of organic matter further reduces water depth. Gradually the water surface is converted to marsh and swamp, then to solid land.

The slow conversion of lake to land is a natural process that may involve plants, animals, and especially people who, by disturbing the surface, contribute to increased rates of sediment flow. Beavers are sometimes responsible for creating fertile mountain meadows by constructing a series of dams and small lakes that fill with sediment. As lakes are filled, beavers abandon their useless quarters to build new ones and start the process over.

Nutrients that are supplied to a lake may also remain within the system. The nutrients are incorporated into local biochemical cycles. They promote rapid plant growth and are locked up for indefinite periods as plant or animal tissue. When organisms die nutrients are released through decay to be recycled through the system.

Nutrient and oxygen content are closely related to lake depth. Light penetration into lakes is limited to a specific depth by turbidity. In shallow lades there is a better chance that light may reach the bottom, thereby promoting rapid growth of vegetation if sufficient nutrients are present. Decomposition of the organic matter in these productive lakes generates heavy oxygen demand and may deplete oxygen supplies, especially near the bottom of the lake where organic matter collects. The oxygen in lake water is absorbed largely from the atmosphere. Therefore, in lakes with little motion to provide mixing of the water, oxygen stratification develops with the upper layer being very rich in oxygen, with the concentration decreasing with depth.

In all lakes some water is output through evaporation, and in hot, dry climates loss by evaporation may exceed discharge to streams. If the water input into the lake contains dissolved salts, those minerals accumulate gradually and the lake waters become saline. Salinity is a major water quality problem in many dry areas. In northwestern Oklahoma and southwestern Kansas, for instance, many of the surface rock formations contain salt. Ground and surface waters dissolve salts and, if the waters are impounded in shallow reservoirs, rapid evaporation causes significant increase in salinity.

Heat Budgets

Annual heat budgets of lakes must balance just as those of land do, otherwise water bodies would grow warmer or cooler over time. In the middle and high latitudes seasonal heat budgets do not balance, and lakes gain heat in summer and lose heat in winter. During fall as surface waters cool they increase in density and tend to settle downward, displacing lighter, warmer water. Warmer water rises to the surface where it cools, absorbing oxygen in the process. As the water throughout the lake cools to the same temperature, wind motion may cause further mixing throughout and helps distribute oxygen. This is known as the **fall overturn**. The fall overturn is sometimes visibly noticeable because of sediment brought to the surface as water rises from the bottom. Once surface waters cool below 4°C (39°F) the surface water becomes less dense and floats. Further cooling may cause the surface to freeze, producing an inverse temperature stratification and preventing any more oxygen exchange. Cold water can absorb more gas than warm water. Carbonated drinks are good examples. The carbonation is due to carbon dioxide gas dissolved in the liquid. The warmer the drink is, the less carbon dioxide it can retain at equilibrium pressure. Thus, the sudden reduction of pressure when the container is opened causes a hot carbonated beverage to "fizz" dramatically.

In the spring increased insolation, warm rain, and runoff cause ice to break up. When the surface temperature reaches 4°C, water grows slightly more dense and a slight convection current may be set up. The high winds of spring may help produce a **spring overturn**, once again causing free mixing of water throughout the lake and replenishing the oxygen supply. Lakes with two overturns each year are called **dimitic lakes**.

As waters warm during the spring and summer, water density decreases, causing warm water to float. A distinct temperature stratification develops in which there is a warm upper layer of water (the **epilimnion**) overlying a cold lower layer (the **hypolimnion**). Temperature differences within each of these layers are minimal, but temperatures change drastically between them. In the zone of rapid temperature change (the **metalimnionor thermocline**) temperature may drop as much as 1°C per meter. The depth of the thermocline increases during spring and summer, but with the approach of fall, rapid heat loss begins and cooling of surface water and the resulting convection destroy the temperature stratification.

The general description we have just presented does not apply equally to all lakes. In very shallow lakes there may be no temperature stratification at all. The epilimnion may reach to the bottom allowing free mixing all summer long. In very deep lakes, on the other hand, there may be a deep zone that never mixes with upper waters. The thermocline does not disappear during overturn, but simply descends to a lower depth. All deep lakes exhibit some temperature stratification during summer, even in tropical areas.

Summary

Most of the world's fresh water is locked up in glaciers. Presently, glaciers occupy about one-third of the area they covered at their maximum extent. Glaciers develop whenever snowfall exceeds snowmelt on an annual basis over a long period of time. Snow is gradually changed into ice and moves in response to internal pressures and gravity. If snowfall at the glacier head exceeds losses through evaporation and melting, the end or terminus, of the glacier moves forward. If output exceeds input, the terminus retreats. Regardless of whether the end of a glacier remains stationary or moves backward or forward, ice constantly moves downslope carrying with it large quantities of debris picked up by the glacier. At the present time northern hemisphere glaciers are melting back, but we do not know how long this cycle will last or if this trend is naturally or anthropogenically induced.

Groundwater systems are made up of water trapped inside the soil or ground. Groundwater systems receive most of their input by infiltration and percolation. Water is naturally discharged from groundwater aquifers into springs, lakes, and streams, but human activities are responsible for upsetting the steady state, or equilibrium, conditions that prevail in natural systems. In addition to using up the water resource rapidly, excessive pumpage creates additional problems. In coastal areas the reduction of fresh water pressure allows salt water to intrude the aquifer. In other locations, withdrawal of fresh water is accompanied by subsidence or collapse of the ground.

Water quality problems, increased production costs, and other undesirable effects are the product of poor management of groundwater resources. During the twentieth century we have frequently exploited groundwater as a separate resource. Effective groundwater management must recognize the interdependencies among all of our hydrologic resources. Waste must be discouraged and beneficial use encouraged while recognizing and reconciling the competition among potential users.

Much of the water that enters stream systems as surface flow is temporarily stored in lakes. In this respect lakes serve as negative feedbacks to maintain steady flow conditions. Lakes are open systems receiving water, sediment, nutrients, salts and gases. Sediments and dissolved materials tend to accumulate within lake systems in many cases.

Middle and high latitude lakes undergo annual cycles of heating and cooling during which lake waters become stratified by temperature. In summer, surface waters are warmer and temperature decreases with depth.

In winter, colder waters collect at the surface and temperature increases with depth, at least for a short distance. Lake waters can mix freely only when they are the same temperature throughout. Surface winds are usually the driving force in this circulation, and many lakes may overturn twice annually, once in the spring and once in the fall.

In this chapter we have examined glaciers, groundwater, and lakes. Each of these acts as a buffer and helps to provide a steady base flow to streams by storing excess water temporarily and releasing it to streams gradually.

Chapter 16

Streams and Overland Flow

Streams are the major means by which excess water falling on continents is returned to the oceans. Precipitation that falls on land represents potential energy with the specific amount of energy dependant upon the amount of precipitation and its elevation above sea level. The flow of water represents the conversion of this potential energy into mechanical work. In this section we will explore the major properties of stream and overland flow systems and examine their behavior with special emphasis on human impacts to runoff and stream flow.

Overland Flow

The first water to find its way into streams following precipitation is that which moves by overland flow. When rain falls at a rate that is in excess of the infiltration capacity, overland flow commences once the surface storage capability has been exceeded. Surface storage has three components, each of which may influence the proportion of rainfall that runs off and the time lag between rainfall and runoff. The three components are interception storage, surface depression storage, and detention storage. **Interception storage** depends on the type and amount of vegetation cover and in heavily forested areas it may amount to as much as 10 to 20 per cent of runoff on an annual basis. **Depression storage**, which begins as soon as rainfall intensity exceeds infiltration capacity, refers to water retained in puddles and other depressions in the soil surface. Surface detention refers to the film of water that covers the soil after depression storage has been exceeded, and once surface detention reaches some initial depth, gravity causes the water to flow downslope as overland flow. Once rainfall ceases overland flow stops soon afterwards, and the remaining surface storage evaporates or infiltrates into the soil.

Most overland flow is assumed to be laminar, at least on upper slopes. In **laminar flow,** water flows in thin layers with the velocity of the layers increasing away from the soil surface (Figure 16.1). If rain is intense or the depth exceeds some critical value, tranquil laminar flow is replaced by a more disorderly form of flow called **turbulent flow**. In a statistical sense turbulence represents the

Figure 16.1 Vertical Velocity Profile

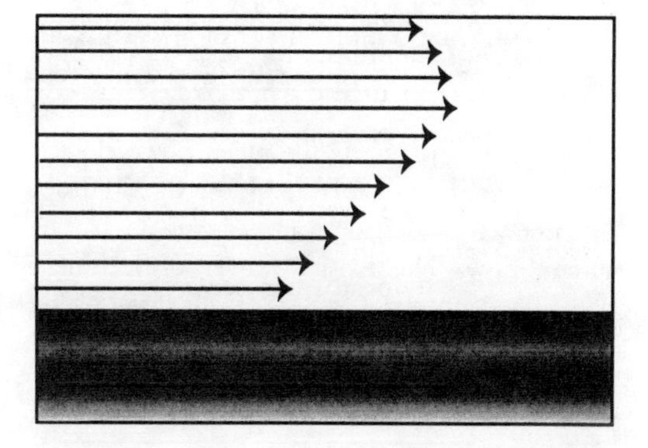

departure from average flow velocity – in other words, those water masses that move forward, backward, up or down with respect to the average downstream displacement of the total water mass. Turbulence is proportional to velocity and roughness, so turbulent flow may prevail on downhill slopes where water is deeper and velocities are greater. Turbulent flow may also occur when rainfall intensities are high or surfaces are very rough.

Flowing water maximizes entropy through turbulence. Without turbulence and boundary friction to help dissipate energy, water flowing down a slope would undergo constant acceleration in accordance with the formula for gravitational acceleration.

$$Ek = \frac{MV^2}{2}$$

where Ek is kinetic energy
M is mass
and, V is velocity if mass is held constant.

Turbulence performs other functions as well. Oxygen is required in lakes and streams for growth and decay of organic matter. As a general rule, the more turbulent the flow of water, the better the water is oxygenated.

Turbulent flow is the primary mechanism whereby flowing water transports sediment. Upward-moving molecules of water lift suspended particles and thereby keep them from settling to the bottom and allowing them to be displaced downslope by the mean flow velocity. In this manner, soil splashed loose by raindrops or entrained by flowing water is carried away by surface flow. The erosive power of turbulent flow is evident in the pattern of rills that quickly develops on the slopes of any newly exposed unvegetated site. Rills can grow quickly into gullies that are nothing more than stream channels that carry no base flow. In fact, rills and gullies are collectors that channel runoff into the stream system.

Streams

Streams receive their inputs from three sources: base flow, throughflow, and overland flow. **Base flow** may be derived from groundwater, lakes, swamps, or glaciers, each of which has sufficient storage capacity to discharge a reliable and fairly steady flow into streams. Base flow contributes a high percentage of dissolved sediment load. **Throughflow**, which flows through the soil, may take anywhere from a few minutes to several days to emerge as stream flow following a precipitation event. **Overland flow** contributes the greatest quantity of water into streams and, except for glacial streams, contributes virtually all of the suspended sediment.

In addition to sediment, non-water inputs into streams include a variety of substances from sewage to toxins, which enter streams either directly or indirectly as a result of human activities. Perhaps the single largest problem is organic pollution such as sewage, and drainage from cattle feeding or hog farming operations. Organic wastes pose two problems. First, their decomposition places heavy demand on a stream's oxygen supply and may almost totally deplete it. In this case all materials do not decompose, and organisms that need oxygen cannot survive. In addition, heavy concentrations of human sewage may pose a health problem. Many diseases may be spread by water. Diseases such as typhoid, paratyphoid and hepatitis may be spread by contaminated water, and in many countries, cholera, amoebic dysentery, schistosomiasis and other diseases debilitate and kill hundreds of thousands of people each year. The effects of other water pollutants such as fertilizers, pesticides, herbicides, and mercury are widely known.

System Structure

Streams are open systems and most natural streams have existed in a stable environment long enough that they have developed a dynamic equilibrium in which a balance has been achieved between input and output and among system variables. As hydraulic systems, streams possess only one completely independent variable, which is the discharge of water. Discharge of sediment in **alluvial streams** is only partially independent because some of the sediment load is derived from bank erosion. An alluvial stream is one that runs on a bed of alluvium rather than bedrock. In general, mountain streams are bedrock streams, whereas streams with large flood plains are alluvial streams.

The components of the stream system are expressed in the basic hydraulic equation:

$$Q = V \bullet D \bullet W$$

where Q is discharge
V is velocity
D is depth
and W is width.

The relationships between discharge and velocity, width, and depth can be observed under three types of conditions: 1) at a single station on a stream as discharge varies over time; 2) at different stations on the same stream as discharge increases in the downstream direction; and 3) on different streams with different discharges.

Although the exact forms of the graphs are different from stream to stream and station to station depending on channel shape, it is apparent from Figure 16.2 that, as at-a-station discharge increases, velocity, depth and width tend to increase also. This is a predictable effect because lag between input and output requires that increased discharge be temporarily accommodated by increased storage which, because of the basic form of stream channels, causes increases in depth and width. Stream channels tend to be narrower at the bed than at the top so that when discharge increases width increases at a faster rate than depth. This means less resistance per unit volume of flow and increased flow velocity. In alluvial channels depth is also increased by scour of the stream bed during high discharge. Sediment input does not increase as rapidly as water input during storms, so streams are capable of transporting more sediment than they receive. Therefore, they tend to scour their beds and banks, but the bed is easier to scour because the bed material is not as cohesive as bank material and is easier to entrain.

The downstream relationships between discharge volume (Q), velocity (V), depth (D), and width (W) are also shown in Figure 16.2. They indicate clearly that channel size is adjusted to discharge. In most streams discharge increases steadily in a downstream direction. As the tributaries add water to the main channel there are corresponding increases in width, depth, and velocity in the downstream direction. The reasons for these changes are essentially the same as those for at-a-station changes – more water requires a bigger channel and/or increased velocity. The slopes of the lines on the graphs indicate that width increases fastest, followed by depth, then velocity. In fact velocity may increase very little in the downstream direction along some streams.

Figure 16.2 At-a-station Discharge Volume vs. Stream Width, Depth, and Velocity

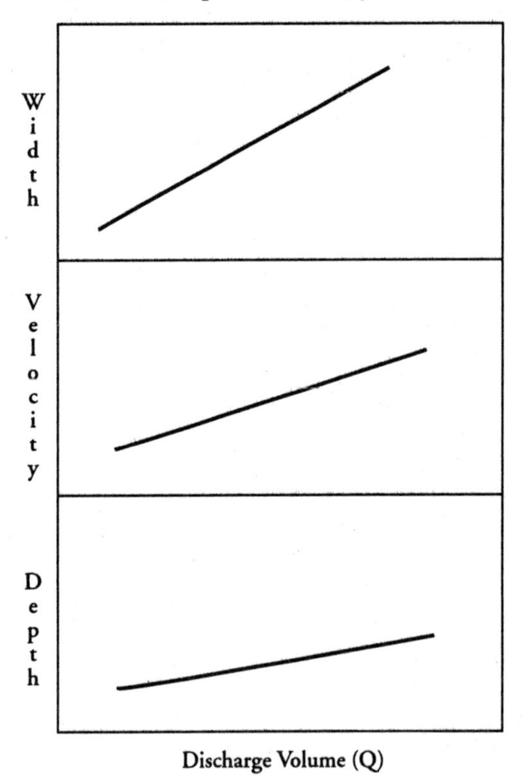

Increased downstream discharge adds kinetic energy to systems by increasing mass and velocity. The kinetic energy of flowing water is equivalent to one half of the mass times the velocity squared. Potential energy depends on mass and elevation.

Ep = mgh
where Ep is potential energy
m is mass
g is the acceleration of gravity (980 cm/sec^2)
and h is elevation above base level.

Potential energy is added to the system wherever tributary streams join the main channel. High velocities and high capacity for erosion may be partly responsible for the typical concave profile of stream gradient from source to mouth that most streams possess (Figure 16.3). Apparently, the concave profile is a necessary condition for streams that have achieved dynamic equilibrium in stream grade. Such streams are stable in that they are neither actively eroding nor depositing in their channels.

Data collected from a variety of streams of different discharges also tend to support the generalization that width, depth, and mean velocity are directly related to discharge and to each other. The mean flow velocity of a stream is an expression of the rate at which potential energy is converted into kinetic energy and dissipated by friction. Thus, velocity is proportional to the head difference divided by the channel length (or channel slope), and inversely proportional to the resistance encountered by a stream as it runs over its bed. Bed resistance depends, in part, on surface area as represented by channel width and depth, but not so obvious is the effective increase in resistance caused by bed roughness and sediment in transport in alluvial streams. An **alluvial stream** is one that runs on a bed of alluvium rather than bedrock. In general, mountain streams are bedrock streams, whereas streams with large flood plains are alluvial streams. A more complete statement of the hydraulic relations of alluvial streams is given by the following formula:

Q = f (E, S, Ssd, D, W)

Figure 16.3 Typical Stream Profile

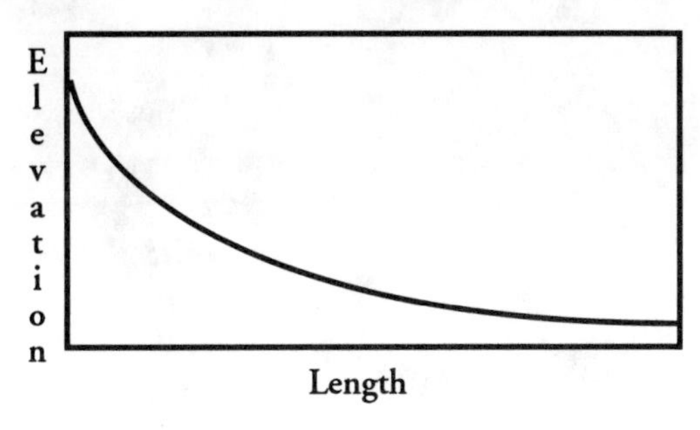

where Q is discharge
f indicates a set of functional relationships
E is the energy gradient, or slope
S is the sediment concentration
Ssd is the sediment size distribution
D is channel depth
and W is channel width.

Two things should be clear from the foregoing discussion. First, for a given stream, flow velocity and channel morphology, as represented by width and depth, are adjusted to discharge. Second, in alluvial streams, width, depth, sediment load, and stream slope are strongly interrelated, and changes in any one of these may induce changes in the others.

Stream Turbulence

Mean velocity is a useful concept for analyzing streams, but when we look at streams, it is readily apparent that flow velocities are quite variable. Eddies and vertical rolls that can be observed near banks and around obstructions in water represent departures from average flow. These eddies, which are especially noticeable in rapidly flowing streams, must be related to velocity and flow resistance. Flow variations are turbulence, and turbulence is greatest when flow velocities are highest and over rough surfaces.

Because the entrainment and transportation of sediment are so dependent upon flow velocity and turbulence we should examine their distribution within streams. Velocity tends to be highest in the deepest portion of the stream just below the surface. The stream bed has the greatest friction and lowest velocities, and friction declines steadily away from the stream bed. The highest velocities are not at the surface because frictional resistance is greater at the air-water interface than it is within the water.

In a rough, irregular channel the pattern of turbulence is largely controlled by obstructions to flow. In a smooth, regular channel maximum turbulence is located near the bottom where the rate of change of velocity is greatest. The drastically different velocities near the bottom cause **shear** (the tendency of molecules to pull at one another), thereby setting up the micro-eddies that are so characteristic of turbulent flow. The upward motions of turbulent flow tend to lift suspended sediment particles at a faster rate than they fall allowing them to be displaced further downslope once they have been picked up.

Stream Pattern

Because of the tendency of water in open channels to flow in a helical or corkscrew manner, streams themselves seldom follow straight paths. Instead, their channels tend to curve in a somewhat regular fashion. Although there is an infinite variety of channel patterns, fluvial geomorphologists recognize three basic categories – straight, meandering, and braided.

Although there are few truly straight stream channels, except where stream location is controlled by geological features such as faults, the term is applied to those streams that are relatively straight. The criterion used to decide whether a stream is straight or meandering depends on the ratio of stream length to down valley distance. Streams with ratios less than 1.5 are classified as straight, and those with values exceeding 1.5 are meandering.

Many bedrock streams in mountainous areas are classified as straight streams. Their very steep gradients and rapid conversion of potential energy to kinetic energy encourage downslope flow more than lateral motion so that these streams seldom develop the regular, curving patterns that are so typical of meandering alluvial streams.

Straight streams that transport a mixture of different sized materials frequently exhibit a pattern of alternating pools and riffles in which the riffles contain coarser bed materials, steeper bed slopes, wider channels, and lower mean flow velocities than pools. Pools tend to alternate from one side of the stream to the other, and are spaced an average of about 5 to 6 channel widths apart. Steep water surface gradients over riffles disappear at high discharges.

Low gradient streams that transport fine-grained sediments flow in a series of sweeping curves spaced at more-or-less regular intervals. These curves are called meanders, and streams with this pattern are called **meandering streams** (Figure 16.4). The course of the lower Mississippi River is one of the best illustrations of a meandering stream (Figure 16.5).

Stream channels are asymmetrical in meander sections because outside banks (the concave banks) are steeper than inside (or convex banks). Meandering streams tend to have more uniform water surface slopes than pool-and-riffle streams do, and this has led to speculation that the curved path of meanders results in a more uniform rate of energy loss.

Figure 16.4 Meandering Streams

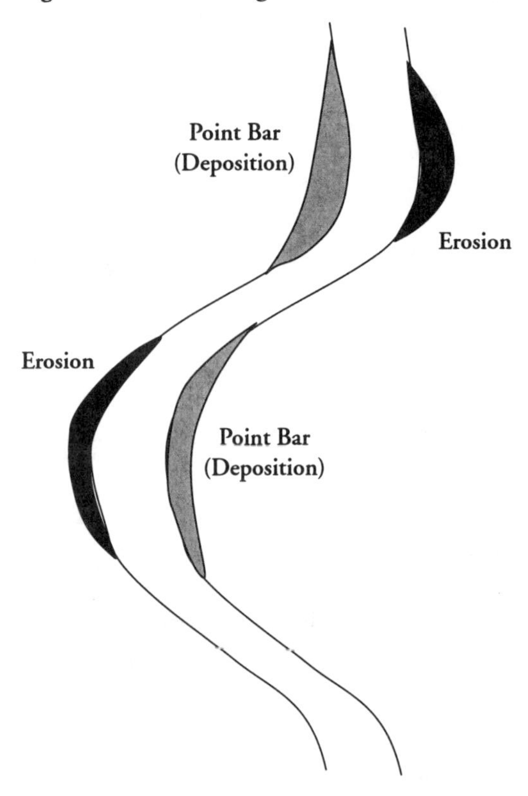

Braided streams have poorly defined main channels. The channels may be split into several distinct streams separated by bars or islands only to converge a short distance downstream or be split into still more channels. Braided streams commonly have steeper gradients, and wider, shallower channels, and transport coarser sediments than meandering streams.

Stream pattern has nothing to do, necessarily, with stream equilibrium, and streams of any type may be capable of transporting as much sediment as they receive so that they are neither building up or eroding their channels. Because they selectively erode their

Figure 16.5 The Meandering Course of the Lower Mississippi River

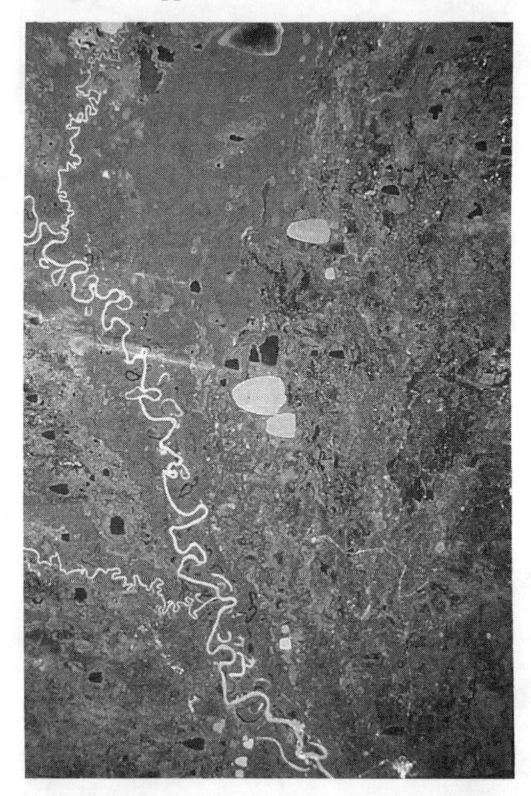

banks and deposit sediment within their channels, streams do move, but stable streams of any pattern do not change their basic form over time.

Properties of streams are derived from energy conversions that occur as water flows through the system. When water is input into the system initially, either as direct precipitation, runoff, or discharge from a spring or lake, it possesses an amount of potential energy proportional to its elevation and its mass. It is equal to the acceleration of gravity times mass times the distance.

Even though this potential energy theoretically can be converted into an equal amount of kinetic energy, in practice much of it is dissipated as heat of friction. This is still another example of system entropy in which friction acts as a negative feedback to restrict velocity, thereby preventing steady acceleration of flow in streams. Experience confirms this. Streams with constant slopes, uniform channel shapes, and no downstream increases in discharge tend to have uniform velocities all along their courses. Such streams are uncommon in nature but are common in man-made canals and channelized streams.

Most natural streams have neither uniform slopes nor uniform channel shapes. If for any reason channel storage is reduced, whether by a decrease in depth, by restriction in channel width, or by obstructions in the channel, water "piles up" and potential energy increases at the expense of kinetic energy. These local increase in water level and potential energy increases water pressure and the stream gradient so that for a short distance below this point velocity is increased. Dams across streams have the same effect, except that potential energy is dramatically increased because of the mass of water stored. This is why, when a hydroelectric facility is built upstream, the generating capacity in downstream facilities is increased. The rapid conversion of potential energy to kinetic energy below dams enables the stream to transport more sediment – thereby increasing its erosive power.

Flow Response

Streams respond to increased inflows of water (and potential energy) by increasing storage. The resulting increases in velocity, depth, and width enable the stream to increase output after a short lag. When input drops, discharge decreases. A graph of stream discharge during and following a period of heavy precipitation would appear similar to Figure 16.6. Prior to the storm the stream is in equilibrium receiving a slow, but steady, input of water called **base flow**. Once rainfall begins, overland flow commences after a short lag. It is slow at first, but as the soil becomes saturated, flow increases rapidly. Discharge rises quickly to a point where output balances input. When input slows or ceases, output exceeds input and discharge declines. Gradually, discharge returns to base flow, but much more slowly than the original increase because of the slow release of water by saturated soil.

A graph like Figure 16.6 is called a **hydrograph**, and the flow from a stream over a long period of time is composed of a sequence of these discharge curves superimposed on base flow. The number, spacing, and height of individual discharge curves depends on the nature, frequency, and amount of precipitation that occur in the basin. The shape of the curve depends primarily on the size, shape and other characteristics of the stream's drainage basin. The more peaked the curve is, the quicker the basin responds to increased input, while flatter curves indicate greater lag.

Figure 16.6 Storm Hydrograph

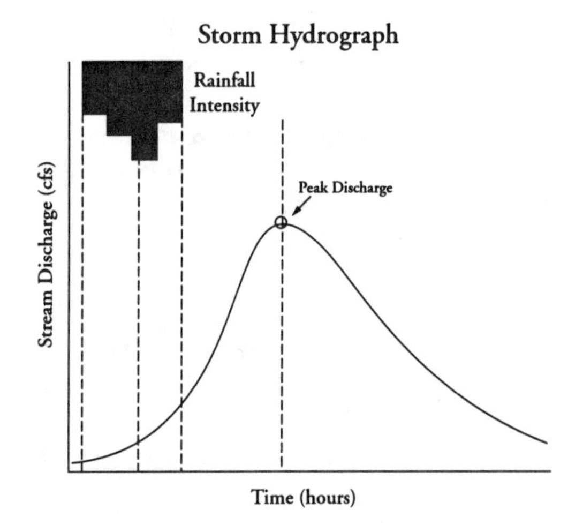

Figure 16.7 Typical Rating Curve

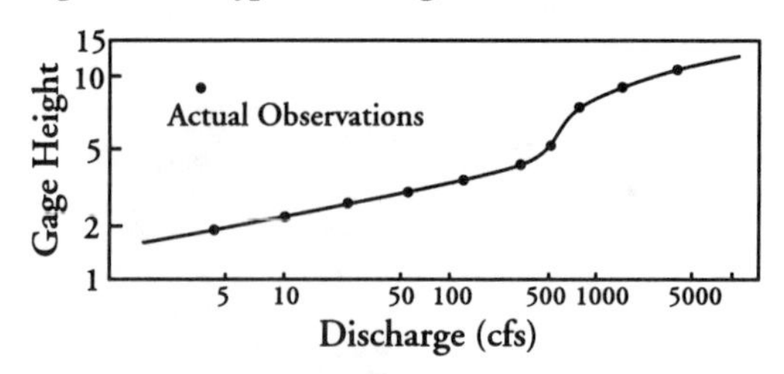

During periods of high flow we are more likely to be interested in the height of the water, or **stage** of the stream, than we are with the volume of water discharged. It is possible to convert discharge data to stage data with a stream rating curve. **Rating curves** show the relationship between stage and discharge for a given stream. They are constructed by measuring stage and discharge under a variety of flow conditions, plotting the data and fitting a line to the points. Then the curve can be used to determine storage for any given discharge (Figure 16.7).

Floods

The problem of flooding in streams is essentially a system storage problem. Water is delivered to the stream subsystem by the surface flow subsystem in large amounts. In order to handle this increased discharge, streams increase their depth, width and velocity. If the increase in volume cannot be accommodated in the normal stream channel, a flood occurs, and the stream overflows into its flood plain.

In general, streams with hydrographs having lag to peak similar to A in Figure 16.8 have serious flood problems. Streams A and B in

Figure 16.8 Hydrographs and Flood Peak

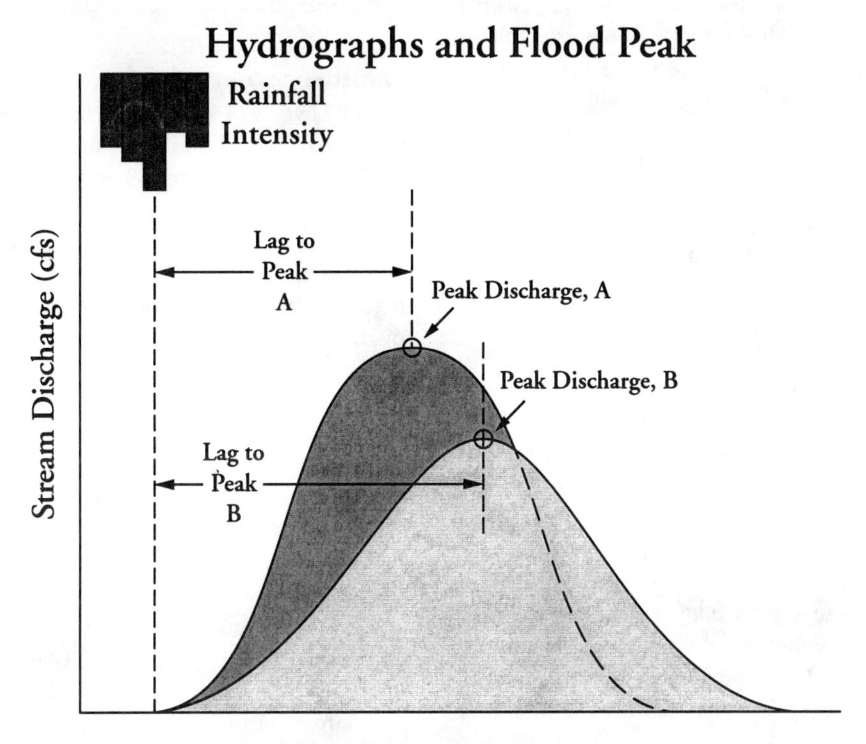

Figure 16.8 carry the same amount of water, yet stream A has a higher peak discharge and bigger floods. Such curves are characteristic of drainage basins that have low infiltration capacities and high surface runoff. These basins do not have the ability to store water in the soil, and thereby increase the lag time until the water is delivered to the stream system. Streams in urban areas tend to have very peaked hydrographs. Destruction of the vegetative cover, large areas covered by impervious surfaces (roofs, streets, parking lots and other surfaces through which water cannot infiltrate), and storm sewer systems designed to carry excess water away as quickly as possible all contribute to steepening of the hydrograph. In undisturbed basins, vegetative surfaces slow surface runoff and increase infiltration capacity and storage capacity of soils. These basins, similar to B in Figure 16.8, are much less likely to flood.

Basin shape also affects the height of the hydrograph. In compact, circular basins with regularly spaced stream networks, all surface runoff is delivered to the stream at the same point at about the same time, thereby increasing the volume of flow that must be accommodated.

In long, narrow basins surface flow gets to all sections of the stream simultaneously, allowing the increased volume to be stored throughout the channel. Basin size affects the height of the hydrograph, but does not necessarily increase the flood problem. Large basins have big streams that can store more water.

Precipitation intensity and amount, the direction that a storm moves across a stream basin, and the condition of the soil (the amount of soil moisture and frozen soil) are some of the other factors that influence the size and shape of the hydrograph. For the most part, these are external factors over which communities and land owners have no influence.

Many of the flood problems we face today are largely our own doing. Destruction of natural vegetative surfaces, use of agricultural practices that promote rapid runoff, and building of cities with their impervious surfaces all contribute to more frequent and larger floods.

Even our flood control techniques contribute to the problems. Construction of levees to confine streams increases the height of floods. We have already seen that volume increases are accompanied by increases in velocity, width, and depth. Thus, if width is restricted velocity and depth have to be greater. Some of the depth increase is accomplished by scouring the channel during the rising flow of a flood, but the height of the water also increases when the channel is artificially restricted by levees.

The practice of channelizing streams may increase the flood problem downstream. When a stream is channelized it normally is cleared, straightened, and given a uniform cross section shape and uniform gradient. These changes reduce channel storage while increasing the rate at which water moves through the system. The net effect is to deliver water downstream faster than it can be accommodated, thus increasing downstream floods.

Channelization and the River System

The hydraulic benefits of channelization are predicated upon the assumption that discharge through a stream section can be increased by manipulating depth, width, or velocity. Deepening and widening increase the channel cross-section, and velocity is altered by increasing the slope, or decreasing the channel roughness.

Few will dispute the fact that drastic changes in one or more of the hydraulic variables – channel width, channel depth, channel slope, sediment concentration or roughness – do, indeed, increase discharge. Typically, proponents of channelization base their arguments upon the ratio of presumed benefits of increased discharge (which may lower flood crests and/or decrease duration of floodstages) and direct costs of channel work. Ignored, however, are the indirect costs that result from changes in dependent hydraulic variables and associated changes in biological systems.

The equilibrium conditions of natural alluvial channels have developed in response to discharge regimes that are a function of the characteristic precipitation regime of the watershed, vegetation, evapotranspiration, and other constant factors that influence the proportion of precipitation that runs off and that which enters the streams as base flow. As urbanization progresses, drainage improvements and changes in land use lead to altered discharge regimes and sediment loads and, consequently, to changes in channel characteristics. In other words, the stream is changed from a system of equilibrium to one of disequilibrium.

Channelization adds to the disequilibrium by further manipulating hydraulic variables so that whatever stability the system possessed is destroyed. Induced changes in the remaining variables represent negative feedbacks by which the stream attempts to reestablish stability.

Relationships among hydraulic variables are manifested in the morphology of the stream channel. Some examples from natural and channelized streams illustrate the direction, if not the magnitude, of changes in channel morphology attributable to feedback adjustments among disturbed state variables.

Hydraulic Effects of Channelization

Because channelization affects one or more of the dependent hydraulic variables – channel slope, depth, width and roughness – and since these variables are not independent of each other, there must be feedback effects that tend to promote a new equilibrium state. The basic hydraulic equations discussed earlier, along with years of evidence from observing channelized streams, provide us with an ample basis to predict the effects of channelization.

Perhaps the most often cited and most obvious effect of channelization is that of shortening the stream length by straightening the channel. The hydraulic effect is increased channel slope. Assuming that discharge of water and sediment remain relatively constant, hydraulic equations reveal that width is proportional to slope raised to some power. Thus when slope is increased, the stream must accommodate the increased slope by widening its channel. System stability is achieved in the short run by increasing sediment discharge through bank erosion, and the attainment of a new equilibrium condition is dependent upon increased channel width. The new wider, shallower channel is more efficient at dissipating energy.

Adjustment of channel width to increased slope can be observed in the field. When natural channels become steeper through such processes as meander cutoff, the cutoff section usually increases its width until it is significantly wider than the segments above and below. Widening can also be observed in many channelized streams that have been steepened through straightening.

Because the variables depth, width, slope, sediment concentration, and roughness are interdependent, seldom, if ever, under natural conditions can one of these change, or be changed, without having some influence on the others. When channels are modified we can artificially constrain the range of these variables, or, in some cases, can hold them constant – as when channels are lined with concrete or when banks are stabilized by riprap. Riprap is the placement of rocks along banks to protect them from scour. When this is done, the stream is denied a source of sediment and a means of attaining equilibrium by channel widening or bed scour.

Another obvious effect of most channelization is removal of vegetative cover within and along the banks of the channel. Vegetation has three effects on streamflow. First, it reduces the effective size of the channel; second, it increases the resistance of banks to erosion; and third, it increases hydraulic resistance. Logically, these effects are more important on small streams than on large ones. Because effects one and three tend to restrict discharge, removal of vegetation in channelization tends to increase discharge through a given reach.

Because vegetation tends to stabilize banks, vegetated stream sections ought to have steeper slopes than strictly alluvial streams with the same discharge. Channel modifications that reduce or eliminate resistance from bank vegetation, release energy associated with above-normal slopes that accompany a vegetative channel. Consequently, elimination of vegetation results in accelerated erosion and a wider channel. In time (years), the over-wide channel is particularly sensitive to encroachment by vegetation that, when reestablished, assists the stream in regaining its original equilibrium.

Widening and deepening channels, and the construction of uniform trapezoidal cross-sections in stream channelization also throw stream systems into disequilibrium. In streams with self-formed alluvial channels the channel width and depth are adjusted to a wide range of frequently occurring discharges. Deepening or widening the channel through dredging upsets the equilibrium and may cause the channel to alternately widen itself through bank erosion during frequent high discharges and deposit in the channel during low flow. Furthermore, unless maintained by constant channel work, uniform trapezoidal channels seldom persist for long. Because of the natural tendency to meander, most streams tend to undercut their banks in an alternating manner in an attempt to reestablish a sinuous course. The ultimate result is to produce a stable point bar, pool-and-riffle sequence. Exceptions to this would be those streams with steep grades and coarse sediment loads that tend to develop a braided channel pattern.

Streams are more than hydraulic systems. They are biological systems as well, with aquatic and terrestrial components. Because fluvial ecosystems are complex and diverse, the major biological impacts of channelization relate to reduction in diversity of habitats and a decrease in the number of desirable habitat spaces. Along channelized sections, grass, weeds, and brush usually increase at the expense of trees, with a corresponding reduction in habitat diversity.

The major hydraulic changes brought about in channelized streams vary with discharge. During low flow, channelized streams tend to be wider and shallower and have decreased velocity than before channelization. During high flow these same streams are usually deeper and faster than for the same prechannelization discharge. The biological impact of these and other hydrologic changes are varied. Maximum velocities may exceed those that can be tolerated by fish, and with the absence of protective cover they may be eliminated. Many fish require pool and riffle habitat, a variety seldom found in wide, shallow channelized streams. Even water temperature may be affected significantly by the removal of vegetation that formerly provided shade as well as cover. Other biological impacts may include changes in nutrient levels, food chain relationships, and reproductive cycles.

The aesthetic appeal of natural streams is reflected in constant literary references that include such phrases as "winding," "wooded streams," and "babbling brooks." Few would deny that aesthetic qualities suffer as much, or more, than the hydraulic or biological, and that aesthetic degradation is certainly more noticeable.

Stream Networks

Streams do not exist as separate, individual entities. The streams in an area form an integrated network with small streams feeding successively larger ones. A map of all streams within a drainage basin appears similar to the limbs on an oak tree, with many tiny streams and only a few large ones. The number and density of streams are a function of the area's climate, its surface characteristics, and its geology. A system of streams may appear **dendritic** (tree like), or it may have peculiar patterns that are controlled by the rock structure.

Stream networks frequently serve as routes of movement for people and animals. Among pre-modern societies boats were often the most frequently and widely used mode of transportation for people and goods. The stream network is an areally efficient system because of the density of the streams that reach into all parts of the region, and all but the uppermost tributaries are frequently navigable by small craft.

Several systems for numbering, or ordering, streams in a drainage system have been developed. The tiny streams with no tributaries are numbered 1. Two 1s coalesce to form a 2, two 2s merge into a 3, etc. Within a particular drainage basin, there is a definite progression in order so that the ratios between successive orders of streams is approximately the same for all orders. For example, there are typically four times as many first order streams as there are second order, and four times as many second order as third order. Similar relationships hold between stream lengths of different orders and stream frequencies and densities (number of streams per unit area and length of streams per unit area). These kinds of stream properties are called **morphometric** properties.

Basin-wide Planning and Management

Events that occur in one part of a stream basin influence events in other parts even though there may be no apparent connection between them. Agricultural practices, such as clear tillage, that reduce infiltration capacity and increase runoff tend to produce quick response systems and increased flood problems downstream along with more erosion and sedimentation. The erosion gully near Greenville, S.C. shown in Figure 16.7 resulted from agriculturally induced increases in runoff.

The solution to many of the problems that arise from patchwork or piecemeal approaches to water resource management is comprehensive basin-wide management. Basin-wide planning and management were advocated long before TVA, yet, we continue to make mistakes because of our failure to treat drainage basins as systems. Reservoirs fill with sediment before half their planned project life has elapsed. Efficient farm and urban drainage systems quadruple the frequency and double the magnitude of downstream floods, and levee systems built to protect the towns further add to the problem.

Multipurpose basin development, incorporating upstream land use management techniques and downstream flood control, and integration of navigation, power generation, recreation, water supply and other uses of water is not an easy task. Streams frequently serve as political boundaries, subdividing the basin among several political units. There are dozens of special interest groups, many of which have conflicting goals and aims that must be arbitrated. And, many stakeholders find it difficult to set aside self-interests and view the issues from the larger perspective of what is best for the community and society at large. Because of these challenges, few multipurpose basin developments have been carried out on large drainage basins. Comprehensive watershed management has been employed with some success, however, in small basins by such agencies as the U.S. Natural Resource Conservation Service.

Figure 16.9 Erosion Gully Near Greenville, S.C.

Water Management

Though all of us recognize the essential nature of water, those of us who live in humid areas where water is plentiful (or has been in the past) may not realize how important it is to those who live in areas where water is in short supply. Moslems believe that all persons should have free access to water, and that to refuse surplus water to one in need of it is a sin against Allah. Scarcity enhances value, and the more valuable water becomes, the greater the conflicts that develop over its use and management become.

In the past, most decisions regarding water use priorities were made independently as though each problem were unique and separate from the others. The result has been a hodgepodge of policies and legislation that varies from one place to another even within the same political unit. Ultimately, all users are competitors because we all depend, in one way or another, on the same sources of supply. Groundwater problems cannot be neatly isolated from surface water problems. Rather, in any sensible management scheme they must supplement and complement one another. Likewise, water problems in Texas cannot be isolated from water problems in Oklahoma when the same aquifers and streams are used as sources.

Good water management must not only integrate all types of water resources and encompass all users, but there must be mechanisms that insure that water is allocated in such a fashion that society receives maximum benefit. In general, this means allocating water to its most productive uses. The mechanism for accomplishing this goal is water law. Nowhere are the contradictions of human actions more apparent than in water law. In the arid western states of the United States (and increasingly in eastern states) water laws can be characterized as "prior appropriation law," whereas the more humid eastern states originally applied riparian doctrine.

Prior appropriation policies treat water as a property item that can be appropriated, bought, and sold by individuals. Under this system a person is free to appropriate and use water as long as it is put to beneficial use. Access to water is not contingent upon ownership of the land on which it is found. There is an hierarchy of uses with domestic use having first priority followed by industrial, agricultural, and less "productive" uses. In general, prior appropriation laws promote more beneficial use of water than riparian doctrine does.

Under **riparian law**, access to water is contingent upon owning land that borders the water. Each owner has equal right to use the water, though no one person can own the water. Ownership is vested in the state. Riparian laws treat water as a common good to which all landowners with water boundaries have equal access. One may use water for disposable purposes, such as irrigation, only so long as such use does not injure other property owners. In effect, riparian policies work to the distinct advantage of land owners along streams by giving them virtual monopoly powers to use water any way they wish. The biggest liability of riparian law is that it does not encourage beneficial use. Common resources are easily exploited because all of society helps bear the cost, even though individuals receive the benefits.

Some economists argue that treating water as a resource subject to market mechanisms insures its efficient use by allowing a price structure to allocate the resource among competing users. Others say that such a completely open market system works to the disadvantage of certain types of users and would not produce the greatest good for the most people. Certainly, just as in the case with real property, individual ownership rights must be subjugated in some cases to insure that problems such as flood control, which involve interdependencies and commonalities with other water management objectives, are developed and managed in a unified manner. In addition, the private sector is ill-equipped to deal with water related problems such as flood control, because flood protection is a common or collective good and cannot be offered or denied on an individual basis.

Summary

The overland flow subsystem includes all surface areas in which water is briefly stored and over which water is transferred except for streams and lakes. The storage capacity of the overland system is low and the response is generally quick. Actual system lag depends on the infiltration capacity of soil and its moisture content, surface variables such as cover and slope, and meteorological conditions (intensity and duration of rain, etc.). Surface flow initially occurs as a sheet-flow, but may become turbulent as water collects in rills and channels. Turbulence is responsible for transporting sediment, oxygenating water, and dissipating energy.

Stream systems collect overland flow and return it to the oceans (or to temporary storage in lakes). Streams also receive discharge from lakes, groundwater, and throughflow. In addition to water inputs, streams receive sediments, organic wastes, and dissolved minerals. Water input into the system represents potential energy that is converted to kinetic energy as water flows downhill. Kinetic energy is ultimately dissipated as heat through friction. Water, sediment, and dissolved minerals are output as discharge from the mouth of the streams. Streams also output water through evaporation and seepage.

The amount of water discharged determines stream size. Width, depth, and velocity all vary directly with discharge. Velocity varies within a stream as a function of friction, decreasing in wide shallow reaches and increasing in deep reaches where friction is less. Turbulence is greatest in rough, rapidly flowing streams, and is concentrated in the deepest part near the bottom where velocity decreases most rapidly. Factors other than turbulence also influence velocity. Sediment increases viscosity and produces lower velocities, and velocity varies with stream slope.

Stream discharge is a product of its velocity times its cross-sectional area (width x depth). Thus, as water is input at faster rates, streams increase their depth, partly by eroding their channel bottoms and also by raising their water surfaces (stages). The relationship between discharge and stage varies from stream to stream and from place to place along a given stream. Following a storm, storage and discharge increase rapidly. If the volume of water exceeds channel storage capacity, a flood occurs. Floods usually fall more slowly than they rise, as the plot of discharge (hydrograph) indicates. Hydrograph shape is indicative of the nature of the flood problem within a stream basin and depends on basin shape, system storage, meteorological conditions, and any anthropogenic impacts upon the drainage system.

Streams are organized into a definite hierarchical structure with many small streams and a few large ones. The pattern is most often a random, treelike pattern, called dendritic, but if strong structural control is present any of several commonly recognized patterns may be found depending on the type of lithologic influence.

The system unity of drainage systems is observed easily when changes that occur in a part of a basin affect processes or conditions elsewhere. Many of these interdependencies are common knowledge, yet we still approach many water management problems as though we were dealing with separate problems, each with its own singular cause-effect relationship.

Chapter 17

Oceans

Introduction

We have found it necessary to mention the ocean several times in prior chapters – as a heat reservoir, as a source of atmospheric moisture, and as a sink for the water and sediments that streams discharge from the continents. But, the ocean is a system in its own right. In fact, it may be considered as several kinds of systems which geographers are interested in. In this chapter we will examine two of these systems, one physical and one biological, and we will amplify some of our earlier discussions concerning exchanges of energy and matter between the ocean and other systems.

Even though ocean waters cover 70 percent of the earth's surface, we are accustomed to viewing maps that emphasize continents and imply that the continents are partitions which subdivide the ocean into separate bodies of water. In fact, these large, open bodies form one interconnected world ocean, but it is divided into separate basins by continents and other features. Like the atmosphere, it is a thermodynamic system driven by the sun – directly through solar radiation and indirectly by atmospheric motion. The ocean may also be viewed as a biologic system which is also driven by solar energy, but subject to nutrient restrictions.

Ocean water is commonly called "soup" by oceanographers. It is a living soup that contains water, salts, solid sediments, gases, and living organisms. More than 65 elements have been detected in seawater, most of them as salts or constituents of organic matter. The most common substances are chloride, sodium, sulfate, magnesium, calcium, potassium, bicarbonate, and bromide, and the nutrient ions nitrogen, phosphorous, silicon, and carbon. The salts and gases are held in solution while the organisms and particulate sediments are in suspension. As we will see in the section on budgets, not all of the constituents of seawater are uniformly distributed, and concentrations vary from one location to another.

The Origin of the Ocean Basins

Not much imagination is required to mentally piece together the opposing continental outlines of South America and Africa. Close examination reveals that other continental outlines bear striking similarities as well. Sir Francis Bacon and others suggested as long ago as the 17th century that perhaps the continents may have been connected at some time in the past, and subsequently broke up and drifted apart.

In the 19th century the rapid advances which occurred in geology provided additional evidence to support the suggestion. Similar fossils, similar rock types, and similar geological structures were detected on several continents. In the early 20th century a German meteorologist by the name of Alfred Wegener assembled all of the available evidence and systematically presented a comprehensive theory of continental drift. He noted the fact that Scandinavia was still rising vertically very slowly in response to the release of pressure associated with the melting of the thick ice sheet that covered Scandinavia during the last glacial stage. He argued that if the earth could flow vertically like this, then surely it could flow horizontally.

When Wegener fit the continental outlines together (not the shorelines) as best he could, he found that many of the similar fossils, rocks, and structures on various continents fit together also. Despite his systematic presentation of the several bits of supporting evidence, Wegener convinced relatively few geologists and geophysicists, mainly because he lacked a convincing explanation to show how the rigid earth crust could move in such a manner.

In the 1950's discovery of several additional bits of supporting information led geophysicists to reexamine the concept of continental drift. The additional evidence was (1) the more complete mapping of the ocean floor, (2) the development of a sensitive network of seismic stations where earthquakes are measured) that allowed for more accurate maps of earthquake locations, and (3) the discovery of a strange pattern of alternating bands of magnetic anomalies in the rocks on the bottom of the ocean.

It had been general knowledge for some time that mountain ridges were located in the ocean near many island chains. Extensive surveys in the 1950's revealed that there was almost a continuous network of ridges that runs through all of the ocean basins. It was apparent from the earthquake maps being generated that these ridges were the locations of most intermediate depth earthquakes that occurred in the ocean. The seafloor investigations also revealed that the basin floor near the ridges was covered with a thin veneer of sediment that increased in thickness toward the continental margins. Nowhere in the Atlantic could bottom sediments be found that were older than 140 million years.

Existing geological theories assumed that the ocean basins were as old as the continents with both having been produced at the same time. If this were so, then why were sediments near the ridges so thin, and why could no sediments be found that were older than 140 million years?

Clearly a new explanation was called for, and independently, two scientists, Harry Hess and Robert Dietz, suggested that the ridges were formed by rising currents of molten rock which cooled and formed the spreading ocean floor. Despite its intuitive appeal, most scientists considered the explanation to be improbable.

At about the same time as these developments were taking place, other scientists were conducting magnetic surveys in an attempt to map the historical migrations of the magnetic north pole. When new rocks are formed from molten material the iron minerals in them are magnetized by the earth's magnetic field. The investigators found that not only did the direction of alignment vary, but, periodically and unpredictably, the polarity was reversed, indicating that the earth's magnetic field reverses itself through time. By examining the polarity of rocks of known age, geophysicists were able to devise a calendar for dating rocks based on the sequence of magnetic reversals.

Paleomagnetic surveys of the ocean revealed a series of alternating zones of magnetic reversals which lay parallel to the oceanic ridges. When the dating technique worked out for terrestrial rocks was applied to the ocean floor basalts, geologists discovered that the rocks nearest the ridges were very young and the oldest rocks were along the margins of the ocean. It was soon obvious that Dietz and Hess probably were correct. The ocean floor might be spreading along the oceanic ridges as molten material rose to the surface. An area where crustal spreading of this type occurs is called a spreading center.

Scientists proceeded to test the idea. Continental outlines were fitted together as accurately as possible. The results were astounding and produced fits with few gaps and minimal overlap. Minor inconsistencies can easily be explained by deformation of continental shapes and by post-rift deposition.

The absolute age of rocks in Brazil and Africa, and sharp lines which separated very old and more recent rocks in Africa could be projected into northeastern Brazil, where identical age sequences were found.

Dating of the rocks from islands in the Atlantic and Pacific Oceans revealed that the ages increase steadily away from the rift zone. The islands that make up the Hawaiian chain reveal this clearly. The large Island of Hawaii, which is the only island in the chain with active volcanoes, is situated near the rift zone. The islands to the west of the Island of Hawaii get progressively older with parts of Kauai ranging between 4.5 and 5.6 million years old. Distances from rift zones enable us to estimate the rates of drift at 2 to 6 cm per year.

In addition to the highly active seismic zones of the oceanic ridge, many earthquakes and volcanoes occur in the mountain belts that border the ocean and along the deep ocean trenches and their associated island arcs. Earthquakes occur when the earth's crust breaks, causing shock waves to be transmitted to the surface. The location of deep and intermediate earthquakes mark the zones along which crustal movements occur. The seismically inactive zones indicate more stable sections of the earth's crust.

Lithologically, the earth is a closed system, and because the crustal surface varies little over time, if surface is being created along the ocean rift zones then crust must be consumed or destroyed elsewhere. Modern theory assumes that this destruction occurs along the ocean trenches and that the earthquakes and volcanoes are by-products of the process. The comprehensive theory that combines continental drift and tectonic processes is **plate tectonics**.

The Ocean Floor

The ocean floor is very different from continental surfaces. It is composed of different materials and its topography is more varied. The basalt which emerges from the spreading zones makes up the bulk of the oceanic crust. The basalt is covered by a veneer of sediments which decreases in thickness from the continental margins to the rift zones where newly formed basalt surfaces may be entirely bare of sediments or contain only scattered, thin patches.

For many years the ocean bottom beyond the continental slopes was thought to be a large, flat plain called the Abyssal Plain. Modern studies reveal a much more complex floor which can be subdivided rather easily into six types of surfaces depending on topography and location: (1) continental shelf and slope, (2) continental rises, (3) ocean basins, (4) volcanoes and volcanic ridges, (5) oceanic rises and ridges, and (6) trenches. The basins account for the largest area (40 percent), followed by rises and ridge areas with 55 percent, and shelf and slope areas with 15 percent. The six physiographic types are perhaps best illustrated in the Atlantic.

Despite the fact that there is one interconnected world ocean, the bottom topography is such that the circulation between these is restricted, and the Atlantic, Pacific, Indian and Arctic oceans have different characteristics. The Pacific Ocean is the largest. It covers over one-third of the earth's surface and accounts for more than 50 percent of the ocean area. The continents that surround the Pacific basins are zones of active mountain building, and the coastal areas in many cases are subduction zones containing deep trenches. This has produced a deep basin, the margins of which are tectonically active. The oceanic ridge, which really is a single feature that runs through all the ocean basins, is located in the eastern part of the Pacific. The western part of the basin contains numerous volcanic ridges and associated islands.

The Atlantic and Indian Oceans are relatively recent features formed by the breakup of the supercontinent Pangea some 200,000,000 years ago. The ridge in the Atlantic occupies a central location from which new ocean floor is being extruded at an estimated 2 to 5 cm per year. The Atlantic is much shallower than the Pacific. It is bordered in many places by gently sloping continental platforms which project into the ocean and form wide continental shelves and shallow coastal areas. Overall, the Atlantic is a shallow ocean, and despite its large boundaries with the Arctic and Antarctic Oceans, it has the warmest and saltiest water.

The Indian Ocean, which is the smallest of the three major ocean basins is connected to the Atlantic and Pacific through broad open seas south of Africa and Australia. The mid-oceanic ridge of the Atlantic continues around South Africa and turns northeastward into the Indian basin. It divides near the center of the ocean where one part goes northward into the Gulf of Aden and the other turns to the southeast toward the Pacific basin.

The Ocean as a Thermodynamic System

The principal thermodynamic properties of the ocean system are temperature, salinity, pressure, viscosity, and density. Biological system properties depend on these same variables plus gases and sunlight. We will examine the system properties before attempting a discussion of dynamic system behavior.

Pressure, Viscosity, and Density

The weight of water causes pressures to increase rapidly with depth. Unlike the atmosphere where pressures change geometrically, water is an incompressible substance and pressure changes linearly with depth. The rate of increase is so rapid that the pressures are measured in atmospheres. The rate of change is one atmosphere for every 10 m (32 feet), and in the deepest areas of the ocean pressures reach 1100 atmospheres.

The viscosity of water, its internal resistance to flow, varies in response to both temperature and pressure. In most liquids viscosity increases with pressure, but in seawater viscosity decreases with pressure and reaches a minimum at depths between 5000 and 10,000 meters (500 to 1000 atmospheres pressure). Apparently, the high pressures help to destroy the remaining structure in the liquid caused by the tendency of hydrogen molecules to bond together, and thereby reduces resistance to motion. Viscosity changes due to temperature variations are minimal compared to those caused by pressure changes. However, viscosities are slightly greater in the colder water of the polar regions.

The density of water helps to support marine organisms, thereby allowing marine animals to attain larger sizes in some cases than land animals. Density does not change with pressure, but does change with temperature as discussed. Cool water is more dense than warm water and will subside, whereas cold water (less than 40°C) and ice are buoyant because of their lower densities. Saline water reaches its maximum density at temperatures lower than 40°C, the maximum density temperature of fresh water. In fact, most seawater freezes before maximum density is attained. Density differences cause water masses to react as units and retard mixing of water of different temperatures and salinities. When ions or particles are added to water, its mass, and, therefore, its density are increased. However, seawater density is more sensitive to changes in temperature than to salinity changes. Nevertheless, more saline water does have a sufficiently greater density which causes it to subside. Dense water will flow downslope along the ocean bottom in the same manner as water on land does. These density currents can erode the bottom in the process. Density differences are also responsible for much of the subsurface circulation in the ocean.

Biosystem Structure

The bulk of the biomass in the ocean is comprised of tiny organisms, both plants and animals, called plankton. Biomass refers to the total mass of biotic organisms existing at any given time. Phytoplankton are basically small, simple-celled plants capable of carrying out photosynthesis. They are free-floating and depend upon their tremendous surface area-to-volume ratios to create sufficient drag to keep them suspended within the **photic zone** where light penetrates the water and makes photosynthesis possible.

Zooplankton are the grazers within the oceanic ecosystem. These organisms have limited mobility and, like the phytoplankton, also depend upon the density of water to keep them suspended and upon currents to move. Both planktonic forms are short-lived, and even though at any one time the existing biomass may be low in comparison to terrestrial ecosystems, annual productivity may be very high.

Free swimming organisms, which include fish, marine mammals and squid, are called **nekton.** They are not restricted, as the plankton are, to the photic zone, but they are found in the highest concentrations in areas with the greatest abundance of plankton which supply the food for the small fish that larger species feed on.

The ocean bottom is home for many creatures. Some plants and animals attach themselves permanently to the bottom, and others live upon the surface or within the bottom mud. These organisms constitute the **benthos.** In shallow coastal areas the benthic and photic zones converge into an especially productive environment, and it is from such areas that most edible fish are caught.

Biosystem Response

One measure of biosystem response is productivity. Productivity refers to the rate at which oceanic plants assimilate carbon by photosynthesis. In the ocean, photosynthesis is limited by sunlight and the available supply of dissolved nutrients. There are several types of conditions and characteristic locations for which the nutrient supply is plentiful and productivity is high.

Upwelling currents bring cold, nutrient-rich bottom water to the ocean surface along the west coasts of most continents in tropical and subtropical areas and around Antarctica. One of the most productive fishing areas in the world is located off the coast of Peru and Equador where the rich supply of upwelling nutrients combines with intense sunlight from cloud-free skies to produce optimal growth conditions. In the past as many as 10 million tons of anchovies per year were harvested from these waters. These anchovies are ground into fish meal and used as a high protein food source for livestock. Like most fishing regions, this area suffers from overfishing and although the tonnage of fish caught is down approximately 10% from peak catches in 1989, it remains one of the most productive grounds. In some years, however, the cold water is replaced by a warm current called El Niño and the disappearance of the anchovies means the loss of a major source of foreign income for Peru and virtual economic disaster for the fishermen involved. The fishing is so important to the nation that Peru has nationalized the industry and now sets annual quotas which are strictly enforced in an attempt to insure sustained production of 5 to 8 million tons per year.

The cold waters off Southern California once supported a sizeable sardine industry that was immortalized in some of John Steinbeck's writings, but overfishing and pollution have reduced it to insignificance. No doubt the Peruvians would like to prevent that as much as possible.

Because of its size and the high dissolved-nutrient content of its upwelling water, Antarctica supports what is probably the single most productive area of the ocean. As in most high latitude areas, production is concentrated in the summer season when sunlight is available for photosynthesis. Until the last quarter century no significant amount of commercial fishing took place in Anarctica because of its remoteness from the major markets in North America, Europe, and Asia.

The world's rivers are a major source of nutrients for the ocean. During the summer season the dissolved load of nutrients delivered to the oceans by rivers is low because indigenous plankton populations consume them before the water is discharged. But in winter, the dissolved load is high. Nutrients are readily recycled in shallow water so that coastal areas with major streams emptying into them may be very productive. Examples of such areas can be found in South and East Asia, and along the European coasts.

The low productivity of mid-oceanic areas contrasts sharply with the highly productive coastal areas and upwelling zones which may be as much as 100 times greater. When one considers that only the upper few hundred feet of the ocean are penetrated by light it is apparent that to view the ocean as an endless reservoir of food resources is erroneous.

The Ocean as a Human Food Resource

People have always looked to the oceans as a source of food. The vastness of the ocean and its productivity, the low fishing pressure, and the simple technology available to fishermen restricted the impact of fishing to certain species and favored fishing grounds until well into the 20th century.

In the last two or three decades, rapid population growth in some nations, and rising standards of living in Europe, North America, and the Pacific Rim nations have caused a dramatic increase in the demand for fish and shell fish.

At the same time that fishing fleets were increasing in size, technological advances improved their efficiency enormously. The effects were predictable. Annual catch figures have leveled off, and in most cases declined;

stocks of the most desirable species have declined in quantity and in quality; and, the costs of harvesting the catch have soared.

The ocean has become a major source of protein to supplement the diets for some cultural groups. There are no proprietary rights to fish other than those that result from the territorial claims of nations. Therefore, fish have traditionally been treated as a common good with the only cost being the cost of catching. Fishermen have tried to maximize individual profits by catching all the fish they could. The effect has been to reduce breeding stocks while fishing pressures have increased. With smaller fish populations and more fishermen, each boat catches fewer fish, but uses more fuel and time to catch them. Increasing costs and fewer fish mean higher prices. The net result is that the richer, more technologically advanced nations, such as Japan, Russia, the United States, and some western European nations catch the bulk of the food fish. Modern technologically sophisticated fishing fleets use radar to navigate in fog, sonar to detect schools of fish, GPS and LORAN to return to highly productive fishery locations. Some operations use aircraft to spot schools, weather radar to anticipate fish migrations, 80 miles of longlines filled with hooks, and even 40 mile long drift nets (mostly illegal).

If we consider the ratio of energy harvested to energy used during the harvest, it is apparent that we are subsidizing our catches by substituting large quantities of inorganic energy in the form of fuel for relatively small quantities of organic energy. Such a system can operate only as long as fuel costs are low, a condition that may not persist much longer.

The financial subsidy to the fishing industry is also substantial. As individual nations attempt to protect their fishing industry through subsidies, harvesting of the world's fish has moved from an economically profitable operation to an uneconomic one sustained artificially by the tragedy of the commons. The annual world fish catch at the end of the 20th century is worth approximately $70 million, but the cost exceeded $120 million. The difference is made up from financial subsidies.

Several suggestions to overcome these problems have been advanced by ocean biologists. To more efficiently use ocean food resources they have proposed that we consume fish that currently are not used as food by grinding them into high-protein fish flour that could be substituted for grain flour in making bread. There has been high consumer resistance to fish flour even in developing nations with meager food resources.

We are forced to conclude that the future potential of the ocean as a food resource is not encouraging. Exploitation of some new supplies in places like Antarctica and the use of some formerly undesirable, unexploited species may increase short-run supply, but there is little reason to expect significant long-term increases.

Energy and Matter Budgets

The world ocean is a system driven directly and indirectly by solar energy and through which a variety of matter cycles. Solar energy in the form of light supplies the energy for photosynthesis that runs the biological system. Solar heat is absorbed and redistributed by the ocean through the movement of ocean currents, and the process of evaporation. Ties between the ocean and the atmosphere are so direct and so important that one cannot be treated adequately without considering the other. The pattern of surface wind systems is duplicated by oceanic surface currents, and transitory winds provide the energy for the waves that shape the beaches and coastlines of coastal areas.

The ocean system continually receives inputs of matter in the form of water, suspended and dissolved sediments, and gases. It outputs all of these substances, too. If the ocean is viewed as a steady state system the processes of which vary little through time, then the rates of output and input of nutrients should be equal. By knowing the concentration of a substance in seawater and the amount of seawater in the ocean, it is possible to calculate its storage. Then, by measuring either rate of input or output, the length of time required to replace completely the amount of any substance in storage can be calculated. Oceanographers refer to this concept as **residence time**. Residence times vary from an estimated 260 million years for sediments to a few hundred years or a few thousand years for materials used up rapidly in biologic processes, or which settle out rapidly as sediments. The water itself has a residence time of about 3500 years.

The Oceanic Water Budget

The origin of the ocean like that of the atmosphere, is directly related to the origin and evolution of the earth itself. Because there is no consensus about details of the earth's origin, we can only engage in informed speculation. Most earth scientists consider that the ocean evolved over millions of years from condensation of water vapor released into the primordial atmosphere by volcanic eruptions. Water-laid sediments 2.5 to 3 billion years old provide the earliest indication of an ocean, but they yield few clues as to its chemical composition. Chances are that its composition may have been very similar to today's ocean with some exceptions. The presence of rich iron deposits in ocean sediments indicates that the ancient ocean contained substantial amounts of dissolved iron. Iron is relatively insoluble in sea water today because of its oxygen content, but the ancient seas contained no free oxygen because there were no plants to produce it.

Today's ocean contains approximately 31,350,400,000 km^3 of water. Of that amount 445,000 km^3 evaporate each year and (to maintain sea level) must be balanced by equal input. Precipitation falling directly on the ocean surface returns 412,000 km^3, and river discharge supplies the bulk of the remainder (29,500 km^3). Groundwater flow and iceberg calving supply the rest (3,500 km^3).

The pattern of input and output is more critical than the amounts involved. The most rapid evaporation occurs in the latitudes equatorward of the subtropical high pressure cells where the warm tradewinds blow. Conditions for evaporation are ideal with warm water, warm dry air, and steady winds. Evaporation is minimal in the polar seas where conditions are most unfavorable. Direct rainfall input is concentrated along the equator and in the subpolar latitudes around 50° N and S. The 33,000 km^3 that is discharged into the oceans as runoff, groundwater, and icebergs is concentrated around the continental margins, especially in the vicinity of large tropical rivers and in polar regions where glacial meltwater is concentrated.

The average ocean surface elevation is constant over time for each location. Therefore, regional oceanic water budgets can be expressed as follows:

$$E = P + R + C$$

where E is evaporation
P is precipitation

R is runoff from river and glaciers
C is transport by ocean currents and may be either negative or positive depending on the balance of input and output.

In those places where evaporation exceeds precipitation, unless rivers input enough to offset evaporative loss, then saline water from currents must balance the equation. The result is a higher than average surface salinity. In places where precipitation and runoff exceed evaporation the water is less saline and the excess is exported by currents.

Heat Budgets

In the atmospheric heat budget approximately 19 percent of the incoming radiation is transferred to the atmosphere by evaporation. A large proportion of this energy originates in the tropics and is transported to higher latitudes, helping to keep temperatures lower in equatorial areas.

A second aspect is the latitudinal variability in heat budgets. The heat budget equation for ocean surface temperature is shown in Figure 17.1.

In general, the greatest amount of incoming shortwave radiation is received around 15° to 25° N and S latitude where evaporation is highest. The additional heat released along the equator in the process of precipitation produces high ocean surface temperatures in the equatorial regions between 20° N and 20° S. Temperatures there are higher than 25°C (77°F)

Figure 17.1 The Ocean Heat Budget

$$Rl\downarrow + Rs\downarrow + LE\downarrow + H\downarrow + C\downarrow = Rl\uparrow + Rs\uparrow + LE\uparrow + H\uparrow + C\uparrow \pm \Delta S$$

where

$\downarrow$ is incoming energy

$\uparrow$ is outgoing energy

Rs is shortwave radiation

Rl is longwave radiation

LE is heat loss by evaporation or gained by condensation

H is sensible heat transfer between the ocean and the air

C is heat imported or exported by ocean currents

and, ΔS is the change in heat stored.

year-round. Poleward of that location the temperatures decline rapidly to 0°C (32°F) at 60°N and S. Both polar regions are cold with much of the Arctic Sea remaining frozen year-round.

Although the ocean exhibits considerable spatial variation in temperature, it has little seasonal variation. The high specific heat of water, its transparency, and the enormous mass of water in the ocean allow it to store large quantities of heat. Thus, even in mid-latitude areas ocean surface temperatures vary less than 10°C (18°F) between winter and summer in most places. Below the surface layer there is even less seasonal variability. Temperatures do decrease vertically, and bottom temperatures in the ocean basins may be as low as 0°C (32°F).

Sediment Budgets

With the exception of gases, salt nuclei, and small quantities of materials removed from the sea through human activities, most nonaqueous matter in the ocean is input and output as sediment. Rivers supply the bulk of sediments in particulate or dissolved form. In addition to the estimated 20 billion tons of particulates and 2 billion tons of dissolved sediments supplied by rivers each year, glaciers supply significant amounts in polar regions, and perhaps as much as 100 million tons of windblown dust settles out of the atmosphere over ocean areas.

Particulate sediments settle out of suspension quickly, so most river sediments are deposited in estuaries and deltas. This is especially true of large-grained sediments which often form beach deposits along coasts. Because few large rivers discharge directly into the open ocean, the bulk of the particulate sediments they discharge into the ocean accumulate in shallow continental shelf areas. Accumulations vary from as much as 15 to 40 cm/l,000 years near the continents to an estimated 10 cm/l,000 years. average for all continental shelf and rise areas. Most of these sediments remain in coastal areas unless transported to the deep ocean by density currents.

By way of contrast some deeper parts of the ocean basins are covered by a thin layer of sediments known as red clays or brown muds that accumulate at rates of approximately 1 cm/ l,000 years. They are fine-grained silicate clays of continental origin which become heavily oxidized in the oxygen-rich environment of the deep ocean.

Biogenous sediments accumulate at rates that vary from one part of the ocean to another. **Biogenous sediments** are those that are biological in origin as opposed to **lithogenous sediments** which originate from continents. The silica and calcium carbonate needed by marine organisms to grow skeletons settle to the bottom when the organisms die. The accumulation rate depends, therefore, on the amount of plant and animal life, which in turn, depends on nutrient availability and sunlight.

Dissolved Gas Budgets

Oxygen (0_2), nitrogen (N_2), and carbon dioxide (CO_2) are the most abundant gases in the ocean. Atmospheric gases diffuse into the sea in much the same manner that water molecules evaporate. The surface layer is nearly always saturated and concentrations decrease with depth as the gas is distributed through turbulent mixing and diffusion.

Saturation level is different for each gas and depends on temperature, salinity, and pressure. Cold temperatures, low salinity, and high pressures encourage the solution of gas.

Peak dissolved oxygen concentrations occur just below the surface. This is the photic zone where aquatic plants carry out photosynthesis and release oxygen in the process. Mixing and photosynthesis keep concentrations high to a depth of about 200 meters (650 feet). Concentrations are low from about 200 meters to 800 meters (650 feet to 2600 feet), but increase again below that depth. Because cold, dense water in polar regions tends to subside and move toward the equator, deep waters have been at the surface more recently and the cold temperatures, high pressures. and lack of respiration help to maintain high oxygen levels.

Carbon dioxide is consumed in photosynthesis, but the level of biological activity has little effect on carbon dioxide concentrations. When large amounts are produced, carbonic acid forms and calcium carbonate is dissolved, and when small amounts are present calcium carbonate precipitates. Both of these negative feedbacks cause the concentration of carbon dioxide to remain nearly constant, between 45 and 54 ml/l. There is some slight increase with depth due to biological decomposition of falling organic matter.

Since nitrogen is a relatively inert gas (does not readily combine with other substances), its dissolved concentration in sea water is almost totally controlled by temperature, pressure, and salinity; and, therefore, it is at saturation levels nearly everywhere.

Oceanic Circulation

Water movements in the ocean are basically of two types — currents and waves. Ocean currents are large-scale displacements of water that are driven primarily by wind or density differences, and which are important in heat transport. In deep water wave motion there is little forward displacement of water as the wave form is transmitted. The general circulation is examined in this section.

Surface Currents

Surface currents are primarily wind driven. Wind is the driving force and the overall pattern of surface currents shows striking resemblance to the general atmospheric circulation. The flow pattern is dominated by a series of gyres, or closed elliptical cells. The largest of these gyres are centered in each ocean basin at about 30° north and south latitudes. This is the same location as the subtropical high pressure cells. The pattern is responsible for several recognized currents — a westerly equatorial drift on either side of the equator, currents paralleling the east coasts of continents that flow toward the poles, and currents that flow toward the equator along continental west coasts. The east coast currents transport warm water northward while the west coast currents bring cold water southward. Smaller gyres are found at about 50° - 60° north and south which circulate in a direction opposite to those centered at 30°. These gyres, which display clockwise rotation in the northern hemisphere are driven by the winds that spiral around the subpolar low pressure cells. In the equatorial zone an easterly counter current helps to offset the effect of the westerly flowing equatorial drifts

The basic gyres described above are semipermanent circulation features, since the winds and pressure systems that cause them are semipermanent atmospheric features. There are some minor seasonal variations, however, due to the latitudinal shifts in the pressure cells between winter and summer. In the Indian Ocean more drastic seasonal effects are evident. The seasonal shifts in the monsoon winds cause northeasterly currents in winter and southwesterly currents in the summer.

The principal forces that drive surface currents are wind direction and the Coriolis effect. Once water is set in motion by wind,

Coriolis causes the movement to be deflected to the right in the northern hemisphere and to the left in the southern. The current actually flows at an angle to the direction of the wind. Surface waters tend to drag subsurface waters along. But, Coriolis is also a factor in this case because subsurface movement is deflected to the right of surface waters in the northern hemisphere. With increased depth, movement is slower and is increasingly deflected, producing a spiraling effect. The characteristic pattern is called an **Ekman Spiral**. Along west coasts where winds blow toward the equator, Ekman flow causes offshore movement of water below the surface to a depth of 200 meters or thereabouts. Water from below moves upward to replace water that moves away from the coast. The phenomenon, known as **upwelling**, is very noticeable because of the cold temperature of the water. Upwelling brings abundant supplies of nutrients to the surface, so that these areas are noted for their plentiful fish and bird life. The cold water temperature has a strong influence on the climates of these coastal areas.

Vertical movements of ocean water occur in response to density differences when more dense subsiding surface waters displace less dense water below. Recalling that both temperature and salinity may cause density variations, we might logically expect more saline and cooler waters to sink. This is generally what happens. In the areas of high evaporation, the more dense saline water that accumulates slowly settles downward to displace less saline water. Likewise, warm saline water that is exported to high latitudes by the warm currents and west wind drifts cools by conduction and radiation as it reaches the cooler midlatitudes. The combined effect of higher salinity and cooling causes the water to grow more dense, and it slowly subsides. It is replaced, in part, by water from polar seas. The subsiding water itself turns equatorward and flows along the bottom. In the polar regions surface waters subside as they lose heat. They are replaced by less dense fresh water from streams and glaciers. It is this fresh water with its lower freezing temperatures that freezes into the pack ice in the Arctic Ocean during the winter.

Density variations are primarily responsible for the subsurface current pattern. Cold Antarctic water flows northward along the bottom, penetrating well into the northern hemisphere. A wedge of water from the north Atlantic flows over the Antarctic water, and it in turn, is overlain by a wedge of Antarctic intermediate water that manages to push past the equator.

These layers constitute the basis for classifying water masses. The concept is similar to that of air masses except that water masses are bodies of water with similar temperature and density characteristics. They originate in areas where salinity and temperature densities are uniform over large areas, allowing large water masses to acquire similar characteristics.

The major water masses which occur in each of the oceans are the central water mass, the intermediate water mass, deep water mass, and bottom water mass. The central water mass is located between the subtropical convergences and it extends down to the permanent thermocline. The salinity of the central water mass is high in the Atlantic Ocean but low in the Pacific. The central water mass is underlain by intermediate water masses, formed in the Arctic and Antarctic convergences. Intermediate water extends to about 1500 km in depth. The water in this mass originates from the cooling of subarctic waters during the winter.

The deep water mass is confined largely to the Atlantic Ocean. The highly saline Gulf stream waters mix with subarctic surface water.

During the winter these waters grow very dense as they cool, sinking to the bottom and flowing southward. They follow the bottom until they meet the denser bottom water of the Antarctic water mass which causes the deep water mass to rise over it and wedge its way between the dense bottom water and the less dense intermediate water.

The Antarctic bottom water forms in the vicinity of Antarctica. As surface water freezes the brines left behind are very dense and cold. The brines settle to the bottom where they mix with cold surface waters, forming a very cold (-0.5°C) and dense water mass. The mass then flows slowly northward along the ocean floor.

Waves

Ocean waves are not important in the transport of water or sediment except in nearshore areas where shallow depths distort wave forms. Although the wave form is transmitted by water particles, the water particles themselves are not significantly displaced in the process. They may move up and down or in a circular path as the wave passes, but they return to their approximate original positions.

Progressive Waves

There are two basic kinds of wave motion that can be observed in the ocean, progressive waves and standing waves. In **progressive waves** the wave motion transmits energy. The water in the wave crest possesses maximum potential energy. An object floating on the water surface undergoes the same motion that water particles at the surface do. In this example the floating object sits in the low part of the wave, the trough. As the crest approaches the object is drawn backward and upward and gains potential energy equivalent to its upward displacement. When the wave crest passes the object, it possesses maximum potential energy, then it moves forward and downward to complete its journey. The path described is a circular one, and the diameter depends on wave height. Water particles below the surface inscribe progressively smaller circles to a depth of approximately one-half the wave length (wave length is the distance from wave crest to wave crest). Below that point the water is relatively undisturbed.

In shallow water, less than a half-wave length deep, the water particles are subjected to bottom friction and move in ellipses rather than circles. The closer to the bottom, the flatter the ellipse. In deep water, waves move at a speed proportional to their length, causing longer waves to move faster. The formula is

$$V = \frac{gL}{2}$$

where V is velocity
g is the acceleration of gravity
L is wave length.

The speed of shallow water waves depends on the water depth. As progressive deep-water waves move into shallow water both the speed of the waves and their length decrease. Since momentum must be conserved, the height of the wave increases at a rate proportional to the decrease in length. It is this increase in wave height accompanying the decrease in wave length that causes wind waves to become unstable and break near shore and which causes the great damage by larger catastrophic waves.

Standing Waves

Standing waves occur in semi-enclosed basins where the width of the basin is a multiple of wavelength. Under these circumstances the waves which develop do not move progressively. Rather, the water surface oscillates or moves up and down. The effect is similar to what happens when children play on a seesaw. The board tilts first in one direction, then another, and the children move up and down. If they push upward as the board starts upward, it moves higher. Likewise, if new energy impulses are added to a basin containing a standing wave at just the right time, the effect is cumulative and the wave grows higher. In a similar fashion the wave can be destroyed if the energy impulse occurs at the wrong time. Standing waves are familiar to us in music also. Guitars, fiddles, and other such instruments have chambers designed to produce standing waves so that the vibrations of the strings add energy impulses periodically and the sound is amplified. The point about which the wave oscillates is called the node, and a standing wave may have one or several nodes. Standing waves in the ocean are caused by storm surges, sudden barometric pressure changes, earthquakes, tides, and other energy inputs.

Wind Waves

When wind blows across a water surface, friction between the wind and the water surface causes ripples, called **capillary waves**, to develop on the water surface. These waves create a rougher surface that makes the wind even more effective in pushing water forward. The continued input of energy by the wind has a cumulative effect which causes the wave to grow taller. The amount of energy, and therefore, the height of the wave, is dependent on wind speed and duration, water depth, and **fetch** (the distance over which it blows). For a given wind speed and duration, however, there is a limit to the height of waves that are produced. A condition of dynamic equilibrium is attained.

Waves generated in windy or stormy areas move outward as progressive deep-waves. Because the speed with which progressive deep-waves travel depends on their wave length, the higher, longer waves move faster than the slower, shorter ones. By this process of wave dispersion, waves sort themselves by size as they move, producing clusters of uniformly sized waves.

Bottom friction causes waves to slow down as they enter shallow coastal zones. Most waves approach coastlines at an angle, and the portions that enter shallow water first slow down first. This causes the wave front to bend, or refract, and align the waves somewhat with the shoreline, although they still do not normally attain a perpendicular alignment. Along irregular coasts with bays and headlands wave energy is concentrated on the headlands while the bays receive least. Thus, headlands are zones of excess energy and erosion, while bays are zones of energy deficiency and accumulation.

As wind waves become shorter and steeper near the coast they also become top heavy and unstable. Waves in shallow water break toward the shore, either because the top part moves faster than the bottom part which dissipates energy through bottom friction, or because onshore winds may blow the oversteepened wave over.

There are several types of breaking waves, depending on the slope of the coast and the wavelength. As waves enter shallow waters the energy they possess becomes concentrated in a shorter, higher wave. This energy must either be dissipated as the wave breaks, or it must be transmitted to the shore as **swash**, the water that runs up onto the beach, or as overwash which washes completely over the beach. Wave height increases gradually for small waves and shallow coasts, and the energy is dissipated gradually as the crest of the wave spills down the front while it moves shoreward. For larger waves or steeper coasts wave height increases more quickly and the energy may be dissipated more quickly by a plunging breaker in which the top curls over leaving a trapped air pocket inside. On even steeper coasts (and on coasts with large waves) huge collapsing breakers may develop in which the top portion breaks completely over the lower portion. The work done by waves in coastal areas depends, in part, on the types of breakers that dominate the surf during different times of the year. Spilling breakers, for example, tend to entrain sand in the surf and move it shoreward to build gently sloping beaches whereas plunging or collapsing breakers and surging waves which break directly on shore may tend to erode beaches and produce steeper profiles. The surf is the nearly continuous belt of breaking waves that borders a shore.

Breaking waves deliver water and energy to the shore in the form of swash, the churning, turbulent motion that transports sand and water up onto the beach. Swash water returns to the ocean as backwash, a gravity flow that develops because of the beach slope. Incoming swash is directed perpendicular to the wave front and usually is at an angle to the coast. Backwash, on the other hand, is directed directly down-slope which is usually perpendicular to the coast. Debris which is moved back and forth by the swash and backwash is gradually moved along the coast. Material moved in this manner is called drift, and the current that develops in response to this net force is called a **longshore** or **littoral current**.

Breaking waves generally transport more water to shore than can return as backwash. Thus, some means of transporting this excess water back to the oceans is a necessity if equilibrium conditions are to be maintained in coastal areas. This excess water is carried back to the ocean by **rip currents** which are narrow zones where water moves seaward at velocities of as high as 6 km per hour (3.7 mph). Rip currents, which are often called undertows, may be 25 m (82 feet) wide and may extend as far as 1,000 m (3,280 feet) from shore. They are frequently distinguishable by their discoloration from sediment or foam.

Catastrophic Waves

Catastrophic waves are generated when unusual events input enormous amounts of energy into the ocean, as may happen with earthquakes, volcanic eruptions, faulting, or landslides. Such catastrophic waves are called **tsunamis**, and they behave as shallow sea waves.

Their wavelengths vary from 100 to 700 km (62 to 430 miles), and in the deep ocean they may move at speeds of up to 740 km per hour (460 mph). The damage done by tsunamis results from the speed at which they travel and the height attained by the waves in shallow water. A tsunami may travel from Alaska to Hawaii in 5 hours, and as the wave is slowed in the shallow coastal waters its height may increase by a factor of ten or more, changing a harmless 30 centimeter (one foot) wave into a 30 to 45 meter (100 to 150 foot) monster that is capable of doing millions of dollars of damage.

Tides

Tides are undulations in the ocean surface caused by the gravitational attraction of the moon, sun and other planets. Tides affect not only the ocean but all of the earth. However, the effects are most noticeable with the ocean.

The major energy component in the tidal system is the moon. As the moon circles the earth the two bodies pull on each other. Since the earth and moon do not move closer together, there must be some force that exactly balances the attractive forces. That force is the centrifugal force that arises from the rotation of the earth-moon pair about their common center of gravity, which lies inside the earth's sphere. Because the earth rotates as it revolves around this center, the centrifugal force which is produced is the same everywhere on earth. This is not true for the force resulting from the gravitational attraction of the moon. That force is greater on the side nearest the moon, and least at the point on the surface of the earth that is exactly opposite the moon. The earth's gravity acts toward the center of the earth and attracts all surface locations approximately equally. Addition of all the force vectors yields the resultant force vectors which are called the tractive force (Figure 17.2). **Tractive forces** are the net forces which act on water at each latitude. The water particles respond by moving in the direction of the arrows, causing a slight bulge at the point on either side of the earth underneath the moon.

Because of their wavelength (one-half circumference of the earth) tides move as shallow waves even in the middle of the ocean. Unlike deep-water waves whose velocity is a function of wavelength, the velocity of a shallow wave is determined by the depth of the water. In mid-ocean the wave height is very minor, but in shallow coastal areas wave height is increased so that the ebb and flow of tides may cause water levels to vary from less than a meter (three feet) to more than 15 meters (49 feet). The exact range depends on the configuration of the coast.

When the moon is in the plane of the ecliptic, both high tides and both low tides are the same. High tides are spaced 12 hours and 25 minutes apart, and the semi-diurnal pattern they produce is called the equatorial tide. Because of its orbit which is tilted 5°9' with respect to the plane of the ecliptic, the moon is in the equatorial position only twice each lunar month (29.53 days). Recalling that the earth itself is tilted 23.5° to the plane of the ecliptic, the moon declination varies from 28°45' on either side of the equator. Thus, when the moon is at its maximum declination the high tides are unequal. This tidal pattern is called a **tropical tide**.

Actual tidal patterns may be vastly different from these idealized forms. Some coastal areas have virtually no tidal range at all, while others have, for all practical purposes, a single daily tide. In addition, the effects of the sun and the other planets are superimposed on the lunar tides to produce many complex tidal configurations.

Tidal Currents

Tidal ranges and tidal currents depend upon the configuration of coastal areas. Seemingly tides ought to flood (rise) for 12 hours and 25 minutes and ebb (fall) for an equal time, and the range between high and low tides ought to be a function of water depth. In actual fact the situation is more complicated, for in addition to simple wave components the tide may possess properties of standing waves or progressive waves. Depending on the geometry of the ocean basin and the configuration of the coast, the tidal range may vary from barely noticeable to several meters, reaching its greatest extremes in funnel-shaped bays such as the Bay of Fundy with a tidal range of more than 15 meters.

Tidal currents depend largely upon the tidal range and the size of the opening that the water must flow through. The highest currents may reach 3 cm per second (6 ft/sec), and generally occur in narrow constrictions which separate semi-enclosed estuaries from open ocean. Such currents are a significant erosional force and may be capable of scouring channels and transporting sediment several miles seaward. Estuaries, in a broad sense, are tidal basins which contain saline

Figure 17.2 Centrifugal Vectors and Gravitational Vectors = Tractive Forces That Cause Tides

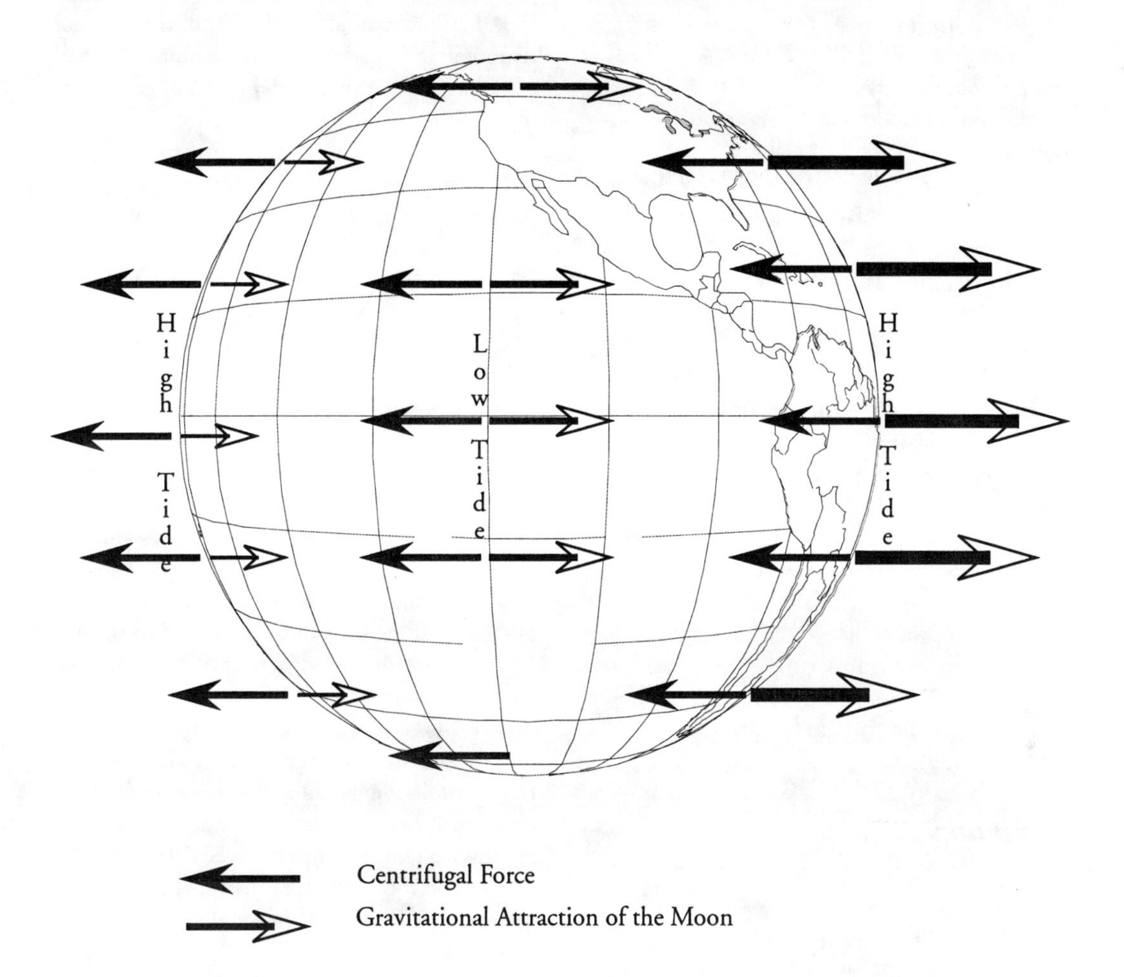

water and which are semi-enclosed and connected to the open sea. Because streams discharge fresh water into the estuaries there is more water to flow out on the ebb tide than flows in on the flood tide, so ebbing flow usually lasts longer.

Tidal currents in estuaries may be very complicated, particularly if the estuary is large and irregularly shaped. In the Chesapeake Bay, for example, the tidal current moves upstream as a progressive wave at a speed of 60 kmph (37 mph). In funnel-shaped bays and river mouths the tidal wave is steepened and may become asymmetric, moving upstream as a wall of water called a **tidal bore**.

In most estuaries the less dense fresh water discharged by rivers spreads out over the salt water below it. The sharp density change between the salt and fresh water retards mixing. The seaward-moving surface currents pull some salt water upward and carry it seaward. Salt water flowing in to replace the lost salt water sets up a current along the bottom, and the amount of salt water flowing in is greater than the amount of fresh water flowing out.

Summary

Like the atmosphere the ocean is a thermodynamic system driven by energy from the sun. It is also a biological system that derives its energy from the sun and its nutrients from the salts, sediments, and gases that are dissolved or suspended in ocean water.

The age of the ocean is a matter of some debate, but it is generally agreed that the modern-day ocean basins are less than 200 million years old, and that they have been formed through continental rifting and continual sea floor spreading.

Temperature, salinity, pressure, viscosity and density are the major properties of the physical ocean system, they are the variables of the thermodynamic system, and they exhibit temporal and spatial patterns that are the result of variations in inputs and outputs of energy and mass.

The biological system consists of the phytoplankton and other plants that carry out photosynthesis, and both herbivorous and carnivorous species of zoo plankton, fish and other types of animal life. Productivity is highest in surface waters which are high in nutrients and which receive large concentration of solar radiation.

The water budget indicates that the inputs and outputs of the 445,000 km^3 of water that cycles through the atmosphere each year are unevenly distributed. Rapid evaporation occurs in the tradewind zones and minimal amounts evaporate at the poles. Inputs of fresh water are greatest along the equator, around the margins of the continents, and in polar areas. The variability in input and output of cyclic water is the major factor responsible for variations in surface salinity.

Ocean heat budgets vary with latitude. High insolation within the tropical latitudes along with heat released by condensation causes the surface temperatures between 20°N and S to be high, 25°C (77°F) or more. Temperatures decline rapidly toward the subpolar areas where average temperatures fall to 0°C. Unlike land areas ocean surface temperatures show only moderate seasonal viability. They vary little in tropical and polar areas, and normally vary less than 10°C (18°F) between winter and summer in the midlatitudes.

Rivers and glaciers are the source of most oceanic sediments. Consequently, suspended concentrations are high only in the vicinity of the continental margins, and most sediments are deposited near their sources. Dissolved silica and calcium carbonate are removed from solution by marine organisms, and these sediments accumulate on the ocean floor when the organisms die.

Oxygen, nitrogen, and carbon dioxide are the most abundant gases in the ocean. Saturation levels vary for different gases and with temperature, salinity and pressure. Oxygen levels are highest just below the surface, decline gradually between 200 and 800 meters, and increase below that point. Carbon dioxide concentrations are relatively constant, varying between 45 and 54 ml/l everywhere. Nitrogen levels are almost totally controlled by temperature, pressure and salinity.

The circulation systems in the ocean depend on inputs of energy from winds, solar energy, and gravitational energy. The basic surface circulation system closely parallels atmospheric circulation. Closed circulation cells are well developed in each of the major ocean basins in the vicinity of the permanent atmospheric pressure systems. The major gyres are associated with the subtropical highs and with smaller cells associated with the subpolar lows. Some seasonal variability is evident, especially in the Indian Ocean. The poleward moving currents found on the western sides of the ocean basins are mostly warm currents while those moving toward the equator on the eastern sides are cold currents. Coriolis force deflects the cold currents away from the continents causing cold water to upwell along these coasts.

Mechanical energy is exchanged between the atmosphere and the ocean when wind friction creates waves. In the open oceans the longer a wave is, the faster it moves, so waves approaching coastal areas tend to be of uniform size and spacing. As waves enter shallow coastal waters they slow down, increase in height, and break toward shore. Breaking waves create swash which runs up onto the beach and which returns as backwash. When waves strike the shore on angle, the back-and-forth motion of swash and backwash produces a longshore current and moves sand along the beach.

Tides are shallow-water waves produced by the combined forces of centrifugal force created by rotation of the earth-moon pair, the gravitational force of the earth, and the gravitational attraction of the bodies in the solar system. Theoretically, high tides are experienced twice each day (12 hours and 25 minutes apart) with low tides occurring in between. Actual tidal patterns and the length of time that waters flood and ebb depend on coastal conditions. The currents that are generated by flooding and ebbing tides are a function of the size of the opening the water must pass through and the volume of water that moves through the opening.

This chapter on oceans concludes our discussion of the hydrosphere. In the three preceding chapters we have examined hydrological processes and patterns with a twofold purpose in mind. On the one hand, we have tried to emphasize the unity of the whole system by examining the flows of energy and water which tie the subsystems together. On the other hand, we have stressed the subsystems themselves, examining each one individually and trying to understand how they function and how they are related to the other earth systems.

Appendix

A

Atlas of Placenames and Regions

Maps are essential tools in meteorology, climatology, and physical geography. They are basic tools for analysis and a valuable aid in the presentation of information. In most physical geography classes, students are confronted with numerous maps and place names. Extensive use is made of thematic maps. In the typical geography class, the instructor — either by sweeping motions of the hand or pointer — makes frequent reference to wall maps, chalkboard outlines, slide or overhead map images. Several regional maps are presented here as a convenient reference for students with less than perfect mental maps. We encourage you to use these maps in class, to organize your review materials, or just as study aids.

Map 1 North America - Political

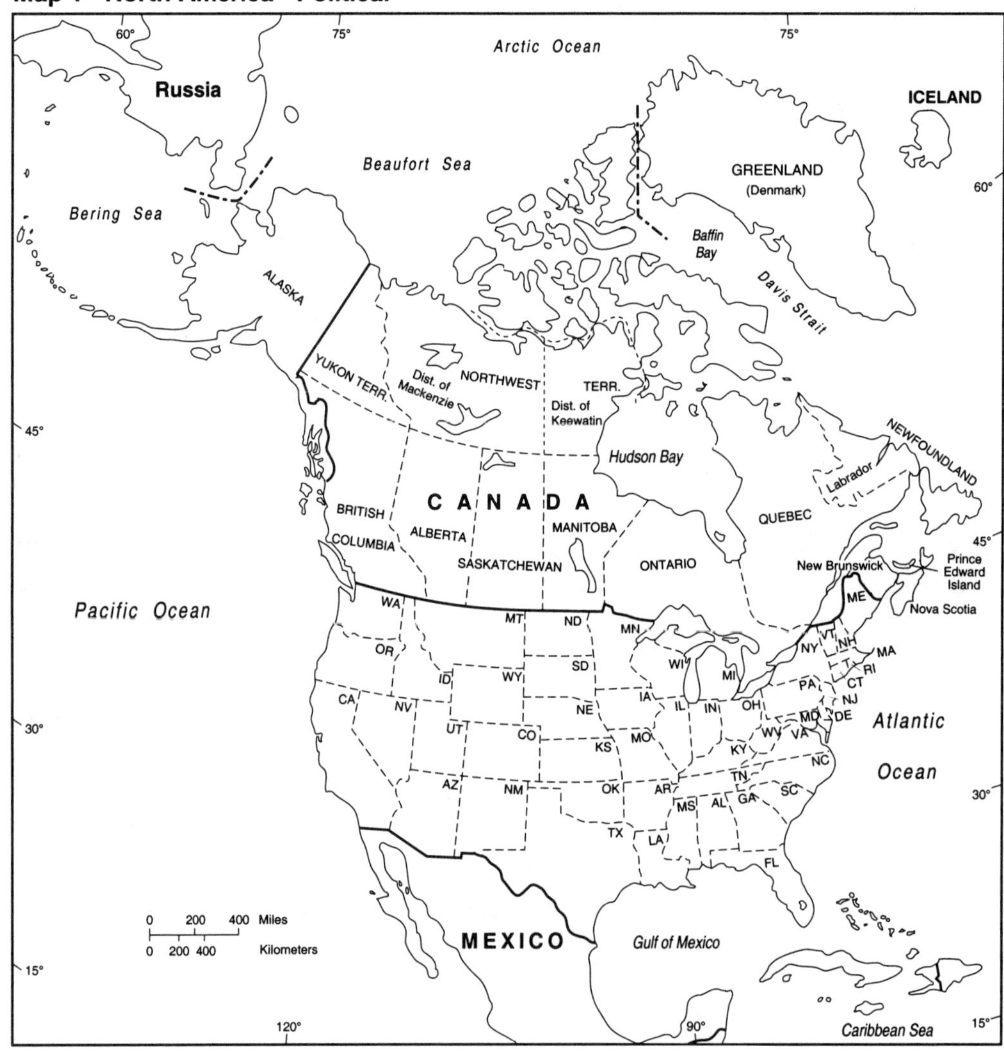

Map 2 North America - Physical

Map 3 United States

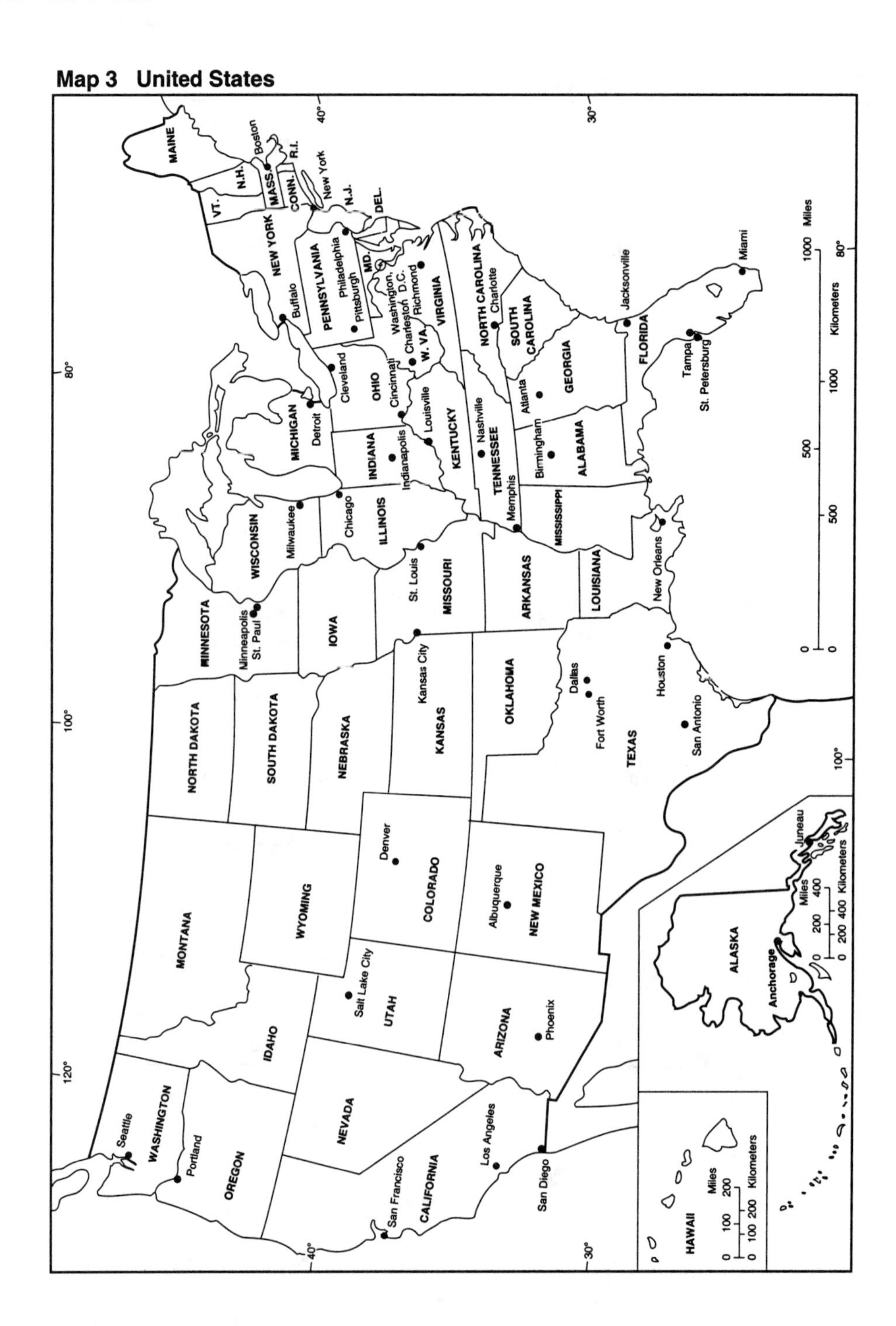

Map 4 Europe

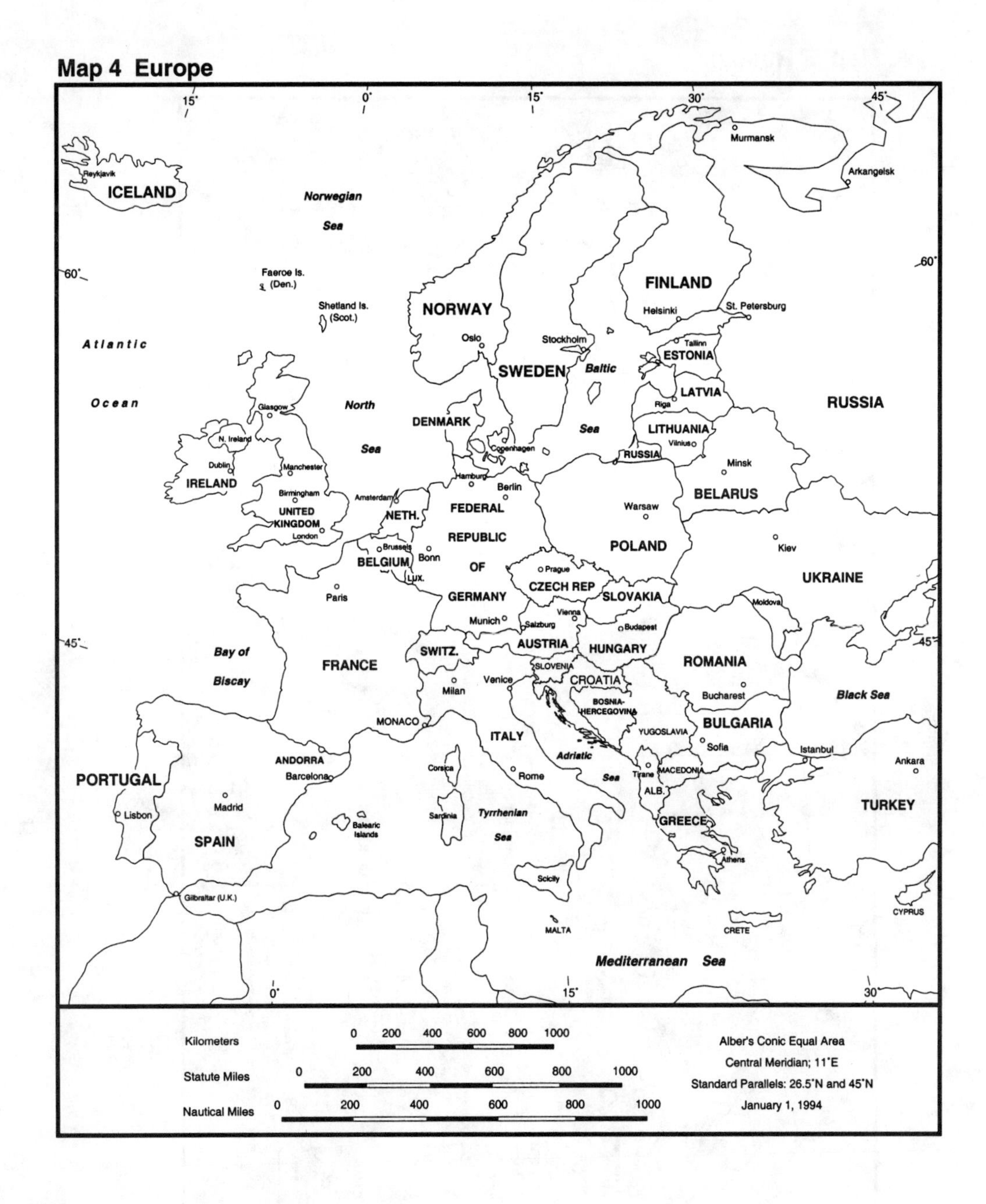
ICELAND
Reykjavik
Norwegian Sea
Faeroe Is. (Den.)
Shetland Is. (Scot.)
Atlantic Ocean
North Sea
NORWAY
Oslo
SWEDEN
Stockholm
Baltic Sea
FINLAND
Helsinki
St. Petersburg
Murmansk
Arkangelsk
RUSSIA
ESTONIA
Tallinn
LATVIA
Riga
LITHUANIA
Vilnius
RUSSIA
BELARUS
Minsk
DENMARK
Copenhagen
IRELAND
Dublin
N. Ireland
Glasgow
Manchester
Birmingham
UNITED KINGDOM
London
Amsterdam
NETH.
BELGIUM
Brussels
LUX.
Hamburg
Berlin
Bonn
FEDERAL REPUBLIC OF GERMANY
Munich
POLAND
Warsaw
Kiev
UKRAINE
Moldova
Prague
CZECH REP
SLOVAKIA
Vienna
Salzburg
AUSTRIA
Budapest
HUNGARY
ROMANIA
Bucharest
Paris
FRANCE
Bay of Biscay
SWITZ.
Milan
Venice
SLOVENIA
CROATIA
BOSNIA-HERCEGOVINA
YUGOSLAVIA
BULGARIA
Sofia
Istanbul
Black Sea
Ankara
TURKEY
MONACO
ITALY
Adriatic Sea
Rome
Tirane
MACEDONIA
ALB.
GREECE
Athens
ANDORRA
Barcelona
PORTUGAL
Lisbon
Madrid
SPAIN
Corsica
Sardinia
Balearic Islands
Tyrrhenian Sea
Sicily
MALTA
CRETE
CYPRUS
Gibraltar (U.K.)
Mediterranean Sea
60°
45°
15°
0°
30°
Kilometers
0 200 400 600 800 1000
Statute Miles
0 200 400 600 800 1000
Nautical Miles
0 200 400 600 800 1000
Alber's Conic Equal Area
Central Meridian; 11°E
Standard Parallels: 26.5°N and 45°N
January 1, 1994

Map 5 Russia

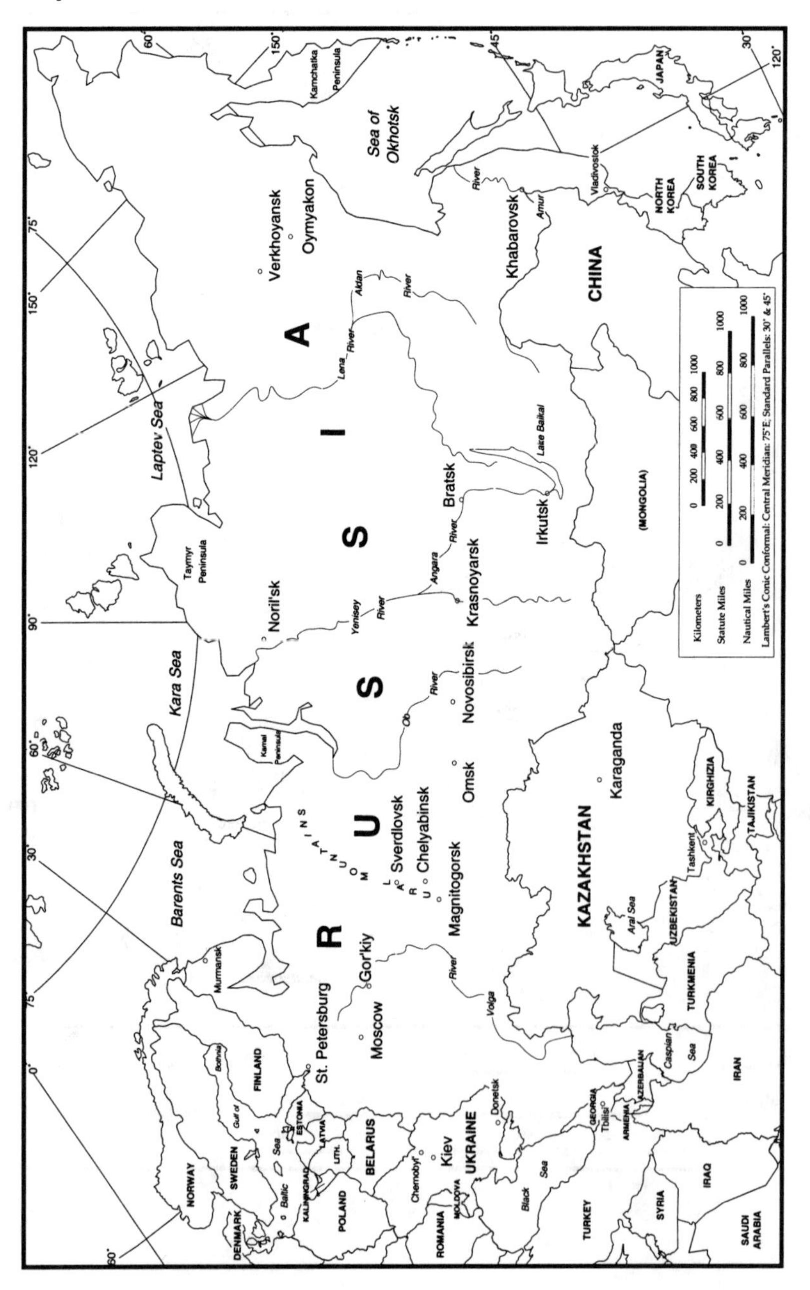
RUSSIA
ASIA
URAL MOUNTAINS
Barents Sea
Kara Sea
Laptev Sea
Sea of Okhotsk
Kamchatka Peninsula
Taymyr Peninsula
Kamal Peninsula
Verkhoyansk
Oymyakon
Khabarovsk
Vladivostok
Noril'sk
Bratsk
Irkutsk
Krasnoyarsk
Novosibirsk
Omsk
Sverdlovsk
Chelyabinsk
Magnitogorsk
Gor'kiy
St. Petersburg
Moscow
Murmansk
Kiev
Chernobyl'
Donetsk
Tbilisi
Tashkent
Karaganda
Lena River
Aldan River
Amur River
Angara River
Yenisey River
Ob River
Volga River
Lake Baikal
Aral Sea
Caspian Sea
Black Sea
Baltic Sea
Gulf of Bothnia
CHINA
(MONGOLIA)
JAPAN
NORTH KOREA
SOUTH KOREA
KAZAKHSTAN
UZBEKISTAN
TURKMENIA
KIRGHIZIA
TAJIKISTAN
IRAN
IRAQ
SYRIA
SAUDI ARABIA
TURKEY
GEORGIA
ARMENIA
AZERBAIJAN
UKRAINE
MOLDOVA
ROMANIA
BELARUS
POLAND
LITH.
LATVIA
ESTONIA
KALININGRAD
FINLAND
SWEDEN
NORWAY
DENMARK
Kilometers 0 200 400 600 800 1000
Statute Miles 0 200 400 600 800 1000
Nautical Miles 0 200 400 600 800 1000
Lambert's Conic Conformal: Central Meridian: 75°E; Standard Parallels: 30° & 45°

Map 6 Asia

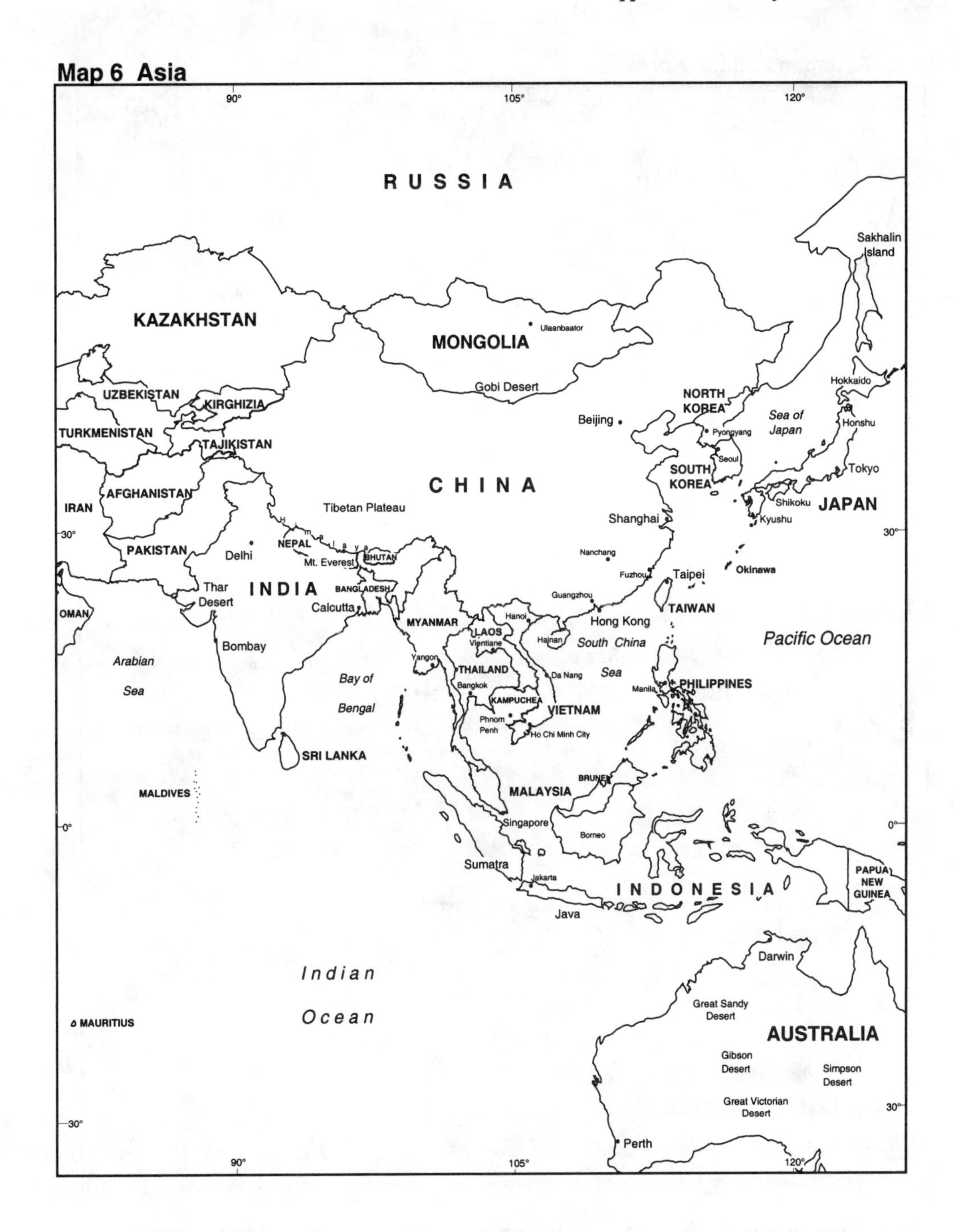

90°
105°
120°
RUSSIA
Sakhalin Island
KAZAKHSTAN
MONGOLIA
Ulaanbaatar
Gobi Desert
Hokkaido
UZBEKISTAN
KIRGHIZIA
NORTH KOREA
Sea of Japan
TURKMENISTAN
Beijing
Pyongyang
Honshu
TAJIKISTAN
Seoul
SOUTH KOREA
Tokyo
CHINA
AFGHANISTAN
IRAN
Tibetan Plateau
Shikoku
JAPAN
Shanghai
Kyushu
30°
NEPAL
PAKISTAN
Delhi
Nanchang
BHUTAN
Mt. Everest
Okinawa
Fuzhou
Taipei
Thar Desert
INDIA
BANGLADESH
Guangzhou
Calcutta
TAIWAN
OMAN
MYANMAR
Hanoi
Hong Kong
LAOS
Vientiane
Hainan
South China Sea
Pacific Ocean
Bombay
Arabian Sea
Yangon
THAILAND
Da Nang
Bay of Bengal
Bangkok
Manila
PHILIPPINES
KAMPUCHEA
VIETNAM
Phnom Penh
Ho Chi Minh City
SRI LANKA
BRUNEI
MALDIVES
MALAYSIA
0°
Singapore
Borneo
Sumatra
Jakarta
INDONESIA
PAPUA NEW GUINEA
Java
Darwin
Indian Ocean
Great Sandy Desert
MAURITIUS
AUSTRALIA
Gibson Desert
Simpson Desert
Great Victorian Desert
-30°
Perth

Map 7 Japan and the Koreas

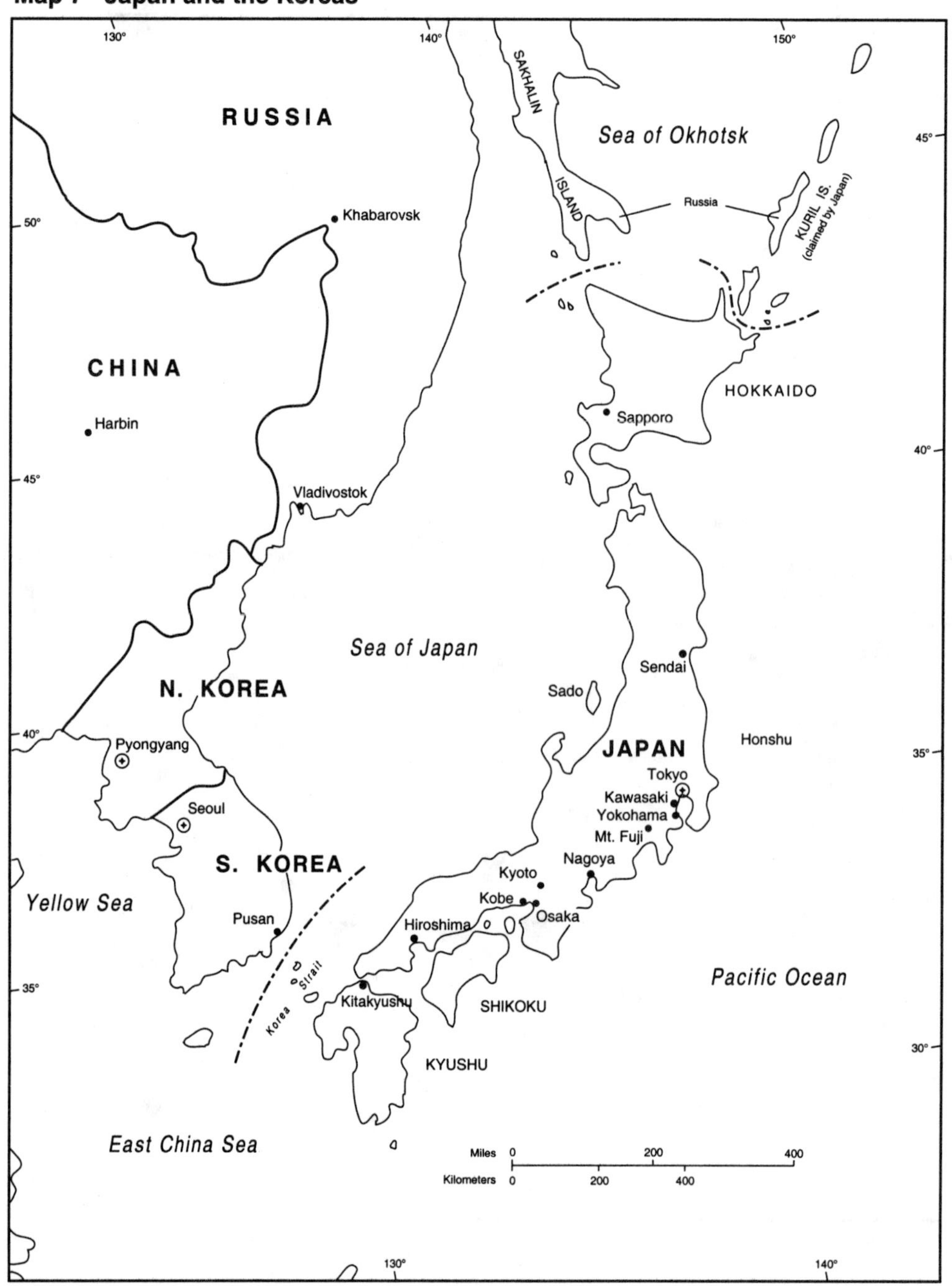

Map 8 Australia

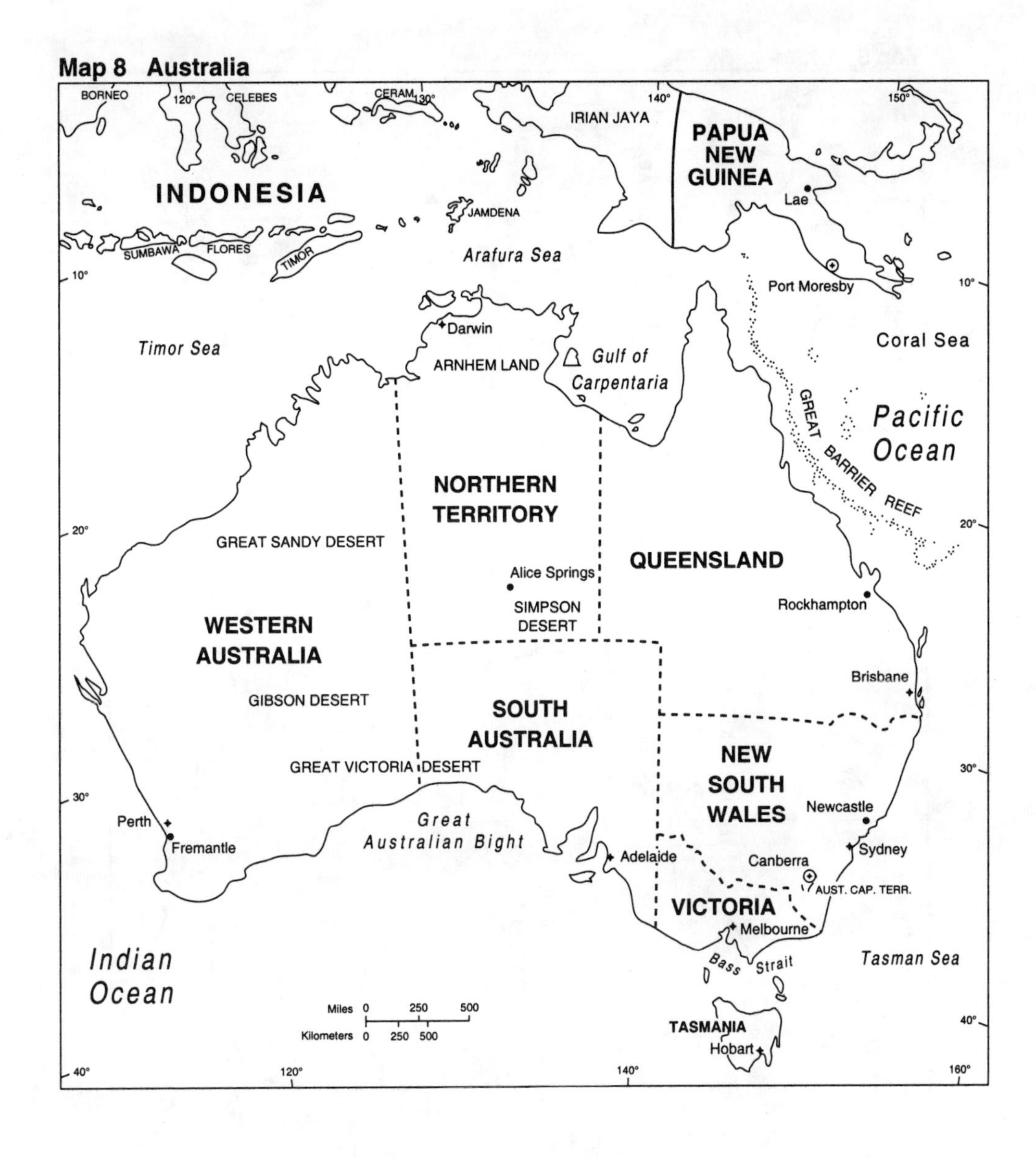

Map 9 China

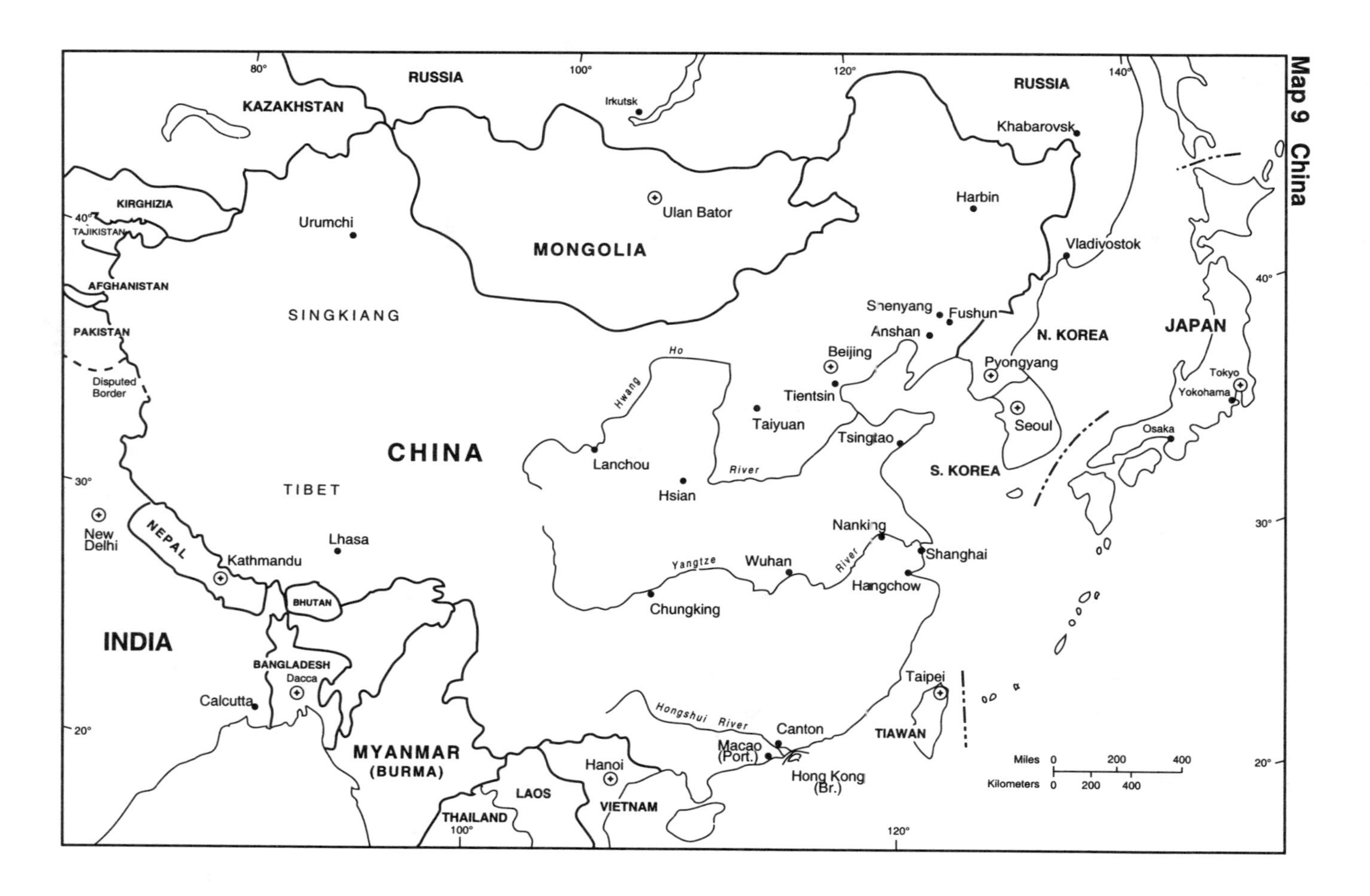

RUSSIA
KAZAKHSTAN
Irkutsk
RUSSIA
Khabarovsk
KIRGHIZIA
TAJIKISTAN
Urumchi
Ulan Bator
MONGOLIA
Harbin
Vladivostok
AFGHANISTAN
SINGKIANG
Shenyang
Fushun
Anshan
N. KOREA
JAPAN
PAKISTAN
Ho
Beijing
Pyongyang
Disputed Border
Hwang
Tientsin
Tokyo
Yokohama
Taiyuan
Seoul
Osaka
Tsingtao
CHINA
Lanchou
River
S. KOREA
Hsian
TIBET
Nanking
New Delhi
NEPAL
Lhasa
Shanghai
Kathmandu
Yangtze
Wuhan
River
Hangchow
BHUTAN
Chungking
INDIA
BANGLADESH
Dacca
Taipei
Calcutta
Hongshui River
Canton
TIAWAN
MYANMAR
(BURMA)
Macao
(Port.)
Hanoi
Hong Kong
(Br.)
Miles
0
200
400
Kilometers
0
200
400
LAOS
VIETNAM
THAILAND
80°
100°
120°
140°
40°
30°
20°

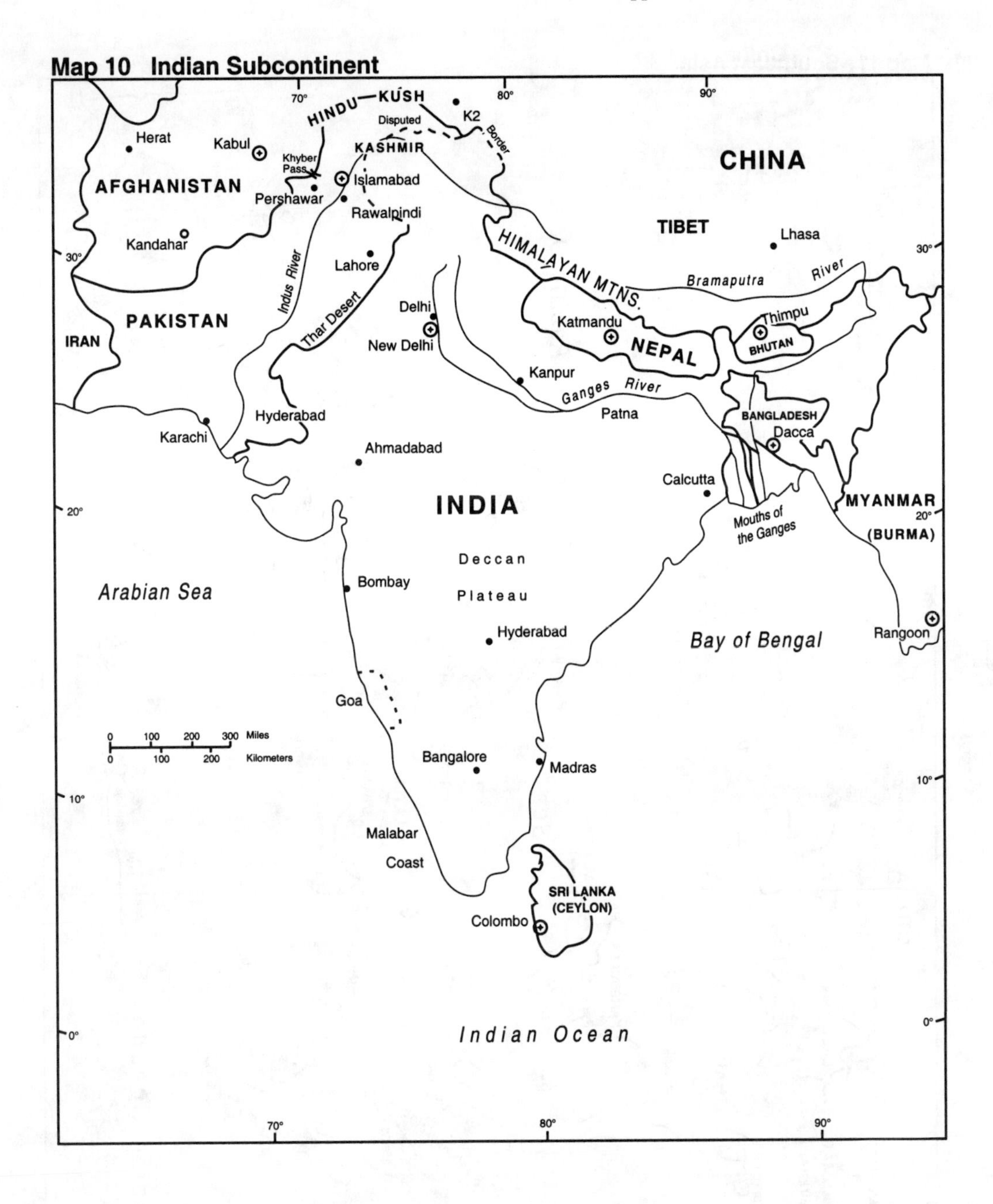
Map 10 Indian Subcontinent
HINDU—KUSH
K2
Disputed
Border
KASHMIR
CHINA
Herat
Kabul
Khyber Pass
Islamabad
AFGHANISTAN
Pershawar
Rawalpindi
TIBET
Lhasa
Kandahar
HIMALAYAN MTNS.
Lahore
Indus River
Bramaputra River
Thar Desert
Delhi
PAKISTAN
IRAN
New Delhi
Katmandu
NEPAL
Thimpu
BHUTAN
Kanpur
Ganges River
Patna
Hyderabad
Karachi
BANGLADESH
Dacca
Ahmadabad
Calcutta
INDIA
MYANMAR
(BURMA)
Mouths of the Ganges
Deccan
Plateau
Arabian Sea
Bombay
Hyderabad
Bay of Bengal
Rangoon
Goa
Miles
Kilometers
Bangalore
Madras
Malabar
Coast
SRI LANKA
(CEYLON)
Colombo
Indian Ocean

Map 11 Southeast Asia

Map 12 Middle East

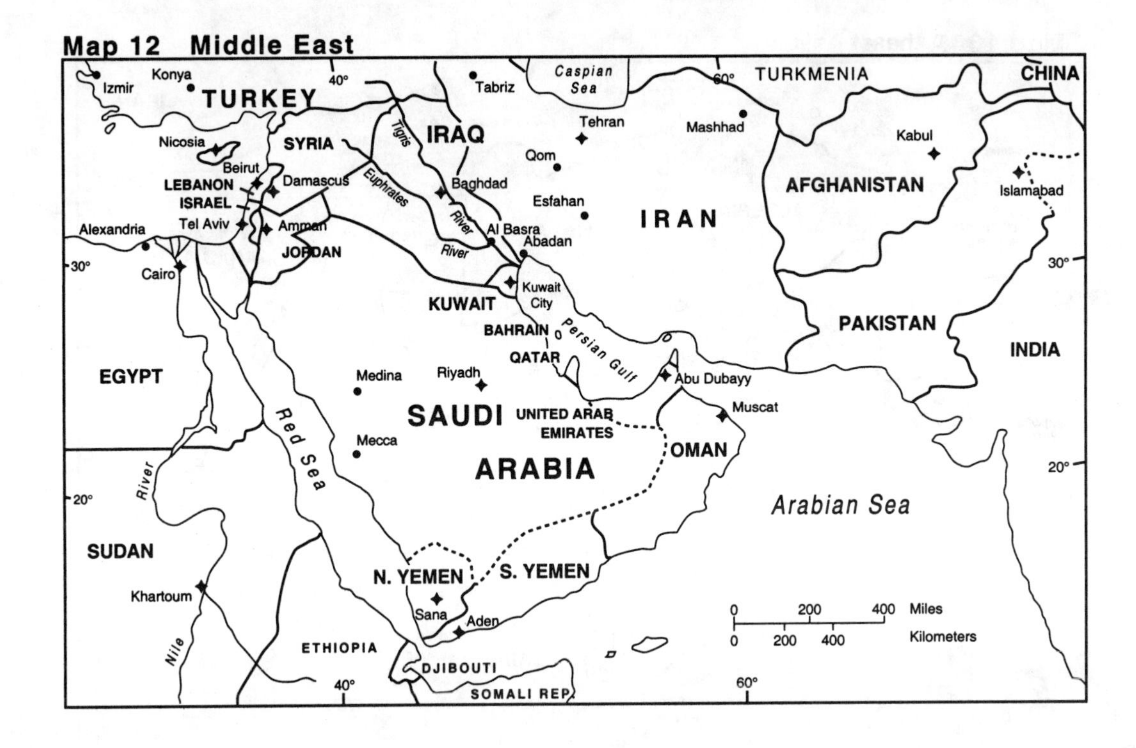

Map 13 Africa

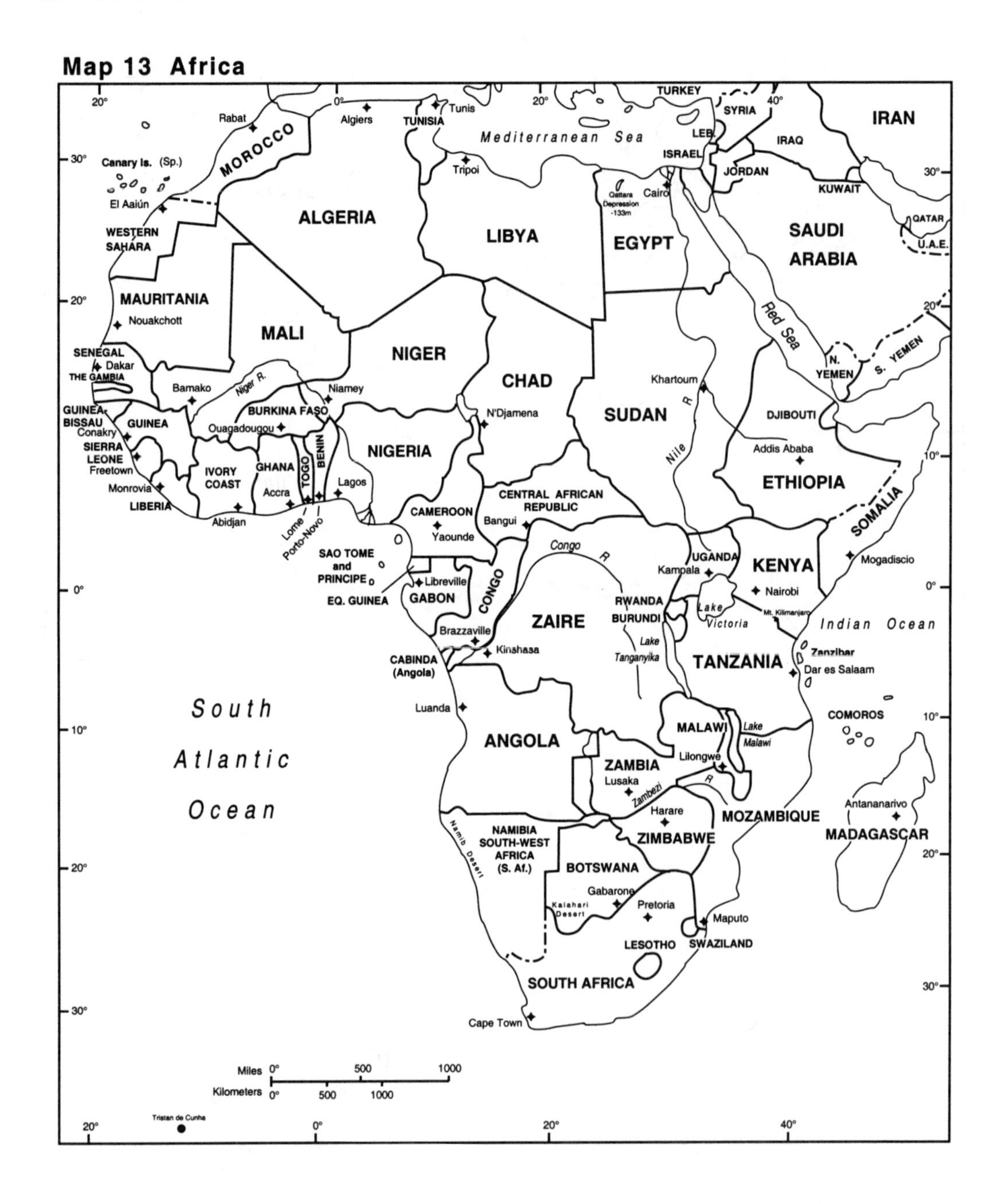

Map 14 Mexico

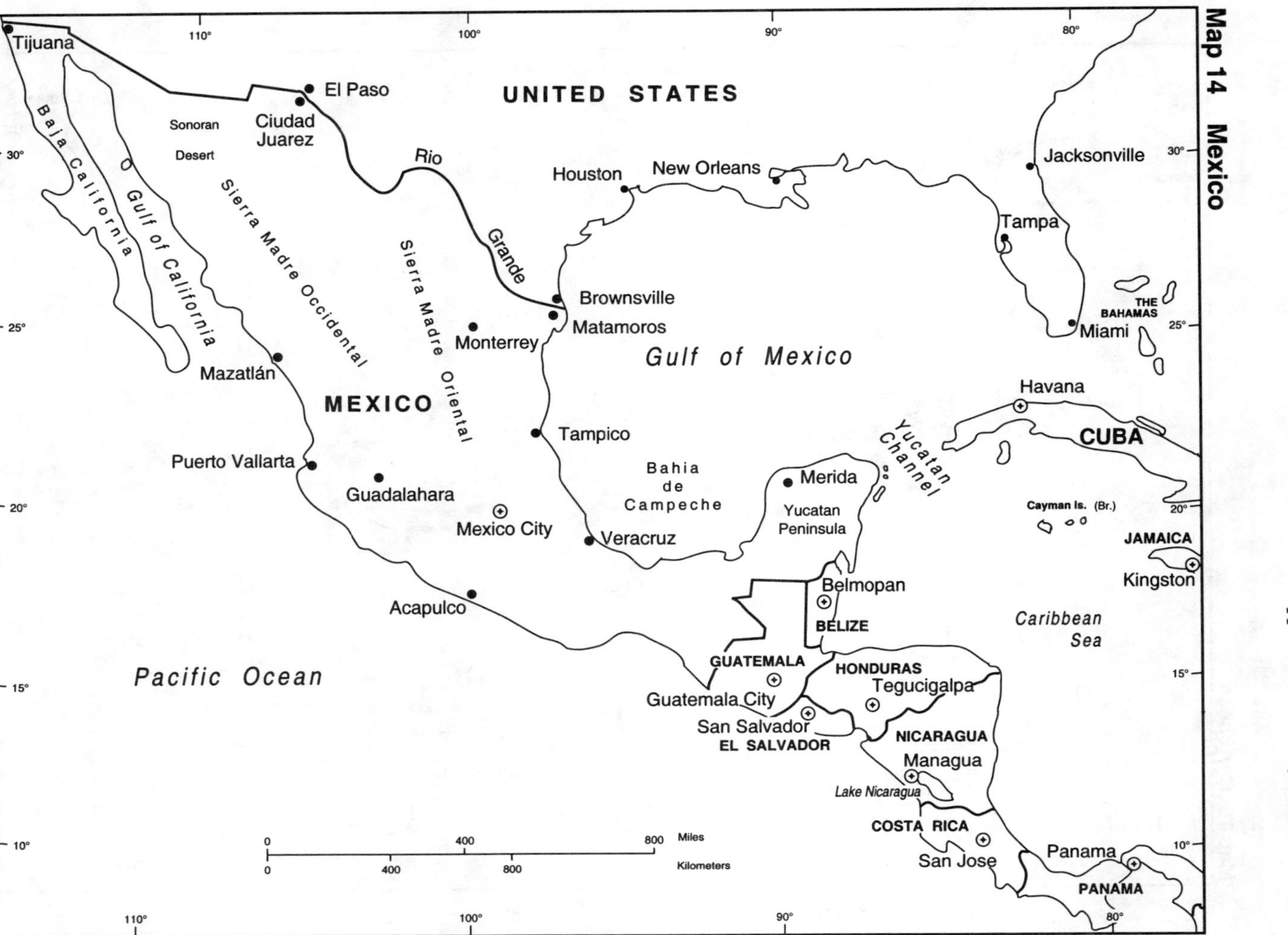

Map 15 Caribbean

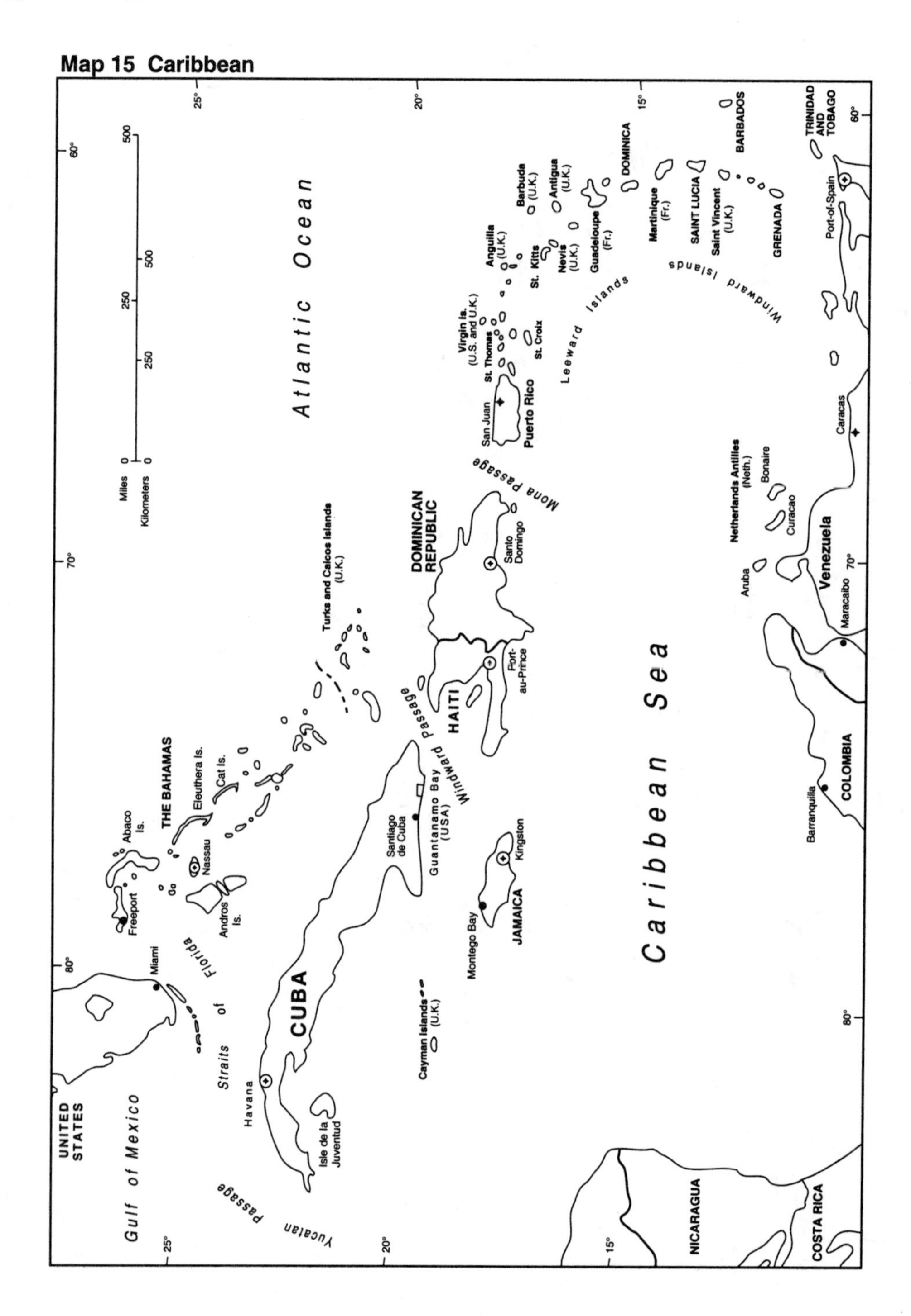

Map 16 South America

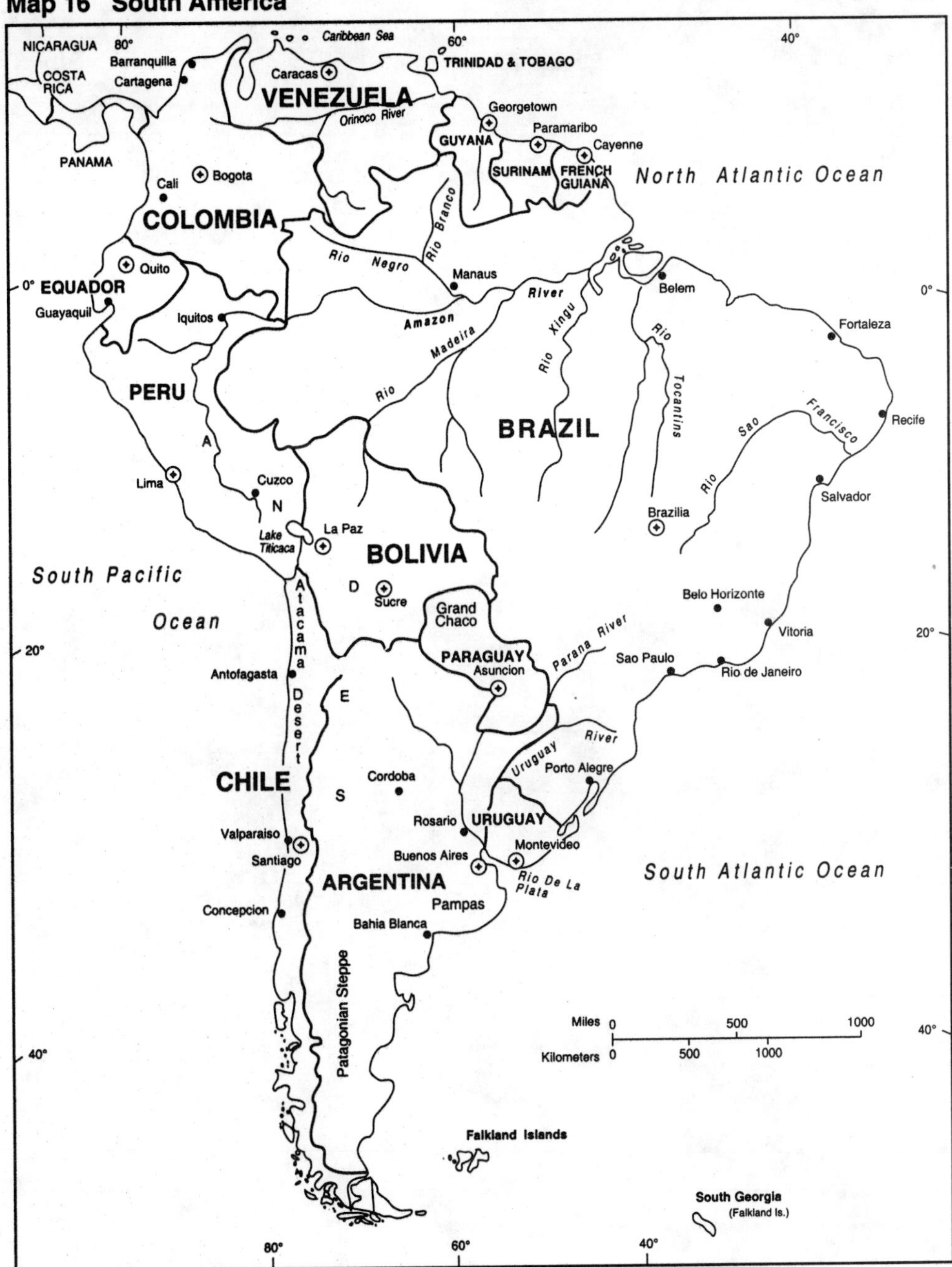

Appendix B: Outline Maps

North America

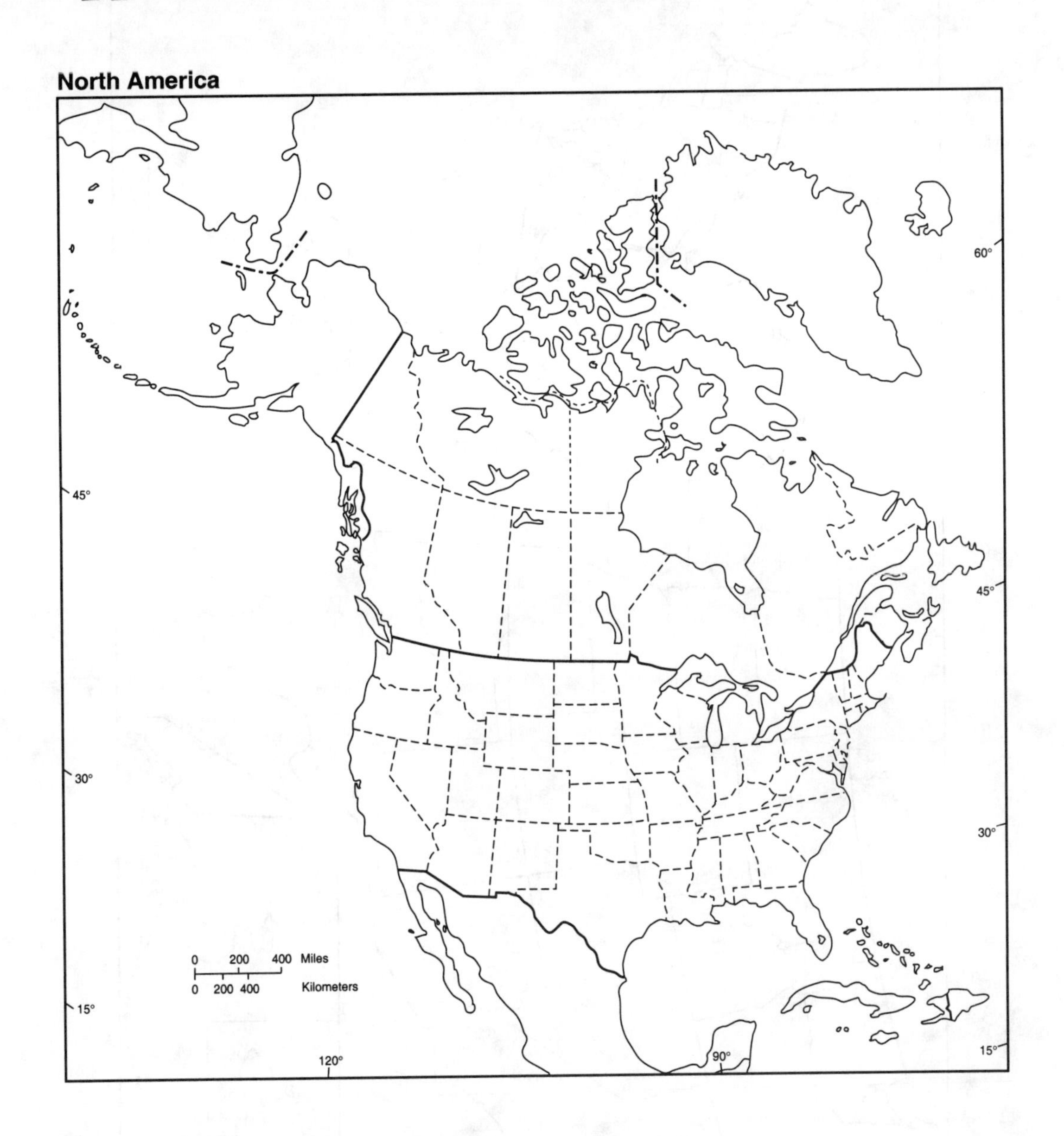

United States

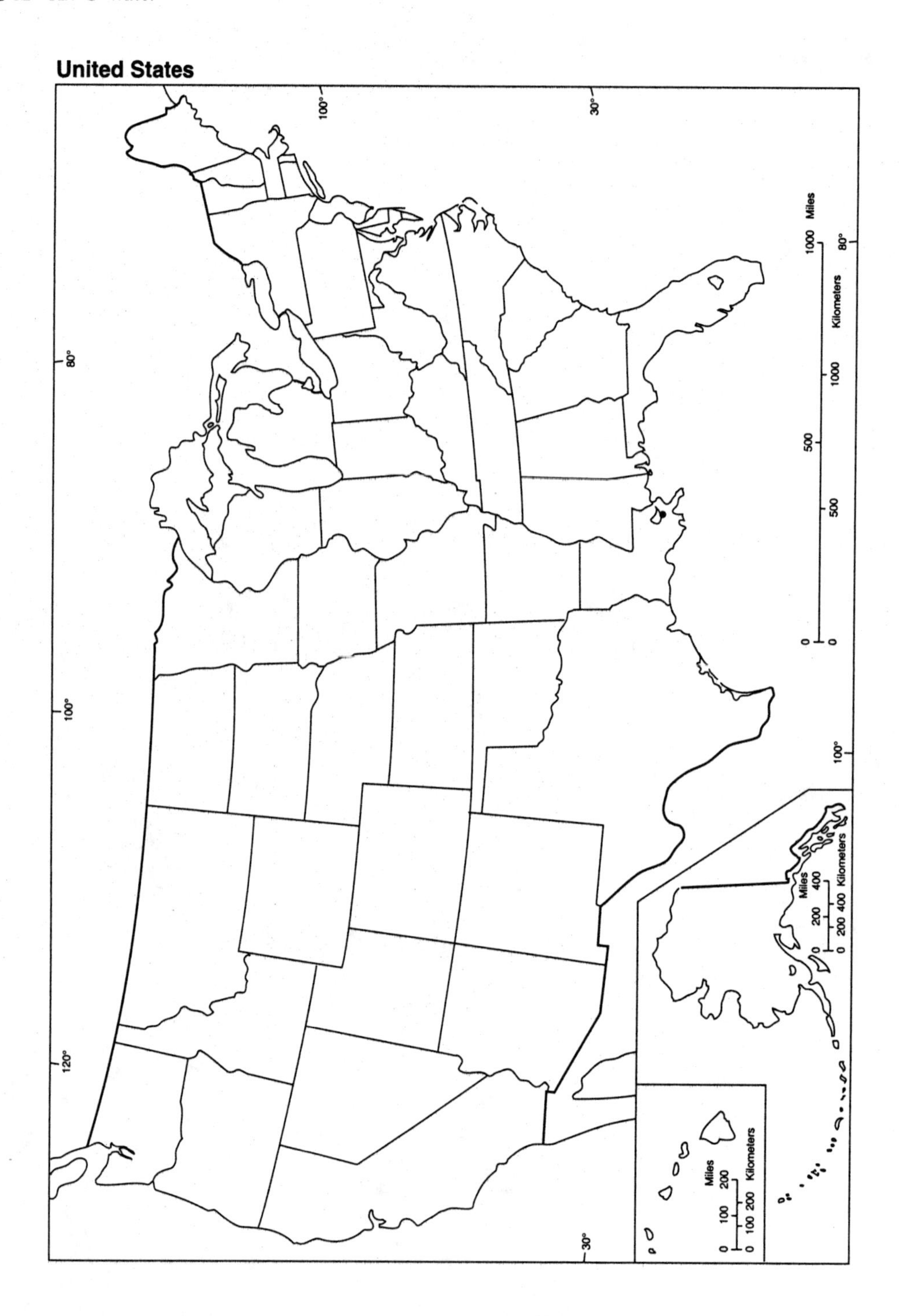

United States Metropolitan Areas

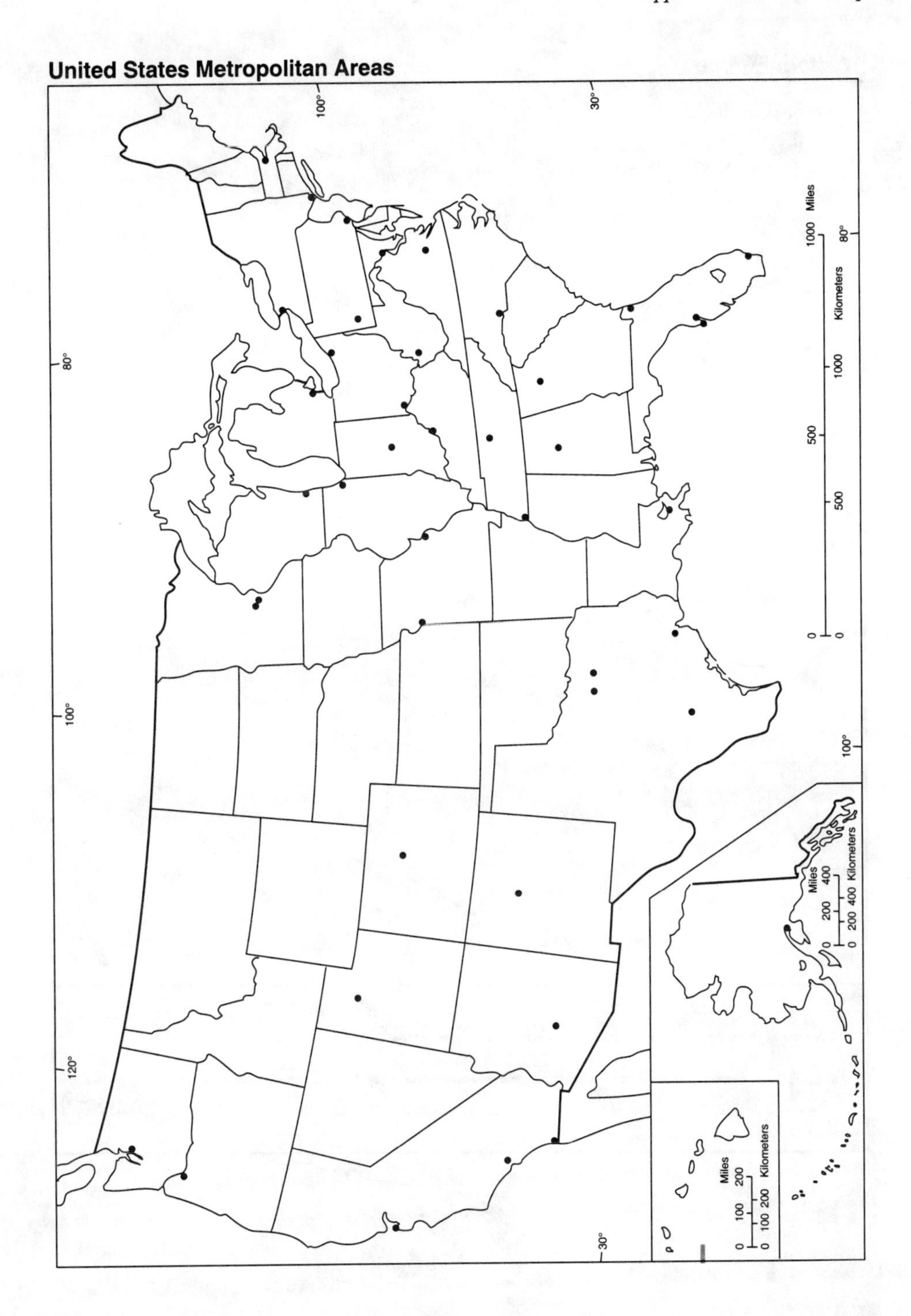

Western Europe

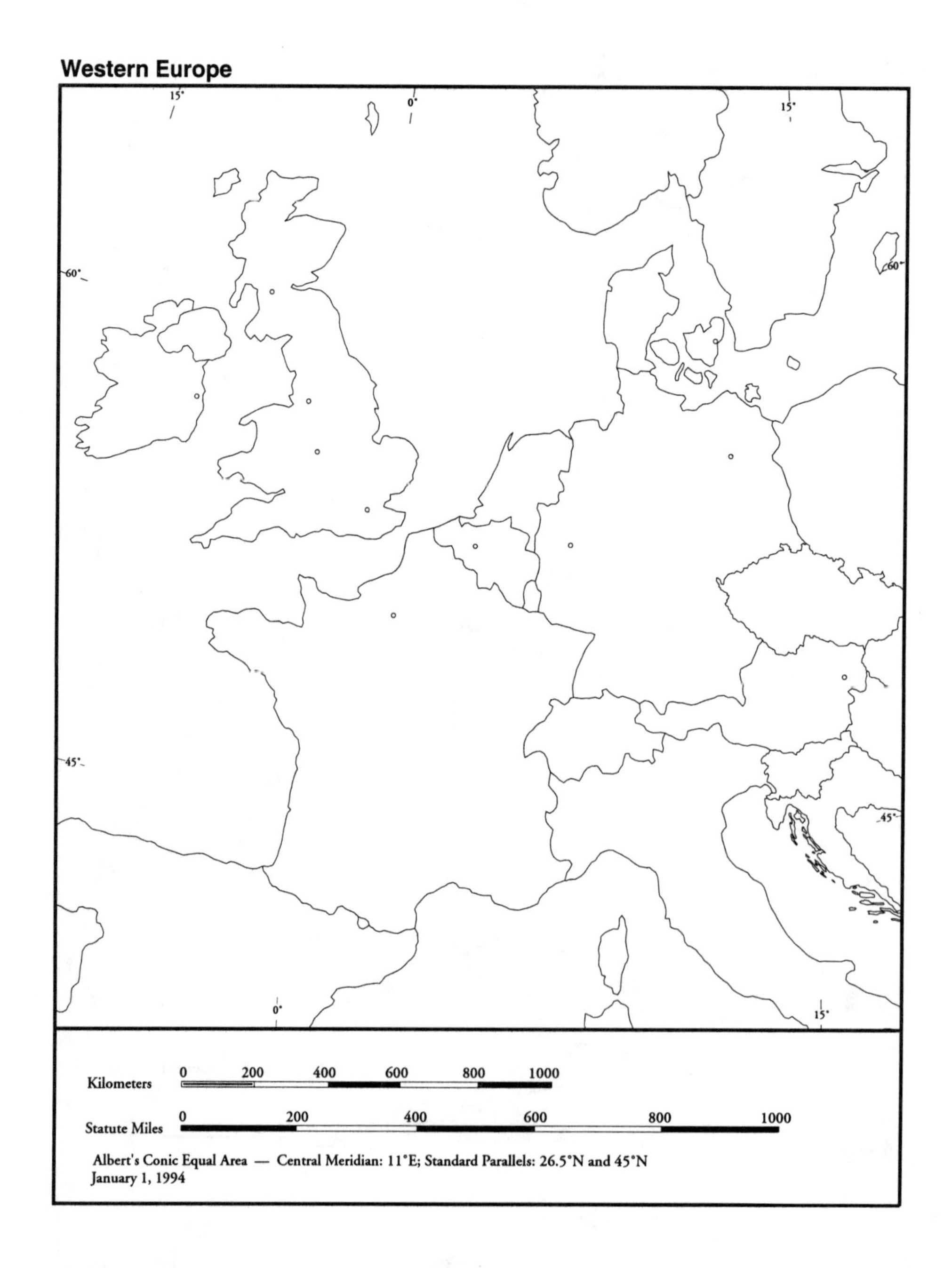

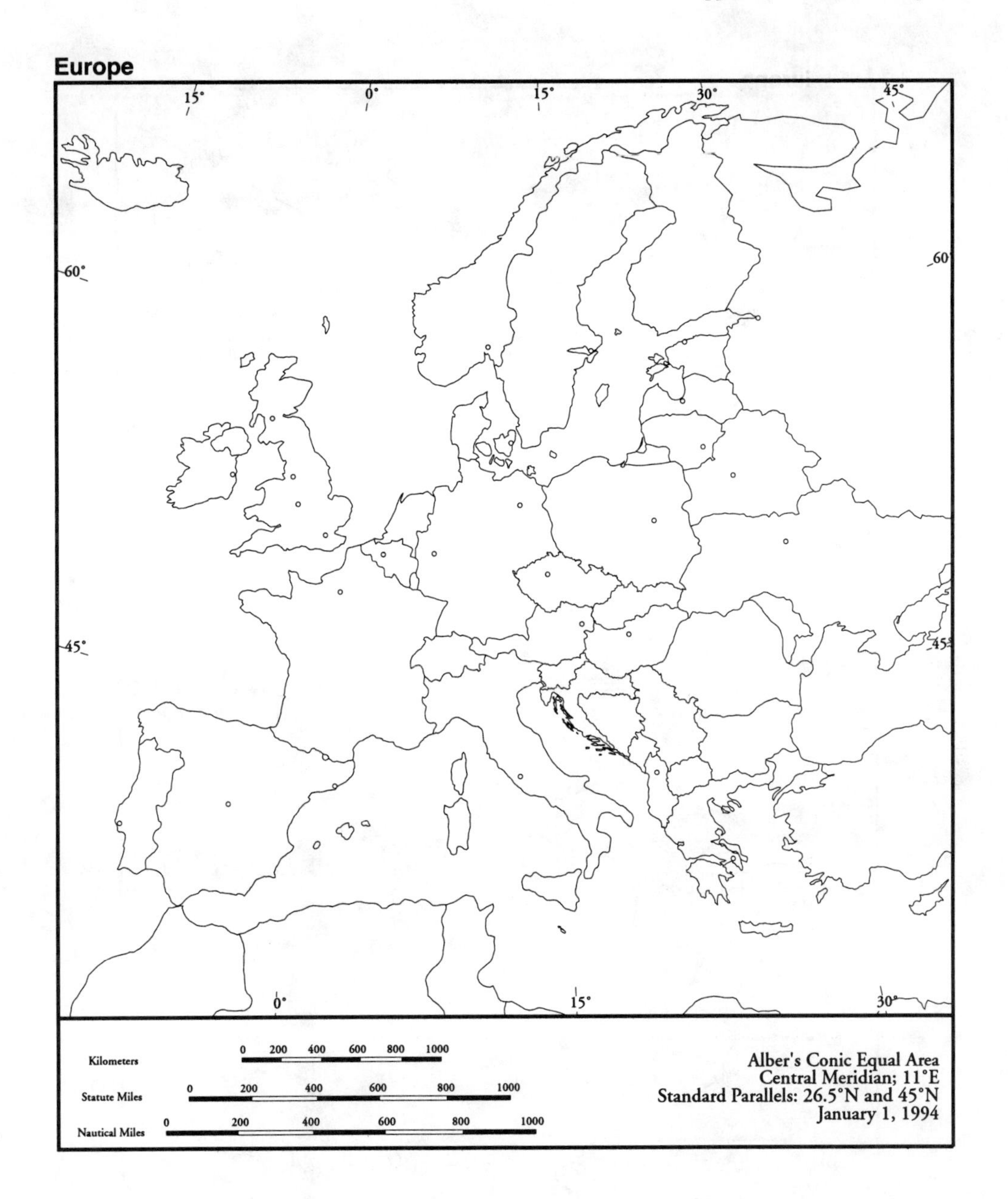
Europe
15°
0°
15°
30°
45°
60°
60
45°
45
0°
15°
30°
Kilometers
0 200 400 600 800 1000
Statute Miles
0 200 400 600 800 1000
Nautical Miles
0 200 400 600 800 1000
Alber's Conic Equal Area
Central Meridian; 11°E
Standard Parallels: 26.5°N and 45°N
January 1, 1994

Russia

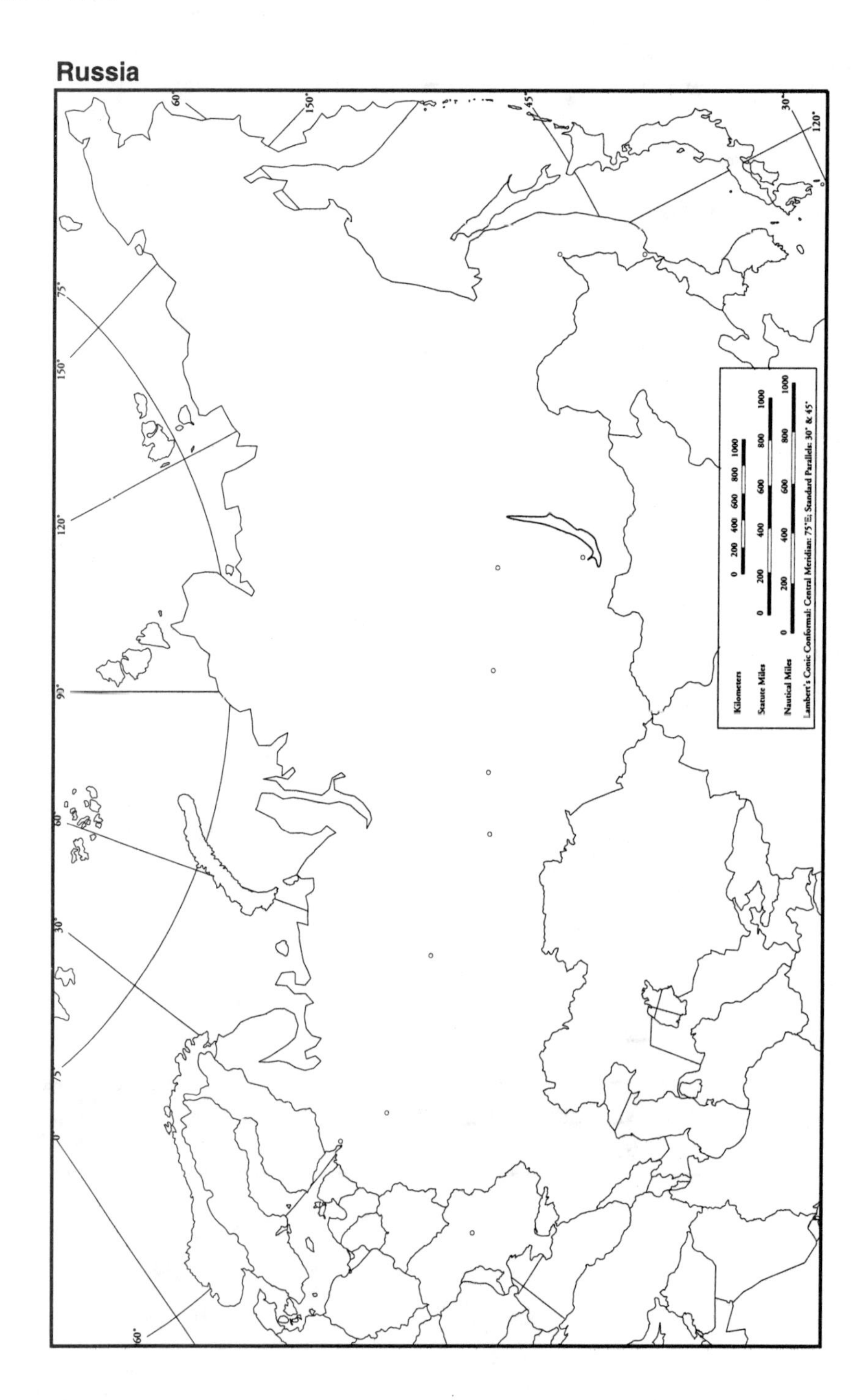

Kilometers 0 200 400 600 800 1000
Statute Miles 0 200 400 600 800 1000
Nautical Miles 0 200 400 600 800 1000
Lambert's Conic Conformal: Central Meridian: 75°E; Standard Parallels: 30° & 45°

Africa

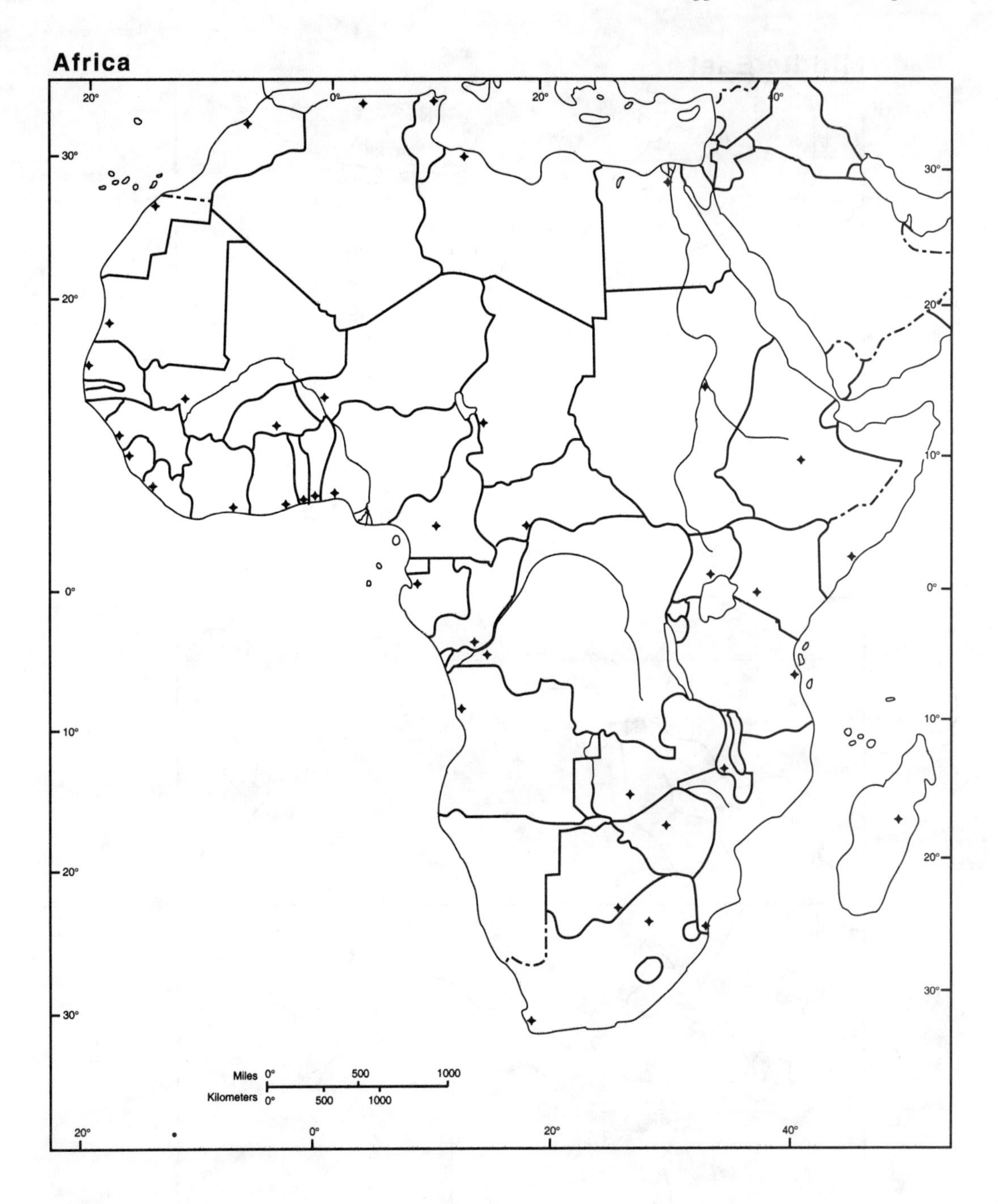
20°
0°
20°
40°
30°
20°
10°
0°
10°
20°
30°
30°
20°
10°
0°
10°
20°
30°
Miles 0° 500 1000
Kilometers 0° 500 1000
20°
0°
20°
40°

Middle East

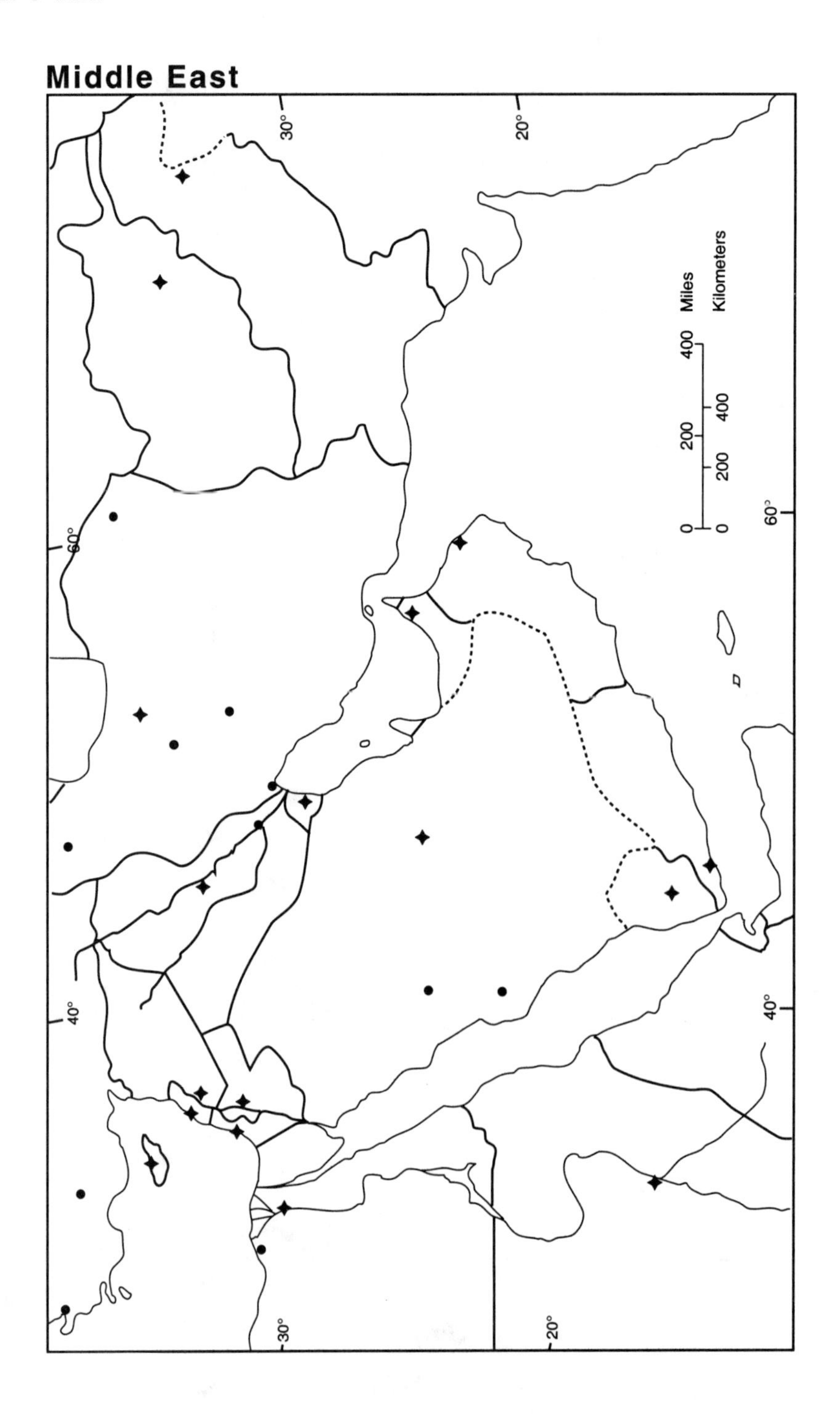

Indian Subcontinent

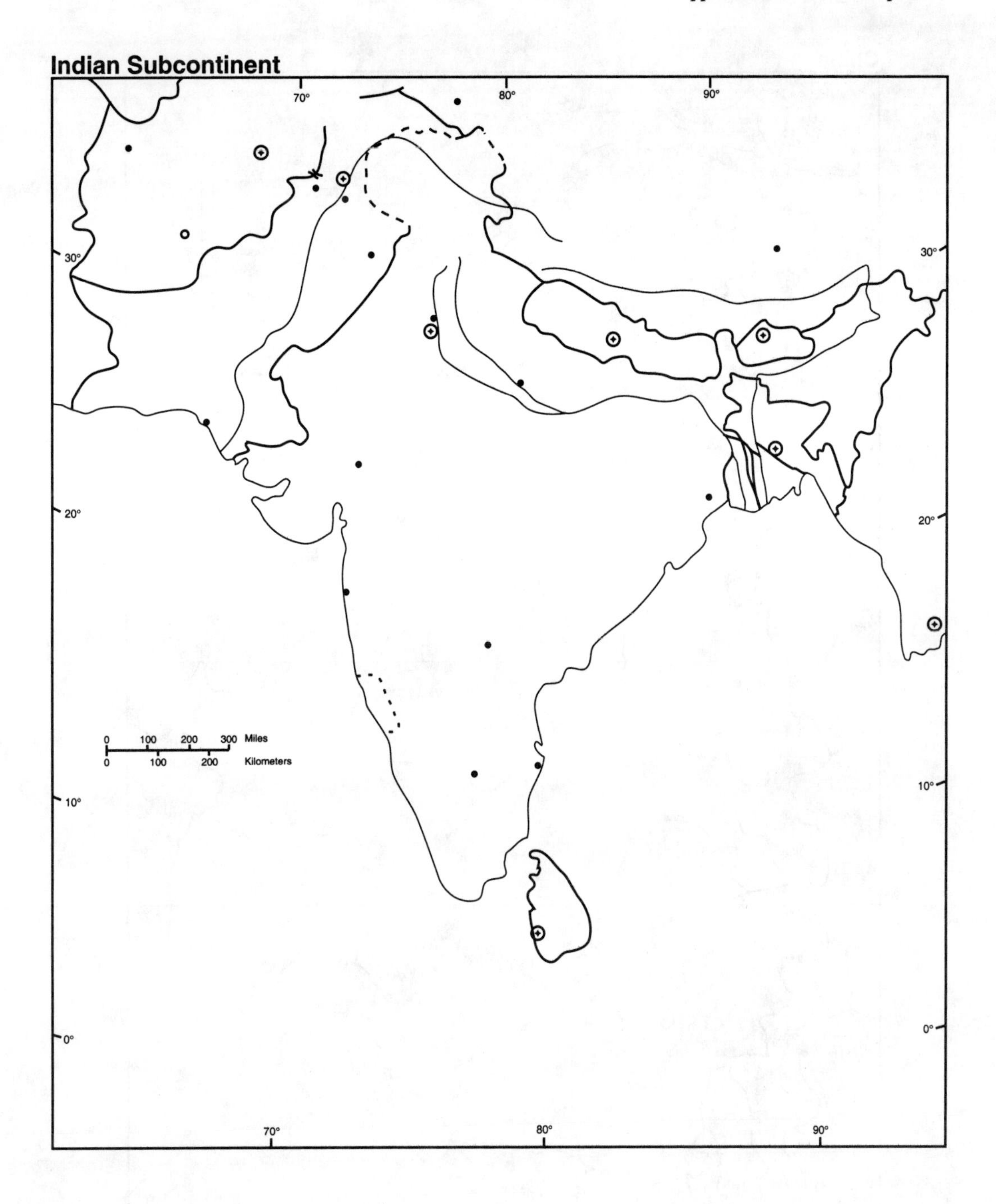

China

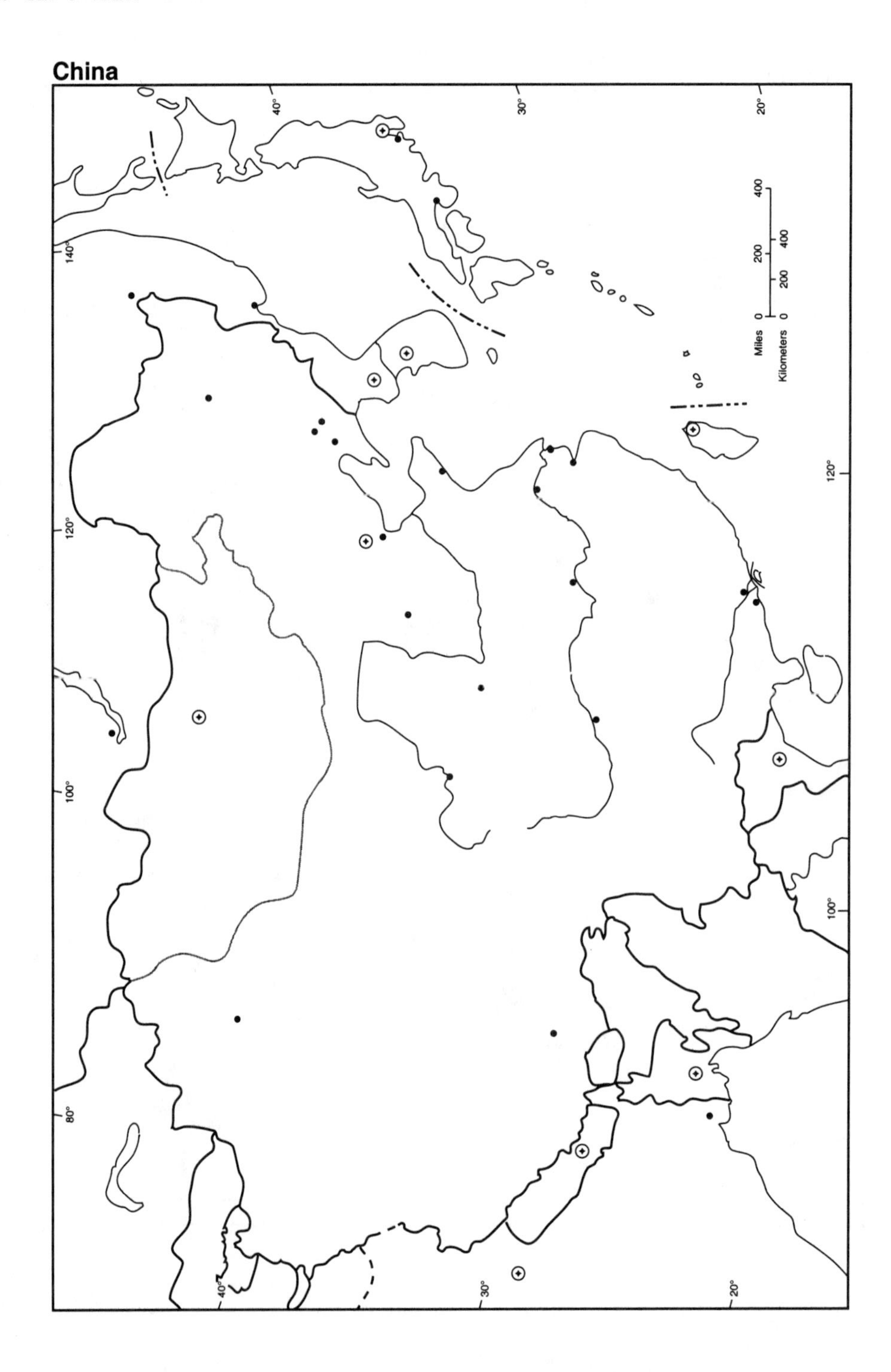
40°
30°
20°
140°
120°
100°
80°
40°
30°
20°
120°
100°
Miles
Kilometers
0
200
400
0
200
400

Japan

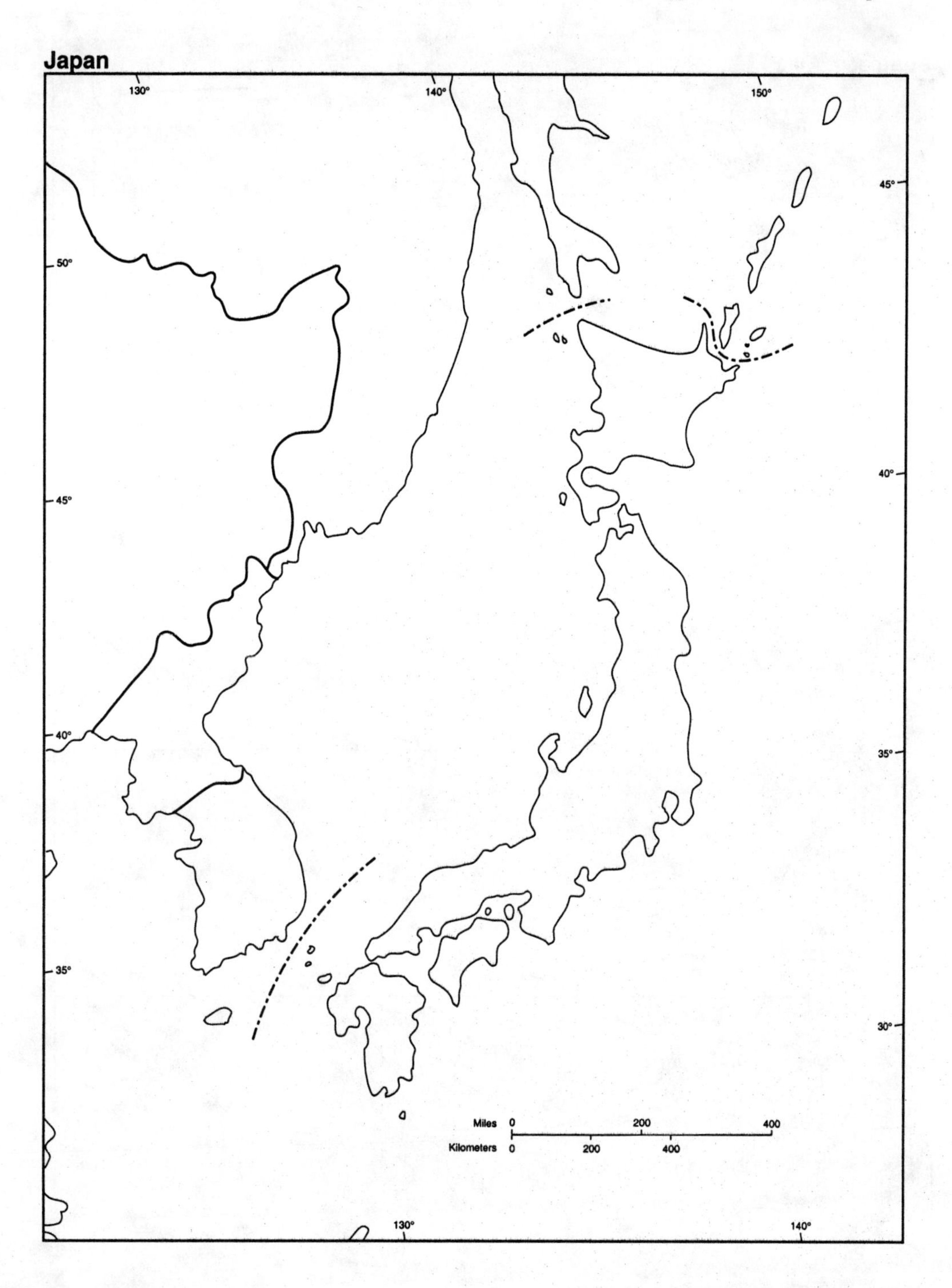
130°
140°
150°
45°
50°
40°
45°
35°
40°
35°
30°
Miles 0 200 400
Kilometers 0 200 400
130°
140°

Southeast Asia

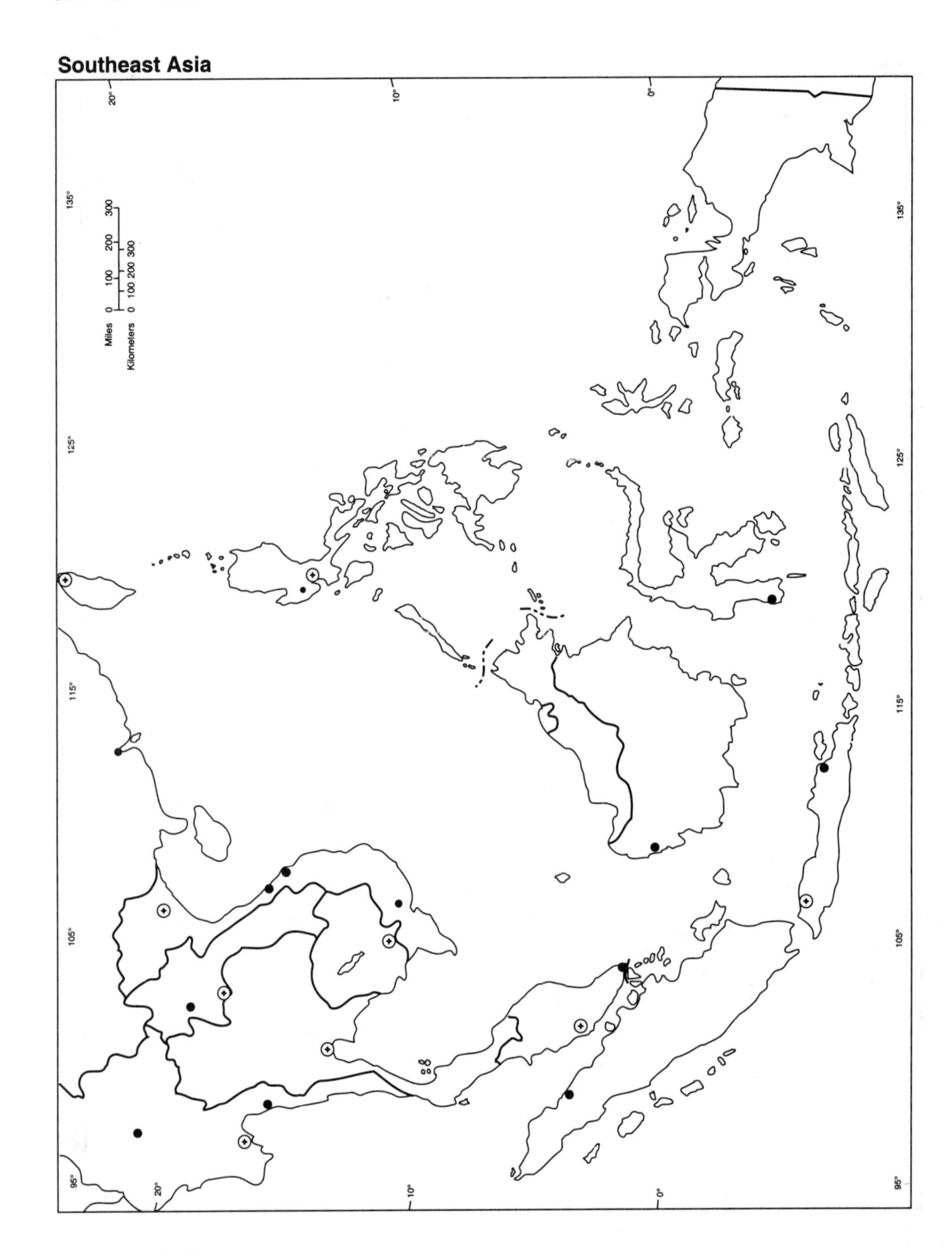

Australia

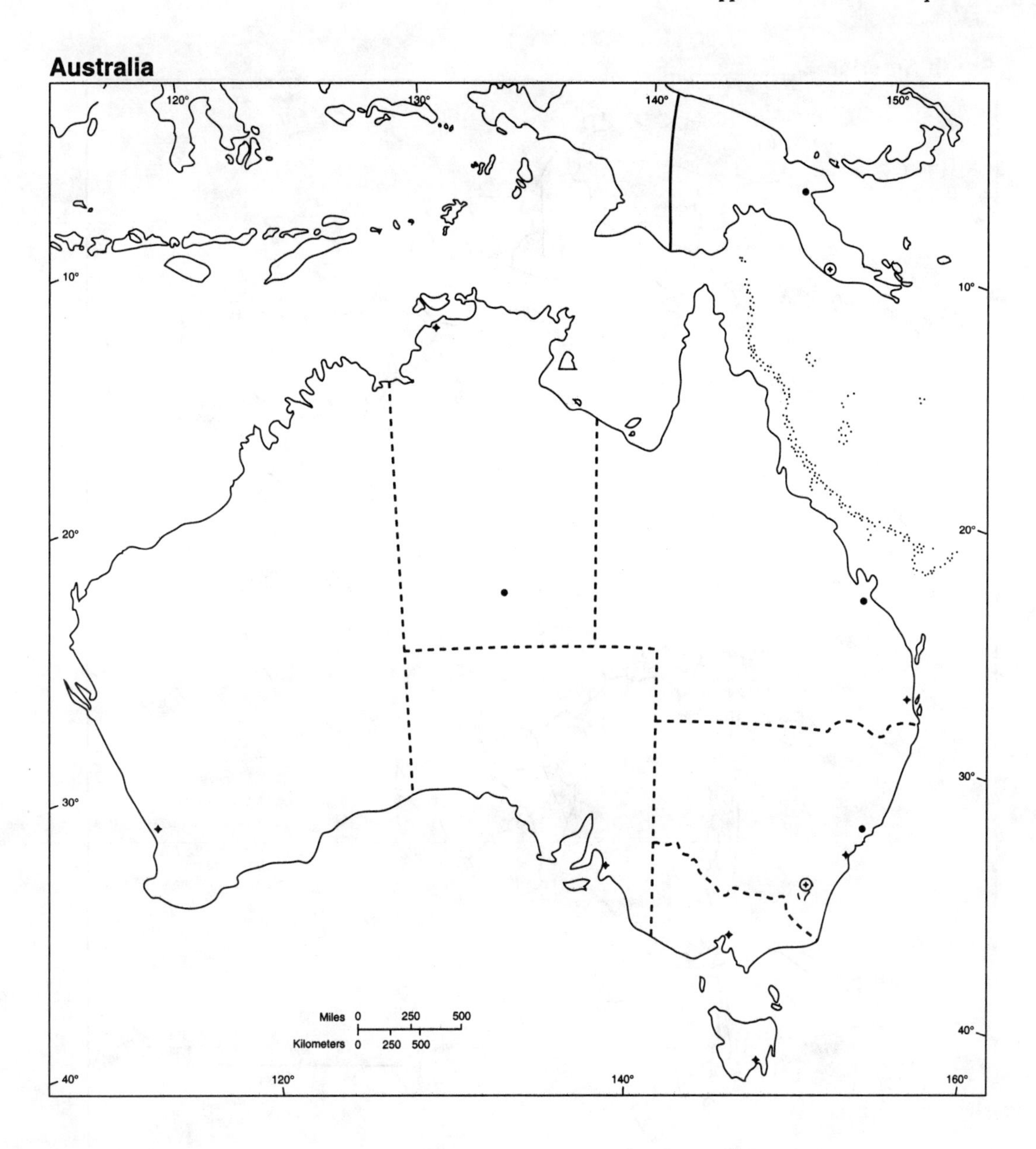
120°
130°
140°
150°
10°
10°
20°
20°
30°
30°
40°
40°
120°
140°
160°
Miles 0 250 500
Kilometers 0 250 500

South America

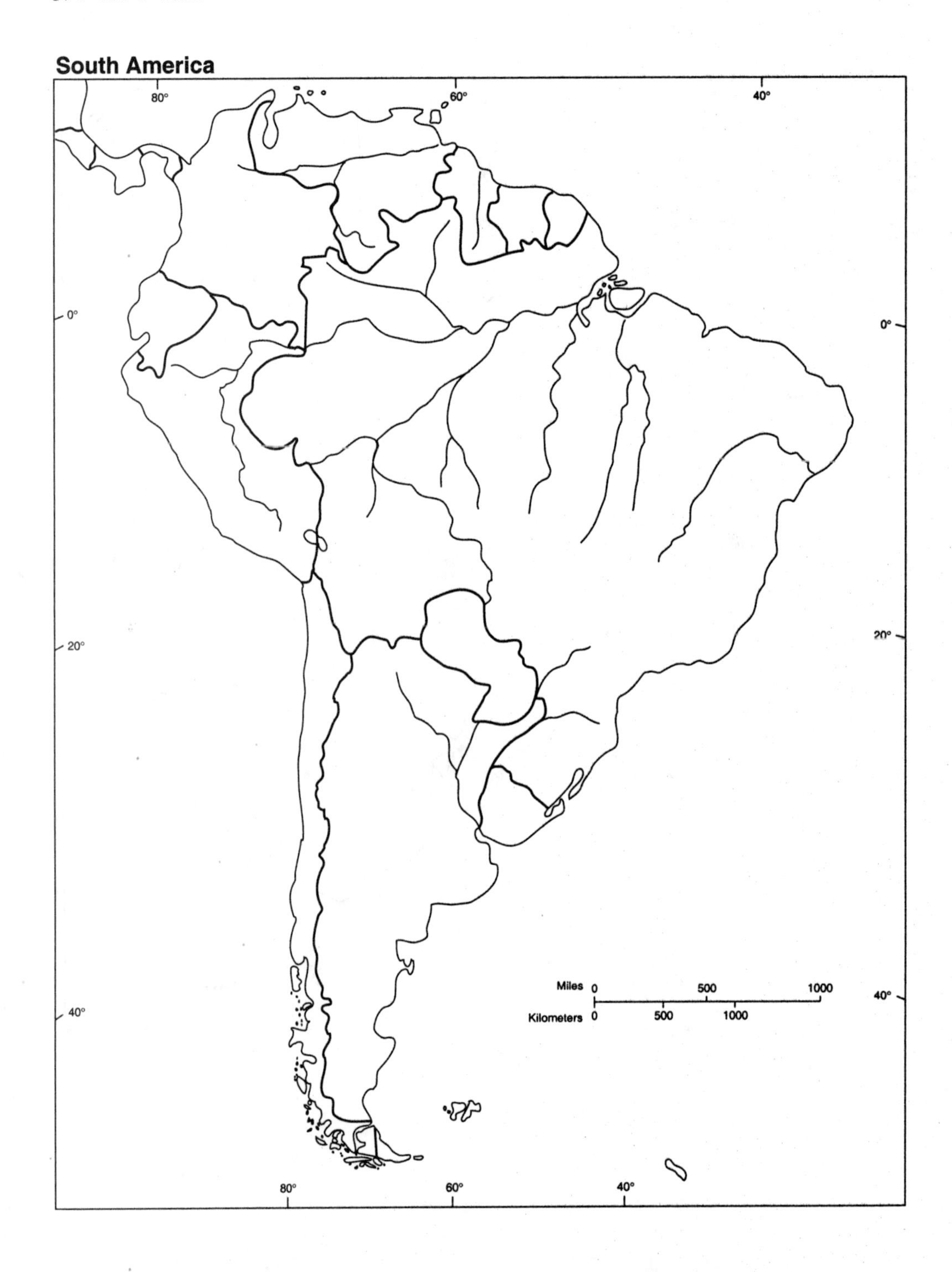

Mexico

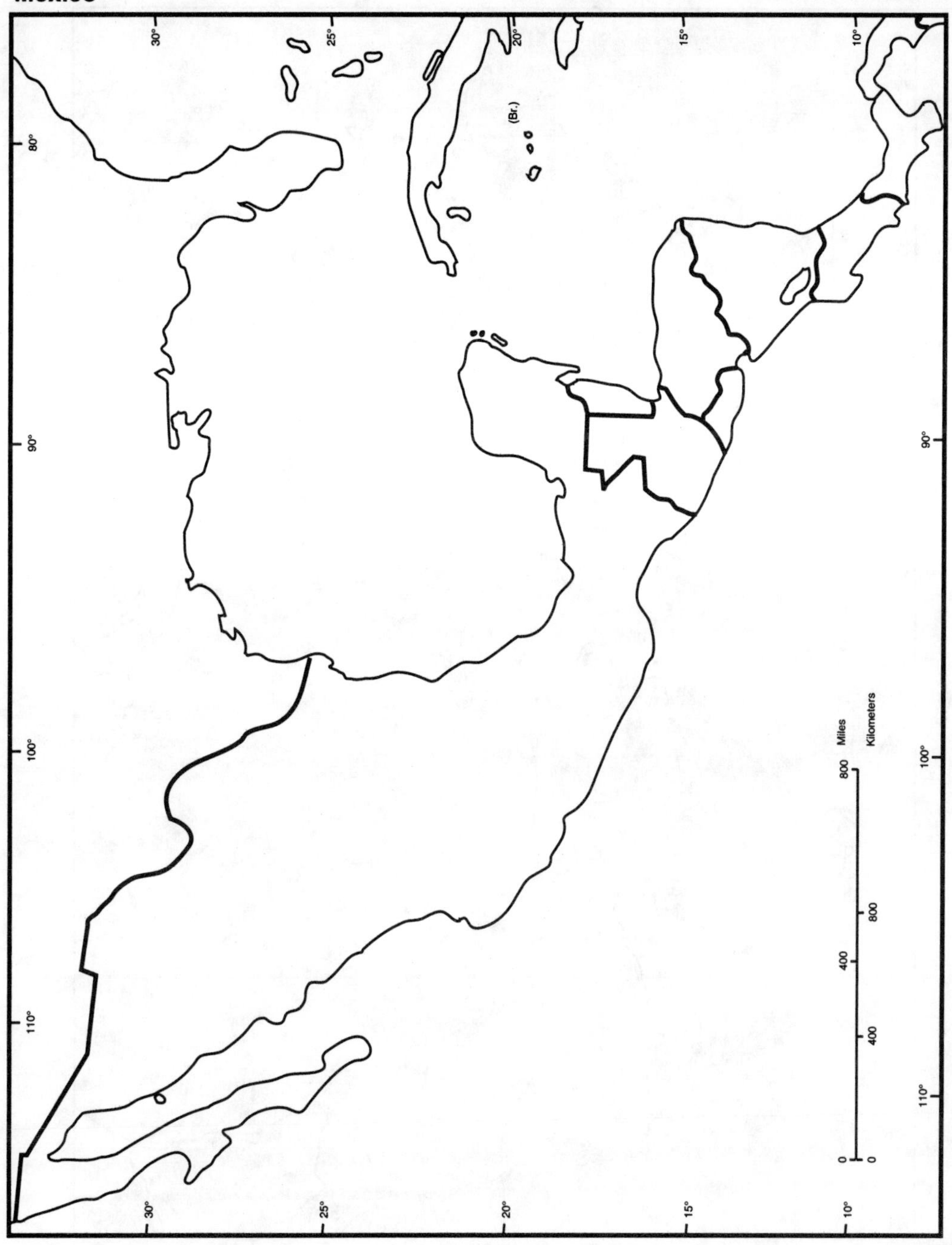
30°
25°
20°
15°
10°
80°
90°
100°
110°
(Br.)
Miles
Kilometers
0
400
800
0
400
800

Caribbean

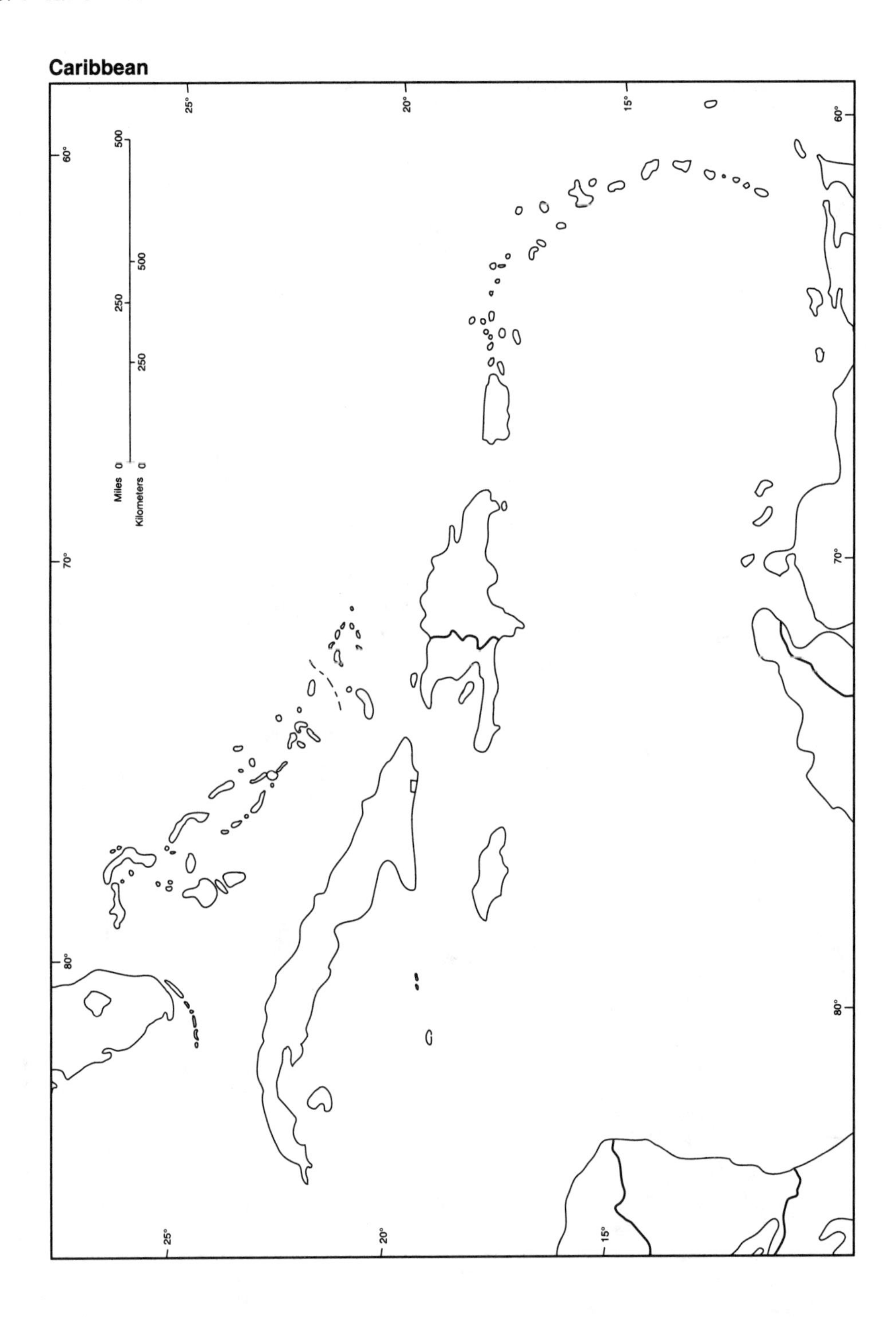

World Map

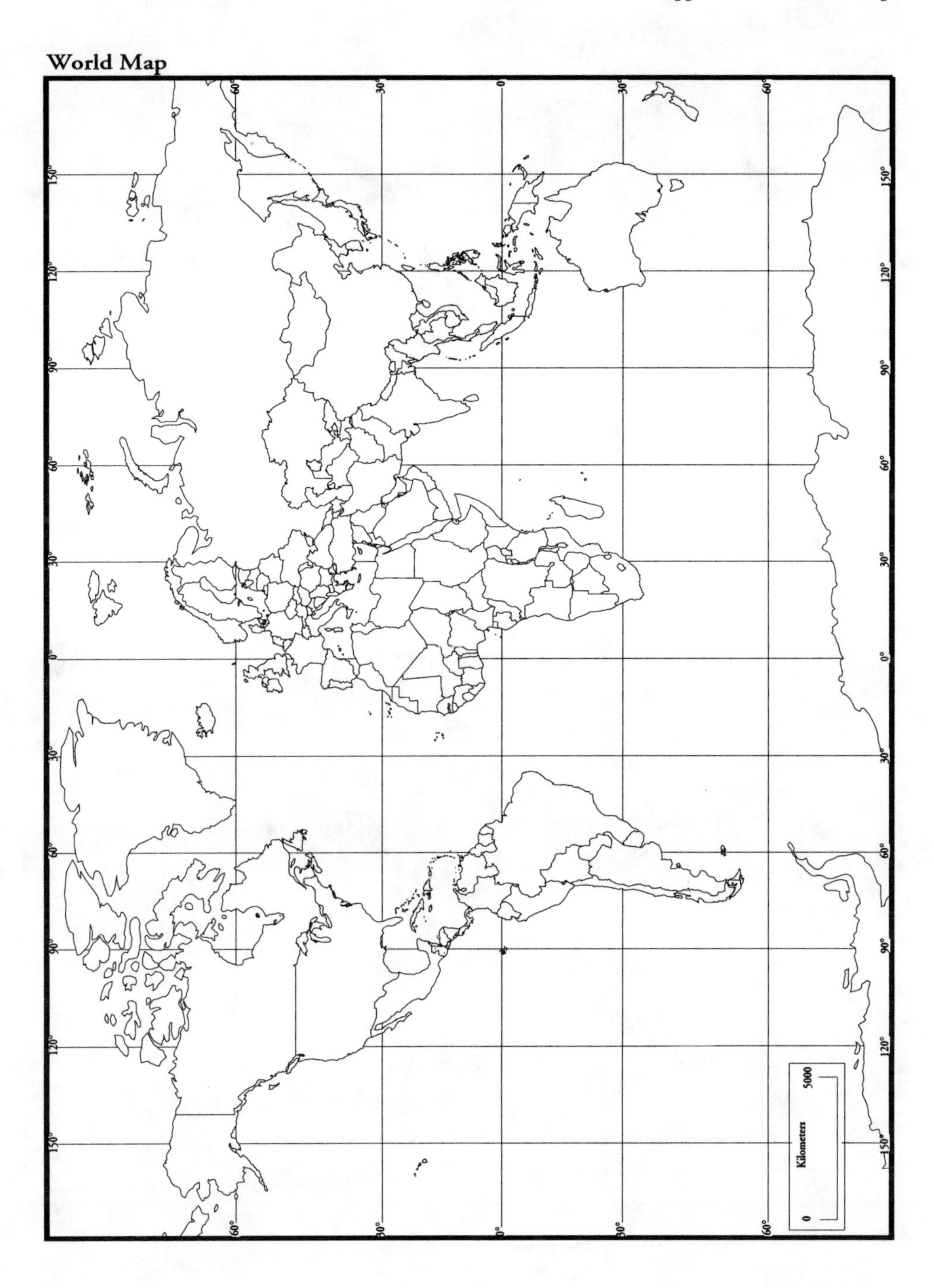

Placename Index

Location	Map #
Kashmir	10
Kathmandu	10
Kawasaki	7
Kazakhstan	5
Kentucky	1, 3
Kenya	13
Khabarovsk	7
Khar'kov	5
Khartoum	12, 13
Khyber Pass	10
Kiev	4, 5
Kingston	14
Kinshasa	13
Kirghizia	5
Kishinev	5
Kitakyushu	7
Kobe	7
Konya	12
Kuala Lumpur	11
Kuang-Chou	11
Kuril Islands	7
Kuwait	6, 12, 13
Kuybyshev	5
Kyoto	7
Kyushu	7
La Paz	16
Lae	8
Lagos	13
Lahore	10
Lake Baikal	5
Lake Balkhash	5
Lake Erie	2
Lake Huron	2
Lake Malawi	13
Lake Michigan	2
Lake Ontario	2
Lake Superior	2
Lake Tanganyika	13
Lake Titicaca	16
Lake Victoria	13
Lanchou	9
Laos	6, 11
Latvia	5
Lebanon	17, 19, 13
St. Petersburg	4, 5
Lesotho	13
Lhasa	9, 10
Liberia	13
Libreville	13
Libya	4, 13
Lilongwe	13
Lima	16
Lisbon	4
Lithuania	5
Lome	13
London	4
Louisiana	1, 3
Luanda	13
Luang Prabang	11
Lusaka	13
Luxembourg	4
Luzon	11
Macao	6, 9, 11
Madagascar	13
Madras	10
Madrid	4
Magnitogorsk	5
Maine	1, 3
Makasar	11
Malabar Coast	10
Malawi	13
Malaya	11
Malaysia	6, 11
Maldives	6
Mali	13
Malta	4
Managua	14
Manaus	16
Mandalay	11
Manila	11
Manitoba	1
Maputo	13
Maracaibo	15
Martinique	15
Maryland	1, 3
Mashhad	12
Massachusetts	1, 3
Matamoros	14
Mauritania	13
Mecca	12
Medan	11
Medina	12
Mediterranean Sea	4
Melbourne	8
Memphis	3
Merida	14
Mexico	1, 14
Mexico City	14
Miami	3, 14, 15
Michigan	1, 3
Mindanao	11
Minneapolis	3
Minnesota	1, 3
Minsk	4, 5
Mississippi	1
Missouri	1, 3
Missouri River	2
Mogadiscio	13
Moldavia	5
Molucca Sea	11
Mona Passage	15
Mongolia	6
Monrovia	13
Montana	1,4
Monterrey	14
Montevideo	16
Montreal	3
Morocco	4, 13
Moscow	4, 5
Moulmein	11

Location	Map #
Rockhampton	8
Rocky Mountains	2
Romania	4
Rome	4
Rosario	16
Rwanda	13
South Yemen	6, 12, 13
Sabah	11
Sado	7
Saint Christopher	15
Saint Lucia	15
Saint Vincent	15
Sakhalin Island	7
Salt Lake City	3
Salvador	16
Samar	11
San Antonio	3
San Diego	3
San Jose	14
San Juan	15
San Salvador	14
Sana	12
Santiago	16
Santiago de Cuba	15
Santo Domingo	15
Sao Paulo	16
Sao Tome and Principe	13
Sapporo	7
Sardinia	4
Saskatchewan	1
Saudi Arabia	6, 12, 13
Sarawak	11
Sea of Japan	7
Sea of Okhotsk	5, 7
Seattle	3
Senegal	13
Seoul	7
Shanghai	9
Shenyang	9
Shikoku	7
Sicily	4
Sidney	8
Sierra Leone	13
Sierra Nevada Mts.	2
Simpson Desert	8
Singapore	6, 11
Singkiang	9
Snake River	2
Sofia	4
Somalia	6, 13
Sonaran Desert	2
South Africa	13
South Australia	8
South Carolina	1, 3
South China Sea	11
South Dakota	1, 3
Falkland Islands	16
South Korea	6, 7
Spain	4
Sri Lanka (Ceylon)	10
St. Louis	3
St. Petersburg	3
Strait of Malacca	11
Straits of Florida	15
Sucre	16
Sudan	12, 13
Sulu Sea	11
Sumatra	11
Sumbawa	8, 11
Surabaja	11
Surinam	16
Sverdlovsk	5
Swaziland	13
Sweden	4
Switzerland	4
Syria	6, 12, 13
Tabriz	12
Tajikistan	5
Taipei	9, 11
Taiwan	6, 11
Taiyuan	9
Tallinn	4, 5
Tampa	3, 14
Tampico	14
Tanzania	13
Tashkent	5, 10
Tasman Sea	8
Tasmania	8
Tbilisi	5
Teguclgalpa	14
Tehran	12
Tel Aviv	12
Tennessee	1, 3
Texas	1, 3
Thailand	11
Thimpu	10
Tientsin	9
Tigris River	12
Tijuana	14
Timor	8, 11
Timor Sea	8
Tirane	5,7
Togo	13
Tokyo	7
Trinidad and Tobago	15, 16
Tripoi	13
Tunis	4, 13
Tunisia	4, 13
Turkey	4, 6, 12, 13
Turkmenia	5
Caicos Islands	15
Russia	4, 5
Siberia	2, 5, 6
Uganda	13
Ukraine	5
Ulan Bator	9
United Arab Emirates	12
United Kingdom	4
United States	6, 14, 15
Uruguay	16

Appendix

C

Symbols, Units, and Conversions

Temperature

°C = .55 [degrees Fahrenheit - 32]

°F = 1.8 degrees Celsius + 32

K = degrees Celsius + 273.15

Celsius is sometimes designated centigrade

Kelvin (K) is sometimes designated absolute temperature

Heat and Power

calorie (cal) = the amount of heat necessary to raise the temperature of 1 gram of water from 14.5°C to 15.5°C. 1 calorie = 0.23892 joule. The calorie is sometimes designated as gram calorie.

kilocalorie (kcal) = the amount of heat necessary to raise the temperature of 1 kg of water from 14.5°C to 15.5°C. Kilocalorie is sometimes designated as food calorie or Calorie.

British Thermal Unit (BTU) = the amount of heat necessary to raise the temperature of 1 pound of water from 63°F to 64°F. 1 BTU = 0.0040 calorie.

Watt (W) is a unit of power. Power is work done per unit of time. 1 watt = 1 joule/second = 0.29308 BTUs per hour.

Joule (J) is a unit of energy. 1 Joule = 1 $kg \cdot m^2/sec^2$

Pressure

1 atmosphere (atm) = 76 cm of mercury
14.7 lb/square inch = 1013.25 millibars

1 bar = 0.9869 atmospheres

1 millibar (mb) = 100 Pascals (Pa) = 1 hectopascal (hPa)

1 Pascal (Pa) = 1 N/m^3 = 1 J/m^3

Trigonometric Functions

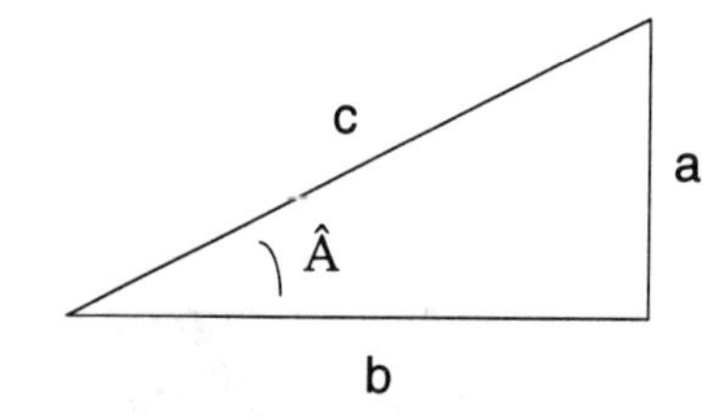

Sine Â = a/c   Cosine Â = b/c   Tangent Â = a/b

Beaufort Wind Scale

No.	Wind Effect	Knots
0	Smoke rises vertically	< 1
1	Direction shown by smoke but not wind vane	1-3
2	Wind felt of face; leaves rustle; vane moved by wind	4-6
3	Leaves and small twigs in constant motion	7-10
4	Raises dust and loose paper; small branches are moved	11-16
5	Small trees in leaf begin to sway	17-21
6	Large branches in motion; umbrellas used with difficulty	22-27
7	Whole trees in motion; walking against wind difficult	28-33
8	Breaks twigs off trees	34-40
9	Slight structural damage occurs	41-47
10	Seldom experienced inland; some trees overturned	48-55
11	Very rarely experienced; widespread damage	56-65
12	Hurricane force	>65

Angles

1 radian = 57° 18′
1 degree = 60 minutes (′) of arc
1 minute = 60 seconds (″) of arc

Time

1 year = 3.1536×10^7 seconds = 8.76×10^3 hours
1 day = 1440 minutes

Length

1 nanometer = 10^{-9} meters
1 micrometer (μ) = 10^{-6} meters
1 millimeter = 10^{-3} meters
1 centimeter = 0.39 inches
1 inch = 2.540 centimeters
1 meter = 39.37 inches = 3.28 feet
1 kilometer = 0.62 miles = 3,281 feet
1 mile = 5280 ft = 1.609 kilometers (km)

Area

1 square centimeter (cm^2) = 0.15 in^2
1 square inch (in^2) = 6.45 cm^2
1 square meter (m^2) = 10.76 ft^2
1 square foot (ft^2) = 0.09 m^2
Area of a circle = $\pi r^2 = 0.7854 d^2$
Area of a sphere = $4 \pi r^2 = 12.566 r^2$

Volume

1 cubic centimeter (cm^3) = 0.06 in^3
1 cubic inch (in^3) = 16.39 cm^3
1 liter (l) = 1000 cm^3 = 0.264 gallon (gal) U.S.
1 cubic foot of water = 62.4283 pounds (weight)
1 cubic meter (m^3) = 264.2 gallons

Mass

gram = .002204 pounds

1 kilogram (kg) = 2.204 pounds

1 pound = 453.592 grams

1 pound = 0.453592 kg

Speed

1 knot - 1 nautical mi/hr = 1.15 statute mi/hr = 0.51 m/sec = 1.85 km/hr

1 mile per hour (mi/hr) = 0.87 knot = 0.45 m/sec = 1.61 km/hr

1 kilometer per hour (km/hr) = 0.54 knot = 0.62 mi/hr = 0.28 m/sec

1 meter per second (m/sec) = 1.94 knots = 2.24 mi/hr = 3.60 km/hr

Force

1 dyne = 1 gram per centimeter per second^2 = 2.2481 x 10^{-6} pounds force

1 pound force= 0.2248 Newtons

1 Newton (N) = 1 kilogram per meter per secon^2 = 105 dynes

Powers of 10

	Prefix	Symbol
10^{-18} = 0.000000000000000001	atto	a
10^{-15} = 0.000000000000001	femto	f
10^{-12} = 0.000000000001	pico	p
10^{-9} = 0.000000001	nano	n
10^{-6} = 0.000001	micro	μ
10^{-3} = 0.001	milli	m
10^{-2} = 0.01	centi	c
10^{-1} = 0.1	deci	d
10^{1} = 10	deka	da
10^{2} = 100	hecto	h
10^{3} = 1,000	kilo	k
10^{6} = 1,000,000	mega	M
10^{9} = 1,000,000,000	giga	G
10^{12} = 1,000,000,000,000	tera	T
10^{15} = 1,000,000,000,000,000	penta	P
10^{18} = 1,000,000,000,000,000,000	exa	E

Geophysical Parameters

Mass of the earth = $5.98 \cdot 10^{24}$ kg

Mass of the atmosphere = $5.14 \cdot 10^{18}$ kg

Mass of the oceans = $1.4 \cdot 10^{21}$ kg

Mass of living organisms (dry weight) = $1.3 \cdot 10^{15}$ kg

Polar radius = $6.36 \cdot 10^{6}$ m

Equatorial radius = $6.38 \cdot 10^{6}$ m

Average radius of the earth = $6.37 \cdot 10^{6}$ m

Average acceleration due to gravity = 9.80665 m/sec^2

Acceleration due to gravity at the poles = 9.83 m/sec^2

Acceleration due to gravity at the equator = 9.78 m/sec^2

Solar constant = 1370 $W \cdot m^{-2}$

Rotation of the earth = $7.27 \cdot 10^{-5}$ radians $\cdot s^{-1}$

Stefan-Boltzmann constant = $5.669 \cdot 10^{-8}\ W \cdot m^{-2} \cdot K^{-4}$

Density of water:

at 0°C	999.87 kg/m^3
at 3.98°C	1,000 kg/m^3
at 25°C	997.07 kg/m^3

Latent Heat of vaporization:

at 100°C	$2.258 \cdot 10^{6}$ J/kg
at 100°C	$539.6 \cdot$ cal/g
at 17°C	$2.459 \cdot 10^{6}$ J/kg
at 0°C	$2.499 \cdot 10^{6}$ J/kg

Greek Letters

Α	α	alpha
Β	β	beta
Γ	γ	gamma
Δ	δ	delta
Ε	ε	epsilon
Ζ	ζ	zeta
Η	η	eta
Θ	θ	theta
Ι	ι	iota
Κ	κ	kappa
Λ	λ	lambda
Μ	μ	mu
Ν	ν	nu
Ξ	ξ	xi
Ο	ο	omicron
Π	π	pi
Ρ	ρ	rho
Σ	σ	sigma
Τ	τ	tau
Υ	υ	upsilon
Φ	φ	phi
Χ	χ	chi
Ψ	ψ	psi
Ω	ω	omega

Appendix D

Hurricane Tracking Chart

This chart is provided for your convenience should you wish to plot the course of a hurricane across the Atlantic or Gulf of Mexico. Hurricanes frequently display erratic storm tracks and recording their path can be fun if they are not headed your way. Satellite imagery can easily spot the eye of the hurricane but precise location of the storm center is still determined by aircraft traverses. Bulletins locating the storm center are periodically broadcast and can be easily plotted on this chart to the nearest half degree of latitude and longitude. If the storm's center is reported to be near 41.5° North and 63° West, read up the page to 41.5° N and left to 63°W. The example is plotted.

To find the path of any North Atlantic hurricane between 1871 - 1989 and to learn more about these furious storms, see Pielke, Roger A. *The Hurricane* , New York; Routledge, 1990.

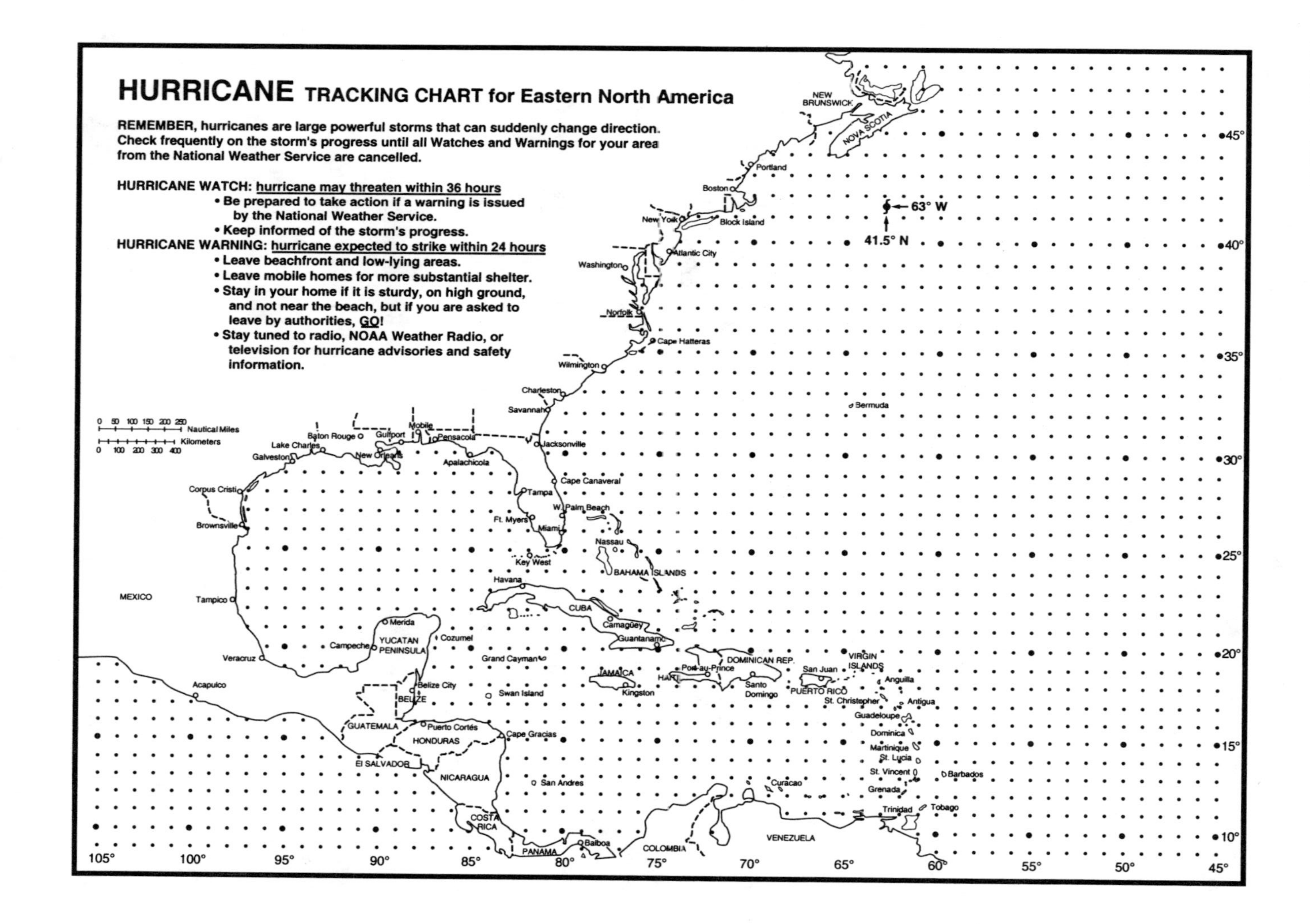

HURRICANE TRACKING CHART for Eastern North America
REMEMBER, hurricanes are large powerful storms that can suddenly change direction. Check frequently on the storm's progress until all Watches and Warnings for your area from the National Weather Service are cancelled.
HURRICANE WATCH: hurricane may threaten within 36 hours
• Be prepared to take action if a warning is issued by the National Weather Service.
• Keep informed of the storm's progress.
HURRICANE WARNING: hurricane expected to strike within 24 hours
• Leave beachfront and low-lying areas.
• Leave mobile homes for more substantial shelter.
• Stay in your home if it is sturdy, on high ground, and not near the beach, but if you are asked to leave by authorities, GO!
• Stay tuned to radio, NOAA Weather Radio, or television for hurricane advisories and safety information.
0 50 100 150 200 250 Nautical Miles
0 100 200 300 400 Kilometers
63° W
41.5° N
NEW BRUNSWICK
NOVA SCOTIA
Portland
Boston
Block Island
New York
Atlantic City
Washington
Norfolk
Cape Hatteras
Wilmington
Charleston
Savannah
Bermuda
Jacksonville
Cape Canaveral
Tampa
W. Palm Beach
Ft. Myers
Miami
Key West
Nassau
BAHAMA ISLANDS
Baton Rouge
Gulfport
Mobile
Pensacola
Lake Charles
Galveston
New Orleans
Apalachicola
Corpus Cristi
Brownsville
MEXICO
Tampico
Veracruz
Acapulco
Havana
CUBA
Camagüey
Guantanamo
Merida
YUCATAN PENINSULA
Campeche
Cozumel
Grand Cayman
JAMAICA
Kingston
HAITI
Port-au-Prince
DOMINICAN REP.
Santo Domingo
San Juan
PUERTO RICO
VIRGIN ISLANDS
Anguilla
St. Christopher
Antigua
Guadeloupe
Dominica
Martinique
St. Lucia
St. Vincent
Barbados
Grenada
Trinidad
Tobago
Curacao
Belize City
BELIZE
Swan Island
GUATEMALA
Puerto Cortés
HONDURAS
Cape Gracias
El SALVADOR
NICARAGUA
San Andres
COSTA RICA
PANAMA
Balboa
COLOMBIA
VENEZUELA
45°
40°
35°
30°
25°
20°
15°
10°
105°
100°
95°
90°
85°
80°
75°
70°
65°
60°
55°
50°
45°

Glossary of Selected Terms

Ablation. — The process by which glaciers waste away. The process includes melting, evaporation, and sublimation.

Abrasion. — Grinding action of one material by another.

Absolute Humidity. — The weight of moisture in the air per unit volume of air. The weight of the water vapor in a given volume of air.

Absolute instability. — The condition of a parcel of moist air, the lapse rate in which is greater than the dry adiabatic rate.

Absolute Zero. — Temperature at which all random molecular motion stops. 0° on the Kelvin scale, minus 273° on thc Celsius scale.

Acceleration. — The rate of change of velocity. Acceleration is velocity divided by time.

Adiabatic Cooling or Heating. — Change of temperature in a gas due to compression or expansion and taking place without gain or loss of heat from the outside. Thermodynamic process in which no heat is exchanged with the surrounding environment. In the atmosphere, when the air is compressed, the air molecules are moved closer together. The work done on them adds energy to the system. When the air expands the air molecules move further apart and the energy consumed by the work of molecular displacement consumes energy.

Adiabatic Temperature Lapse Rate. — Rate at which air cools or heats when it is lifted or descends in altitude.

Advection. — Horizontal flow of air at the surface or aloft; one of the means by which heat is transfered from one region to another.

Advection Fog. — Fog produced by condensation of a moist air layer moving over a cold land or water surface.

Aeolian. — Pertaining to the action or effect of the wind.

Aerosol. — Particles that remain suspended in the air for extended periods of time.

Air Current. — A stream of air moving in any direction other than the horizontal, especially in the vertical.

Air Drainage. — The flow of air down a slope or channel. Air tends to flow downhill, due to gravity, when its density is greater than that of the adjacent air at the same level.

Air Mass. — Large body of air within which the vertical gradients of temperature and moisture are relatively uniform. An extensive body of air that has essentially uniform conditions of humidity and temperature along a horizontal plane.

Air Pressure. — Barometric pressure.

Albedo. — The reflectivity of a surface.

Aleutian Low. — The semi-permanent cyclone or low that is usually located near the Aleutian Islands.

Alluvial Fan. — Gently sloping fan-shaped accumulation of course alluvium formed below the point of emergence of a channel from a narrow canyon or gorge.

Alluvial Stream. — A stream that flows through an alluvial deposit, i.e. the lower Mississippi.

Alluvium. — Stream-laid sediment deposit found in a stream channel.

Alpine Glaciers. — Glaciers that are confined to valleys in mountainous areas.

Altimeter. — An aneroid barometer graduated to show height instead of pressure.

Altitude. — Vertical distance above the earth. The angle of an object above the horizon.

Alto Clouds. — Clouds of the middle height range, 2 to 6 km (6,500 to 20,000 ft).

Anabatic Wind. — A "valley breeze or wind" that occurs when there is uneven heating of the air in mountain areas. The air near the slope is heated faster than air at an equal altitude over the valley thus creating a convectional cell.

Anemoneter. — An instrument for measuring the speed or force of the wind.

Aneroid. — Applied to a barometer which contains no liquid.

Aneroid Barometer. — An instrument for measuring atmospheric pressure, first built in 1843 by Vidie.

Angle of Incidence. — The angle at which an object strikes a surface measured from a line perpendicular to that surface. The angle at which the sun's rays strike the surface of the atmosphere or earth.

Angle of the Sun. — The angle or arc of the sun above the horizon. A complementary angle to the angle of incidence.

Anticyclone. — A region of high pressure with winds that spiral out from the center in a clockwise direction in the Northern Hemisphere and a counterclockwise direction in the Southern Hemisphere.

Aphelion. — The point of farthest distance between the earth and the sun. In the elliptical orbit around the sun, the earth is farthest from the sun on approximately July 4.

Aquifer. — Rock mass or layer of sediment that readily transmits and holds ground water. A rock formation that allows water to move through it at a greater rate than the adjacent rock.

Aquifuge. — A substance that has a very low ability to store or transmit water, i.e. clay.

Arctic Sea Smoke. — A fog that occurs when cold air moves over much warmer water.

Arid. — Dry; lacking any significant amount of moisture.

Artesian Flow. — Spontaneous rise of water in a well or a fracture in the earth's crust that lies above the level of the surrounding water table.

Atmosphere. — The gaseous material that surrounds the earth. A measure of pressure - 1 atmosphere = 14.7 p.s.i. = 1013 millibars.

Atmospheric Pressure. — The force per unit area exerted by the atmosphere in any part of the atmospheric envelope.

Atom. — The smallest part of an element taking part in a chemical reaction.

Autumnal Equinox. — That moment following summer and preceeding winter when the subsolar point strikes the equator. Occurs on or about September 21 in the northern hemisphere.

Available Soil Water. — The difference between a soil's field capacity and wilting point water storage values.

Azimuth. — The arc of the horizon intercepted between a given point and an adopted zero point. North is usually the zero point.

Bar. — A unit of pressure equal to the mean atmospheric pressure at about 100 meters above mean sea level. Standard atmospheric pressure of 760 mm, 29.921 inches, or 1,013,250.144 dynes/sq. cm.

Barograph. — A barometer that makes a continuous record of pressure changes.

Barometer. — An instrument for measuring atmospheric pressure.

Base Flow. — The stream flow when there is a steady input of water.

Bases (Base Cations). — Certain cations present in the soil solution that are important plant nutrients.

Batholith. — Massive body (perhaps part of a solidified magma chamber) of igneous rock that can extend over hundreds of square kilometers.

Bauxite. — The ore of aluminum.

Beaufort wind scale. — A system of estimating wind velocities, originally based (1805-1808) by its inventor, Admiral Beaufort of the British Navy, on the effects of various wind speeds on the amount of canvas which a full-rigged frigate of the early nineteenth century could carry.

Bed Load Of Streams. — Heavy matter that is rolled or pushed along the stream channel bottom.

Bedrock Stream. — A stream that flows over bedrock, i.e. the upper reaches of the Rio Grande.

Bergeron Theory. — A theory that uses the concept of differential vapor pressure to explain the formation of precipitation. The vapor pressure over ice crystals is less than that of water droplets. Thus, water vapor is attracted to the ice crystals until the crystal is no longer buoyant and falls from the sky.

Bermuda High. — The name often given to the high pressure cell usually found over the Atlantic Ocean near the Bermuda Islands, though it varies in position and intensity.

Biota. — Life forms.

Bituminous Coal. — Soft coal; coal that is high in carbonaceous and volatile matter.

Black Body. — An object that absorbs all the radiation that falls upon it. It neither reflects nor transmits any of the incident radiation.

Black Box. — A term used in model building to avoid detailed description or in cases where the process is not completely understood. A situation where the system is treated as a unit without consideration for the objects of their interactions.

Blowouts (Deflation Hollows). — Broad, shallow depressions formed by the action of wind removing loose surface particles.

Boreal Forest. — Expansive needleleaf evergreen forest of the earth's subarctic regions.

Boundary. — Plane established by a state at the limit of its territory.

Boundary Layer. — The contact zone between two materials of dissimilar natures, i.e. the atmosphere and the earth's surface, cold and warm air, water and oil, etc.

Boyle's Law. — A thermodynamic law that states that the volume of a gas varies inversely to its pressure with temperature constant.

Braided Stream. — A stream that is carrying an overload of sediments, thus deposition is taking place in the channel creating an interwoven pattern of the stream and deposits.

British Thermal Unit. — The BTU is a unit of heat equal to 0.01 of the quantity required to raise the temperature of one pound of water from the melting point to the boiling point.

Broadleaf. — Leaf form that is wide in relation to its length, and is thin and comparatively large.

Buoyancy. — The tendency of a body to float or rise in a fluid.

Buys Ballot's Law. — The principle governing the relation of wind direction to pressure distribution: "If one stands with their back to the wind, the pressure on their left side is lower than on their right." Thus stated, the law applies to the northern hemisphere.

Calcic Horizon. — Subsurface soil horizon of carbonate enrichment.

Calorie. — The amount of energy required to raise one gram of water from 14.5°C to 15.5°C at sea level.

Capillary Water. — Water that clings to solid surfaces by means of the force of capillary or surface tension.

Carbohydrate. — Compounds made of carbon, hydrogen and oxygen that constitute a major class of food.

Ceiling. — The lowest height above ground at which all clouds at and below that level cover more than one-half of the sky.

Celsius Scale. — A temperature scale proposed in 1742 by Andres Celsius, Professor of astronomy at Upsala, with 0° as the boiling point of water and 100° as the melting point of ice, just the reverse of the centigrade scale. Today the term Celsius is used to describe the centigrade scale that was an adaptation.

Centigrade Scale. — A temperature scale with 0° as the melting point of ice and 100° as the boiling point of water. Probably derived from the Celsius.

Channelized Stream. — A stream that has been straightened and has a uniform grade and cross section. This is used to move water rapidly through an area.

Chemical Weathering. — A chemical change that takes place in rock minerals through exposure to atmospheric conditions in the presence of water; a process of decomposition.

Chinook. — Name given in North America to a warm, dry, wind along the eastern slopes of the Rockies. Identical to the European Foehn.

Cinder Cone Volcano. — Steep-sided conical-shaped volcano built of coarse ejecta.

Cirque. — An amphitheater, or bowl-shaped depression formed by mountain glaciers.

Cirrus Clouds. — High clouds formed of ice and shaped into delicate white filaments, streaks, or narrow bands.

Climate. — Generalized statement of weather conditions for a given location over a long period of time.

Climatology. — The science closely linked with meteorology and geography that seeks to understand the long term condition of the atmosphere at various scales from local to global.

Climax Vegetation. — Stable community of plants and animals reached at the end of a series of plant succession stages.

Climograph. — Graph upon which monthly temperature and precipitation are plotted.

Cloud. — A visible mass of water droplets and ice crystals formed by condensation in the atmosphere.

Clouds. — Dense concentrations of suspended water or ice particles.

Coal. — Solid, combustible, organic material formed by the decomposition of plant material without free access to air.

Cold Fog. — A fog that is made up primarily of ice crystals.

Comfort Energy. — Any form of energy whose end use is the heating and cooling of buildings and homes.

Condensation. — A change in the state of matter from gas to liquid.

Conduction. — The trasference of heat within and through a substance by means of internal molecular activity and without any obvious external motion.

Conformal. — Map property of preserving accurate shape between the map and the earth's surface.

Cone Of Depression. — Area in which the water table is depressed, usually because of excess pumping of groundwater.

Coniferous. — Cone bearing.

Conservation Of Angular Momentum. — A law of physics that states the momentum of a rotating body equals mass times angular velocity times radius of rotation.

Continent. — Large body of land that stands above sea level.

Continentality. — A term that describes the location of an area on a land mass and the resulting dryness and seasonal temperature variations that occur because of the location.

Convection. — Process by which heat is transferred by moving matter.

Convective Clouds. — Clouds formed from the lifting of warm air above denser surrounding air.

Continentality. — A measure of the degree to which continental interior locations heat and cool more rapidly and to a greater extent than do coastal or insular locations.

Coriolis Effect. — Effect of the earth's rotation tending to turn freely moving objects toward the right in the Northern Hemisphere and to the left in the Southern Hemisphere.

Convection. — Circulation resulting from a non-uniform temperature within a fluid owing to differences in density acted upon by gravity.

Convergence. — The increase of mass within a layer of the atmosphere when there is a net horizontal inflow of air to that layer.

Coriolis Force. — See Coriolis Effect.

Counter Radiation. — Reradiation. Heat radiation that is radiated back to the earth by the atmosphere.

Cultural Landscape. — Natural landscape as transformed by mankind, acting through culture, to create a manmade landscape.

Cumuliform Clouds. — Clouds of globular shape, often with extended vertical development.

Cumulonimbus Clouds. — Large, dense cumuliform clouds yielding precipitation.

Cumulonimbus Mammatus Clouds. — Cumulonimbus which displays a downward rolls or pouches indicative of extreme instability usually associated with severe weather.

Cyclone. — A region of low pressure with winds that spiral toward the center in a counterclockwise wise direction in the Northern Hemisphere and a clockwise direction in the Southern Hemisphere.

Debris Slide. — Small areas of detached, fast moving unconsolidated crustal debris and soil.

Deciduous. — Tree or shrub that sheds its leaves on a seasonal basis.

Declination. — The angular distance of a celestial body north or south of the celestial equator. Also, the angle between true north and magnetic north.

Degree Day. — A departure of one degree of temperature per day from a reference temperature that is used computing heating and air conditioning loads.

Deflation. — Process by which loose surface particles may be lifted or rolled along to be removed from an area by the action of wind.

Deflation Hollows. — See Blowouts.

Delta. — Low-lying wedge of land formed of alluvium and projecting into the sea.

Dendritic. — A random tree-like stream pattern.

Dendritic Stream Pattern. — Branching treelike arrangement of streams that converge into a single channel outlet that are formed on surfaces comprised of relatively homogeneous materials.

Density. — The amount of mass contained in a unit volume. Mass divided by volume.

Deposition. — A change in the state of matter from gas to solid.

Depression Cone. — The drawdown of the water table at a well.

Depression Storage. — Water that is stored in natural surface depressions.

Desert Pavement (Reg). — Desert surface blown free of fine mineral particles by the wind.

Dew. — Moisture that condenses on a cool body.

Dew Point. — Temperature at which the air is fully saturated and below which condensation will occur. The temperature at which saturation pressure = vapor pressure.

Diastrophism. — Reorganization of solid rock materials by forces within the earth.

Diffusion. — To become less concentrated. Dispersal.

Dike. — Platelike layer of igneous rock formed by intrusive igneous activity, often found in a near vertical position and typically cutting across the strata of older rock formations.

Dimitic Lake. — A lake that overturns in the spring and fall.

Dissolved Stream Load. — Mineral matter in a soluble state carried along by a stream.

Diurnal. — Daily.

Dog Days. — A period of hot weather supposedly extending from late July to early September.

Doldrums. — Belt of calm and variable winds located in the vicinity of the equator.

Drainage Basin. — Total surface area occupied by a drainage system and bounded by a drainage divide. The area that is drained by a stream and its tributaries.

Drainage System. — Branched network of stream channels and their adjacent land, bounded by a drainage divide.

Drift. — Glacial debris that is transported by the glacier, and deposited in back of the terminal moraine.

Drizzle. — Form of precipitation comprised of water droplets less than 0.5 mm (0.02 in) in diameter.

Drumlin. — Lens-shaped hill of glacial till formed by plastering of till beneath moving, debris-laden glacial ice.

Dune. — Sandy deposit formed when rapidly moving air, laden with sand particles, encounters an obstruction that causes the wind to lose velocity and deposit its load.

Dust Bowl. — Name given early in 1935 parts of Colorado, Kansas, New Mexico, Texas, and Oklahoma then afflicted with droughts and dust storms.

Dynamic Equilibrium. — A system in which the average input equals the average output and in which there is constant motion.

Dynamic Meteorology. — A branch of meteorology that applies mathematics and physics to explain motions and heat transfers in the atmosphere.

Dyne. — Unit of force required to accelerate one gram one centimeter in one second.

Earthflow. — Moderately rapid downhill flow of masses of water-saturated soil or crustal debris.

Earthquake. — Trembling or shaking of the ground resulting from a shock wave associated with movements of the earth's crust.

Easterly Waves. — A wavelike disturbance in the tropical easterlies.

Ecliptic. — The plane defined by the motion of the center of earth as it orbits the center of the sun.

Ecumene. — Permanently inhabited portion of the earth.

Effluent Stream. — A stream which receives water from groundwater discharge. Also called a gaining stream.

Ekman Spiral. — A theoretical change of wind direction and velocity with height from the surface to the gradient wind (~500 meters).

Electromagnetic Radiation. — Energy waves given off by atoms when electrically charged particles are accelerated, as when an atom's electrons move from one energy level to another. Electromagnetic radiation travels at the speed of light. Light, heat, and radio waves are all examples of electromagnetic radiation.

Electromagnetic Spectrum. — An array of all electromagnetic radiation usually with the short waves to the left and the long waves to the right.

Electron. — A negatively charged atomic particle that orbits the nucleus of an atom.

Elevation. — Vertical distance or height above mean sea level.

Endangered Species. — Life forms facing extinction.

Energy. — A quantity having the dimensions of a force times a distance which is conserved in all interactions within a closed system. The ability to do work.

Entropy. — The energy in a system that is unavailable for work. Entropy has to do, not with the amount of energy in a system, but with its distribution.

Environmental Determinism. — A philosophical doctrine developed in the late nineteenth and early twentieth centuries that stated cultural characteristics are primarily determined by the environment, especially the physical and biotic environment.

Environmental Lapse Rate. — The rate at which the temperature will increase or decrease as one ascends or descends in the atmosphere. The average rate of change is 0.5°C/100M or 3.5°F/1000 Feet.

Environmental Perception. — The concept which holds that people of different cultures perceive, interpret, and use similar natural environments differently.

Epeirogeny. — A term meaning "continent making."

Ephemeral Plants. — Small desert plants that complete a life cycle very rapidly following desert rainfall.

Epilimnion. — A warm circulating layer of water near the surface of the lake.

Epiphytes (Air Plants). — Plants that live above the ground out of contact with the soil.

Equatorial Trough. — Low-pressure trough centered more or less over the equator and lying between the two belts of trade winds.

Equinox. — Days on which the illumination period equals the darkness period, approximately March 21 and September 23. An event that occurs twice a year when the sun's rays cross the equator and the nights and days are of equal length, 12 hours each.

Erg. — A sandy desert.

Esker. — A narrow, often sinuous ridge of coarse gravel and boulders deposited in the bed of a meltwater stream that was enclosed within a tunnel in a stagnant ice sheet.

Evaporation. — A change in the state of matter from liquid to gas. The process by which liquid is changed into vapor.

Evapotranspiration. — The combined water loss to the atmosphere by evaporation from the soil and by transpiration from plants. The sum of the processes of evaporation and transpiration, or the volume of water contributed to the atmosphere by transpiration from plants and evaporation from water and soil for a given area.

Evergreen. — A tree or shrub that holds most of its green leaves throughout the year.

Extratropical Cyclone. — A wave cyclone tracking eastward along the polar front.

Extrusive Vulcanism. — A form of vulcanism wherein fluid rock moves from beneath the surface to points above the surface.

Fault. — A sharp break in rock with displacement of the block on one side with respect to the adjacent block.

Fahrenheit. — Temperature scale named after Daniel Fahrenheit (1686-1736) who was the first to use mercury in a thermometer. Generally the lower limit (0) was as estimation of the temperature at which a solution of water and salt would freeze and the upper limit (100) was based on an approximation of body temperature.

Federation. — A form of political organization that features a considerable measure of local autonomy.

Feedback. — The property of a system which leads the effect of change back to the initial variable. See negative feedback and positive feedback.

Ferrell Cell. — A cell that occurs between 30° and 60° latitude to help explain general atmospheric circulation patterns.

Field Capacity. — The maximum capacity of soil to hold water against the pull of gravity. The amount of water held in the soil after the gravitational water is removed.

Fiord. — A narrow, deep ocean embayment that occupies a glacial trough.

First Law Of Thermodynamics. — Law of Conservation of Energy. In all chemical and physical changes, energy can not be created nor destroyed but can be transformed from one form to another.

Fission. — The splitting of heavy nuclei into two lighter nuclei with the release of large amounts of energy and one or more neutrons.

Flood. — Any high water flow that causes the stream to rise above the natural or artificial banks of the channel.

Floodplain. — An area of low, flat ground that is present on one or both sides of a stream channel and subject to flooding about once annually.

Fluid Drag. — The resistance (friction) a body incurs as it moves through a liquid or gas.

Fluorine. — A chemical element of the halogen family.

Flurocarbons. — Inert chemical compounds composed primarily of fluorine and carbon.

Fog. — A cloud layer in contact with a land or sea surface. A cloud formed near the surface of the earth when humid warm air is cooled to the point of condensation.

Folds. — Wavelike patterns in rock, resulting from crustal compression.

Force. — The cause of movement or acceleration of a object. Magnitude of force equals mass times acceleration.

Fossil Fuels. — Any naturally occurring fuel such as coal, crude oil, or natural gas, formed from the fossil remains of organic materials.

Frequency. — The number of wave crests that pass a given point per unit time, usually per second.

Friction. — The resistance encountered when two bodies are in contact, or the force that offers resistance to motion between two bodies that are in contact.

Friction of Distance. — Resistance to circulation, generally applied to the movement of economic goods, and figured in terms of costs.

Front. — Surface of contact between two unlike air masses.

Frontal Clouds. — Clouds formed when relatively warm and/or moist air is forced to rise over denser, cooler air masses.

Frost. — The direct freezing of condensation on terrestrial objects.

Fusion. — The formation of a heavy nucleus from lighter nuclei, such as hydrogen isotopes, with an attendant release of energy.

Gamma Rays. — A high energy form of electromagnetic radiation.

General Systems Theory. — An inductive approach that considers the properties and interrelationships of the system to create a general body of theory.

Geography. — A field of study that seeks to provide scientific description of the earth as the home of humankind through the accurate, orderly, and rational description, analysis, and interpretation of physical and cultural gradients.

Geostrophic Wind. — Frictionless air in which pressure gradient force and the Coriolis effect are in balance. An upper altitude wind that flows parallel to the isobars.

Geothermal Energy. — The heat energy in the earth's crust whose source is the earth's interior.

Geyser. — A fountain of heated groundwater and steam that normally spouts at irregular intervals.

Ghost Balloons. — Balloons that are released into the atmosphere and tracked in order to gain more information on air movement at different altitudes.

Glacial Moraine. — A till deposit formed at the terminus of a glacier.

Glacial Plucking. — A removal of masses of bedrock from beneath a glacier as the ice moves forward.

Glacial Trough. — A deep, steep-sided valley of a U-shaped cross-section formed by mountain glacier erosion.

Glaciers. — A large natural accumulation of land ice that shows evidence of present or past flow.

Gradient Wind.— Steady, horizontal wind that blows parallel to curved isobars with centrifugal, Coriolis, and pressure gradient forces in balance.

Gram. — Originally a unit of mass equal to 1cc of water at 4°C, but now defined as 0.001 of the standard kilogram.

Graupel. — Soft hail.

Gravity. — Force that attracts objects to the earth's surface equal to an acceleration of 980.665 centimeters per second.

Gravitational (Free) Soil Water. — Surplus precipitation that percolates through the solum to lower levels under the influence of gravity.

Greenhouse Effect. — An atmospheric phenomenon that is analogous to a greenhouse. The short energy waves readily pass through the atmosphere and heat the earth. The longer energy waves emitted by the earth are trapped in the lower atmosphere by water vapor and condensation. If there is little water vapor or CO2 then a large amount of energy is reradiated to outer space creating a wide range in day and night time temperatures. If there is a significant amount of water vapor and CO2 present the reradiated energy is trapped in the lower atmosphere creating a narrow range between night and day temperatures.

Ground Fog. — A shallow but often dense radiation fog caused by a temperature inversion.

Ground (Terrestrial) Radiation. — Long-wave energy emitted from the earth's surface.

Ground Water Table. — The top of the ground water zone.

Ground Water Zone. — Water that is found in the zone of saturation.

Groundwater. — Subsurface water that occupies a saturated zone of loose earth material beneath the crustal surface.

Growing Season. — The length of time between the last frost of spring and the first frost of fall.

Gully. — A narrow shallow gorge that is made by running water.

Hadley Cell. — A model that describes the convectional cell that exists in the atmosphere between 0° and 30°N and 30°S Latitude. This general closed circulation is caused by differential heating and cooling.

Hail. — A form of precipitation consisting of pellets or spheres of ice with a concentric layered structure.

Hail. — Precipitation that is usually spherical in shape with more or less concentric layers of ice.

Halo. — A group of optical effects that appear as colored or whitish rings and arcs about the sun or moon when seen through thin ice crystal clouds.

Halophyte. — A plant tolerant of relatively high quantities of mineral salts in the soil.

Hanging Valley. — A stream valley that has been truncated by erosion so as to appear stranded above the main valley into which it formerly flowed.

Hard Water. — Water that has a large amount of dissolved mineral in it, especially calcium, bicarbonate and magnesium.

Hardwood. — A broad-leaved, generally deciduous tree, as distinguished from a coniferous tree.

Heat. — The kinetic energy of the random translational motions of molecules of a substance.

Heat Sink. — An area, such as the polar zone, that loses more heat than it gains, thus creating a heat deficit.

Heat Source. — An area that receives a great deal of heat energy, thus gaining a heat surplus. This area exports heat energy, e.g. the equatorial region.

High Pressure Ridge. — A high pressure area at the 500 millibar level.

Hoar Frost. — A frost that occurs when water vapor sublimates on terrestrial objects.

Horse Latitudes. — see Subtropical High Pressure System.

Humid Climate. — A moist climate in which surplus precipitation occurs in at least one season of the year.

Humidity. — The amount of water vapor present in the air.

Humus. — The more or less stable fraction of soil organic matter that remains after the major portion of plant and animal residues has decomposed.

Hurricane. — A large tropical storm that generally has a diameter of several hundred kilometers and winds over 75 m.p.h. See Tropical Cyclone.

Hydraulic Action. — The excavation of unconsolidated materials (gravel, sand, silt, and clay) by the force of flowing water exerting an impact and drag effect upon the bed and banks of the stream channel.

Hydraulic Gradient. — A pressure gradient, in which the water pressure depends on the degree and length of the slope of the water surface. Expressed as dh/dl, where dh equals the change in the height of the fluid (head) and dl equals the flow length being considered.

Hydraulic Head. — The potential energy of a ground water system. The total hydraulic head is equal to the sum of the pressure head (P), the elevation head (Z), and the velocity head (v). In considerations of groundwater, velocity head is normally negligible and is ignored.

Hydrologic Cycle. — A model of moisture movement into and out of the atmosphere. The earth subsystem in which the movement of water is described.

Hydrostatic Pressure. — The pressure of a fluid which acts equally in all directions.

Hygroscopic Nuclei. — Small particles that are suspended in the air and serve as condensation surfaces for water vapor.

Hygroscopic Water. — Soil water that is tightly bound to mineral matter within the soil. Water that has adhered to soil particles and is unavailable to plants.

Hypolimnion. — A cold non-circulating layer of water near the bottom of the lake.

Iceberg. — A mass of ice floating in the ocean and derived from the terminal end of a glacier.

Ice Fog. — Fog formed under frigid conditions by sublimation most frequently in higher latitudes. Also known as frost in the air, frozen fog, and pogonip.

Ice Storm. — Glazing by freezing rain.

Iconography. — The spirit or traditions of a group tending to hold it together.

Igneous Rocks. — Rocks that have solidified from a high-temperature molten state.

Impervious. — Impenetrable. In hydrology, the property that prohibits any substance from absorbing water. A surface through which water will not move.

Influent Stream. — A stream from which water flows downward through the ground towards a lower elevation water table. Also called a losing stream.

Infrared. — Electromagnetic waves that have wave lengths between .02 and .00007 cm. These waves are longer than the visible light and shorter than radio waves.

Insolation. — The portion of solar radiation that the earth intercepts. The amount of solar radiation received by the earth.

Instability. — Vertical distribution of temperature such that air, if given either an upward or downward impulse, will tend to move away with increasing speed.

Intercepted Water (Rain). — Falling water that is stopped in its journey to the ground (earth) when it lands on an object in its path, i.e. a leaf, a building, etc.

Interfluve. — Upland area between streams.

Intertropical Convergence Zone (Doldrums). — A belt that lies close to the equator where the trade winds converge to form a zone of calm to light and variable winds.

Intrusive Vulcanism. — A form of vulcanism wherein magma is forced into rock formations that lie beneath the surface of the earth.

Inversion. — A stable atmospheric condition that occurs when warm air is found above cooler air.

Ion. — An atom or group of atoms that carries an electrical charge. If an electron is lost, the atom has a positive charge and is called a cation. If an electron is gained, the atom has a negative charge and is called an anion.

Ionization. — The chemical process by which bases, acids and salts are broken up into positively and negatively charged particles.

Ironstone. — Plinthite that has irreversibly hardened.

Island. — A small body of land that stands above sea level.

Isobar. — A line on a map connecting points of equal barometric pressure.

Isohyet. — A line on a map connecting points of equal precipitation.

Isolated System. — Closed system: A system with clearly defined boundaries across which no transportation of energy or matter occurs.

Isotherm. — A line on a map connecting points of equal temperature.

Isothermal. — Equal temperature or no temperature change.

Jet Stream. — A rapidly moving stream of air that generally has a core of 200 km or less and is about 3 km thick. This stream of air meanders through the atmosphere between 25,000 and 40,000 feet with speeds that are often over 100 miles per hour. Each hemisphere has 2 jet streams, one at approximately 25° and one that ranges between 40° and 60°.

Joint. — A fracture in a body of rock wherein there is no displacement of the rock on either side of the fracture.

Joule. — A practical but very small unit of work equal to less than 0.001 of a BTU.

Jungle. — A type of tropical forest, consisting mainly of low, dense plants.

Katabatic Winds. — Air flowing down an incline such as a mountain wind caused by the cooling of surface air on a slope which in turn becomes more dense and moves down slope.

Kelvin Temperature Scale. — Absolute temperature scale proposed by Kelvin (1848) wherein a single unit is the 0.01 of the distance between freezing (273.13K) and boiling (373.13K) water.

Kinetic Energy. — The energy a body has due to its motion.

Krakatoa. — Volcanic island that in 1883 produced the most violent eruption in recorded history spreading dust around the earth.

Laccolith. — A lens-shaped body of igneous rock formed by intrusive igneous activity.

Lag. — A delayed response to an input in a system. A delay of specific duration.

Laminar Flow. — A sheetlike flow.

Lamprey. — A slender eel-like fish widely distributed in both fresh and salt water.

Land Breeze. — Local wind along costs that blow from land to water during the night.

Landslide. — See Rock slide.

Lapse Rate (Environmental). — The rate at which the temperature will increase or decrease as one ascends or descends in the atmosphere. The average rate of change is 0.5°C/100M or 3.5°F/1000 feet.

Latent Heat. — Energy that is absorbed and held in storage when water is converted from a liquid to a gas, or from a solid into a gas.

Latent - Heat Of Sublimation. — The amount of heat absorbed by a material without it becoming hotter, i.e. ice at 0°C requires an addition of 80 calories/gram of ice to become water.

Latent - Heat Of Vaporization. — The amount of heat absorbed by a material without it becoming hotter, i.e. water at 100°C requires an addition of approximately 540 calories/gram to become water vapor at 100°C and more at lower temperatures.

Latitude. — Angular distance north or south of the equator.

Lava. — Magma that has emerged onto the earth's surface.

Lava Flow Plateau. — A volcanic tableland formed where fluid magma reaches the surface along extensive fractures in the crust.

Leachate. — Substances that are dissolved by water as it percolates through the soil.

Leader Stroke. — The first of the component strokes of a lightning discharge.

Levee. — The banks of a stream which are higher than the surrounding floodplain. These may be natural or man-made.

Liana. — Thick, woody vines.

Lifting Condensation Level. — The level air becomes saturated when it is lifted adiabatically.

Lightning. — Enormous electrical gradients are set up between clouds and between clouds and the ground as water droplets stratify charge between the top, middle, and base of cumulonimbus clouds.

Lignite. — A brown low-grade coal of recent geological origin.

Linear Velocity. — Straight line speed.

Loess. — Wind deposited sediments made up chiefly of silt.

Long Wave Radiation. — Energy of long wavelength form emitted from bodies of relatively low temperature.

Longitude. — Angular distance east or west of the Prime Meridian.

Low Pressure Trough. — A low pressure area at the 500 millibar level.

Lumped System. — A system where similar objects are analyzed together to form generalization about the system, i.e. biosphere, soils of tall grass prairies, etc.

Lux. — A practical metric unit of illumination equal to one lumen per square meter, or one candlepower per square meter.

Magma Chamber. — A source region at considerable depth below the surface of the earth from which fluid rock moves in the processes of vulcanism.

Map Projection.— A systematic representation of latitude and longitude lines on a plane surface.

Mass. — The quantity of matter in an object.

Mass Wasting. — The spontaneous downward movement of soil and crustal debris under the influence of gravity.

Matter. — Anything that has weight and occupies space.

Meanders. — Arcuate or looping bends in a stream. More or less regularly spaced curves in a stream.

Meandering Stream. — A sinuous stream channel in which the bends tend to be regularly spaced.

Mei Scattering. — Scatter the results when a wavelength strikes a molecule that has a diameter equal to the wavelength.

Mesopause. — Upper limit of the mesosphere, approximately 80 km (48 mi) above the earth's surface. The outer layer of air in the lower atmosphere.

Mesosphere. — The atmospheric zone that is between 50 and 85 kilometers above the surface of the earth.

Metalimnion. — Zone of rapid temperature change - transition zone between the epilimnion and hypolimnion.

Metamorphic Rock. — Rock that has been either physically or chemically altered in the solid state by the action of heat, pressure, shearing stress, or infusion of elements, all taking place at a depth substantially beneath the surface.

Meter. — The primary standard of length originally 0.0000000001 of the distance between the equator and the north pole along the Paris meridian, but now the length of a platinum-irridium bar at 0°C.

Microwaves. — Electromagnetic waves with a wavelength between 1 cm and 100 cm.

Millibar. — A unit of pressure equal to one-thousandth of a bar. A millibar equals a force of 1,000 dynes per square centimeter.

Momentum. — The length of time required to bring a moving object to rest under the action of a constant force. The product of mass times velocity.

Monsoon. — Seasonal winds that blow seaward in winter and inland in summer.

Monsoon Winds. — Winds that seasonally reverse their direction of flow.

Mountain Sickness. — Ailment experienced by persons subjected to air pressures much less than that to which they are accustomed. Symptoms include exhaustion, nausea, headache, sleeplessness, nose bleeds, lapse of memory, and inability to think clearly.

Mudflow. — A very fluid and rapid downhill flow of masses of saturated soil and crustal debris.

Natural Gas. — A naturally occurring mixture of hydrocarbons found in porous geologic formations under the earth's surface, often in association with petroleum.

Needleleaf. — A leaf form that is very narrow in relation to its length.

Negative Feedback. — A system feedback that dampens the original action on the system.

Normal Fault. — A fault, generally steeply inclined, along which the hanging-wall block has moved relatively downward.

Normal Temperature Lapse Rate. — The long-term average decrease in temperature as altitude increases.

Nuclear Energy. — The energy released during reactions between atomic nuclei.

Occlusion or Occluded Front. — A condition that occurs when the cold front of a midlatitude cyclone overtakes the warm front and forces the warm air aloft.

Ocean Conveyor. — The path of a global oceanic circulation system that includes a deep cold salty current flowing from the south Atlantic through the Indian to the south Pacific and a warm shallow return current.

Oil Shale. — A sedimentary rock containing a solid organic material called kerogen.

Open Systems (Non-Isolated). — A system in which energy and matter are constantly being supplied and removed across the boundary.

Opportunity Costs. — Profits available from competing alternatives.

Orogeny. — A major geologic episode in which land masses are deformed by folding and faulting, and resulting in mountain formation.

Orographic Clouds. — Clouds formed when air is forced to ascend the slopes of a mountain range.

Outwash Plain. — A flat, gently sloping plain formed by the deposition of sand and gravel by the meltwater streams in front of the margin of an ice sheet.

Overland Flow. — Broad sheets of water that are eventually carried to a stream from interfluve areas.

Oxic Horizon. — A subsurface soil feature of tropical soils that consists of a mixture of oxides of iron and/or aluminum, quartz sands, and clay.

Ozone. — Gas molecules made up of three oxygen atoms.

Ozone Layer. — A region of the atmosphere, 15 km (9 mi) to 55 km (35 mi) in altitude, wherein ozone is concentrated.

Paddy. — A flooded field as used in rice farming in southern and eastern Asia.

Palmen Model. — An atmospheric model that explains vertical and horizontal distribution of air circulation. This model shows turbulent circulation from 60° - 90° latitude.

Parallelism Of Axis. — As the earth orbits the sun, its axis remains parallel at all times. This plus an axis angle of 66 1/2° from the plane of the ecliptic combine to form the seasons.

Peneplain. — A land surface of slight relief and low elevation.

Percolation. — One of the processes by which water moves down through the soil.

Permafrost. — Permanently frozen subsoil.

Photochemical Reaction. — A chemical reaction that is triggered by light, usually sunlight.

Photon. — A particle of light. A subatomic particle with zero rest mass.

Photosynthesis. — A process whereby plants utilize solar energy to produce chemical energy in the form of carbohydrates.

Photosynthesis. — A biological process that takes place in green plants where carbon dioxide and water, in the presence of chlorophyll and sun energy, are transformed into carbohydrates and oxygen.

Physical Weathering. — The breakup of massive rock into small particles (a process of disintegration) through the action of physical forces acting at or near the earth's surface.

Physiological Density. — The number of people per unit of arable land.

Piezometric Surface. — The top of the water table.

Plane of the Ecliptic. — A theoretical plane in space which coincides with the earth's orbit around the sun.

Plankton. — Small marine organisms that float freely in the water.

Plant Communities (Plant Associations). — A group of plants that coexist in a fashion that appears to be mutually beneficial to one another.

Plant Succession. — A predictable evolutionary sequence in the occupance of a site by plants.

Plate Tectonic Theory. — A means to explain the processes that move the earth's crust about and change the relative positions of land masses.

Plateau. — An upland area, more or less flat and horizontal.

Plinthite. — An iron-rich mixture of clay with quartz that commonly occurs as dark red mottles within the soil.

Plucking. — An erosion process associated with glaciation where the glacial ice freezes to surface material and pulls it loose as it moves.

Polar Easterlies. — System of easterly surface winds located at high latitudes. Winds that blow from the high pressure at the poles to the low pressure zones that exist at approximately 60° latitude. The winds are deflected to the right in the Northern Hemisphere and to the left in the Southern Hemisphere.

Polar Front. — Front lying between cold polar air masses and warm tropical air masses. A mid-latitude front that divides the polar and tropical air masses. This front is best developed during the winter when it can dip as far equatorward as 30° latitude.

Polar High. — A persistent center of high atmospheric pressure over cold surfaces at high latitudes.

Polders. — A tract of low land reclaimed from the sea by the use of dikes and dams.

Pool. — A deepened portion of a sinuous stream.

Porosity. — The ratio of pore space to the total volume of a given unit.

Positive Feedbacks. — A system feedback that reinforces the original action on the system.

Potential Energy. — The energy a body has due to its position.

Potential Evapotranspiration. — The maximum amount of evapotranspiration that will occur provided there is never a deficit of water.

Potential Temperature. — Temperature of air if it were reduced adiabatically to surface pressure (1,000 mb).

Pound. — Pound is used to denote both mass and force. When used to denote mass, it describes the amount of material contained in an object which is a constant. When used to denote force, it is a measure of weight which is the force exerted by that object as a result of gravitational forces acting on that object. The metric system denotes mass in kilograms and force in newtons.

Precipitation. — The falling to earth of various forms of condensation.

Pressure. — Force per unit area. Force divided by area.

Pressure Gradient Force. — Change of atmospheric pressure along a horizontal line. A force that exists due to the location of atmospheric pressure systems. The greater (steeper) the millibar surface, the greater the gradient force and vice versa. A force that is proportional to the difference and distance between two pressure systems.

Prevailing Westerlies. — Surface winds blowing from a generally westerly direction in the midlatitudes, but varying greatly in intensity and direction.

Pseudoadiabatic.— A non-reversible process wherein condensation occurs and latent heat is released into the atmosphere.

Quasi-biennial oscillation. — Alternating periods of easterly or westerly wind flow in the equatorial stratosphere having a periodicity of 24 to 30 months between successive regimes.

Radial Stream Patterns. — A pattern of stream channels formed where an elevated structure, such as a volcano or dome, exists.

Radiation. — A process by which energy is emitted either as tiny particles (alpha rays) or as electromagnetic waves (X-rays, gamma rays).

Radiation Fog. — Fog produced by radiational cooling of the air in contact with the earth's surface. A fog that is formed when rapid terrestrial radiation cools the contact layer. If there is sufficient water vapor present fog will occur. See Ground Fog.

Radiation Pollution. — Occurs when levels of radioactivity exceed the planetary norm.

Radioactivity. — A naturally occurring process wherein atoms of certain substances undergo spontaneous nuclear disintegration and emit both high-speed particles and penetrating electromagnetic rays.

Rain. — A form of precipitation composed of water droplets more than 0.5m (0.02 in) in diameter. Precipitation that falls in liquid form.

Rain Shadow. — A dry area that occurs on the lee side of a barrier, usually a mountain. Area of dry climate to the lee of a mountain barrier, produced as a result of adiabatic warming of descending air.

Recharge. — The water that is put back into the ground water system.

Refraction. — Change in direction of propagation that occurs when sound or light waves pass obliquely from one medium to another of different density.

Reg. — See Desert pavement.

Region. — An area defined and delimited on the basis of criteria that provide it with homogeneity. A relatively homogeneous area usually defined on the basis of matter and energy transfers.

Regional Method. — An analytic approach toward determining the character of area that requires an assumption that some part of the earth contains homogeneous landscape elements, and then proceeds with an examination of the area's component factors to explain the interrelationships providing for the region's identity.

Relative Humidity. — Ratio of water vapor present in the air to the maximum quantity it could hold at the same temperature. The ratio of the amount of water vapor in a given volume of air to the amount of water vapor that could be held by the same volume at the same temperature. The ratio between actual vapor pressure and saturation vapor pressure.

Resultant Wind. — Vectorial average of all wind direction and speeds at a given place for a certain period.

Reverse Fault. — A fault, generally steeply inclined, along which the hanging-wall block has moved relatively upward.

Rhumb Line. — Navigational term defined as a curved line on the earth's surface crossing all meridians at a constant angle.

Riffle. — A shallow rapids in a stream. Riffles usually alternate with pools.

Rill. — A very small gully.

Rock Slide (Landslide). — The rapid sliding of large masses of bedrock from steep mountain slopes, from high cliffs, and from unstable slopes of stream banks and shorelines.

Rossby Wave. — Very slow moving undulations in the upper air westerlies. Also known as longwaves.

Runoff. — A term encompassing the physical movement of water, under the influence of gravity, toward lower elevations, and ultimately the sea.

Saltation. — A leaping, impacting, and rebounding motion of materials moved by the wind or water.

Saturation. — An atmospheric moisture condition such that no additional net moisture can be added to the atmosphere. The state reached when all the voids between soil particles have been filled with liquid.

Saturation Vapor Pressure. — The pressure exerted by water vapor when the atmosphere is saturated. See Supersaturation.

Scalar. — Having only magnitude as opposed to vectors that have magnitude and direction.

Scatter, Nonselective. — Reflection of light in all directions from large spherical surfaces.

Sea Breeze. — Local wind along coasts that blow from sea to land during the daylight hours.

Seasonal Aridity. — A climatic characteristic of a region wherein a given season normally receives scant amounts of precipitation.

Second Law Of Thermodynamics. — All systems spontaneously go from a state of order to disorder. Systems spontaneously increase entropy.

Sedimentary Rock. — Rock formed from the accumulated sediments derived from preexisting rocks.

Selva. — The tropical rainforest vegetation.

Semideciduous. — Plants that shed their leaves at intervals not in phase with a season.

Sensible Temperature. — Apparent temperature indicated by the sensations of the human body.

Shield Volcano. — A domelike accumulation of lava flows.

Short Wave Radiation. — Energy of short wavelength form emitted from very hot radiating surfaces, such as the sun.

Sill. — An intrusive igneous rock in the form of a plate and created where magma was forced into a natural parting in the bedrock.

Site. — The immediate area of a specific place.

Situation. — The regional or relative position of a place with respect to other places or regions.

Sleet. — A form of precipitation consisting of ice pellets, which may be frozen raindrops. Frozen precipitation, grains of ice.

Slurry. — A solid suspended in a liquid so it can be pumped through a pipeline.

Smog. — Term coined in 1905 by Des Voeux to signify a mixture of smoke and fog, but today loosely used to refer to air pollution generally.

Snow. — A form of precipitation consisting of ice particles.

Snow Blindness. — Temporary blindness caused by the glaring light reflected from snow surfaces.

Softwood. — A coniferous (cone-bearing) tree.

Soil. — A natural layer at the outermost part of the earth's crust containing living matter and supporting or capable of supporting plant life.

Soil Creep. — The extremely slow downhill movement of soil and crustal debris as a result of continued agitation and disturbance of the surface particulate matter by activities such as frost action, temperature change, or the wetting and drying cycles of the soil.

Soil Profile. — A vertical representation of the soil from the surface downward through the solum.

Soil Reaction. — The degree of acidity or alkalinity of a soil, usually expressed as a pH value.

Soil Structure. — The manner in which soil particles are clustered together to form aggregates or lumps.

Soil Texture. — The proportion of mineral particles of given sizes that comprise the soil.

Solar Constant. — The amount of solar energy received on a plane perpendicular to the sun's rays at the outer edge of the atmosphere.

Approximately 1.95 Langleys per square centimeter. 1.95 calories per square centimeter per minute. An amount of energy reaching the earth's atmosphere when it is at an average distance from the sun. A value ranging from about 1,358 to 1,380 watts per square meter.

Solar Energy. — The electronmagnetic radiation transmitted by the sun.

Solar Radiation. — Radiant energy from the sun. See Electromagnetic Radiation.

Solar System. — A celestial family of which the earth is a member.

Solstice. — Moment of most poleward solar declination.

Solum. — Soil.

Specific Heat. — The amount of heat required to raise the temperature of a given volume of a substance a given number of degrees. For water, usually 1 gram and 1 degree Celsius.

Specific Humidity. — The weight of moisture in the air per unit weight of air.

Spodic Horizon. — A subsurface soil horizon in which aluminum, iron, and organic matter have accumulated.

Stable Air. — Air that resists lifting.

Standard Time. — Time used at a place in accordance with international agreement in 1884 linking time zones with standard meridians.

Steady State. — A situation in which output or system response remains constant.

Stefan's Law. — A fundamental law of radiation that states a black body will emit radiation from its surface in proportion to the fourth power of its absolute temperature.

Stemflow. — Water that flows to the ground along the stems of a plant.

Steppe. — Term used to identify a semiarid climate region or the type of vegetation it contains.

Strata. — Layers.

Stratified Drift. — A glacial deposit of materials sorted by glacial melt waters prior to their being deposited.

Stratiform Clouds. — Clouds of layered, blanket-like forms.

Strato Clouds. — Cloud types of the low-height family, below 2 km (6,500 ft.), formed into dense, dark gray layers.

Strato Volcano. — A volcano constructed of alternate layers of lava flows and coarse blocky volcanic ejecta.

Stratopause. — The upper reaches of the stratosphere, approximately 50 km (30 mi) from the earth's surface. The transition zone between the stratosphere and the mesosphere.

Stratosphere. — A strata of air that surrounds the troposphere and extends to an altitude of 50 km (30 mi). The part of the atmosphere between the troposphere and the mesosphere. The temperature changes very little with altitude in the stratosphere.

Stream. — A form of channelized flow of any size, including small brooks and large rivers.

Stream Abrasion. — The grinding action of streams as they strike suspended rock fragments against their channel walls.

Stream Corrosion. — The process by which soluble rock materials in stream valleys are dissolved by their streams.

Strike-Slip Fault. — A fault on which displacement has been horizontal.

Strip Mining. — Surface, or open-pit, mining.

Sublimation. — A change in the state of matter from solid to gas. A process by which a solid is converted to a vapor without becoming a liquid first or vice versa.

Subsolar Point. — The point where the sun's rays are perpendicular with the earth's surface.

Subtropical. — Pertains to the climate found at the tropical margins of the temperate zones.

Subtropical High Pressure (Horse Latitudes). — A high pressure region centered around the 30°N and 30°S. This is an area of descending and diverging air. Areas of persistent high atmospheric pressure, trending east-west and centered about 30° N and S.

Summer. — The time of year when the latitude of the subsolar point is closest.

Summer Solstice. — The date when the angle of the sun is greatest. In the northern hemisphere, that moment when the subsolar point reaches the Tropic of Cancer (23.5°N), its most poleward migration into the northern hemisphere. Occurs on or about June 21.

Sunspots. — Dark areas on the surface of the sun. Periods of great sunspot activity are associated with minute increases in insolation.

Supersaturation. — A condition that exists when a fluid holds more of a substance than is needed for 100% saturation. This can occur in the atmosphere and in liquids.

Surface Detention. — Water that is held to soil particles as a thin film.

Surface Tension. — That property by which the surface molecules of all liquids tend to exert mutually attractive forces so that a film with the least surface area is produced.

Suspended Stream Load. — Materials carried by a stream in floatation.

Syncline. — A troughlike fold in rock strata wherein the limbs of the fold are oriented upward.

Synoptic. — Atmospheric conditions existing at a given time over an extended region drawn from simultaneous observations.

Synoptic Disturbances. — Daily weather disturbances.

Synoptic Hours. — 0000, 0600, 1200, 1800 Greenwich mean time.

Systematic Method. — An analytical approach toward determining the character of area that involves examining the areal variation of individual landscape elements (whether cultural, economic, or physical).

Systems Analysis. — An examination and an understanding of the components of a system and their relationships to each other in order to understand the functioning of the entire system.

Taconite. — A hard flint-like rock that serves as an ore of iron.

Taiga. — See Boreal forest.

Tar Sand. — A sandy geologic deposit in which oil is found.

Temperature. — The average kinetic energy of the molecules of a substance.

Temperature Range. — Difference between the maximum and minimum temperatures.

Terminal Velocity. — The maximum speed obtained by an object falling freely through a fluid.

Terrestrial Radiation. — See Ground Radiation.

The Green Revolution. — A term applied to efforts to increase food production in the Third World by introducing high-technology methods (high-yield plants, fertilizers, and pesticides).

Thermal Equilibrium. — A situation where heat input equals heat output - a steady state.

Thermal Gradient. — In earth or water, the rate at which temperature increases or decreases with depth below the surface. A temperature difference between two bodies or areas which causes heat to flow. The intensity of the gradient is directly related to the closeness of the points and the difference in temperature. Heat always flows from hot to cold.

Thermals. — Updrafts of heated air. Convective cells established because of differential heating and cooling.

Thermodynamic Temperature. — Temperature measured as average kinetic energy of molecules. If molecules are dispersed heat content may be low despite high thermodynamic temperature.

Throughflow (Interflow). — Downslope movement of water at or just below the surface.

Thunderstorm. — An intense, local convectional storm yielding heavy precipitation along with lightning and thunder. Local atmospheric disturbance that is accompanied by thunder and lightning. Thunderstorms usually develop in association with warm, humid, unstable air. These storms can be locally severe.

Tides. — Periodic fluctuations in the elevation of the sea's surface.

Till. — A glacial deposit of an unsorted mixture of clays, sand, cobbles, and boulders.

Tornado. — The smallest and most violent cyclonic storm having intense winds and extremely low air pressure. A small, severe storm that occurs when a vortex extends toward the earth usually from a thunder cloud.

Trade Winds. — Surface winds of the low latitudes. Easterly winds that blow between 30°N and 30°S and the equator.

Transient System Responses. — System responses produced by infrequent, large or irregularly spaced inputs, i.e. floods, drought, forest fire.

Transpiration. — The loss of water to the atmosphere from the leaf pores of plants. Loss of water vapor by plants, primarily through openings in the leaves called stomata.

Tree Farm. — Where trees are raised as a crop.

Trellis Stream Pattern. — A system of stream channels formed in surface materials made up of parallel bands of rock that contrast in their resistance to erosion.

Tropic of Cancer. — 23 1/2° North Latitude - The line that marks the most northern limit of the sun's perpendicular rays.

Tropic of Capricorn. — 23 1/2° South Latitude - The line that marks the most southern limit of the sun's perpendicular rays.

Tropical Cyclone (Hurricane). — An intense traveling cyclone of tropical and subtropical latitudes, accompanied by high winds and heavy rainfall.

Tropopause. — Upper limit of the troposphere, averaging 20 km (12 mi) in altitude at the equator and 10 km (6 mi) at the poles. The zone that separates the troposphere from the stratosphere.

Troposphere. — Layer of atmosphere in contact with the earth's surface, wherein temperatures normally decrease with increased altitude. The lowest level of the atmosphere. The troposphere is the zone of active weather.

Tundra. — Climate and associated vegetation between 0°C in the north and 10°C in the south. Only occurs in the northern hemisphere.

Turbid. — A cloudy or muddy condition in fluids caused by suspended particulates.

Turbulence. — A disturbance in flow, or an uneven flow. Random departures from the mean flow in a fluid. A heterogeneous flow.

Typhoon. — See Hurricane.

Ultraviolet. — Electromagnetic waves that are shorter than visible light but longer than X-rays.

Unstable Air. — Air that spontaneously rises because it is warmer than surrounding air.

Upwelling. — The rising of water from deep layers of the ocean to the surface.

Vadose Zone. — The layer of ground above the zone of saturation consisting of soil, rock, air, and water. Also called the zone of aeration.

Vapor Pressure. — The pressure exerted by water vapor in a given volume of air.

Vapor Pressure Gradient. — A gradient created by the difference in vapor pressure between two points or areas.

Vector. — An arrow which describes both a direction and magnitude of force.

Velocity.— Distance covered per unit time. Distance divided by time.

Vegetation. — A mosaic of plant assemblages in which the individual plant is the basic structural unit.

Ventifacts. — Wind-blasted cobbles that have faceted surfaces joined in sharp edges.

Vernal Equinox. — Occurs on or about March 21. See Equinox.

Viscosity. — The resistance of a fluid to internal flow.

Volcanic Ash. — Finely divided igneous rock blown under gas pressure from a volcano.

Volcano. — A conical, circular structure built of accumulations of lava and volcanic ash.

Vorticity. — Rotational circulation of air about an axis that is usually vertical but may be oriented in any direction.

Vulcanism. — The redistribution of fluid rock.

Warm Front. — Line of discontinuity along the earth's surface where the forward edge of an advancing mass of relatively warm air is replacing a retreating colder air mass.

Warm Fog. — A fog that is made up primarily of water droplets.

Water Budget. — A model that is used to account for water in the earth system and how it moves from one subsystem to another.

Water Table. — The upper limit of the saturated zone of groundwater.

Wavelength. — The distance from wave crest to the adjacent wave crest or from wave trough to adjacent wave trough.

Weather. — The physical state of the atmosphere at a given place and time.

Weathering. — The total of all processes acting at or near the earth's surface to cause physical disruption and chemical decomposition of rock.

Westerlies. — Prevailing winds that are found in the regions from 30° to 60° North and South Latitude.

Wien's Displacement Law. — The wavelength of maximum radiation intensity is inversely proportional to the absolute temperature of the radiating body.

Willy - Willy. — See Hurricane.

Wilting Point. — The point at which the quantity of water stored in a plant is insufficient to prevent a plant from wilting (unless it is adapted to dry conditions). The point at which no soil water is available to plants.

Wind. — The horizontal movement of air.

Wind Shadow. — An area behind an obstacle wherein wind speed is reduced.

Winter. — The time of the year when latitude of the subsolar point is most distant.

Winter Solstice. — The date when the angle of the sun is smallest. In the northern hemisphere, that moment when the subsolar point reaches the Tropic of Capricorn (23.5°S), its most poleward migration into the southern hemisphere. Occurs on or about December 21.

Woodland. — A form of forest in which the trees are widely spaced, the canopy cover being only 25 to 60 percent.

Work. — The displacement of an object by a force. The energy expended in moving a mass. Work equals Force times Distance.

X-Rays. — High energy shortwave electromagnetic radiation.

Xerophyte. — A plant adapted to a dry environment.

Zonal Index. — Measure of the intensity of the west-east component of the general circulation, usually between 35° and 55°N. High index numbers (+15) indicate strong west-east flow, and low index numbers (+5 to -2) indicate greater meridional flow.

Further Readings

Adams, R.M., Fleming, R.A., Chang, C.C., McCarl, B.A. and Rosenzweig, C., 1995. A Reassessment of the Economic Effects of Global Climate Change on US Agriculture. *Climatic Change*, 30(2): 147-168.

Adams, R.M. et al., 1990. Global climate change and US agriculture. *Nature*, 345(6272): 219-223.

Alavaja, M.C.R. et al., 1994. Residential Radon Exposure and Lung Cancer Among Nonsmoking Women. *Journal of the National Cancer Institute*, 86(24): 1829-1837.

Albers, S.C., McGinley, J.A., Birkenheuer, D.L. and Smart, J.R., 1996. The local analysis and prediction system (LAPS): Analyses of clouds, precipitation, and temperature. *Weather and Forecasting*, 11(3): 273.

Aldhous, P., 1991. Climate Change. *Nature*, 349(6306): 186.

Alexander, M.A., 1992. Midlatitude Atmosphere Ocean Interaction During El Niño .2. The Northern Hemisphere Atmosphere. *Journal of Climate*, 5(9): 959-972.

Anderson, D.G., 1970. Effects of Urban Development on Floods in Northern Virginia. Water Supply Paper 2001-C, USGS.

Arkin, P. and Janowiak, J., 1991. Analyses of the Global Distribution of Precipitation. *Dynamics of Atmospheres and Oceans*, 16(1-2): 5-16.

Atkinson, B.W., 1981. Meso-scale Atmospheric Circulations. Academic Press, New York.

Avila, A., 1996. Time trends in the precipitation chemistry at a mountain site in Northeastern Spain for the period 1983-1994. *Atmospheric Environment*, 30(9): pp. 1363-1373.

Bakun, A., 1990. Global Climatic Change and Intensification of Coastal Ocean Upwelling. *Science*, 247: 198-201.

Balling, Jr., Robert C. 2000. The Geographer's Niche in the Greenhouse Millennium. *Annals of the Association of American Geographers* 90 (1): 114-122.

Barber, R.R. and Chavez, F.P., 1983. Biological Consequences of El Niño. *Science*, 222(December 16, 1983): 1203-1210.

Bardach, J.E., 1989. Global Warming and the Coastal Zone, Some Effects on Sites and Activities. *Climate Change*, 15(1-2): 117-150.

Barry, R.G., 1969. Evaporation and Transpiration. In: R.J. Chorley (Editor), *Water, Earth and Man*. Methuen and Co., London, pp. 169-184.

Beggs, P., 1998. Pollen and pollen antigen as triggers of asthma - what to measure ? *Atmospheric Environment*, 32(10): 1777-1783.

Bell, G.D. and Bosart, L.F., 1994. Midtroposheric Closed Cyclone Formation Over the Southwestern United States. *Monthly Weather Review*, 122(5): 791-813.

Bigg, G.R., 1997. Currents of change: El Niño's impact on climate and society. *Atmospheric Environment*, 31(10): 1583-1583.

Bosart, L.F. et al., 1996. Large-scale antecedent conditions associated with the 12-14 March 1993 cyclone ("superstorm '93") over eastern North America. *Monthly Weather Review*, 124(9): 1865.

Brauer, M. et al., 1995. Measurement of Acidic Aerosol Species in Eastern Europe: Implications for Air Pollution Epidemiology. *Environmental Health Perspectives*: 103:482-488.

Brecque, M.L., 1989. Detecting Climate Change I: Taking the world's shifting temperature. *Mosaic*, 20(4): 2-9.

Brecque, M.L., 1989. Detecting Climate Change II: The impact of the water budget. *Mosaic*, 20(4): 10-17.

Budyko, M.I. et al., 1962. The Heat Balance of the Surface of the Earth. *Soviet Geography: Review and Translation*, 3.5: 3-16.

Businger, S., 1991. Arctic Hurricanes. *American Scientist*, 79(1): 18.

Businger, S., Knapp, D.I. and Watson, G.F., 1990. Storm Following Climatology of Precipitation Associated with Winter Cyclones Originating over the Gulf-of-Mexico. *Weather and Forecasting*, 5(3): 378.

Canby, T.Y., 1984. El Niños Ill Wind. *National Geographic*, 171(2): 144-183.

Cane, M.A., 1986. El Niño. *Ann. Rev. Earth Planet Sci.* 1986, 14(1986): 43-70.

Cao, M., Gregson, K. and Marshall, S., 1998. Global methane emission from wetlands and its sensitivity to climate change. *Atmospheric Environment*, 32(19): pp. 3293-3299.

Chang, J.-H., 1968. *Climate and Agriculture*. Aldine Pub. Co., Chicago.

Chapman, E.G. and Luecken, D.J., 1993. Associations Between Pollutant Emissions and Precipitation Chemistry - An Empirical Analysis. *Journal of Geophysical Research*, 98(D1): 1113.

Charles, C.D., Hunter, D.E. and Fairbanks, R.G., 1997. Interaction between the ENSO and the Asian monsoon in a coral record of tropical climate. *Science*, 277(5328): 925 (Reports).

Chen, T.C., Yen, M.C. and Schubert, S., 1996. Hydrologic processes associated with cyclone systems over the United States. *Bulletin of the American Meteorological Society*, 77(7): 1557.

Chu, P.S., Yu, Z.P. and Hastenrath, S., 1994. Detecting Climate Change Concurrent with Deforestation in the Amazon Basin- Which Way Has It Gone? *Bulletin of the American Meteorological Society*, 75(4): 579-584.

Clark, K., Nadkarni, N., Schaefer, D. and Gholz, H., 1998. Cloud water and precipitation chemistry in a tropical Montane Forest, Monteverde, Costa Rica. *Atmospheric Environment*, 32(9): pp. 1595-1603.

Cody, R.P., Weisel, C.P., Birnbaum, G. and Lioy, P.J., 1992. The Effect of Ozone Associated with Summertime Photochemical Smog on the Frequency of Asthma Visits to Hospital Emergency Departments. *Environmental Research*, 58: 184-194.

Cohen, A.J. and III, C.A.P., 1995. Lung Cancer and Air Pollution. *Environmental Health Perspectives*, 103(Supplement 8): 219-224.

Crosson, P., 1989. Climate Change and Mid-Latitude Agriculture: Perspectives on Consequences and Policy Responses. *Climatic Change*, 15(1-2): 51-74.

Dansereau, P., 1957. *Biogeography; An Ecological Perspective*. Ronald Press, New York.

Delfino, R.J., Becklake, M.R. and Hanley, J.A., 1994. The Relationship of Urgent Hospital Admissions for Respiratory Illnesses to Photochemical Air Pollution Levels in Montreal. *Environmental Research*, 67: 1-19.

Dockery, D.W. and III, C.A.P., 1994. Acute Respiratory Effects of Particulate Air Pollution. *Annual Review Public Health*, 15: 107-32.

Dolph, J. and Marks, D., 1992. Characterizing the Distribution of Observed Precipitation and Runoff over the Continental United-States. *Climatic Change*, 22(2): 99.

Dooge, J.C.I., 1992. Hydrologic Models and Climate Change. *Journal of Geophysical Research - Atmospheres*, 97(D3): 2677-2686.

Dorling, S.R., Davies, T.D. and Pierce, C.E., 1992. Cluster Analysis - A Technique for Estimating the Synoptic Meteorological Controls on Air and Precipitation Chemistry - Method and Applications. *Atmospheric Environment*, 26(14): 2575.

Dunne, T. and Leopold, L.B., 1978. *Water in Environmental Planning*. W. H. Freeman, New York.

Easterling, W.E. et al., 1993. Agricultural Impacts of and Responses to Climate Change in the Missouri-Iowa-Nebraska-Kansas (MINK) Region. *Climatic Change*, 24(1-2): 23-62.

Ebert, P. and Baechmann, K., 1998. Solubility of lead in precipitation as a function of rain drop size. *Atmospheric Environment*, 32(4): 767-771.

Elsner, J.B., Lehmiller, G.S. and Kimberlain, T.B., 1996. Objective classification of Atlantic hurricanes. *Journal of Climate*, 9(11): 2880.

Emanuel, K.A., 1991. Carbon Dioxide and Hurricanes - Implications of Northern Hemispheric Warming for Atlantic Carribbean Storms - Comments. *Meteorology and Atmospheric Physics*, 47(1): 83-84.

Evans, J.L., 1993. Sensitivity of Tropical Cyclone Intensity to Sea Surface Temperature. *Journal of Climate*, 6(6): 1133-1140.

Fairley, D., 1990. The Relationship of Daily Mortality to Suspended Particulates in Santa Clara County, 1980-1986. *Environmental Health Perspectives*, 89: 159-168.

Fernandez Paartagas, J. and Diaz, H.F., 1996. Atlantic hurricanes in the second half of the nineteenth century. *Bulletin of the American Meteorological Society*, 77(12): 2899.

Fleagle, R.G. and Businger, J.A., 1963. An Introduction to Atmospheric Physics. Academic Press, New York.

Fokianos, K., Kedem, B. and Short, D.A., 1996. Predicting precipitation level. *Journal of Geophysical Research - Atmospheres*, 101(D21): 26473.

Fontham, E.T.L. et al., 1994. Environmental Tobacco Smoke and Lung Cancer in Nonsmoking Women. *Journal of the American Medical Association*, 271: 1752-1759.

Fosdick, E.K. and Smith, P.J., 1991. Latent Heat Release in an Extratropical Cyclone that Developed Explosively over the Southeastern United-States. *Monthly Weather Review*, 119(1): 193.

Frederick, K.D., 1993. Climate Change Impacts on Water Resources and Possible Responses in the MINK Region. *Climatic Change*, 24(1-2): 83-116.

Frederikson, J.S. and Frederikson, C.S., 1993. Monsoon Disturbances, Intraseasonal Oscillations, Teleconnection Patterns, Blocking, and Storm Tracks of the Global Atmosphere During January 1979 - Linear Theory. *Journal of the Atmospheric Sciences*, 50(10): 1349-1372.

Freydier, R., Dupre, B. and Lacaux, J., 1998. Precipitation chemistry in intertropical africa. *Atmospheric Environment*, 32(4): pp. 749-765.

Fulton, T. and Braestrup, P., 1981. The New Issues: Land, Water, Energy. *Wilson Quarterly* (Summer): 120-137.

Geiger, R., 1959. *The Climate Near the Ground*. Harvard University Press, Cambridge, Massachusetts, p. 494.

Germehl, P., Dammrath, W. and Gross, H., 1980. *Physical Geography*. Saunders College, Philadelphia.

Gibbs, W.W., 1996. Where the wind blows. *Scientific American*, 275(No. 6): 44.

Giorgi, F., Marinucci, M.R. and Visconti, G., 1992. A 2XCO2 Climate Change Scenario over Europe Generated Using a Limited Area Model Nested in a General Circulation Model 2. Climate Change Scenario. *Journal of Geophysical Research*, 97(D9): 10011.

Glantz, S.A. and Parmley, W.W., 1995. Passive Smoking and Heart Disease. *Journal of the American Medical Association*, 273(13): 1047-1053.

Goldenberg, S.B. and Shapiro, L.J., 1996. Physical mechanisms for the association of El Niño and west African rainfall with Atlantic major hurricane activity. *Journal of Climate*, 9(6): 1169.

Goldsmith, J.R., Griffith, H.L., Detels, R., Beeser, S. and Neumann, L., 1983. Emergency Room Admissions, Meteorologic Variables, and Air Pollutants: A Path Analysis. *American Journal of Epidemiology*, 118(5): 759-778.

Gordon, A.H., Bye, J.A.T. and ByronScott, R.A.D., 1996. Is global warming climate change? *Nature*, 380(6574): 478.

Graedel, T.E. and Crutzen, P.J., 1989. The Changing Atmosphere. *Scientific American*, 261(3): 58-68.

Graf, W., 1985. *The Colorado River*. Assoc. of American Geographers, Washington, D. C.

Greene, J.S., 1996. A synoptic climatological analysis of summertime precipitation intensity in the eastern United States. *Physical Geography*, 17(5): 401.

Guetter, A.K. and Georgakakos, K.P., 1996. Are the El Niño and La Nina predictors of the Iowa River seasonal flow? *Journal of Applied Meteorology*, 35(5): 690.

Gutowski, W.J., Gutzler, S.D. and Wang, W.C., 1991. Surface Energy Balances of 3 General Circulation Models-Implications for Simulating Regional Climate Change. *Journal of Climate*, 4(2): 121.

Hagelberg, T. and Mix, A.C., 1991. Climate - Long-Term Monsoon Regulators. *Nature*, 353(6346): 703-704.

Handler, P., 1990. USA Corn Yields, the El Niño and Agricultural Drought - 1867-1988. International *Journal of Climatology*, 10(8): 819.

Hanna, E., 1996. The role of Antarctic sea ice in global climate change. *Progress in Physical Geography*, 20(4): 371.

Hare, F.K., 1960. The Westerlies. Geographical Review, 50: 345-367.

Hare, F.K., 1989. Climatology and Meteorology. In: J.G. Henry and G.W. Heinke (Editors), *Environmental Science and Engineering*. Prentice Hall, Englewood Cliffs.

Harwell, M.A., 1993. Assessing the Effects of Global Climate Change - The Pan-Earth Project Series. *Climatic Change*, 23(4): 287-292.

Hegerl, G.C. and North, G.R., 1997. Comparison of statistically optimal approaches to detecting anthropogenic climate change. *Journal of Climate*, 10(5): 1125.

Henderson, K.G. and Vega, A.J., 1996. Regional precipitation variability in the southern United States. *Physical Geography*, 17(2): 93.

Hendersonsellers, A. et al., 1993. Tropical Deforestation - Modeling Local-Scale to Regional-Scale Climate Change. *Journal of Geophysical Researc - Atmospheres*, 98(D4): 7289-7317.

Hodanish, S. and Gray, W.M., 1993. An Observational Analysis of Tropical Cyclone Recurvature. *Monthly Weather Review*, 121(10): 2665-2689.

Hoggan, 1989. *Computer–Assisted Floodplain Hydrology and Hydraulics*. McGraw–Hill, New York, 518 pp.

Holdridge, L.R., 1947. Determination of World Plant Formations from Simple Climate Data. *Science*, 105: 367-368.

Holland, G.J. and Lander, M., 1993. The Meandering Nature of Tropical Cyclone Tracks. *Journal of the Atmospheric Sciences*, 50(9): 1254-1266.

Houghton, J. T., Meira Filho, L. G., Callander, B. A., Harris, N., Kattenberg, A.; and Maskell, K., eds. 1996. *Climate Change 1995: The Science of Climate Change*. Cambridge: Cambridge University Press.

Houghton, R.A., 1991. Tropical Deforestation and Atmospheric Carbon Dioxide. *Climate Change*, 19(1-2): 99-119.

Houghton, R.A. and Woodwell, G.M., 1989. Global Climatic Change. *Scientific American*, 260(4): 36-44.

Hromadka II, T.V., McCuen, R.H. and Yen, C.C., 1987. *Computational Hydrology in Flood Control Design and Planning*. Lighthouse Publications, Mission Viejo, California, p. 510.

Hubbert, G.D., Holland, G.J., Leslie, L.M. and Manton, M.J., 1991. A Real-Time System for Forecasting Tropical Cyclone Storm Surges. *Weather and Forecasting*, 6(1): 86.

Hulme, M., 1992. A 1951-80 Global Land Precipitation Climatology for the Evaluation of General Circulation Models. *Climate Dynamics*, 7(2): 57-72.

I.P.C.C., 1995. *Climate Change 1994*. Cambridge University Press, Cambridge, p. 339.

Ishikawa, Y. and Hara, H., 1997. Historical change in precipitation pH at Kobe Japan: 1935-1961. *Atmospheric Environment*, 31(15): 2367-2369.

Ito, K., Kinney, P.L. and Thurston, G., 1995. Variations in PM-10 Concentrations Within Two Metropolitan Areas and Their Implications for Health Effects Analyses. *Inhalation Toxicology*, 7: 735-745.

Jamieson, D., 1996. Ethics and intentional climate change. *Climatic Change*, 33(3): 323.

Jauregui, E. and Romales, E., 1996. Urban effects on convective precipitation in Mexico city. *Atmospheric Environment*, 30(20): 3383.

Jones, R.L. and Mitchell, J.F.B., 1991. Climate Change - Is Water Vapour Understood. *Nature*, 353(6341): 210-211.

Kane, R.P., 1997. Relationship of El Niño Southern oscillation and Pacific sea surface temperature with rainfall in various regions of the globe. *Monthly Weather Review*, 125(8): 1792.

Karl, T.R., Groisman, P.Y., Knight, R.W. and Heim, R.R., 1993. Recent Variations of Snow Cover and Snowfall in North America and Their Relation to Precipitation and Temperature Variations. *Journal of Climate*, 6(7): 1327-1344.

Keim, B.D., Fairs, G.E., Muller, R.A., Grymens, J.M. and Rohi, R.V., 1995. Long-Term Trends Of Precipitation and Runoff in Louisiana, USA. *International Journal of Climatology*, 15(5): 531-542.

Knappenberger, P.C. and Michaels, P.J., 1993. Cyclone Tracks and Wintertime Climate in the Mid-Atlantic Region of the USA. *International Journal of Climatology*, 13(5): 509-532.

Köppen, W., 1900. Versuch einer Klassifikation der Klimate. *Geographishe Zeitung*, 6: 593-611.

Köppen, W. and Geiger, R., 1930. *Handbuck der Klimatologie*, 1-5. Gebruüder Borntraeger, Berlin.

Krzysztofowicz, R. and Sigrest, A.A., 1997. Local climate for probabilistic quantitative precipitation forecasting. *Monthly Weather Review*, 125(3): 305.

Küchler, A.W., 1964. *Potential Natural Vegetation of the Coterminous United States*. Special Publication No. 36. American Geographical Society.

Kumar, A. and Hoerling, M.P., 1997. Interpretation and Implications of the observed inter-

El Niño variability. Journal of Climate, 10(1): 83.

Leopold, L.B., 1974. Water, A Primer. W. H. Freeman and Co., San Francisco.

Lewandrowski, J.K. and Brazee, R.J., 1993. Farm Programs and Climate Change. *Climatic Change*, 23(1): 1.

Lhermitte, R., 1990. Attenuation and Scattering of Millimeter Wavelength Radiation by Clouds and Precipitation. *Journal of Atmospheric and Oceanic Technology*, 7(3): 464-479.

Li, Y. and Roth, H.D., 1995. Daily Mortality Analysis by Using Different Regression Models in Philadelphia County, 1973-1990. *Inhalation Toxicology*, 7:45-58 (Roth Associates, Inc.).

Lippmann, M., 1993. Health Effects of Tropospheric Ozone: Review of Recent Research Findings and their Implications to Ambient Air Quality Standards. *Journal of Exposure Analysis and Environmental Epidemiology*, 3(1): 103-129.

Lynch, J.A., Grimm, J.W. and Bowersox, V.C., 1995. Trends in Precipitation Chemistry in the United States: A National Perspective, 1980-1992. *Atmospheric Environment*, 29(11): 1231-1246.

Manning, J.C., 1987. *Applied Principles of Hydrology*. Merrill Publishing Co., Columbus.

Marinucci, M.R. and Giorgi, F., 1992. A 2XCO2 Climate Change Scenario over Europe Generated Using a Limited Area Model Nested in a General Circulation Model .1. Present-Day Seasonal Climate Simulation. *Journal of Geophysical Research*, 97(D9): 9989.

Mayewski, P.A. et al., 1993. Greenland Ice Core 'Signal' Characteristics - An Expanded View of Climate Change. *Journal of Geophysical Research - Atmospheres*, 98(D7): 12839-12848.

Meggers, B.J., 1994. Archeological Evidence for the Impact of Mega-Nimo Events on Amazonia During the Past Two Millenium. *Climate Change*, 28(4): 321-338.

Melillo, J.M. et al., 1995. Vegetation ecosystem modeling ans anlysis project; Comparing biogeography and biogeochemistry models in a continental-scale study of terrestial ecosystem responses to climate change and CO2 doubling. *Global Biogeochemical Cycles*, 9(4): 407.

Melillo, J.M. et al., 1993. Global Climate Change and Terrestrial Net Primary Production. *Nature*, 363(6426): 234-239.

Mircea, M. and Stefan, S., 1998. A theoretical study of the microphysical parameterization of the scavenging coefficient as a function of precipitation type and rate. *Atmospheric Environment*, 32(17): 2931-2938.

Mohnen, V.A., Goldstein, W. and Wang, W.-C., 1993. Tropospheric Ozone and Climate Change. *Journal of Air & Waste Management Association*, 43(October): 1332-1334.

Monk, G.A., 1992. Synoptic and Mesoscale Analysis of Intense Mid-Latitude Cyclones. *Meteorological Magazine*, 121(1445): 269.

Montgomery, M.T. and Farrell, B.F., 1993. Tropical Cyclone Formation. Journal of the Atmospheric Sciences, 50(2): 285-310.

Moolgavkar, S.H. and Luebeck, E.G., 1996. A Critical Review of the Evidence on Particulate Air Pollution and Mortality. *Epidemiology*, 7(4): 420-428.

Moolgavkar, S.H., Luebeck, E.G., Hall, T.A. and Anderson, E.L., 1995. Particulate Air Pollution, Sulfur Dioxide, and Daily Mortality: A Reanalysis of the Steubenville Data. *Inhalation Toxicology*, 7: 35-44.

Myers, N. and Goreau, T.J., 1991. Tropical Forests and the Greenhouse Effect - A Management Response. *Climatic Change*, 19(1-2): 215-227.

Nicholson, S.E. and Kim, E., 1997. The relationship of the El Niño Southern oscillation to

African rainfall. *International Journal of Climatology*, 17(2): 117.

Nielsen, J.W. and Dole, R.M., 1992. A Survey of Extratropical Cyclone Characteristics During GALE. *Monthly Weather Review*, 120(7): 1156.

Nobre, C.A., Sellers, P.J. and Shukla, J., 1991. Amazonian Deforestation and Regional Climate Change. *Journal of Climate*, 4(10): 957-989.

Nordlund, G. and Tuomenvirta, H., 1998. Spatial variation in wet deposition amounts of sulphate due to stochastic variations in precipitation amounts. *Atmospheric Environment*, 32(17): pp. 2913-2921.

Nunnally, N.R., Brehob, K.R. and Gunter, J.D., 1974. Public Policy and Private Enterprise in the Development of Flood Plains: A Laboratory Exercise in Physical Geography. *Journal of Geography* (October): 12-25.

Obrien, S.T., Hayden, B.P. and Shugart, H.H., 1992. Global Climatic Change, Hurricanes, and a Tropical Forest. *Climatic Change*, 22(3): 175.

Oke, T.R., 1978. Boundary Layer Climates. Methuen, New York.

Oliver, J.E., 1973. *Climate and Man's Environment.* John Wiley & Sons, New York.

Oppenheimer, M., 1989. Climate Change and Environmental Pollution: Physical and Biological Interactions. *Climatic Change*, 15(1-2): 255-270.

Ott, W.R. and Roberts, J.W., 1998. Everyday Exposure to Toxic Pollutants. *Scientific American*, 277(2): 86-91.

Pastor, J. and Post, W.M., 1993. Linear Regressions Do Not Predict the Transient Responses of Eastern North American Forests to CO2-Induced Climate Change. *Climatic Change*, 23(2): 111-120.

PateCornell, E., 1996. Uncertainties in global climate change estimates - An editorial essay. *Climatic Change*, 33(2): 145.

Perry, A.H. and Walker, J.M., 1977. *The Ocean-Atmosphere System.* Longman, London.

Pitelka, L.F. et al., 1997. Plant migration and climate change. *American Scientist*, 1997(85): 464.

Pope III, C.A., Bates, D.V. and Raizenne, M.E., 1995. Health Effects of Particulate Air Pollution: Time for Reassessment? *Environmental Health Perspectives*, 103(5): 472-480.

Postel, S., 1984. *Water: Rethinking Management in an Age of Scarcity.* Worldwatch Institute, Washington, D. C.

Price, M., 1985. *Introducing Groundwater.* George Allen and Unwin, Boston.

Puxbaum, H., Simeonov, V. and Kalina, M., 1998. Ten years trends (1984-1993) in the precipitation chemistry in Central Austria. *Atmospheric Environment*, 32(2): 193-202.

Ramanathan, V., 1988. The Greenhouse Theory of Climate Change: A Test by Inadvertant Global Experiment. *Science*, 240: 293-299.

Rasmusson, E.M., 1985. El Niño and Variations in Climate. *American Scientist*, 73: 168-177.

Reason, C.J.C., 1992. On the Effect of ENSO Precipitation Anomalies in a Global Ocean GCM. *Climate Dynamics*, 8(1): 39.

Reisner, M., 1985. Here's Mud in Your Eye: The Best and the Worst Tap Waters in the Nation. *Connoisseur*(July): 100-104.

Reisner, M.A., 1986. *Cadillac Desert.* Viking, New York, p. 582.

Rind, D. et. al., 1990. Potential Evaporation and the Likelihood of Future Drought. *Journal of Geophysical Research - Atmospheres*, 95(D7): 9983-10004.

Rind, D., Suozzo, R., Balachandran, N.K. and Prather, M.J., 1990. Climate Change and the

Middle Atmosphere. Part I: The Doubled CO2 Climate. *Journal of the Atmospheric Sciences*, 47(4): 475.

Robock, A. et al., 1993. Use of General Circulation Model Output in the Creation of Climate Change Scenarios for Impact Analysis. *Climatic Change*, 23(4): 293-336.

Rogers, P., 1986. Water: Not as Cheap as You Think. *Technology Review* (November/December): 30-43.

Rogers, R.F. and Davis, R.E., 1993. The Effect of Coastline Curvature on the Weakening of Atlantic Tropical Cyclones. *International Journal of Climatology*, 13(3): 287-300.

Romieu, I. et al., 1995. Effects of Urban Air Pollutants on Emergency Visits for Childhood Asthma in Mexico City. *American Journal of Epidemiology*, 141(6): 546-553.

Rosenberg, N.J., 1992. Adaptation of Agriculture to Climate Change. *Climatic Change*, 21(4): 385.

Rosenberg, N.J., 1993. A Methodology Called MINK for Study of Climate Change Impacts and Responses on the Regional Scale - An Introductory Editorial. *Climatic Change*, 24(1-2): 1-6.

Rowe, C.M., 1993. Global Land-Surface Albedo Modelling. *International Journal of Climatology*, 13(5): 473-496.

Rozanski, K., Araguasaraguas, L. and Gonfiantini, R., 1992. Relation Between Long-Term Trends of Oxygen-18 Isotope Composition of Precipitation and Climate. *Science*, 258(5084): 981.

Salati, E. and Nobre, C.A., 1991. Possible Climatic Impacts of Tropical Deforestation. *Climate Change*, 19(1-2): 177-197.

Schemenauer, R.S., Banic, C.M. and Urquizo, N., 1995. High Elevation Fog and Precipitation Chemistry in Southern Quebec/Canada. *Atmospheric Environment*, 29(17): p. 2235.

Schimmelpfennig, D., 1996. When we don't know the costs or the benefits: Adaptive strategies for abating climate change. *Climatic Change*, 33(2): 235.

Schneider, S.H., 1989. The Changing Climate. *Scientific American*, 261(3): 70-79.

Schonwiese, C.D., Stahler, U. and Birrong, W., 1990. Temperature and Precipitation Trends in Europe and Their Possible Link with Greenhouse-Induced Climatic Change. *Theoretical and Applied Climatology*, 41(3): 173.

Scudlark, J., Russell., K., Galloway., J., Church., T. and Keene, W., 1998. Organic Nitrogen in Precipitation at the mid-Atlantic U.S. Coast: Methods evaluation and preliminary measurements. *Atmospheric Environment*, 32(10): 1719-1728.

Seilkop, S.K. and Finkelstein, P.L., 1987. Acid Precipitation Patterns and Trends in Eastern North America, 1980–84. *Journal of Climate and Applied Meteorology*, 26(No. 8): 980–994.

Sellers, W.D., 1965. *Physical Climatology*. University of Chicago Press, Chicago.

Sevruk, B., 1996. Adjustment of tipping-bucket precipitation gauge measurements. *Atmospheric Research*, 42(1-4): 237.

Sinha, A. and Toumi, R., 1997. Tropospheric ozone, lightning and climate change. *Journal of Geophysical Research - Atmospheres*, 102(D9): 10667.

Sinha, A. and Toumi, R., 1997. Tropospheric ozone, lightning, and climate change. *Journal of Geophysical Research - Atmospheres*, 102(D9): 10667.

Sirois, A., 1993. Temporal Variation of Sulphate and Nitrate Concentration in Precipitation in Eastern North America - 1979-1990. *Atmospheric Environment*, 27(6): 945-964.

Skaugen, T., Creutin, J.D. and Gottschalk, L., 1996. Reconstruction and frequency estimates of extreme daily areal precipitation. *Journal of Geophysical Research - Atmospheres*, 101(D21): 26287.

Smith, J.B. and Tirpak, D.A., 1989. The Potential Effects Of Global Climate Change On the United States - Appendix A Water Resources. EPA-230-05-89-051, U. S. (United States Environmental Protection Agency).

Smith, K. and Tobin, G.A., 1979. *Human Adjustment to the Flood Hazard*. Longman Group Limited, London.

Somerville, M.C., Shadwick, D.S., Meldahl, R.S., Chappelka, A.H. and Lockaby, B.G., 1992. Use of a Non-Linear Model in Examining Growth Responses of Loblolly Pine to Ozone and Acid Precipitation. *Atmospheric Environment*. Part A, General Topics, 26(2): 279-287.

Stull, R.B., 1988. *An Introduction to Boundary Layer Meteorology*. Kluwer Academic, Boston.

Syrett, W.J., Albrecht, B.A. and Clothiaux, E.E., 1995. Vertical cloud structure in a midlatitude cyclone from a 94-GHz radar. *Monthly Weather Review*, 123(12): 3393.

Thornthwaite, C.W., 1948. An Approach Toward a Rational Classification of Climate. *Geographical Review*, 38: 55-94.

Trewartha, G.T., 1943. *Introduction to Weather and Climate*. McGraw-Hill, New York.

Wallace, L. A., 1995. Human Exposure to Environmental Pollutants: A Decade of Experience. *Clinical and Experimental Allergy*, 25(1): 4-9.

Wanielista, M., 1990. *Hydrology and Water Quantity Control*. John Wiley & Sons, New York, p. 565.

Ward, G.F.A., 1995. Prediction of Tropical Cyclone Formation in Terms of Sea-Surface Temperature, Vorticity and Vertical Wind Shear. *Australian Meteorological Magazine*, 44(1): 61-70.

Weber, G. R., 1990. North Pacific Circulation Anomalies, El Niño and Anomalous Warmth Over the North American Continent in 1986-1988: Possible Causes of the 1988 North American Drought. *International Journal of Climatology*, 10(3): 279-290.

Webster, P.J. and Yang, S., 1992. Monsoon and ENSO - Selectively Interactive Systems. *Quarterly Journal of the Royal Meteorological Society*, 118(507): 877.

Wilks, D.S., 1992. Adapting Stochastic Weather Generation Algorithms for Climate Change Studies. *Climatic Change*, 22(1): 67.

Williams, P., 1989. Adapting Water Resourses Management to Global Climate Change. *Climatic Change*, 15(1-2): 83-94.

Willmott, C.J., Robeson, S.M. and Janis, M.J., 1996. Comparison of approaches for estimating time-averaged precipitation using data from the USA. *International Journal of Climatology*, 16(10): 1103.

Yasunari, T., 1991. The Monsoon Year - A New Concept of the Climatic Year in the Tropics. *Bulletin of the American Meteorological Society*, 72(9): 1331-1339.

Yin, Z.-Y., 1993. Streamflow Regimes in Urban and Rural Basins of the Middle-Upper Chattahoochee River Basins, North Georgia. *Southeastern Geographer*, 33(1): 44-64.

Yu, Q., 1992. Numerical Silumation of Pollutants Removal by Precipitation. *Journal of Atmospheric Chemistry*, 14(1-4): 143-152.

Zhao, W.N. and Khalil, M.A.K., 1993. Downscaling of Global Climate Change Estimates to Regional Scales - An Application to Iberian Rainfall in Wintertime. *Journal of Climate*, 6(6): 1161-1171.

Subject Index

A

B

C

D

E

F

G

H

I

J

K

L

M

N

O

P

Q

R

S

T

U

V

W

X

Y

Z